Reliability and Degradation

The Wiley Series in Solid State Devices and Circuits

Edited by

M. J. Howes and D. V. Morgan

Department of Electrical and Electronic Engineering, University of Leeds

Microwave Devices

Variable Impedance Devices

Charge-coupled Devices and Systems

Optical Fibre Communications

Large Scale Integration

Reliability and Degradation

Reliability and Degradation

Semiconductor Devices and Circuits

Edited by

M. J. Howes
D. V. Morgan
Department of Electrical and Electronic Engineering, University of Leeds

A Wiley–Interscience Publication

JOHN WILEY & SONS
Chichester · New York · Brisbane · Toronto

British Library Cataloguing in Publication Data:

Reliability and degradation.—(The Wiley series
in solid state devices and circuits)
1. Semiconductors
I. Howes, M. J. II. Morgan, D. V.
621.3815'2 TK7871 80-42310

ISBN 0 471 28028 3

Typeset by Preface Ltd, Salisbury, Wilts. and printed at the Bath Press, Avon

Contributors

A. CHRISTOU	*US Naval Laboratory, Washington DC, USA*
J. E. DAVEY	*US Naval Laboratory, Washington DC, USA*
W. GOODBLOOD	*Philips Research Laboratory, Eindhoven, Netherlands*
G. KERSUZAN	*Philips Research Laboratory, Eindhoven, Netherlands*
J. W. MAYER	*Material Science Dept, Cornell University, Ithaca, New York, USA*
D. V. MORGAN	*Department of Electrical and Electronic Engineering, The University of Leeds, UK*
S. D. MUKHERJEE	*National Research and Resource Facility for Submicron Structures, Cornell University, Ithaca, New York, USA*
D. H. NEWMAN	*British Telecom Research Centre, Martlesham Heath, Ipswich, UK*
G. OTTAVIANI	*Institute di Fisica, Modena, Italy*
S. RITCHIE	*British Telecom Research Centre, Martlesham Heath, Ipswich, UK*
A. G. VAN NIE	*Philips Research Laboratory, Eindhoven, Netherlands*
J. WOOD	*ITT Standard Telecommunications Laboratories Ltd, Harlow, Essex*

Series Preface

The Oxford Dictionary defines the word revolution as 'a fundamental reconstruction'; these words fittingly describe the state of affairs in the electronic industry following the advent of solid state devices. This 'revolution', which has taken place during the past 25 years, was initiated by the discovery of the bipolar junction transistor in 1948. Since this first discovery there has been a worldwide effort in the search for new solid state devices and, although there have been many notable successes in this search, none have had the commercial impact which the transistor has had. Possibly no other device will have such an impact; but the commercial side of the electronics industry stands poised, awaiting the discovery of new devices as signicant perhaps as the transistor.

Research and development in the field of solid state devices has concerned itself with two important problems. On the one hand we have device physics, where the aim is to understand in terms of basic *physical concepts* the mode of operation of the various devices. In this way one seeks to optimize the technology in order to achieve the best performance from each device. The second aspect of this work is to consider the important contribution of the circuit to the operation of a device. This problem has been called *device circuit interaction*. It is a great pity that in the past these two major aspects of the one problem have been tackled by separate groups of scientists with little exchange of ideas. In recent years, however, this situation has been somewhat remedied, the improvement being due directly to the very rigorous system specifications demanded by industry. Such demands constantly require greater performance from devices, which can only be brought about by coordinated team work.

The objective of this new series of books is to bring together the two aspects of this problem: device physics and device circuit interactions. We hope to achieve, by coordinated co-authorship of leading experts in the respective fields, a varied and balanced review of past and current work. The books in this series will cover many aspects of device research and will deal with both the commercially successful and the more speculative devices. Each volume will be an in-depth account of one or more devices centred on some common theme. The level of the text is designed to be suitable for the graduate student or research worker wishing to enter the field of research concerned. Basic physical concepts in semiconductors and elementary ideas in passive and active circuit theory will be assumed as a starting point.

University of Leeds
December 1980

M. J. HOWES
D. V. MORGAN

Preface

Since their conception solid-state semiconductor devices have always been regarded as having the potential of high reliability. This expectation was not diminished by the fact that early devices were characterized by lower lifetimes since it was quickly recognized that the sources of device failure were the result of the relatively primitive technology. Problems identified were poor crystal quality and undesirable chemical reactions between the metal–semiconductor and metal–metal interfaces which constitute major features of device structures. The exacting performance specifications impose great demands on the architecture of modern devices. Device structures are more complex, much smaller, may transport high current densities and in some cases may be expected to withstand high interfacial electric fields. Furthermore the device may have to operate at elevated temperatures.

Much of the early work on device reliability consisted of life testing and visual inspection of the degraded structures which provided semi-quantitative indirect evidence of possible degradation mechanisms. However, in recent years the development of a number of sophisticated near surface analytical techniques has shed new light on the physical processes involved. These techniques are discussed in detail in Chapter 1. Chapters 2 and 3 cover the basic interactions between metal films and silicon surfaces and between metal films and the two important III–V semiconductors GaAs and InP. Chapers 4 and 5 extend these studies to the degradation of Si, GaAs and InP device structures. The special cases of optoelectronic devices such as lasers and light emitting diodes are dealt with in Chapter 6 whilst the final chapter considers the increasingly important microwave integrated circuit.

The editors wish to thank most warmly the authors who have contributed to this volume and are particularly indebted to Dr. J. E. Davey (Naval Laboratory, Washington), Professor J. W. Mayer (Cornell University), Dr. J. Poate (Bell Laboratories), and J. Wood (ITT) for the valuable advice and constructive criticism.

University of Leeds
December 1980

D. V. MORGAN
M. J. HOWES

Contents

Reliability and Degradation
Edited by M. J. Howes and D. V. Morgan

CHAPTER 1

Interactions of metal films on semiconductors

S. D. MUKHERJEE

1.1 INTRODUCTION

All semiconductor devices need electrical accessibility and therefore some form of metal contact is required. The contact may be rectifying or Ohmic, but the metals used are often unsuitable for making connections to the outside world. Invariably, therefore, layered metallizations are used for making the top metallic layer suitable for wire or tape ultrasonic–thermocompression bonding usually involving aluminium or gold, or soldering using lead–tin alloys.

When a device is fabricated, it goes through a number of processes[1,2] at the end of which it is metallized, passivated, encapsulated, or packaged or any combination of these. The device engineer knows the structure of the device which will include a number of semiconductor–metal, insulator–metal, and metal–metal interfaces. In order to ascertain the operational reliability of the device accelerated life-tests and pre-delivery burn-in or screening (or both) based on such life tests are often carried out. The stressing involves operating the device either at a higher temperature or current density in normal atmospheric, corrosive or highly irradiated environments, or in environments consisting of combinations of these.

The different interfaces in the device may change their characteristics through materials transfer by various means[3,4] at the different stages mentioned above. The interfacial changes and any resultant alterations in the bulk of the constituting materials invariably alter the electrical or mechanical performances (or both) of the device and the device is said to have degraded. More importantly, operation of the device at high power-levels or at high ambient temperatures for long periods of time reduces its useful life of operation as specified by the required electrical tolerances, while unnecessarily high stressing causes thermal runaway and device failure. It is therefore very important that the device engineer be aware of the various possible ways in which the different interfaces alter themselves and the bulk materials that they separate, both during processing and the actual operations of the device in adverse conditions, and bears in mind the likely degradation or failure mechanisms caused by such interactions.

In this chapter we shall look into the fundamentals of these interactions. We shall also explore a number of macro- and microscopic analytical techniques used for studying them. In effect, we shall attempt to create a ground work necessary for the proper understanding of the in-depth reviews that appear later in this volume.

1.2 PROPERTIES OF THIN METALLIC LAYERS

1.2.1 Deposition techniques

Thin layers of metals can be deposited onto a semiconductor in a number of different ways,[5,6] the most important of which are:

(i) Thermal evaporation from a source heated resistively or by an electron beam.[5–7]
(ii) Molecular beam epitaxy—(MBE)[1,8] from Knudsen effusion sources, the process being similar in many respects to thermal evaporation.[9]
(iii) Sputtering from a large, planar metallic source[6,10] using non-reactive, high atomic weight gas-ions such as Ar^+, by DC–RF diode and triode cathodic sputtering,[5,6,11] magnetron sputtering, both planar[12–14] and cylindrical,[15] and large diameter ion-beam sputtering with[16] and without[17,18] ion neutralization facility.
(iv) Plating from aqueous solution,[5,6] electron assisted[19,20] and electroless.[21,22]
(v) Chemical vapour deposition (CVD) from an organometallic carrier gas,[6,23–27] deposited either pyrolitically or by chemical reactions.

Table 1.1 shows some relevant parameters of deposition techniques performed in vacuum.

Thermal evaporation from filaments or boats may involve vaporization from the liquid state, e.g., for Al, Au, Pd, Ni etc., while some metals, such as Ti, sublime from the solid state. Metals evaporated from the liquid state by resistive means are susceptible to bubbling and spitting which may degrade the uniformity of the deposited layer. Controlled heating by magnetically or electrostatically deflected electron beams reduce such problems.

For filament, boat or crucible evaporation, the melting point T_m limits the highest temperature achievable: Al_2O_3(2020 °C), BeO(2520 °C), Mo(2620 °C), Ta(2980 °C), W(3410 °C) and graphite (3700 °C). Appropriate metal–source combination is necessary. Al reacts with all refractory metals (W, Ta, Mo); Ti reacts with W but not with Ta, although Ta burns out before sufficient Ti vapour-pressure is generated; Au wets and reacts with Ta,

Table 1.1 Physical parameters associated with the different methods of metal depositions performed in vacuum

	Filament and boat evaporation	Molecular beam epitaxy	Electron beam evaporation	DC–RF sputtering	Planar magnetron sputtering
Chamber pressure (torr)	10^{-7}–10^{-5}	10^{-11}–10^{-8}	10^{-6}–10^{-5}	10^{-3}–10^{-1}	10^{-5}–10^{-2}
Deposition rate (nm min^{-1})	10–500	10–100	50–5000	20–200	20–5000
Ratio of gas to deposited atom collision rate on substrate	10^{-4}–10^{-2}	10^{-8}–10^{-5}	10^{-5}–10^{-2}	10–100	10^{-2}–1
Energy of deposited atoms on substrate (eV)	0.1–0.5	0.05–0.5	0.1–0.3	0.1–20	0.1–5
Irradiation of substrate by	IR and visible photons	IR and visible photons	X-rays, IR and visible photons, some ions as well	Electrons, UV and gas ions	UV photons and very few ions
Thermal loading of the substrate (W cm^{-2})	0.1–10	0.01	0.1–1.0	5–10	1
Substrate heating or cooling and temperature (°C)	Heated or cooled, 50°–300 °C	Temperature controlled 100°–700 °C	Heated, 200°–300 °C can be cooled in research units	Cooled, 50°–250 °C	Cooled or heated 50°–250 °C

but only partially wets W and Mo; Ni alloys with Mo, Ta, and W to form eutectics, but if Al_2O_3 coated W boats are used there is always the possibility of evaporating Al_2O_3 as well. BeO crucibles are better; but it is hard to evaporate Pt from filaments or boats.

In electron-beam evaporation, the charge (large lumps of high purity metals) is melted *in situ* inside a water-cooled copper crucible in high vacuum. Only the molten metal at the highest temperature, where the electron-beam strikes the melt, actually acts as an efficient evaporation source. The temperature diminishes radially and since the copper crucible is cooled, no mass transfer takes place there. In this way any metal can be evaporated easily and in a controlled fashion, with no contamination from the container.

In DC and RF sputtering a negative potential is developed on the metal target plate by directly coupled DC or capacitatively coupled RF sources, and a plasma is generated in the low pressure argon atmosphere. Ar^+ ions bombard the plate, dislodge metal atoms by kinetic energy transfer, which are deposited onto the facing substrate kept at ground or slightly positive potential. Electrons and some Ar^- ions bombard the substrate and so do energetic metallic ions, increasing the substrate temperature[28] and damaging electrically active layers up to a depth of 10 nm. As a result Schottky barrier height and other characteristics on GaAs are adversely affected.[29,30]

In magnetron sputtering a crossed magnetic field compels the electrons to move in cycloidal paths, increasing gas ionization probability and nearly eliminating electron bombardment of the substrate. This not only reduces the operating pressure providing a 'clean' environment, but also allows the sputtering to be used in Si–MOS technology reducing charge trapping on SiO_2. Sputtering at not very low pressures often causes gas occlusion in deposited metals that may result in porosity.[31]. Ion beam sputtering can be achieved in a differentially pumped deposition chamber operating at a much lower pressure, ensuring even cleaner environment. In all these physical deposition processes, where atoms are stripped off their neighbours in a solid state, part of the substrate heating is caused by the latent heat of condensation of these atoms[28] (cohesive energy), which on principle, cannot be eliminated.

Although it is generally believed that plating from solutions invariably incorporates impurities in the metal film, this is not so in every case. In well controlled aqueous environment less than 100 or 10 ppm or impurities can be achieved.[19] These figures are, at times, better than similar figures for physical vapour deposited layers. In plating, as well as in CVD, non-metallic atoms that are part of the electrolyte or the carrier gas may be deposited as well. For example electroless deposited Ni[22] and CVD Pt[23] may contain phosphorus while CVD Cr[24] and Ti[27] often have carbon as impurities. In some cases such impurities may be useful, e.g. for making a stuffed diffusion barrier (Section 2.5.3). Electroplating has been used for Schottky barrier diodes (SBDs)[32] and is currently used extensively for Au and Ag heat sinking of high power devices, such as IMPATTs,[33,34] TRAPATTs and TEDs (Gunn diodes).

1.2.2 Grain size and epitaxiality: grain boundary structure

Three types of solid surfaces exist: crystalline (e.g. (100) and (111) Si and (110) GaAs), polycrystalline (e.g polycrystalline silicon and metallic layers, SiO_2 and Si_3N_4) and amorphous (e.g. fused quartz, borosilicate, and phosphosilicate glasses and amorphized metal or semiconductor) and we need to deposit metals on all these surfaces for different applications. Solid state interactions are strongly dependent upon the structures of both the interface and the dissimiliar solids on the two sides of this interface. We therefore set out to find what the structure of the deposited metal is going to be like.

When a metal is deposited on a low Miller index plane on a crystalline material, oriented growth (heteroepitaxy) may occur if the misfit $\leqslant 7\%$. The misfit is defined as $100(b - a)/a$ per cent where a is the lattice parameter of the semiconductor and b, the network spacing of the metal which is some multiple of its lattice parameter that allows parallel matching of two major crystal planes of the metal and the semiconductor.[35] Such epitaxy occurs due to the lowering of interfacial potential energy when atoms on either side fall into the periodic potential wells caused by the crystallinity of the other, with that of the substrate playing the decisive role since it is thicker.[36] Any misfit in a and b can be accommodated in two ways. Firstly by the generation of strain. When the deposited layer is very thin, such as a few monolayers, the substrate forces the area of the basal plane to decrease (or increase) with a corresponding elongation (or compression) of the lattice in a vertical direction to maintain unit cell volume. This is called 'pseudomorphism' and the phenomenon is responsible for heteroepitaxy, even when misfit >10 or 20 per cent, for a larger number of cases.[37] Secondly, for thicker layers pseudomorphism cannot be maintained and the strain caused by mis-registration is reduced through the formation of narrow stripes of poor-fit, called interfacial 'misfit dislocations'[35] because of their close analogy with crystal dislocations. For a start these misfit dislocations form closed loops starting from and ending on the interface. Often they climb up to meet the free surface and threading dislocations are produced. Heteroepitaxial layers usually have high dislocation densities, 10^9 to 10^{10} cm^{-2}, i.e. at least one dislocation for every 0.1 to 1 μm of traverse on the surface.

Growth often follows a layer by layer or islands on a layer sequence when the surface energy (or the surface tension) of the deposit is very much lower than that of the substrate and when the substrate temperature during deposition is high.[38–40] If the converse is true or if misfit $\leqslant$ 7–10 per cent, individually nucleated islands are formed first. These move and enlarge in size and may coalesce to form a single crystal layer when misfit $\leqslant 7$ per cent at a high enough temperature, with threading misfit dislocations forming at or near the coalescing side faces. Or, at intermediate temperatures, for misfit <7 per cent, nucleated single crystal islands coalesce to form a texturized film having grains with a common low Miller index axis normal to the surface by being joined to one another side by side by low or medium angle grain boundaries. Alternatively, they form a polycrystalline layer when misfit >7 to 10 per cent and at lower temperatures with high or low angle grain boundaries separating the coalesced islands. Subsequent deposition may increase the sizes of these crystallites in a vertical direction following homoepitaxy giving rise to columnar growth, or new crystallites are formed and a layer is produced with grain sizes smaller than its thickness.

In most practical cases metals are deposited onto crystalline semiconductors at temperatures much lower than that necessary for creating epilayers for fear of interfacial reactions or device or equipment constraints or both. As a result the film has a polycrystalline structure and the higher the melting point of the metal the smaller the grain size in the film.[41] Dislocation densities in these polycrystals are high, 10^9–10^{12} cm^{-2}, i.e. even 100–500 nm^2 diameter grains have at least one threading dislocation in them.

Metal epitaxy on useful semiconductors has been observed in a number of cases,[42–44] a few of the earlier cases being: Ag and Ni on (111) Si by MBE; Co, Ni, and Ag on (111) Si by evaporation; Ge on (111) and (001) Si by MBE, CVD and evaporation; Ag and Pt on (111) Si by sputtering; Ag on (001) Si by MBE; Al on (001), (110) and (111) GaAs by MBE; Zn on (001) and (111) GaAs by MBE (but Zn is a dopant); Ge on (001), (110), and (111) GaAs by MBE, CVD, LPE and evaporation; Ag on (001) and (110) GaAs by MBE and Fe on (100) GaAs by evaporation.[44] This partial list has been increased considerably in recent years.

The rise in surface temperature during sputtering in a plasma hinders rather than facilitates epitaxial growth. This is because most of the energy is deposited as kinetic energy of the incoming energetic metal atoms and ions and a few gas ions, and it disrupts the small crystallites. Therefore, not only are sputtered epitaxial layers rare but also grain size becomes much smaller than in an evaporated film, and especially so if adsorbed gases are occluded in its structure.[31a] Controlled ion-beam sputtering under UHV conditions, however, makes epitaxy possible.[31b]

Plating on the other hand, invariably creates large, faceted grains of crystallite and it is necessary for structural uniformity to reduce grain size by agitation, lower current density and higher solution temperature. Plating is often initiated from a vacuum deposited thin layer of the same metal. Homoepitaxial grain growth is the starting point and the grains become larger with increasing thickness.

In some cases epitaxial growth is also possible on amorphous[45] or liquid surfaces.[46] If the interaction of the deposit with the substrate is weak, the crystal orientation is decided by nucleation facilitated by cohesive energy, e.g. (111) Si on SiO_2. In this case thermal activation is essential. If, on the other hand, surface-atom interaction is strong, initial orientation follows the short range order of the amorphous body and growth occurs accordingly, e.g. vari-

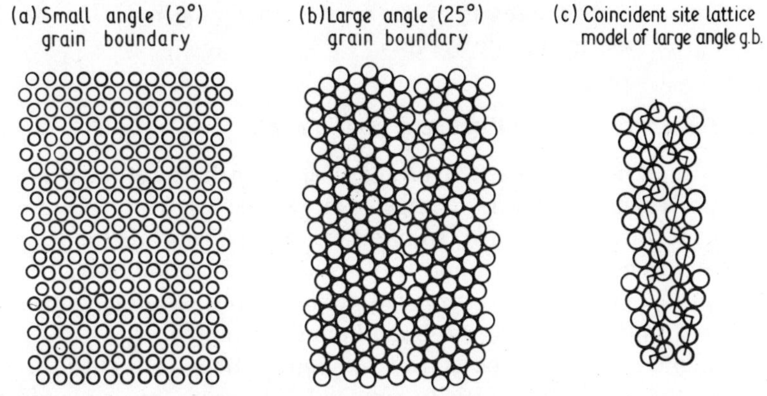

Figure 1.1. Grain boundary structures: (a) and (b) after Lomer and Nye[47] and (c) after Aaron and Bolling[48]

ous metals on liquid mercury.[46] The main driving force dictating the selection of orientation in both the cases is the tendency of the crystal to approach a minimum free energy configuration.

Grain boundaries play decisive roles in both macro- and microscopic properties of thin films and in their interactions. Therefore we shall explore their structures.

Grain boundaries (GBs) are conveniently separated with two groups:[47] small angle (~1°–5°)GBs, often called sub-grain boundaries, and large angle (15°–30°)GBs. The angle is measured between the two most highly packed crystal planes belonging to the two grains, making the smallest angle to the boundary. A small angle GB is composed of a spatially well separated (e.g. 10–50 atomic spacings) array of edge dislocations (see Figure 1.1(a)). The core structure of each is very nearly the same as that of an isolated dislocation—it has up to five or even seven strings of atoms slightly displaced from their equilibrium positions constituting heavily strained regions with a neighbouring relatively relaxed region having larger area per atomic row. As the angle increases to ~7°–12°, the number of strained atomic rows diminishes at the expense of the formation of open spaces, separated by 4–7 atomic spacings. A large angle (~15°–30°)GB has an entirely different structure: local fit and misfit with large open spaces in the misfit regions which is equivalent to an alternately space-filled (zig-zag gapped) structure with a tendency for filling up as much void space as possible (Figure 1.1(b)). The coincident site lattice (CSL) model[48] describes such a boundary as one which has, at the fit regions, atomic rows belonging to both the crystallites and, at the misfit regions, atoms stuck primarily to one of the grains and separated from the other grain by a large gap (Figure 1.1(c)). Although the gap promotes some atomic relaxation within it, disturbances of the crystal lattice are restricted to a zone no greater than one or two atomic spacings. This has been verified by field ion microscopy (FIM).[49]

1.2.3 Stored energy in thin films and grain growth

Energy is stored in a deposited layer in at least four different ways. For both epi- and polycrystalline-layers, energy is stored as: (i) free surface energy (equivalent to surface tension), (ii) interface energy, which is the lowest for well-lattice-matched epilayers, and (iii) biaxial (in directions parallel to the substrate surface) tensile of compressive strain. In polycrystalline films energy is also stored as (iv) grain- or sub-grain-boundary energy.

A crystalline surface can be considered conceptually as a cut in a larger crystal, and the two pieces spatially removed. Atoms on this cut surface are bonded onto the solid on one side and not the other, and are thus relaxed. They therefore sit at potential minima higher up than at lattice sites, where they are normally surrounded by neighbouring atoms, and this extra potential energy constitutes surface energy or tension.

For small angle GBs the slight displacements of atomic rows near the dislocation cores constitute strain and energy is associated with it. At the open spaces at large angle GBs lattice atoms are relaxed in the same way as on the surface. The extra separation causes some atoms to sit at potential minima which are higher than those at lattice sites and thus energy is stored.

But for deposited polycrystalline thin metal films, the more relevant aspect is its biaxial strain energy. Atoms which condense from the vapour state in physical vapour deposition deposit at least the sum of latent heats of vaporization and solidification on to the surface. In dynamic equilibrium the surface temperature is higher than the substrate temperature during deposition. For deposition onto cold substrates, this rise in temperature is inadequate for substantial atomic movements within the time scale of deposition, and so smaller grains are formed. When cooled, the layer contracts and tensile (pull-type) biaxial stress is developed,[50] the stress being larger the higher the melting point of the metal. For example,[51] tensile stress of $1–5 \times 10^9$ dynes cm^{-2} for Au (m.p. 1063 °C) to $5–20 \times 10^9$ dynes cm^{-2} for Ni (m.p. 1455 °C) are common, corresponding to linear strains of 0.5–2.0 per cent which are almost equal to or higher than the elastic limits for some single crystal metals.

Most thin film stress is tensile[50,51] although compressive stress can occur when reactive metals, such as Al, Ti, Ta, V, etc. absorb gases during or after deposition[51] or when columnar grains are formed during sputtering[52] at oblique angles.

Smaller defects, such as vacancies, voids, and gas occlusions, inside epilayers[53] or the grain crystallites give rise to microstrains. All these different kinds of thin film stresses can be measured by a number of optical, electronic and X-ray techniques (see Section 1.7.1, 1.7.5 and 1.7.7).

When a single layer of deposited metal is annealed, strain relief occurs through restructuring, mainly by grain growth. In the bulk of a thin film, that is for grains far away from the interfaces and the surfaces, grain growth occurs in a way similar to the growth of soap bubbles through GB motion.[54] The tendency is to form planar GBs from curved ones, to reduce GB areas and smaller grains are consumed by the larger ones to reduce GB area to volume ratio. Also, equilibrium angles are formed at GB triple points, i.e. where three grains meet (see Figure 1.2(a)).

Metals with high melting points have inherent higher stress in films and their surface–GB energy density is also higher. Consequently, the small grains in high m.p. metallic films grow faster and bigger.[41]

However, in thin films, grains often intersect the surface. Under the action of tensile stress vacancies flow from GBs to the surface with an equal and opposite flow of atoms relieving stress by film elongation (Figure 1.2(b)). Apart from grain growth, atom-transfer from the surface to GBs create shallow grooves where the GBs intersect the surface and in this way annealing induced surface roughness is formed.[55]

The presence of impurities in the film hinders grain growth in two ways. Firstly, reactive gases that accumulate at grain boundaries reduce free energy

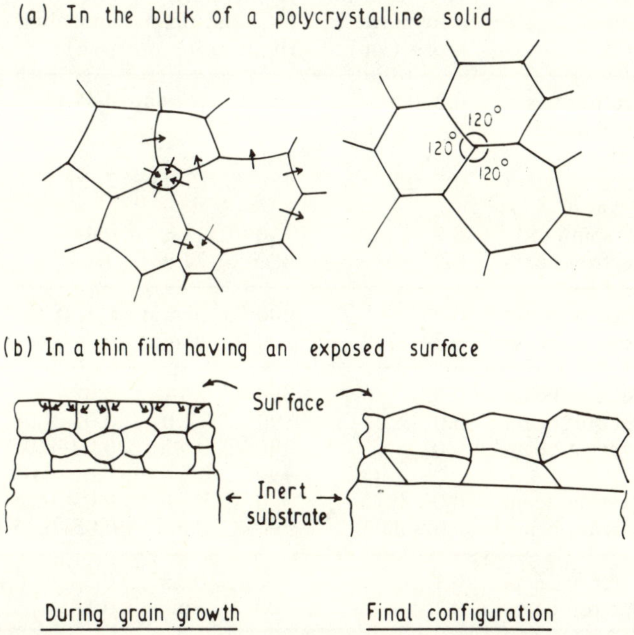

Figure 1.2. Grain growth (a) in the bulk, and (b) near the surface of a polycrystalline thin film, see text: Section 1.2.3

by both compound formation and strain relief (see Section 1.3.4). The driving force for grain growth, viz. a reduction in stored energy, is therefore reduced considerably.[56] Secondly, precipitates present at GBs stop GB motion in a way similar to the pinning down of moving dislocation by point defects in single crystals. Grain sizes of even unannealed thin metal films are therefore not only a function of substrate temperature, energy of deposition of atoms and surface disruption during sputtering but also the absorption and inclusion of impurities present in the deposition environment and outside. Table 1.2, which gives some examples of grain size and growth, shows that metallic interdiffusion can reduce grain growth to relatively smaller sizes.[59]

It is possible to facilitate recrystallization of polycrystalline films on amorphous–polycrystalline substrates (e.g. SiO_2 and Si_3N_4) locally. This is accomplished by sweeping a thermal gradient across the surface by using e.g. a laser or electron beam so as to cause the growth to commence at one side

Table 1.2 Grain size and grain growth data of physical vapour deposited thin metal films

Metal	Deposition technique and substrate temperature (°C)	Grain size for as-deposited layers (μm)	Anneal temperature (°C) and time (h, min)	Grain size after annealing (μm)	Reference
Al	electron-beam unheated SiO_2	0.2–0.5	450 °C, 15 min	0.4–3.0	57
	electron-beam, 200 °C SiO_2	0.5–0.8	450 °C, 15 min	2–5	
Au	RF sputtered	0.2	450 °C, 24 h	0.4	
	electron-beam	0.2	450 °C, 24 h	1.1	
Cu	electron-beam	0.05	200 °C, 20 min	0.1–0.4	58
Pb	electron-beam	0.2–0.3	200 °C, 20 min	0.2–0.4	
Ti	electron-beam	0.02	390 °C, 0.5–1 h	0.05	59
Pd	electron-beam	0.02	390 °C, 0.5 h	0.05–0.08	[a,b]
Rh	electron-beam	0.01	390 °C, 0.5 h	0.03–0.05	
			490 °C, 0.5 h	0.05–0.08	
Au	electron-beam	0.02–0.15	390 °C, 0.5 h	0.05–0.25	
Pt	electron-beam	0.02–0.03	390 °C, 0.5 h	0.05–0.25	
Ag	electron-beam	0.1	300 °C, 0.5 h	0.3	60
Cu	electron-beam	0.1	500 °C, 5 min	0.15	[b]

[a] Annealing performed in air.
[b] Interdiffusion occurred in all cases.

and proceed across the surface. The concept has been known for some time[61] and recently long grains 2–3 × 25–100 μm^2 have been grown in solid and liquid phases.[62–64] In some cases[65,66] the preferential crystal faceting on the amorphous surface[45] has been used for growing nearly single crystal films and the process has been named 'graphoepitaxy'.

1.2.4 Equilibrium phase diagrams for binary substitutional alloys

Binary phase diagrams tell us how two elements would interact chemically in the solid state at any proportion at one atmosphere of pressure as a function of temperature, provided sufficient time is given for the mixture to come into equilibrium.[67,68] A detailed description of phase diagrams is beyond the scope of this work, but we shall sketch the most relevant points, with examples from Figure 1.3.[69]

Metals that alloy together in a single crystal structure in all proportions (such as Ag–Pd and Ni–Pd) are said to be isomorphous. Isomorphous systems not only need identical crystal structure of the elemental phases, the atomic volumes and electronegativities should also be very similar. Limited solubilities up to about 5 per cent occur when the electronegativities χ do not differ by more than about ±0.2 V and the atomic radii r_a are within ±15 per cent, the χ *vs.* r_a plots being called Darken–Gurry plots.[70,71] Below a liquidus line in a phase diagram liquid and solid coexist; below a solidus the phase is entirely solid. At any temperature the relative concentrations on the solidus and liquidus refer to alloys having the lowest free energy at liquid and solid states respectively. The temperature at which the free energy curves for liquid and solid states coincide at their extrema along a common horizontal tangent (i.e. along an isotherm) is called the congruent point, where freezing occurs without any composition change (e.g. the Ni–Pd system).

In a eutectic system a given alloy composition remains in a liquid state down to the lowest possible temperature, called the eutectic temperature, where it solidifies. A eutectic system may have very low, or finite, solid solubilities at the end phases as in Ag–Si and Al–Si respectively. Any deviations from the eutectic composition is accommodated by the rejection into the solid phase (in a region enclosed by the liquidus and solidus) of the excess pure metals (Au–Si) or dilute alloys (Al–Si). The eutectic point is a special case of congruent point.

When the interacting elements are rather dissimilar, different intermetallic compounds are seen to occur in the phase diagram, which themselves act as end phases in eutectic systems (as in Au–Ga, Al–Au, Pd–Si, Al–Pd and Zr–Al). The end phases may have very low solubilities (e.g. in the entire Pd–Si system, and Al–$AuAl_2$, AuGa–$AuGa_2$, $ZrAl_2$–$ZrAl_3$ systems) or have finite solubilities (e.g. in PdAl–Pd_2Al, Pd_2Al–Pd systems).

In peritectic systems, solidification occurs with a complete change in crystal

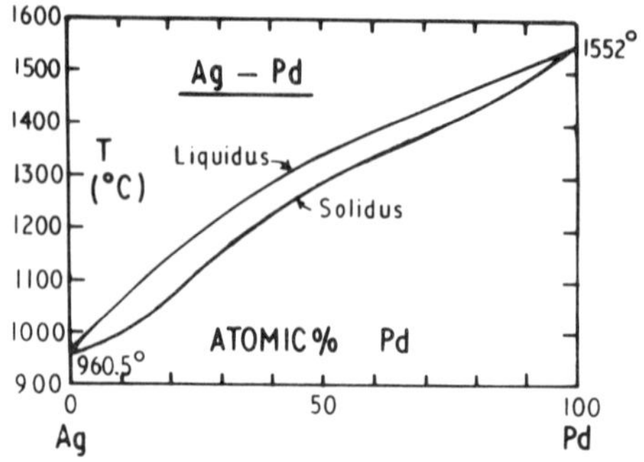
Ag – Pd
T (°C)
Liquidus
Solidus
1552°
960.5°
ATOMIC % Pd
Ag
Pd

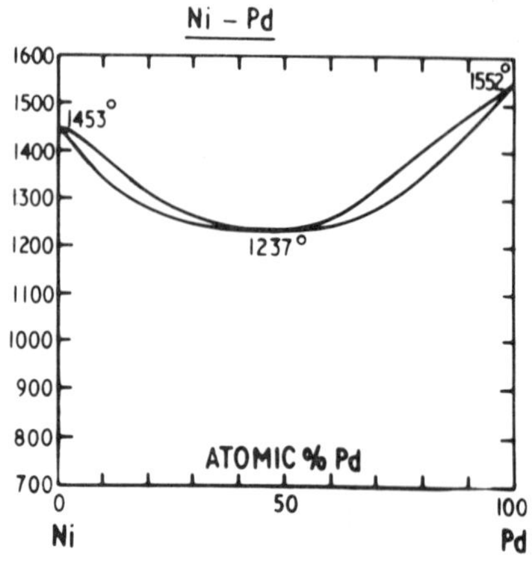
Ni – Pd
1453°
1552°
1237°
ATOMIC % Pd
Ni
Pd

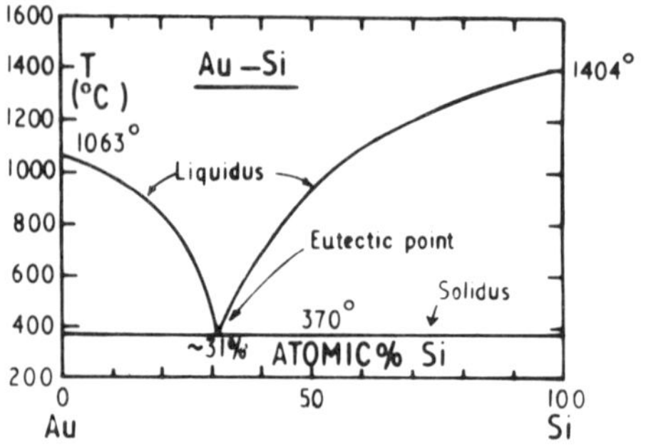
Au – Si
T (°C)
1063°
1404°
Liquidus
Eutectic point
Solidus
370°
~31%
ATOMIC % Si
Au
Si

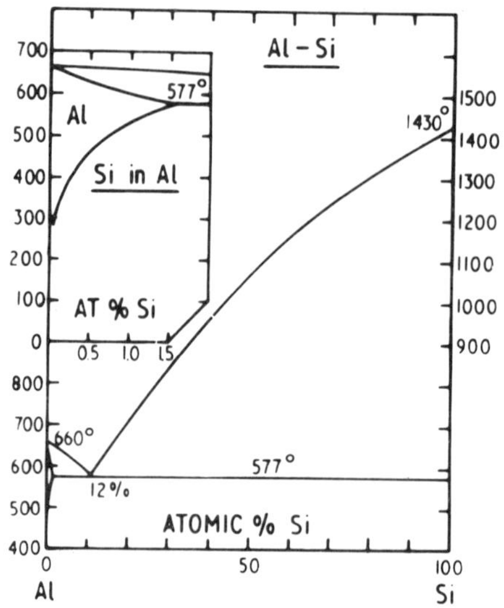
Al – Si
Al
Si in Al
577°
AT % Si
1430°
660°
12%
577°
ATOMIC % Si
Al
Si

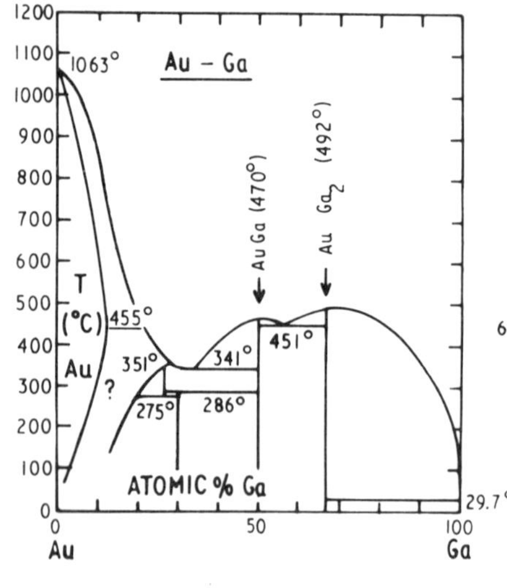
Au – Ga
1063°
T (°C)
Au
?
455°
351°
275°
341°
286°
451°
Au Ga (470°)
Au Ga2 (492°)
29.7°
ATOMIC % Ga
Au
Ga

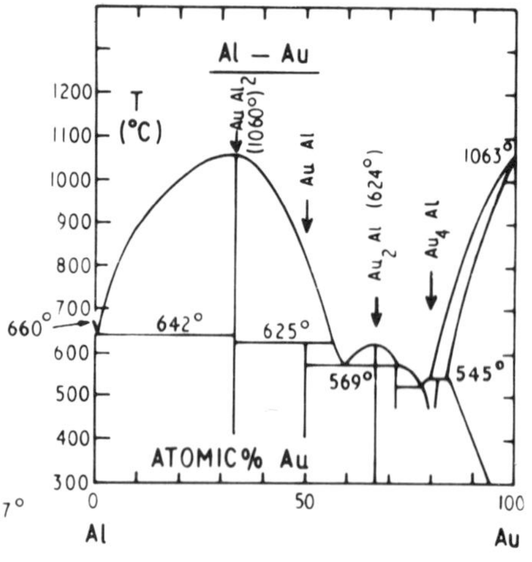
Al – Au
T (°C)
Au Al2 (1060°)
Au Al
Au2 Al (624°)
Au4 Al
1063°
660°
642°
625°
569°
545°
ATOMIC % Au
Al
Au

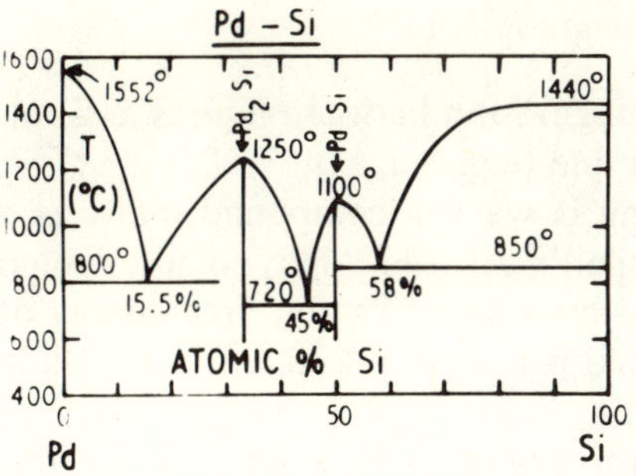

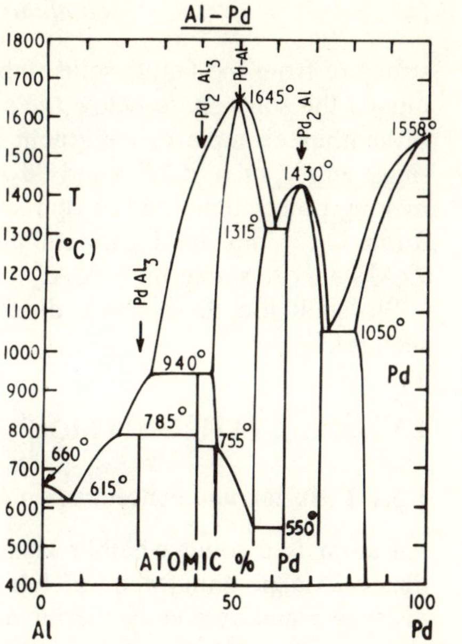

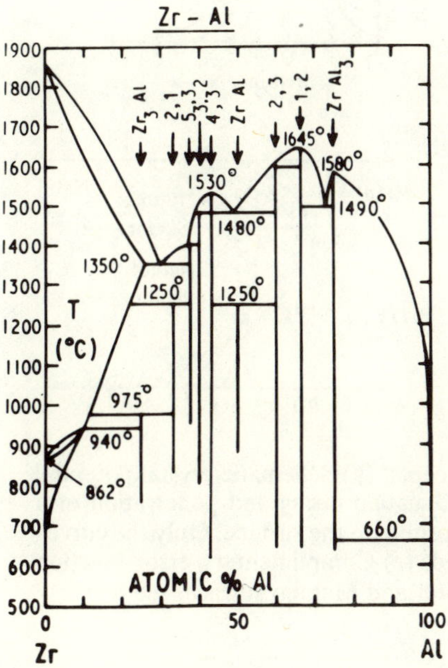

Figure 1.3. Binary phase diagrams of a few general cases and those relevant to device metallization. Ag–Pd: isomorphous, without a congruent point. Ni–Pd: isomorphous, with congruent point. Au–Si: single eutectic with very low end phase solid solubilities. Al–Si: single eutectic with finite solid solubility in one end phase. Au–Ga: compound forming, multi-eutectic with a large end phase solid solubility. Al–Au: multicompound forming, single peritectic, multi-eutectic with a large end phase solid solubility. Pd–Si: multi-compound and eutectic forming with very low mutual solid solubilities. Al–Pd: multicompound and eutectic forming with very large intermediate solid solubilities. Zr–Al: multicompound and eutectic forming with practically no intermediate but one end-phase solid solubilities. (From Hansen and Anderko[69])

structure from the liquid–solid stable configuration between liquids and solidus to the solid phase below the solidus line (e.g. Au_4Al).

Another example of congruent melting is when a compound melts as a single entity, such as PdAl and Pd_2Al, while Pd_2Al_3 and $PdAl_3$ do not. A high melting point compound is characterized by a high negative free-energy of formation,[72] implying higher stability. PdAl is more stable than Pd_2Al, while $ZrAl_3$ is more stable than $PdAl_3$.

We shall use these phase diagrams for examples in a number of later sections.

1.3 SOLID STATE DIFFUSION

1.3.1 Diffusion and concentration equations

An atom is a solid, whether crystalline or amorphous, sits inside a three dimensional potential well which is separated from its neighbouring wells by a distance *a* and an energy barrier ΔE (Figure 1.4(a)). Because of its thermal

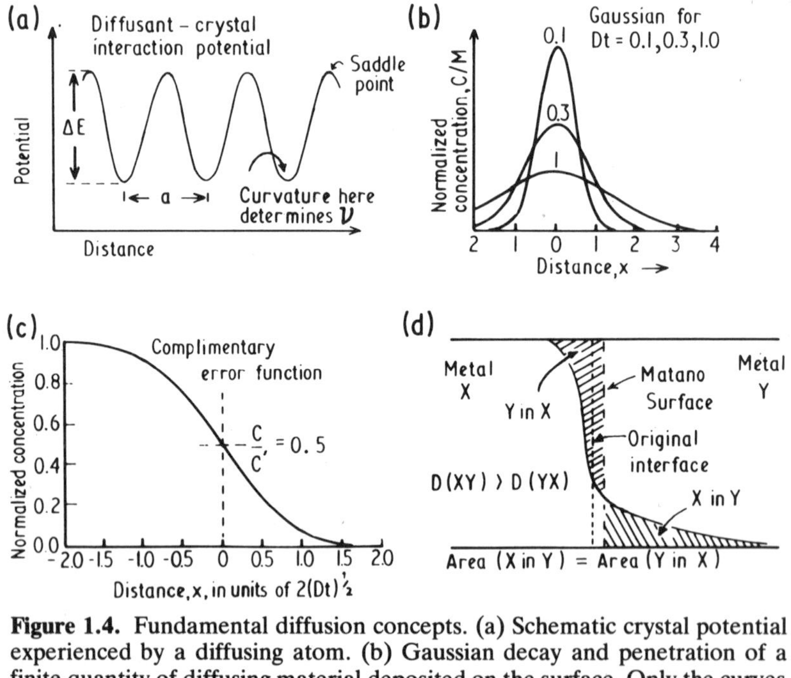

Figure 1.4. Fundamental diffusion concepts. (a) Schematic crystal potential experienced by a diffusing atom. (b) Gaussian decay and penetration of a finite quantity of diffusing material deposited on the surface. Only the curves RHS of the origin should be considered. (c) Complimentary error function distribution. (d) Concept of interdiffusion and Matano surface, see text

energy at a temperature T K it oscillates with a mean frequency ν and has a mean thermal energy of kT, where k is the Boltzmann constant. From time to time the fluctuating thermal energy exceeds ΔE and the atom makes a transition into a neighbouring equilibrium site on surmounting the potential barrier with Boltzmann's probability $\nu \exp(-\Delta E/kT)$.

If the atom is an impurity, and its concentration C varies with position in the crystal, the random walk of these atoms constitutes an effective diffusive flux $\vec{J}$ in a direction of maximum gradient (Fick's law):[73]

$$\vec{J} = -D \text{ grad } C(x, y, z) \tag{1.1}$$

where D (in $cm^2 sec^{-1}$) is the diffusivity or the diffusion coefficient, which varies with T as (Arrhenius relationship):

$$D = D_0 \exp(-\Delta E/kT), \qquad D_0 = \nu a^2 \tag{1.2}$$

ΔE is often expressed in eV when $k \simeq 8.62 \times 10^{-5}$ eV atom^{-1} K^{-1} or it may be expressed in either kcal mol^{-1} or kJ mol^{-1} in which case k is replaced by the ideal gas constant R = 1.986 cal mol^{-1} K^{-1} or 8.31 J mol^{-1} K^{-1} respectively. [Multiply kcal mol^{-1} K^{-1} by 0.0434 to obtain eV atom^{-1} K^{-1}.]

Solutions of Fick's law for various boundary conditions are available[74,75] and the simplest ones are when:[76]

(i) A quantity of M atoms of diffusants are deposited per unit area of the substrate as a very thin layer and is allowed to diffuse at temperature T K for t seconds.

$$C(x, t) = M(\pi Dt)^{-1/2} \exp(-x^2/4Dt) \tag{1.3}$$

which is half a Gaussian distribution (Figure 1.4(b)).

(ii) For two semi-infinite materials X and Y separated by an initial planar interface (i.e. $C = C'$ for $x < 0$ and $C = 0$ for $x > 0$) such that the diffusivities of X in Y and Y in X are equal and independent of concentration.

$$C(x, t) = (C'/2)\text{erfc}[x/(4Dt)^{1/2}] \tag{1.4}$$

where erfc z is the complimentary error function[77] (1 − erf z) (Figure 1.4(c)). Note that $C(x, t) = 1$ per cent at $x = 4.3(Dt)^{1/2}$ for a Gaussian and at $x = 3.3(Dt)^{1/2}$ for a complimentary error function concentration profiles. In reality, however, the diffusivity D (XY) of X in Y is different from D (YX). Then x in Equation 1.4 is measured neither from the position of the original interface, nor from the point where $C_X = 50$ per cent, but from the plane which has equal total concentration of the diffusing materials (i.e. X in Y and Y in X) on either side. This plane is called the Matano surface[78] which is characterized by equal areas under the concentration curves (Figure 1.4(d)), and can often move away from the original interface.

Also for binary interdiffusion the quantity $\bar{D}$ (XY) is often called the interdiffusion coefficient and is effectively the diffusivity of X (or Y) in a matrix of X and Y. This quantity is concentration dependent[78,79]

$$\bar{D}(XY) = D(XY)C_Y(x) + D(YX)C_X(x) \quad (1.5)$$

and is often the only experimentally measurable quantity. Suitable methods exist for extracting $D(XY)$ and $D(YX)$ from $\bar{D}(XY)$.[78,79]

Equations 1.3 and 1.4 hold only when X and Y are single crystals and only for one type of diffusion mechanism. In the next section we explore the different mechanisms for diffusion in single crystals and later we shall consider situations, such as for diffusion in polycrystals, where these equations do not hold at all.

1.3.2 Lattice diffusion

1.3.2.1 Substitutional diffusion

In a dilute solution of metal X in the crystalline matrix of Y the X atoms are partitioned between substitutional and interstitial sites. Impurity atoms at substitutional sites may diffuse either by interchanging positions with a neighbouring host atom or by the 'ring mechanism' (Figure 1.5(a)). However, the local distortion of the lattice caused by the interchanging atoms sitting at saddle points is so high that as much as five to ten eV of energy is needed to create such elastic strain. It is now well accepted that these mechanisms occur,

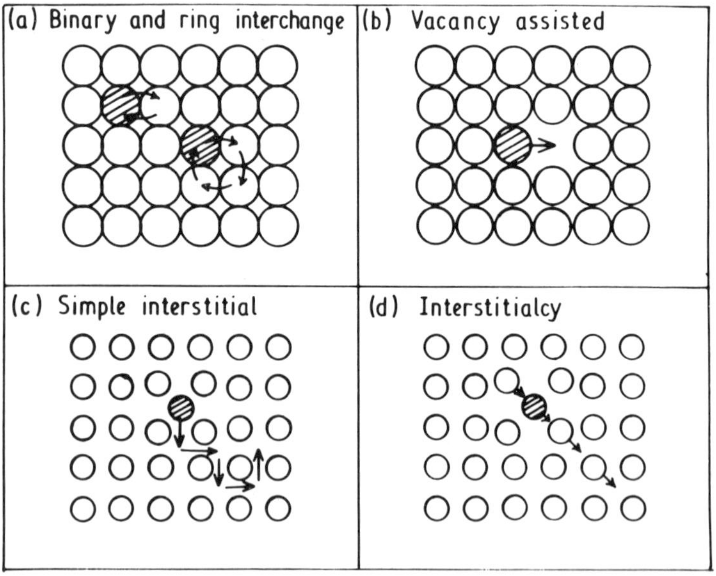

Figure 1.5. Mechanisms of diffusion in a crystalline solid

if at all, only at very close to T_m. In all f.c.c., diamond structure and some other cubic crystals as well, substitutional diffusion occurs mostly through the assistance from vacancies, requiring less energy.

By its very nature the flow of vacancies through the lattice requires an equal and opposite flow of atoms and this constitutes the diffusive flux of the solute (Figure 1.5(b)). The total free energy (Gibb's function, $G = H - TS = E + PV - TS \simeq E - TS$ for small P) needed for a single jump to occur is the sum of the free energy of the formation of a vacancy and the energy barrier encountered by the diffusant in jumping into the vacant site:

$$D = \nu a^2 \exp[(\Delta S_v + \Delta S_j)/k]\exp[-(\Delta E_v + \Delta E_j)/kT] \tag{1.6}$$

where ΔS_v and ΔE_v are the entropy and energy of formation of a vacancy, ΔS_j is the jump strain entropy change and ΔE_j is the jump barrier energy.

Vacancies are created by dislocation climb, by grain boundary migration and from the surface. We have already seen an example where the production of vacancies near the GB–surface intersection causes film strain relief (Section 1.2.3). Vacancies are also created by irradiation and then diffusion is enhanced.[80,81]

In binary interdiffusion, if diffusion occurs by the vacancy assisted mechanism and if $D(XY) > D(YX)$, more X atoms diffuse into Y than Y atoms into X. Thus matrix Y is compressed while matrix X has to absorb all the vacancies and expands as vacancies combine together to form voids having characteristic shapes.[82] The vacancies then do not help the outdiffusion of Y atoms any more. In this way the Matano surface is forced to move towards matrix Y if the expansion of X due to void formation is more than that of Y due to higher X-atom concentration and *vice versa* (Figure 1.4(d)). The generation of voids in X is referred to as the formation of Kirkendall porosity.[83]

1.3.2.2 Interstitial diffusion

Interstitial atoms can diffuse by pushing lattice atoms along a closely packed axis and by moving half of lattice distance—the mechanism being called the interstitialcy mechanism[84] (Figure 1.5(d)). More important is the process where the interstitial atom jumps into another neighbouring interstitial position (Figure 1.5(c)). The crystal strain is very much less and since the process does not depend upon the availability of a vacancy with its associated energy of formation, the activation energy for this is very much lower than any other mechanism. Consequently, the diffusivities are very large[85,86] and are comparable to those in the liquid state, $D \simeq 10^{-5}$ to 10^{-4} cm^2 sec^{-1} [87,88] which are about 10^5 to 10^9 times the self diffusivity of the host.

The partitioning of solute atoms between substitution and interstitial sites is a function of the relative sizes of solute and solvent atoms. If the former is smaller, more solute atoms sit at interstitial sites and *vice versa*. Often a solute atom sitting at a substitutional site 'dissociates' itself by the available thermal energy to move out and sit at an interstitial site, diffuses with a high diffusivity, meets a vacancy, falls into it and becomes substitutional again. Through this dissociative–interstitial mechanism rapid transport occurs even when a very small fraction of the diffusants sit at interstitial positions.

Table 1.3 gives a few representative values of lattice diffusivities of metals in metals and semiconductors according to the mechanisms by which they diffuse.

Table 1.3 Lattice diffusivities of some metals in metals and semiconductors

Diffusant	Matrix	D_0 ($cm^2\ sec^{-1}$)	ΔE (eV)	In temperature range (°C)	D at around 300 °C ($cm^2\ sec^{-1}$)	Reference
Vacancy-assisted self-diffusion						
Ge	Ge	20	3.0	~800	10^{-24}	89
Si	Si	5000	5.0	~1200	10^{-39}	89
Ga	GaAs	10^7	5.6	~750	10^{-40}	90
In	InP	10^5	3.9		10^{-28}	90
Au	Au	0.1	1.8	~800	10^{-16}	89
Ag	Ag	0.4	1.9	~800	10^{-17}	89
Vacancy-assisted impurity-diffusion						
P, As, B	Si	~10	3.4–4.1	~1000	10^{-31}	91
Ni, Pd, Pt	Ag	~10	2.3–2.4	~750	10^{-15}	92
Cr	GaAs	4000	3.4	~800	10^{-25}	90
Pt	Au	0.1–10	2.3–2.6	~900	10^{-20}	94
Interstitial fast diffusion in semiconductors and low T_m metals						
Cu	Ge	4×10^{-3}	0.33	~500°	10^{-5}	89, 93
	Si	5×10^{-3}	0.43	~500°	10^{-6}	89, 93
	GaAs	0.03	0.53	~400°	10^{-6}	90, 93
	InP	4×10^{-3}	0.69		10^{-8}	89
Au	Si	3×10^{-4}	0.40	~500°	10^{-7}	89
	InP	10^{-5}	0.48		10^{-10}	89
Ag	InP	5×10^{-4}	0.59		10^{-8}	89
Fe	Ge	0.1	1.08		10^{-10}	89
	Si	5×10^{-3}	0.85		10^{-10}	89
Li	Ge	5×10^{-3}	0.50		10^{-7}	89
	Si	2×10^{-3}	0.70		10^{-9}	89
	GaAs	0.53	1.0		10^{-9}	89
Cu	Pb, Sn	5×10^{-3}	0.35		10^{-5}	86
Au	In	10^{-2}	0.3		10^{-5}	86
	Pb	5×10^{-3}	0.4		10^{-6}	86
Ag	Sn, In	10^{-2}–10^{-1}	0.5		$10^{-7} - 10^{-6}$	86

1.3.3 Diffusion through extended defects

1.3.3.1 Fundamentals and general behaviour

We have seen in Section 1.2.2 that relatively open spaces exist adjacent to strained and un-open regions respectively at a dislocation core and a GB (Figure 1.1(b)). Diffusion along these regions occurs by the vacancy-exchange mechanism and the diffusivities are enhanced due to three main reasons:[95] (i) lower potential due to relief of strain energy, (ii) availability of a large concentration of vacancies, and (iii) a decrease in the potential barrier caused by the large separation of atoms at the dislocation cores and GBs. Diffusion along a crystal surface is also fast, the diffusion channel being enclosed by the spatially oscillating crystal potential field on one side and the potential barrier of the vacuum on the other.

Faster diffusion along dislocation pipes have been known to occur for dopants in silicon.[95,96] For arsenic $D_1 \simeq 0.50 \exp(-3.6/kT)$ cm^2 sec^{-1} is much lower than the dislocation diffusivity $D_d \simeq 10 \exp(-2.56/kT)$ cm^2 sec^1.[97] As a result short circuit occurs between the emitter and collector when As pipe diffuses through the narrow base of high frequency bipolar transistors.

Diffusion spikes, which are different from pipes, occur through low angle tilt boundaries (which are similar to small angle GBs, Section 1.2.2) in that they are two-dimensional, or, planar, in structure. Dopants reach much deeper into the crystal through these spikes due to the one dimensional diffusion away from the tilt boundary being less effective than the radial diffusion away from an isolated dislocation in dispersing the diffusive flux.[95]

A GB has an associated width, say δ, of the order of 0.5 to 1 nm. A dislocation has an associated radius, δ', or an area $A_d \simeq 1$ nm^2. Often diffusivities along these are expressed as $D_{gb} \cdot \delta$ and $D_d \cdot \delta'$ or $D_d \cdot A_d$. Self diffusivities of various f.c.c metals have been found to follow a common pattern according to their melting points, T_m.[55.98] The pre-exponent remains roughly in the range ~0.1 to 10, if we take $\delta = 1$ nm and $A_d = 1$ nm^2, and ΔE for D_l, D_d and D_{gb} respectively are 17.0, 12.5, and 8.9 times kT_m. Near T_m the GB self-diffusivity approaches that in a liquid (~10^{-5} cm^2 sec^{-1}) implying the formation of a rather disordered region around the GBs. Surface self diffusivities also show similar behaviour: $D_0 \simeq 0.014$ cm^2 sec^{-1}, $\Delta E = 6.5kT_m$ for f.c.c and $D_0 \simeq 1.0$ cm^2 sec^{-1}, $\Delta E = 9.3kT_m$ for b.c.c. metal surfaces.

Since D_0 values are essentially similar, $D_{gb} > D_d > D_l$ at all temperatures for self diffusion. Relative fluxes through a thin layer depend, however, not only upon the relative magnitude of the D's but also how many short circuit paths exist per unit area. Figure 1.6(a) shows a map of relative flux plotted against ρ_d (dislocation density) and r (grain size) as a function of T/T_m.[55] The shaded region indicates ranges of ρ_d and r values commonly found in deposited thin layers of metals. Note how the transition triple point changes position as T increases.

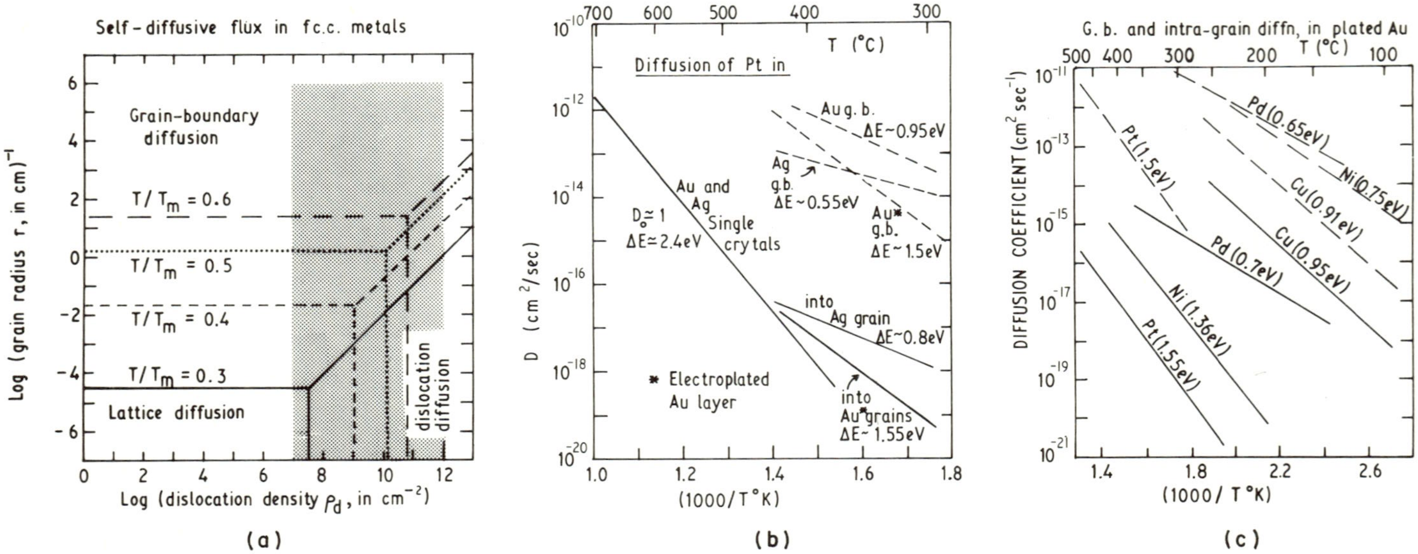

Figure 1.6. Diffusion through dislocations and grain-boundaries. (a) Relative self-diffusive flux of f.c.c. metals as a function of dislocation density, ρ_d (x-axis), grain size, r (y-axis) with T/T_m (T_m = melting point) as the changing parameter. The shaded region indicates possible ρ_d and r values in thin films. (After Balluffi and Blakely[55]). (b) Lattice, grain-boundary and intra-grain diffusivities of Pt in Au and Ag thin films. (Data from Tables 1.3 and 1.4). (c) GB and intra-grain diffusivities of various metals in electroplated gold layers. (After Hall and Morabito[113])

While substitutional lattice diffusion for impurities in metals has D_0 and ΔE values very similar to self diffusivities (Table 1.3) such a generalization does not hold at all for impurity diffusion through extended defects. Often ΔE for GB diffusion is a third or even a quarter of that for lattice diffusion and D_0 is different by many orders of magnitude. In Figure 1.6(b) are plotted D_l and D_{gb} values for the diffusion of Pt in Au and Ag. In Figure 1.6(c) are shown the GB (broken lines) and intra-grain (solid lines) diffusivities of metals in electroplated Au.[113] Table 1.4 gives some representative values of D_{gb} and D_l' (intra-grain) in thin films and it includes the cases shown in Figures 1.6(b) and (c).

The temperature at which lattice and GB diffusivities become equal is called the Tamman temperature, T_T. For self-diffusivities in f.c.c. metals and also for a number of other cases (Figure 1.6(c)) T_T is non-existent. If, however, we take the practical case of diffusive flux through a deposited thin layer, the Tamman temperature can be regarded as that below which the flux is dominated by GB diffusion which, in most cases, lies between $T_m/3$ and $T_m/2$ (Figure 1.6(a)).

Let us now consider the structure: inert substrate, metal X, metal Y, and vacuum, where X and Y are thin deposited layers, and follow the diffusion of X in Y as the temperature is raised suddenly to $T < T_m/3$. We take it that no interdiffusion had taken place during deposition or during storage at room temperature.

Within a very short period of time, such as a few tens of seconds to a few minutes, X atoms would flow into the GBs of Y and fill them up to almost saturation. This would happen due to two main reasons. Firstly, the open

Table 1.4 Grain-boundary and intra-grain impurity diffusivities in thin films

		Grain-boundary diffusion		Intra-grain diffusion		
Diffusant	Matrix	D_0(cm^2 sec^{-1})	ΔE (eV)	D_0(cm^2 sec^{-1})	ΔE (eV)	Reference
Cu	Al	10	0.81			102
Pt	Cr	10^{-2}	1.69			102
Pt	Ag	10^{-11}	0.55	10^{-11}	0.8	100
Pt	Au	10^{-5}	0.95			79
		4×10^{-2}	1.5	3×10^{-6}	1.55	113
Cr	Au	10^{-3}	1.09			102
		10^{-4}–10^{-6}	1.1–1.2			79
Cu	Au	10^{-5}	1.05			79
		10^{-4}	0.91	10^{-5}	0.95	113
Ag	Au	10^{-5}	0.63			102
Pd	Au	5×10^{-6}	0.65	5×10^{-10}	0.70	113
Ni	Au	4×10^{-5}	0.75	7×10^{-6}	1.36	113

spaces in the GBs would act as an abundant source for the supply of vacancies for the diffusants to flow unabated and secondly, the tensile stress in layer Y would be relieved as X fills up these gaps and elongates the film in a similar way as grain growth relieves stress (Section 1.2.3). This is analogous to the phenomenon of effusion, rather than diffusion, when a gas at a higher pressure fills a volume at a lower pressure. Although this has been observed in a number of cases, e.g. for the diffusion of Au in Pt,[99] Pt in Ag,[100] and Ag in Cu[60] a theoretical understanding is still lacking.

After this very fast initial penetration a steady state is reached and the diffusant starts to diffuse into the grains from the boundaries, provided a solid solution is permitted in the phase diagram. If we take $y_{gb} \simeq (D_{gb} \cdot t)^{1/2}$ as the distance of GB penetration of the diffusant and $y_l \simeq (D_l \cdot t)^{1/2}$ as the penetration distance into the grains, then three types of diffused conditions exist according to Harrison.[101,102]

Type A:	$y_l \gtrsim r$,	grain penetration is considerable
Type B:	$y_l < y_{gb}$,	some interface penetration into the grains,
and,	$y_l < r/10$,	small grain penetration from the GBs
Type C:	$y_l < \delta$,	practically no grain penetration.

Types A and B do not generally occur for well annealed large grained layers having very few lattice defects. In layers having much smaller grains, and where a large number of both point defects and dislocations exist within each grain, such as in Au and Ag (Figures 1.6(b) and (c)), lattice diffusivity is increased considerably. For example, for the diffusion of platinum in gold and silver, ΔE decreases from ~2.4 eV or D_l to 0.8 eV in Ag and 0.8 or 1.6 eV in Au for the modified, defect assisted intragrain diffusivity, D'_l. In such cases the diffusing atom X often samples all the three different kinds of diffusion paths for type A diffusion and the effective diffusivity is given by[55,103]

$$D_{eff} = D_l + fD_d \quad \text{or} \quad D_{gb} \tag{1.7}$$

where f is the fraction of atoms in the short circuit paths. Naturally, $D_{eff} \gg D_l$ and the growth of X in Y follows a $t^{1/2}$ dependence, typical of lattice diffusion.[79] When the GBs constitute the dominant short circuit paths in thin films, f in Equation 1.7 $\simeq \delta/r$[55,104] while in single crystal films and in the grains in a thin polycrystalline layer, where diffusion through dislocations dominates, $f \simeq A_d \cdot \rho_d$.[55] The latter is partly responsible for the measured values of $D'_l \gg D_l$ in thin films.

1.3.3.2 Methods for measuring GB diffusivity

In thin films, $r \sim 0.01$ to 0.1 μm and, except for TEM (Section 1.7.5), most analytical techniques such as RBS (Section 1.7.3) and AES–sputtering (Section 1.7.6) sample a large number of grains laterally. So, we are forced to deduce D_{gb} from macroscopic measurements only.[105]

For type B diffusion, if a depth profile of the diffusant is available, the Whipple analysis[106] can be applied

$$D_{gb} \cdot \delta = (\partial \ln C/\partial x^{6/5})^{-5/3}(4D_l/t)^{1/2}K^{5/3} \quad (1.8)$$

either according to Le Claire[107] when $K \simeq 0.78$ or Suzuoka[108] when $K \simeq 0.72\beta^{0.008}$, $\beta = D_{gb} \cdot \delta/2D_l \cdot (D_l t)^{1/2}$. Here x is measured from the original interface and a knowledge of D_l is needed, since the model incorporates a slow loss of diffusants from the GBs into the grains. Ideally, the defect enhanced intra-grain (lattice) diffusivity D'_l should be used instead of D_l. For $\beta = 1$ to 10^8 the two expressions are within ± 10 per cent of each other. The Whipple analysis has been applied to a number of cases.[79,100,102,109]

For type C diffusion the depth profile for the diffusion is either flat (theoretically) or has a very small slope, nearly unmeasurable. In such cases D_{gb} is measured either by observing the first appearance (e.g. a 10 per cent coverage) of X on Y surface[110] in a substrate, X, Y, and surface system or by following the surface build up of X with time.[111,112] In many cases the surface acts as a semi-infinite sink and the diffusant accumulates there in preference to remaining within the solid[112] (see Section 1.3.4). In such cases $D_l \ll D_{gb} \ll D_s$. The accumulated layer of X on Y surface then increases as[111,112]

$$C_s = C_{so}[1 - \exp(-H\tau_b)] \quad (1.9)$$

where C_{so} is the final, saturated surface concentration, $H = l\delta/2\delta'\sigma r$ and $\tau_b \simeq D_{gb}t/h^2$ where h is the thickness of the Y layer, δ' is the free surface effective width for saturation and σ is the surface segregation coefficient. Alternative to expression 1.9 a nomograph based on surface appearance measurement can be used.[110] These methods have also been used in a number of cases.[60,112,113] For type A diffusion it is harder to separate out the components in Equation 1.7. However, at least the lattice diffusivities for the diffusion of X into the Y grains at the X/Y layer interface can be measured assuming an error function profile. A method and a nomograph exist[114] for the fast elimination of errors incorporated into the measured profile from the finite depth resolution of the measuring technique. For short times and low diffusivities $(D_l \cdot t)^{1/2}$ as small as 5 to 50 nm can be isolated. In Figures 1.6(b) and 1.6(c) some of the defect enhanced intra-grain diffusivities D'_l were measured in this way.

This section primarily dealt with cases where metals X and Y form at least limited solid solutions. Compound forming system will be discussed later (Section 1.4.2). However, two important points should be borne in mind. Firstly, it is not always necessary to raise the temperature to observe diffusion. It often occurs at room temperature: Au–Sn,[115] Au–Ga,[116] Ag–In, Ga, Sn, Te[117] and Rh–Au.[59] Secondly, GB diffusion also occurs in cases where either very small or practically no mutual solid solubility exists; in such cases the diffusing material remains mostly at the GBs and does not diffuse into the grains: Cu–Pb,[58] Mo–Au,[118] Au–NiFe,[119] and Au–Rh.[59] This behaviour therefore falls beyond the jurisidiction of binary phase diagrams which only deal with substitutional solids in equilibrium.

1.3.4 Defect-diffusant interactions

In the last two sections we have discussed the mechanisms of transport in both single crystals and polycrystalline materials. In this section we shall explore some of the very large number of cases where the diffusants interact with micro- and macroscopic defects in the solid. In single crystals, vacancies and impurity atoms constitute point defects, while dislocations, stacking faults and low-angle tilt boundaries are regarded as macroscopic defects. The surface and any interfaces present in a multilayer metallization system or in a device structure involving a single crystal semiconductor and a polycrystalline metallic layer are considered as large defects. For solutes present in the grains of a polycrystalline material, the grain boundary behaves somewhat like both an interface and a (semi-)surface and is therefore a macroscopic defect.

In many cases these defects act as low potential regions for the diffusant and, when this occurs, the diffusant finds it eneregetically favourable to remain there. Table 1.5 gives some examples of this.

Often when the temperature is lowered, after a high temperature processing step or stressing, solid solubility of the solute exceeds saturation level in the crystalline matrix.[149] A not-so-fast cooling rate causes excess solute to come out and segregate at dislocations,[150] the surface,[151] and, in a polycrystalline material, at the grain boundaries.[152] In most cases the lowering of two different kinds of energy, e.g. lattice strain energy caused by supersaturation and the dislocation, surface, interaction, or GB energy, acts as the driving force. In some cases, however, the formation of compounds, such as oxides and intermetallics, acts as surface or interface sink and the diffusants are trapped even in the absence of cooling.

Simple concepts based on binary phase diagrams exist which enable one to predict when solute segregation from solid solutions would occur.[152] If the addition of solute causes the freezing temperature to come down, i.e. if the liquidus has a depression with increasing solute concentration, then surface and GB segregation occurs. The amount that would segregate is inversely proportional to the solid solubility in the bulk at the annealing temperature. These rules are based on equilibrium thermodynamics and no exception has been found.

No such simple rules exist, however, for the prediction of surface segregation of material present in the GBs of the host thin film, in the absence of a surface sink, chemical (e.g. oxidation) or physical (e.g. vaporization). While

Table 1.5 Electrical and mechanical effects caused by the interactions of solutes or diffusants with crystal defects

Interaction of solute or diffusant with	Systems	Caused by/driving force for such interactions	Physical effects, electrical and/or mechanical	Examples	Reference
Vacancies	Semiconductor single crystals	Lowered potential at vacancies	Trapping of interstitially diffusing heavy metals	Table III	Section 1.3.2
		Surface acting as vacancy source	Higher concentration near surface, U-shaped concentration profile	Au in Si	120
Vacancies and other impurities	Semiconductor single crystals	Lowering of free energy by complex formation, local strain relief	Compensation in n-type semiconductors	Au–C complex in Si	121
			Recombination or acceptor site changed into a donor site	Metal–O and metal–C complex in Si	122
			Decoration and strain field development	Cu and Li in Si	123, 124
Other similar solute atoms	Semiconductor single crystals	Formation of metallic precipitates due to rejection from super-saturated solid solution and cohesion between metal atoms in the matrix	Local high field induced Zener breakdown in Schottky barrier and p–n diode depletion regions, soft reverse characteristics	Cu, Fe, and Au in Si	125
Dislocations	Semiconductor single crystals	Initial strain relief around dislocation core, subsequent build up of local compressive stress	Decoration of dislocations	Cu in Si	126
		due to precipitation of metal atoms caused by cohesion	Emitter-collector electrical short due to current filament	Cu in Si	127

Table 1.5 *(continued)*

Interaction of solute or diffusant with	Systems	Caused by/driving force for such interactions	Physical effects, electrical and/or mechanical	Examples	Reference
Grain boundaries	Polycrystalline metal films	Reduction in lattice strain and GB	Loss of GB adhesion	O_2 in Mo, W, Ir	128–130
		surface energy	Reduction in GB diffusion	Ta in Au	131
Interfaces	SiO_2–polySi interface in MOS system	Lowering of interface free energy	Shift in MOS flat-band voltage, interface acts as infinite sink	Al	132, 133
	Single crystal InP–Al interface	Oxygen indiffusion and oxidation induced interface sink	Increase in Schottky barrier height due to MIS structure formation	P_2O_5 at interface	134
	Ti–Pd interface	Same	Outdiffusion of Ti into Pd stops	TiO_2 at interface	135
	Single crystal Si–SiO_2 interface	Segregation from super-saturated metal solution in bulk Si	Reduction in MOS inter-face charge and density of states	Ni	136
		Indiffusion through and segregation from SiO_2	Increase in density of states at interface	Au	137
	Single crystal *p*-InP–Au-Mg alloy interface	Interface segregation from bulk and some oxi-dation after deposition	Initial high contact resistance	Mg	138
	Ag–Pb interface in Au–Ag–Pb trilayer	Intermetallic compound formation acting as semi-infinite interface sink	Rapid migration of Au through Ag GBs	$AuPb_2$ at interface	139

Grain boundaries and dislocations	Electroless Ni containing P in supersaturated solid solution	Precipitation of compounds at GBs and dislocation	Developed compressive stress causes layer peel-off and cracking	Ni_3P formation	20
Surface	Polycrystalline metal X–metal Y–surface	Lower adsorption energy than in GBs, often after GB saturation, semi-infinite surface sink	Rapid GB diffusion of X through Y, surface wetting and diffusion	Ag, Cu, Co, Ga on Au	112
				Ag, Au on Ni	
				Au on Ag	
				Ag on Cu	60
				Pt on Ag	100
				Au on Pt	140
				Au, Ag on Rh	59
	Polycrystalline metal X–noble metal Y–surface or Semiconductor–noble metal Y surface	Surface oxidation acting as semi-infinite surface sink	Continuous GB out-diffusion of metal X and eventual total loss, if thin.	Ti on Au	135
				Si on Au–Ag	141
				Fe on Au	142
			Contacting problem due to surface oxide layer.	Cr on Au	143
				Cu on Au	144
			Layer crack up	Ga on Pt, Ni	145
	CVD SiO_2 and Si_3N_4 on GaAs	Diffusion through porous (?) CVD dielectrics and surface segregation	Suitability of encapsulants for annealing after implantation	Ga from GaAs	146
Vacuum	Au and Ag alloys on GaAs and GaAlAs	Diffusion of the components of compound S/C s through metals, evaporation of the volatile material (As) from surface	Preferential intermetallic formation with Ga etc. for ohmic contacts on compound S/Cs	Au	147
				AgGeNi	148
				AgGeNi	
				AgSn	

all the cases in Table 1.5 do follow the rules based on phase diagrams, there are at least two exceptions, Au on Ag[112] and Pt on Ag.[100] Non-equilibrium kinetics and deviations from binary phase diagrams are quite common in thin film interactions (Section 1.3.3) and this is one such example. Thin film contraction during cooling, which might press out materials from the GBs onto the surface, and the effects of surface diffusion on segregation, may also play important roles in this respect.

In a majority of cases these interactions cause reliability problems and device degradation (Table 1.5 here and Table 1.6 in Section 1.5.1). In some cases, however, such behaviour is used to our advantage; these include strain field generation at point defects and dislocations to render them visible by infra-red[126] and X-ray transmission,[123,124] the reduction of GB diffusion[131,152] (see Section 1.5.3 as well) and increase in Schottky barrier height on *n*-InP which generally gives low barriers.[134,153]

Perhaps the most important application of diffusant–defect interactions is the trapping of the fast interstitially diffusing metallic impurities in semiconductors, the process being called 'gettering'. The presence of metallic impurities in the active layer of a device affects device characteristics adversely: they act as generation and recombination centres and hence increase leakage in MOS capacitors in RAMs, decrease minority carrier lifetime and cause soft reverse breakdown in bipolar transistors, trap charge carriers to compensate semiconductors and many others (see Table 1.5). Gettering in silicon has been achieved mostly by generating large crystal defects, such as dislocations and stacking faults in silicon by heavy B or P diffusion,[125,154,156] by the precipitation of oxides,[157,158] by damage created mechanically[159,160] and by ion-implantation.[161,165]

In some cases complex formation in the bulk and segregation of the metals out of the semiconductors into a deposited layer[95,166] have also been considered as possible mechanisms for both Si and GaP. Gettering has enabled the removal of point defect sources, e.g. line defects and voids in reducing Cr out-diffusion from semi-insulating (SI) GaAs substrates to the *n*-GaAs active layers during VPE growth[167] and in removing Cr present in ion-implanted SI–GaAs.[168]

1.4 COMPLEX INTERACTION PROCESSES

1.4.1 Microscopic interactions at solid–solid interfaces

When two dissimilar solids are allowed to make a contact involving interatomic distances they are rarely separated by an abrupt interface. In the absence of interfacial oxides or contaminations they interdiffuse intimately even at room temperature to form a very thin (~2 to 50 monolayers thick) interfacial layer having composition almost equal to the lowest melting point eutectic (if it exists) in the phase diagram and a structure almost totally disordered, and, therefore, metastable.

In the case of covalently bonded elemental semiconductors such as Si, deposited Au forms an interfacial layer with a composition of the Au–Si eutectic[169] with $T_m \simeq 370$ °C (Figure 1.3). In doing so, the covalent bonds of Si are broken and the Si atom behaves like a metal atom with much lower bond strength with other Si or Au atoms. The diffuse interface contains the equivalent of 45 and 20 monolayers of Au respectively on (110) and (111) Si, consistent with the fact that surface energy is higher for (110) than (111) Si. The metallic Si atoms move quickly through the Au overlayer to reach the surface and oxidize even at room temperature (Table 1.5). Au also diffuses into crystalline Si interstitially with $\Delta E \simeq 0.4$ eV (Table 1.3) and moves away from this alloy layer reasonably quickly. In spite of this the loss of Au is slow and so it is believed that a reasonably high potential barrier exists at the interfaces between the Au layer and the interfacial alloy layer, and the alloy layer and bulk Si. The Au and Si atoms are thought to be ejected over these barriers before they start diffusing respectively through the Si crystal and the Au layer.

The diffuse interface hence acts as a separate entity and has the ability to trap its constituents within the two parallel membranes which separate them from pure Au and pure Si regions.

Such behaviour has also been found for a number of compound semiconductor interfaces[170] provided the ionicity of the semiconductors are not very high, e.g. for band gaps $\leqslant 2.5$ V (i.e. when the Fermi level is pinned at the surface or the interface). Room temperature alloying of interfaces has been verified by AES[170] and XPS.[44]

This diffuse and disordered interfacial layer has the composition of the lowest melting eutectic because at this concentration metastable metallic glasses are found to be the most stable in a supercooled state. Because of this, such a membrane is capable of sustaining enormous concentration gradients on its two sides. The immediate and the most useful application of the idea has been the prediction of the composition of the first nucleated phase in metal–silicide formation.[171–173] It is proposed that with the increase of temperature the width of this metastable glassy membrane increases and eventually intermetallic crystallites start to nucleate. The first point to nucleate is the most stable (i.e. of the highest melting point) congruently melting compound adjacent to the lowest temperature eutectic in the phase diagram.[171,172] In the case of silicides no exception to this rule has been found[173] and the idea seems to hold equally well for compound forming metal–metal thin film systems.[174] In the latter case, however, often the compound closest to the lowest eutectic forms first rather than the one that is the most stable,[174] but that the starting phase is a glassy metallic eutectic appears to hold in almost every case.

1.4.2 Multiphase diffusion and compound formation

To elucidate a few fundamental points let us first consider a single-eutectic binary system which forms no compounds but has finite solid solubilities at both end phases α and β. Assuming a planar initial interface and no short-circuit diffusion, after some inter-diffusion at a temperature T', the concentration profile for α would look like Figure 1.7(a). The diffusion of β in α begins with a maximum concentration $(1 - C_{\beta\alpha})$ corresponding to the maximum solid solubility of β in α at T' and, since no intermediate solid solutions exist, an interface $i_{\alpha\beta}$ separates $C_{\beta\alpha}$ from $C_{\alpha\beta}$. Such an abrupt change in concentration needs a glassy membrane having a composition similar to that at the eutectic point (Section 1.4.1). The Matano surface may not coincide with $i_{\alpha\beta}$ and the separation $x'(t)$ follows the time dependence[79,175,176]

$$\left.\begin{aligned} & x'(t) = F_{\alpha\beta} \cdot t^{1/2} \\ & \text{where} \\ & F_{\alpha\beta} = 2[\{(DK)_{\alpha\beta} - (DK)_{\beta\alpha}\}/(C_{\alpha\beta} - C_{\beta\alpha})] \end{aligned}\right\} \qquad (1.10)$$

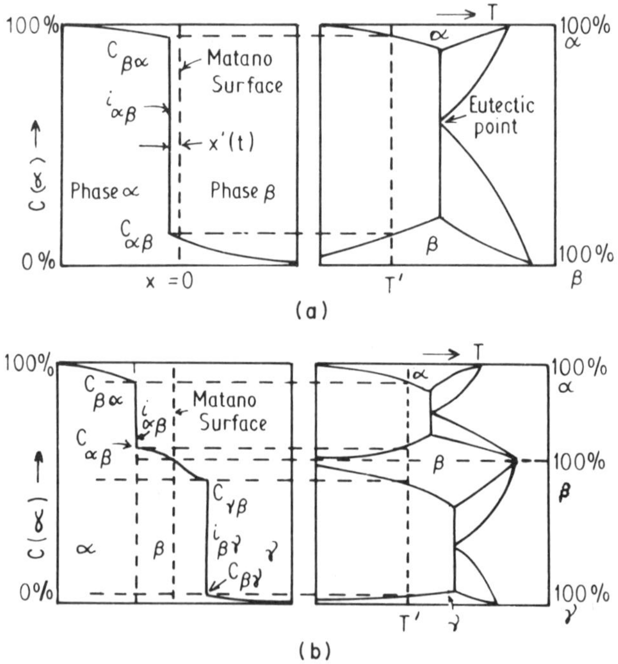

Figure 1.7. Schematic binary interdiffusion profiles for crystalline solids with (a) single eutectic and finite end phase solid solubilities, and (b) one intermediate compound phase and two eutectics with finite solid solubilities at all the three phases

in which $K = dC_{\alpha\beta}/d(x/t^{1/2})$. In a binary system α,γ having an intermediate compound phase β and mutual solid solubilities the concentration profile would look like Figure 1.7(b) and the separation between interface $i_{\alpha\beta}$ and $i_{\beta\alpha}$ is the algebraic sum of their respective distances from the Matano surface[79,175–177]

$$x'_{\alpha\gamma}(t) = (F_{\beta\gamma} - F_{\alpha\beta}) \cdot t^{1/2} \tag{1.11}$$

which is equal to the width of the compound layer. The compound layer thickness will therefore follow a parabolic growth rate with time. Any barrier, such as native oxides on Si and GaAs[178] is broken only at a higher temperature[179,180] the reaction front becomes non-planar and sideways compound growth occurs around the punctured region causing a super linear, t^z, $z > 1$, growth rate.

Because the Ds and Cs are functions of temperature, F has a complicated temperature dependence and may not follow an Arrhenius form, except when the diffusivity of one of the two components is very much larger than the other.

In situations where a number of intermediate phases exist, e.g. in the Al–Au systems (Figure 1.3) the quantity $F_{\alpha\beta} - F_{\beta\gamma}$, which is a measure of the negative divergence of diffusants in the β-layer, can be positive, zero, or negative.[55,176,177] In the two latter cases compounds cannot form at all but when $F_{\alpha\beta} - F_{\beta\gamma}$ is very small, the quantity formed may be undetectable.[55,176] A compound which cannot be observed in a diffusion couple at one temperature may be detected at another, if $F_{\alpha\beta} - F_{\beta\gamma}$ changes sign or becomes larger. Nucleation of some compounds can also become difficult[84] due to high surface tension and lack of stability at low temperature or because of peculiar crystal structure having larger unit cell volume and the strain associated with its formation. In these cases its formation becomes energetically unfavourable, especially when the enthalpy change from one compound to another is very small.[181] In the Al–Au system[55,176] Au_4Al and AuAl intermetallics are rarely observed. Instead, Au_2Al and $AuAl_2$ (the 'purple plague' culprit) are detected, the former preceding the formation of the latter. A number of similar cases exist in the literature.[55,181]

In thin films, however, atomic fluxes necessary for compound formation flow through GBs, and therefore planar reaction fronts may or may not be observed. For B and C type diffusions, compounds form around the GBs. For smaller grains and for type A diffusion the compound formation reaction front may be essentially planar.

Another entirely different method of quasi planar phase growth is when fast interface diffusion occurs[55] (Figure 1.8(a)). If phase α (metal X) is the dominant diffuser and if its diffusion rate along α–β and β–γ interfaces (where β = compound XY and γ = metal Y) is fast then GB diffusion flux will be supplied through the X/XY interface flux and phase growth will be facilitated by the XY/Y interface diffusion of the X-atoms. The kinetics of phase growth would then resemble those in a system where the layer thickness increases purely by lattice diffusion. It is only when both the interface and lattice diffusivities are very much smaller than the GB diffusivity[182] or when the grains are large[183,184] that non-planar growth around the GBs occurs. Compounds having high melting points tend to form around GBs giving non-parabolic, macroscopic growth kinetics. In the case of Ti–Pt interactions[182] (Figure 1.8(b)), TiPt ($T_m \simeq 1830$ °C) forms around the GBs of Pt while almost planar growth occurs for Ti_3Pt ($T_m \simeq 1370$ °C). For Hf–Al reactions[183,184] a fine grain Al layer has a planar and a large grain Al layer has a non-planar reaction front forming $HfAl_3$ (Figure 1.8(c)).

A non-parabolic growth rate also occurs when the phase growth is interface reaction rate controlled rather than controlled by the diffusion rate. On principle, this happens in almost every binary system at very early stages when

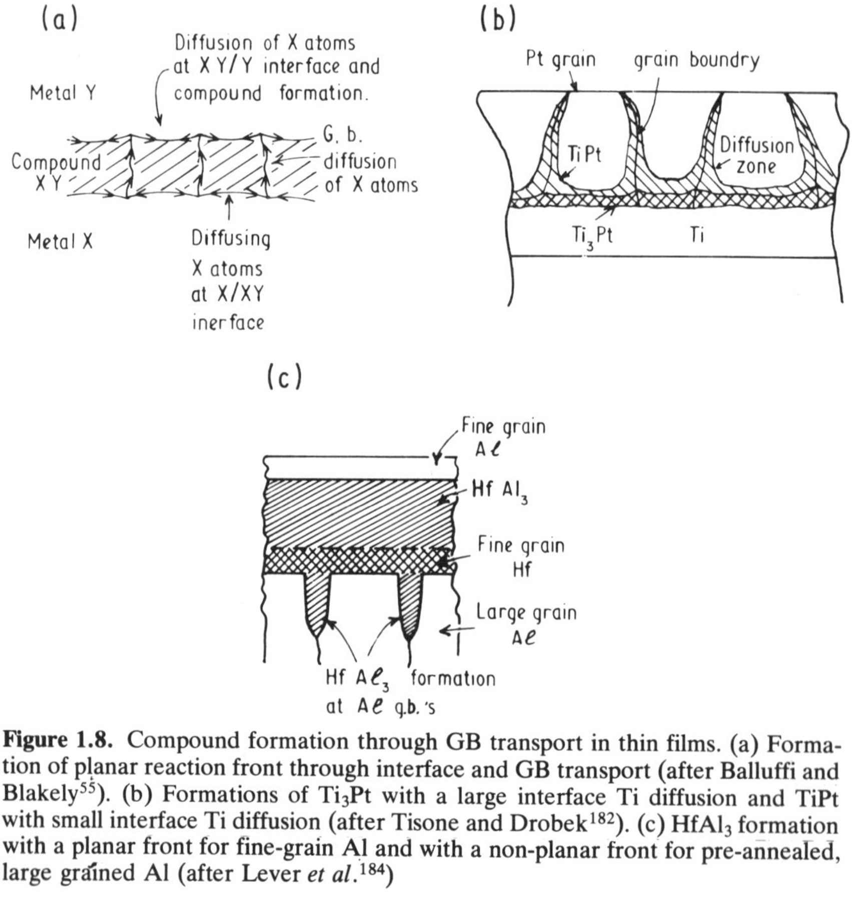

Figure 1.8. Compound formation through GB transport in thin films. (a) Formation of planar reaction front through interface and GB transport (after Balluffi and Blakely[55]). (b) Formations of Ti_3Pt with a large interface Ti diffusion and TiPt with small interface Ti diffusion (after Tisone and Drobek[182]). (c) $HfAl_3$ formation with a planar front for fine-grain Al and with a non-planar front for pre-annealed, large grained Al (after Lever *et al.*[184])

more diffusants are supplied than are consumed by interface reactions to form the compound. When the reaction rate is very slow such behaviour is observed even in laboratory time scales.[79,174,177,178,185] In general, therefore, the equation for the growth of the β-phase in the α–β–γ system looks like[186]

$$x^2 + K_1 x = K_2 t \tag{1.12}$$

where $K_1 \propto$ (interface reaction rate)$^{-1}$ and $K_2 \propto$ effective diffusivities of compound forming elements. At initial stages $x \simeq (K_2/K_1)t$ and when $t \gg K_1^2/K_2$, $x^2 \simeq K_2 t$.

In thin film reactions due to the dominance of GB transport the quantity of unreacted solid in the compound layer may be much larger than that permitted by solid solubility limits, and therefore $C_{\alpha\beta} - C_{\gamma\beta}$ in Figure 1.7(b) may be larger in magnitude. Also because of the non-planar reaction front the vertical idealized concentration profiles invariably have finite slopes, over and above that caused by the finite depth resolution of the depth profiling technique.

We have discussed in Section 1.4.1 the selection rules for the nucleation of the first phase in a binary reacting couple. As reactions proceed other compounds may form and the end phase is determined by the relative atomic proportions of the reacting components. For silicide formation on thick silicon substrates the end product is the most silicon rich compound.[117,185] In interacting metallic bilayers, such as Au–Al,[55,176] if Al is more abundant then a layered structure of Al–$AuAl_2$ forms, and if more Au is present Au_4Al–Au forms. Most metal rich compounds are then parts of the end product. Another useful example of this trend is the case of the most Al rich compound formation in the making of a sacrificial diffusion barrier (Section 1.5.2).

In ternary systems, such as bi-metallic silicide forming layers on silicon (Si–Pd–Ni, Si–Pd–Pt, Si–Pt–Ni, etc.),[187,188] or silicon–silicide–aluminium system (Section 1.5.2) or as in the case of reactive metals on compound semiconductors,[189–191] layered binary compounds form in most cases dictated by the relative diffusivities and formation energies of the components. Ternary compounds form either only at high temperatures such as 600–700 °C[187,188] which are not achieved in the case of GaAs because of thermal dissociation, or in the liquid state, during ohmic contact formation on GaAs.[192] As a first approximation to such problems, therefore, ternary phase diagrams can be disregarded. Silicide formation and metal-compound semiconductor interactions are discussed in detail elsewhere in this volume and so we shall not proceed any further.

1.4.3 Effect of corrosion on multilayer metallization

Corrosion is the destructive attack of a metal by chemical or electrochemical reaction with its environment.[193] Invariably electrochemical cells are formed in which the more electropositive metals,[194,195] such as Al, Ti, Ta, etc. act as the anode (positive electrode) and the more electronegative metals, such as Au, Pt, Cu, etc. form the cathode (negative electrode). The cell consists of a cathode, an anode, an electrically connected external path (circuit) and an internal ion path (electrolyte). In most metallization corrosion in microelectronics, moisture is an essential ingredient in forming the electrolyte while the mobile ions are usually Na^+ (from contamination and trace impurities in metals), H^+ (water, either from leaks in hermetically sealed metal packages or diffusion through plastic encapsulation),[196] Cl^- (from HCl chemical treatment or oxidation of Si slices, $POCl_3$ used in phosphorous doping and CCl_4 used either as a cleaning reagent or in plasma etching of Al and Ti), P^- ions (from PSG–phosphosilicate glass used for passivation,[197] orthophosphoric acid used for etching Si_3N_4 etc.). In corrosion terminology, oxidation involves loss of electrons and although the end product of corrosion degradation may not necessarily be an oxide, it often is.[198] Reduction is the gain of electrons from the external circuit into the electrode.

There are three possible types of corrosion cells:

(i) Galvanic cells—due to dissimilar metals. Examples: Cu segregated in Al GBs for electromigration reduction (Section 1.6.2) acts as a cathode with relation to Al;[199] Mo–Au and Ti–Au metallizations are particularly corrosion prone when Au acts as the cathode.

(ii) Concentration cells—due to differences in electrolytic ion concentration with identical electrodes. This often arises when polarization of ions occurs after a current-flow through the electrolyte is interrupted.

(iii) Anodic cells—due to externally applied voltage. The anode is oxidized which often dissolves in the electrolyte, and at the cathode metal is deposited as in electroplating. Electropolishing of metals and semiconductors and anodic–plasma oxidation of, e.g. GaAs and InP, are examples of macroscopic anodic cells.[200]

As in the case of interfacial compound formation, the rate of corrosion is determined not by equilibrium thermodynamics but by reaction kinetics. Interface or surface barriers, such as Al_2O_3 on Al and Cr_2O_3 on Cr help prevent corrosion. Chloride ions destroy Al_2O_3 to form $AlCl_3$ and corrosion proceeds. $AlCl_3$ breaks up into $Al(OH)_3$ and Cl^- ions in contact with moisture, Cl is thus regenerated and corrosion continues unabated not only for Al but also for Ti and Cr metallizations.[201,202]

The absence of pinholes in, e.g. Au or Ag on Ti or Cr prevents cell formation. But if pinholes are present, the large Au cathode area and the small Ti anode area causes devastating local corrosion at a very fast rate. Presence of an intermediate barrier, such as Pd or Pt, which reduces the number of pinholes, helps reduce such cell formation considerably.[201–203]

In microelectronics, corrosive electrochemical cells are naturally microscopic. In some cases, however, the end result is a macroscopic failure. For example in Al–Cu cell formation[199] at Al GBs, grain-to-grain adhesion is lost and grain separation followed by open circuit results, caused by the large tensile strain commonly found in these layers. In anodic corrosion in Al–Cu, corrosive degradation often occurs at the negative (cathode) side of the conducting stripes. In metallization involving silver, chlorine induced corrosion has often been observed with the formation of AgCl.[100,201]

1.4.4 Interactions in the liquid state

While we did set out to discuss mainly solid state interactions in this article, it would remain incomplete without exploring the degradation mechanisms involving alloys in the liquid state, mainly eutectics. Such interactions have often been called 'liquid metal corrosion' but we prefer to reserve 'corrosion' to cover cases where only electrochemical reactions occur (see the previous section).

Low temperature eutectics are often used in device metallizations such as ohmic contacts to GaAs,[204,205] InP[138] and Si,[206] Pb–Sn solder contacts to IC leads and Au–In preforms for mounting device chips on headers. On the debit side, vertical electromigration or filamentary current (or both) induced liquid metal failure of power microwave devices has remained a perennial problem in both industry and laboratories (Section 1.6.3). Here we shall describe eutectic formation by thermal processes only.

In most cases regardless of what separates the eutectic-forming metal from

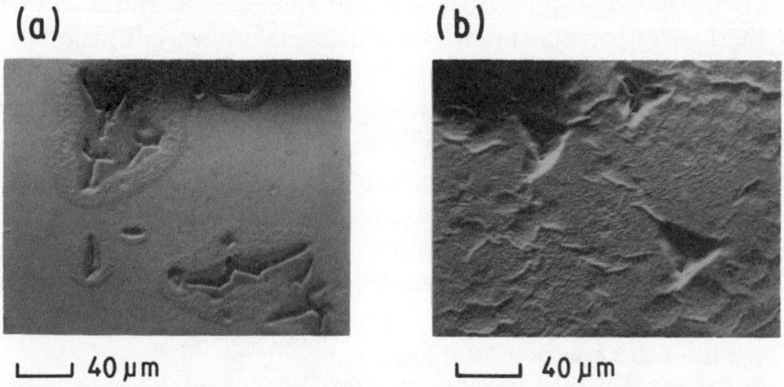

Figure 1.9. Liquid phase Au–Si interaction in (111) Si (substrate) Ti–Pd–Au (each about 100–200 nm thick) structure. (a) After 30 minutes 360 °C anneal and (b) after 30 minutes 410 °C anneal in dried forming gas (10%H_2–90%N_2). Localized gross interactions start occurring at temperature close to Au–Si eutectic temperature (370 °C) and at higher temperatures extensive alloying is observed with possible eutectic penetration and cracking of Si along (110) type planes intersecting along <111> axis which is perpendicular to the (111) type surface, in (b) (from Mukherjee and Morgan[209])

the semiconductor, as in $Si–SiO_2–W–Au$,[207] Si–Cr–Au,[208] Si–Ti–Pd–Au and Si–Ti–Pt–Au,[209] if the temperature is raised above the eutectic point (~370 °C for Si–Au) the metal reaches the semiconductor by some means not clearly understood. It could be simple GB diffusion, however small, or penetration through pinholes or cracks, or even lattice diffusion. Subsequent interaction is rapid when the local liquid formation at the interface adjacent to silicon induces sufficient stress to crack open the intervening layer(s) to accelerate eutectic formation on a massive scale and at a very fast rate, probable involving diffusion in the liquid state (see Figure 1.9).

Solder contacts of 95 per cent Pb–5 per cent Sn on Al are often separated by Cr, Cr–Cu, or Cr–Pd–Cu contacts. If the Pb–Sn alloy comes in contact with Al then Al–Sn eutectic (278 °C) segregates at Al GBs. Either due to tensile strain, inter-grain adhesion is lost and the stripe fails through open circuit (a case of liquid metal embrittlement) or, under current stressing, the liquid migrates fast through the GB network, increasing intergrain resistance which causes further Joule heating, higher migration rate and eventual failure due to thermal runaway.[199]

1.5 DIFFUSION BARRIERS

1.5.1 The need for diffusion barriers in microelectronics

We have seen in the last few sections how corrosion is inhibited by an interposing layer of noble metal and how much we need to safeguard against the direct contact of Au with Si. In Table 1.6 further illustrations are given of some of the many problems in microelectronics that need diffusion barriers as solutions.

While GB diffusion dominates the scene of potential failure or degradation mechanisms, lattice diffusion also plays a devastating role by increasing the electrical resistivity of stripe conductors.[113,199,226,227] In integrally heat sunk (IHS) devices lattice diffusion would cause the thermal impedance to increase as well and, in view of low temperature diffusion of solutes into grains containing a large concentration of defects (Section 1.3.3) such a mechanism is significant. Note also that intermetallics more often than not have higher resistivities than the component elements.[176,199,218]

In many devices multilayer metallizations are used and often three layers are adequate. The first is the adhesion layer (A), generally Ti, Cr, Ta, etc., which are capable of reducing SiO_2, Ga_2O_3, etc. locally[79] to form a strong adhesive bond with the semiconductor surface and the surrounding oxide. At the middle is the barrier layer (B) whose task is to stop solid state interactions as much as possible including the loss of Ti or Cr into the top conducting or contact layer (C) which is often Au (see Table 1.5).

Table 1.6 Some degradation and/or failure oriented problems where one or the only solution is a diffusion barrier

Case	System	Device use	Problem	Caused by	Solution	Reference
1.	Poly-Si–Al	Gate contact in MOS ICs	Post-metallization sintering 30 min at 450 °C changes MOSFET threshold voltage	Almost total loss of poly-Si in the Al GBs and the diffusion of Al in poly-Si GBs to contact SiO_2. Al acts as metal plate with a different work function in MOS system	(i) Deposit Al-4 at % Si (ii) Pre-anneal poly-Si at 900°–1100 °C for 30 min to increase grain size (iii) Use diffusion barrier	132, 210
2.	Poly-Si–Au	Rare	Si dissolves in Au on temperature increase, on cooling large Si crystallites form starting at Au GBs	High GBs surface energy in small grained poly-Si encourages grain growth through solid phase epitaxy	Barrier between poly-Si and Au needed if top contact to be used for Au wire/tape bonding	211, 212
3.	*n*-Si–Al	Al Schottky barrier diodes in bipolar transistor clamping and ICs	Post-metallization sintering results in barrier height increase of Al SBDs. Also soft reverse characteristics and/or lowered reverse breakdown voltage	Si dissolution in Al and Al spiking in Si. Recrystallization of Al-doped Si underneath Al to form *n*-*p*-Al (Camel diode) structure with higher effective barrier height	(i) Deposit Al-4 at % Si (ii) Use transition metal silicides as Schottky barrier (iii) Avoid direct Al contact, use diffusion barrier	199 213–215
4.	Silicide–Al	Silicides for Schottky barriers on n-Si and Al for interconnects in ICs	(i) Barrier height reduces from 0.85V (PtSi) to 0.60 V and ideality factor *n* increases to ~2.0.	Formation of $PtAl_2$ Schottky barrier in parallel with PtSi.	Diffusion barrier needed	199
			(ii) Barrier height increases from 0.66 V (NiSi) to 0.76 V.	NiSi is converted to $NiAl_3$ which forms new Schottky barrier on *n*-Si.	Diffusion barrier needed	216
			(iii) In Si–Pd_2Si–Al structure barrier height increases and *n* increases from 1.04 to 1.33	Al diffuses fast ($\Delta E \simeq 0.5$ eV) through Pd_2Si, forms *p*-layer and increases barrier height. Possible parallel diode formation increases *n*	Diffusion barrier needed	231

Table 1.6 (*continued*)

Case	System	Device use	Problem	Caused by	Solution	Reference
5.	Al–Au contacts	Au or Al wire bonding on to Al or Au pads on ICs and discrete devices	Resistance increase, adhesion loss and bond pull-off	Formation of $AuAl_2$. Commonly called 'purple plague' due to its colour. Adhesion loss due to brittleness, stress and porasity	Diffusion barrier absolutely essential	176, 217, 218
6.	Metal–PbSn solder contacts	Connections in thin or thick film hybrid circuits	Resistance increase and loss of bond strength. Mechanical breakdown of conductors stripe	Formation of AuSn, $PdSn_4$, etc. Al–Sn eutectic formation and liquid metal embrittlement	Diffusion barrier essential	219 Section 1.4.4
7.	Metal–Au	Au plated Cu wires used for thermo-compression bonding in hybrids and discretes	Poor T.C. bonds having low adhesion	Cu diffuses through Au GBs to reach surface and oxidize to form Cu_2O which acts as a barrier against bonding	Perform TC bonding in inert atmosphere. Use diffusion barriers	113, 144, 220
		Au plated Cu blocks for mounting discrete devices and ICs	Device comes off header during mechanical stressing	Extensive Cu diffusion into Au to form CuAu and $CuAu_2$ in inert atmosphere or Cu_2O on Au surface in air (during TC bonding). Much less Au diffn. into Cu causes micro	Diffusion barrier essential	221–223

				and macro-voids (Kirkendall porosity) to form near interface and contact area becomes smaller. Brittle compounds and less contact area causes bonding loss		
8(a)	Semi-conductor–metal–Au.	Au used as a heat sinking material in integral heat sink (IHS) flip-chip microwave power devices (TEDs, IMPATTs, TRAPATTs and FETs)	Device impedance reduction and catastrophic short circuit failure	Metal spiking in active region, short circuiting of the device by liquid eutectic formation	Barrier to Au indiffusion with no pinholes and a substrate with very low dislocation density needed	Section 1.6.3.
8(b)	GaAs–metal–Au	Schottky gate FETs and IMPATTs, variable impedance devices, etc.	Schottky barrier degraded to a lower barrier height and higher ideality factor, *n*	Even the smallest quantity of Au at GaAs–metal interface causes extensive Ga–Au alloy formation and As loss	Barrier essential	224, 225
9.	Semi-conductor beam-lead devices	Hybrid circuits, well isolated and heat sunk devices	Leakage current paths to unwanted parts of the device or circuit	Electromigration of Au to micro-cracked SiO_2–Si_3N_4 regions, formed by moisture-halogen induced corrosion of stressed regions	Anneal out stresses, eliminate moisture, reduce corrosion by barrier metallization	232, 233

The relationships among the three layers should ideally be as follows:[228]

(i) A and C should not diffuse into B but the A–B and B–C interface bonding should be strong.

(ii) B should not diffuse into or form compounds with A and C or do both.

(iii) A–B and B–C interfaces should have very small electrical and thermal contact impedances.

(iv) B should have a uniform structure with no pinholes, low internal stress and should be chemically inert. It should also be malleable enought to withstand biaxial stresses of A and C and should be both electrically and thermally conducting.

(v) A, B and C should not have widely differing thermal expansion coefficients.

(vi) C should protect B and B should protect A from the in-diffusion of reactive gases (H_2, O_2, N_2, etc.) and other chemicals.

(vii) The couples A–B and B–C should not form galvanic cells with relatively large electromotive potentials.

(viii) The combinations A–B, B–C and C–A should not have a low temperature eutectic alloy composition.

In practice not all the conditions are satisfied simultaneously. Often (i) is allowed when the barrier is composed of a single metal such as Pd and Pt.[232] Incorporation of Ni between Cu and Au reduces the possibility of Cu diffusion into Au grains with much less increase in film resistance, although D_{gb} of Ni in Au is higher than that of Cu (Figure 1.6(c)) and the formation of NiO on the Au surface causes more bonding problems than Cu_2O.[113,144,230] In some cases a bilayer is needed as a barrier: to reduce Al–Au interactions the structure Al–Ti–Pt–Au has been suggested.[229] Here Al–Ti intermetallic formation inhibits Al–Pt compound formation while Pt acts as a diffusion barrier for Au which forms intermetallics with Ti, although Au diffuses into Pt somewhat. Preannealing and grain growth of the underlying layers such as Al often helps[132,210,229] as it reduces part of the interdiffusion driving force, e.g. biaxial strain and grain boundary energies.

Intermetallics are also used as barriers because of their relatively higher thermodynamic stability (lower free energy state). But perhaps more fascinating is when foreign atoms can be introduced into the GBs of B to hinder or block the diffusion of A and C. We shall explore these two cases in the following subsections.

1.5.2 Intermetallic and sacrificial diffusion barriers

An intermetallic barrier having a free energy state lower than other possible compounds formed with A and C at least ensures its stability, although it does not preclude the possibility that A and C might diffuse through its GBs. Care should be taken to ensure than no interstitial fast diffusion occurs since some intermetallics have crystallographic structures which have large open spaces to facilitate such a mechanism.[234]

Silicides are rarely very stable, and in any case, metals having higher heats of silicide formation might deplete the barrier. Metal–Si bond energies (in eV per atom) for some silicides are:[235] Cr (4.81)–Ni (4.98)–Pt (5.38)–V (5.41)–Ti (5.45)–Mo (6.10)–Ta (6.75)–W (6.92). Actually, since the first silicide is in contact with silicon, annealing causes Si atoms to diffuse through the silicide to form a new silicide with the top metal[235] and differences in formation energy do not play any relevant role.

The formation of intermetallics with Al, on the other hand is often dictated by the differences in formation energies of the silicides and aluminides which acts as the driving force for such a reduction (Case 4, Table 1.6).

Traditionally Ti_xW_{1-x} has often been used in industry as a barrier in Si metallization technology, as in Si or PtSi–TiW–Al[236] and its use in GaAs devices has also been suggested[237] in the structure GaAs–TiW–Au, although Ti–W is not a compound. W is immiscible in Ti, but Ti has ~12 atomic per cent solubility in W at 600 °C. Generally a bit more Ti is added during deposition and Ti–W acts as a good barrier to Au indiffusion. TiO_2 at the surface acts as a corrosion resistant layer while good Ti–Au or Ti–Al and Ti–O bonds ensure adhesion at the semiconductor and metal interfaces. At higher temperatures, such as 550°–600 °C, Ti reacts with both Si and GaAs and therefore TiW is basically a metastable barrier. Alloys have negative free energy of formation and may be useful as barriers, but a lowering of free energy cannot be satisfied simultaneously for both the A and C layers.

There is an interesting case where the 44 atomic per cent Pd + 56 atomic per cent Ni congruently melting alloy ($T_m \simeq 1237$ °C) was electroplated on GaAs and no reaction between Ni or Pd and GaAs was found after a 300 °C 10 min anneal.[238] In view of the rather reactive nature of the Ni–GaAs[145] and Pd–GaAs[191] interfaces, such an inertness is unusual. It is not clear if plating induced impurities played any role here since no composition analysis is available.

Often an intervening metal starts reacting with either or both of the materials on its two sides but retards their mutual interaction. Ti, V, Cr, etc. act as the adherent as well as a barrier metal between Si and Al[239] and form compounds such as $TiAl_3$, VAl_3, $CrAl_7$, etc. with Al but not with Si at the temperatures involved (~300°–350 °C). Such a compound forming elemental barrier inhibits interdiffusion as long as part of it remains unreacted and is therefore called a 'sacrificial' barrier. In Pd_2Si–Cr–Al system[240] Cr forms $CrSi_2$ with Si and $CrAl_7$ or Cr_2Al_{11} with Al and ceases to be a barrier when consumed.

Among the sacrificial barriers between Si and Al the metals that form low melting aluminides generally do so at a faster rate than those having higher melting points, and with the exception of Ti, are better barriers.[241] Metals belonging to the former group form $PdAl_3$ ($T_m \simeq 785$ °C), $PtAl_3$ (787°), $CrAl_7$ (790°) and $NiAl_3$ (854°), have phase diagrams similar to that of Pd–Al in Figure 1.3 with the exception of Cr, and follow a parabolic growth rate with an almost planar reaction front (Section 1.4.2). Those in the latter category form $TiAl_3$ (1340 °C), VAl_3 (1360°), $HfAl_3$ (1400°), $ZrAl_3$ (1580°) and $TaAl_3$ (1700°), have phase diagrams similar to that of Zr–Al in Figure 1.3 and follow a non-parabolic growth rate since the reaction front is rugged (Figure 1.8(c)). As mentioned in Section 1.2.4, high T_m compounds are more stable and a correlation can be seen.

1.5.3 Stuffed barriers

In many situations the annealing ambient has been found to affect GB diffusion: (i) At 350 °C Au diffuses through Pt GBs at rates 15–30 times faster in N_2 or He ambient than in vacuum.[99] (ii) At 300°–400 °C the diffusion of Co in Au is lower in N_2–H_2 (forming gas) than in pure N_2.[242] (iii) Surface build-up of carbon due to hydrocarbon cracking in vacuum caused by electron beam reduces the surface diffusion of Ag on Au.[243] (iv) In Ti–Rh–Au system Au diffuses into Rh at 3–4 atomic per cent at room temperature, but not so in Ti–Au–Rh system due to unknown gaseous contamination of Rh.[59] (v) Adsorption of hydrogen in Al GBs reduces GB diffusion of Au and hence intermetallic formation.[244]

While the reason for GB diffusivity enhancement is not yet understood, its reduction is attributed to the filling up of the open channels through which the flow of diffusants occurs. The addition of Ta to Au reduces self-diffusion through GBs and sub-GBs (dislocations) while lattice self-diffusivity is increased.[245] It is also well established that the mere addition of a solute is not sufficient for such effects: GB segregation is necessary.[152]

Such a process of 'stuffing' the grain boundaries with impurity atoms reduces GB diffusion of either the host or foreign atoms by one or more of the following mechanisms:[245]

(i) By increasing the base line of the potential minima, especially when a high negative free energy compound is formed at the GBs between the host and the stuffed atoms.
(ii) By increasing the potential barrier ΔE.
(iii) By decreasing ν, by having lower bonding with the diffusant.
(iv) Because of (i) and (ii), by inducing the diffusant to flow along a different short circuit path having lower diffusivity.
(v) Because of chemical bonding and GB pinning, etc., by making vacancy formation more difficult.

The value of a in Equation 1.2 should not be allowed to become larger since this would increase the magnitude of D quadratically.

In Section 1.3.4 we have seen a number of cases where solute metals and oxygen segregate at grain boundaries of the host. In general GB diffusion of oxygen and nitrogen through metals and intermetallics causes oxides and nitrides to form at or near the vicinity of the grain boundaries.[198] In such cases, inter-grain electric current flows by tunneling through the insulating thin oxide, but most transition metals form carbides and nitrides that are extremely good electrical conductors,[228,246] and so film resitivity is rarely increased.

Because of this, thin metal films with segregated metals and nitrogen at GBs and partially or totally nitrided layers sandwiched between layers of the elemental metal such as Ti, Mo, or Ta, have received considerable attention lately as candidates for diffusion barriers. These metals, when deposited under UHV together with subsequent metallizations in a single pumpdown, interact readily with Au and Ag.[118,247,248] If oxygen or nitrogen or both together are allowed to react either during or after thermal evaporation,[249,250] or if the metals are RF sputtered in the presence of a small quantity of N_2 mixed with Ar[251–254] the barrier properties are improved considerably. Even Ti–W barriers have been improved[255] and there are indications that some of the early successes of this barrier alloy were due to the inadvertant inclusion of nitrogen from the deposition ambients. The excess Ti segregated at GBs together with N_2 forms TiN and suppresses GB diffusion.

Grain boundary stuffing has also been made using exotic combinations such as: (i) Au and Ni to stuff Ta_2N layers to stop oxygen indiffusion,[256] (ii) dopants such as P to stuff polySi GBs to stop Si self-migration which is responsible for the loss of Si into Al GBs[257] and (iii) Pt to stuff Ta to improve its ability as a barrier between Si and Au.[258] Also electroplated Pt containing K and Sb impurities from the electrolyte has been found to increase ΔE of GB diffusivity of gold from ~0.95 eV to ~1.65 eV through GB segregation[259] and has improved the barrier properties of Pt at the low temperature range. The rather high ΔE values shown in Figure 1.6(c) for the GB diffusion of Pt in Au and of intra-grain diffusion of Pt and Ni in plated Au grains are also believed to be due to the presence of some impurities deposited from the plating baths.

1.6 ELECTROMIGRATION

1.6.1 General concepts and fundamentals

During thermal diffusion an atom first reaches the saddle point (Figure 1.4(a)). In an isotropic solid such as a cubic crystal (i.e. of f.c.c., b.c.c. or diamond structure) it then has equal probabilities of falling into any one of the potential wells surrounding the saddle point in three dimensions, including its original position. When an electric current flows it has a higher probability, depending upon the electric field or the current density or both, of falling preferentially in a potential well situated towards or against the electron current depending upon whether the atom has a negative or a positive charge state. Atoms therefore migrate in a direction not dictated by the concentration gradient, as in thermal diffusion, but by the direction of electric field or current or both, the phenomenon being called electromigration.[260]

The ion flux $J = Nv$, where N is the number density of the solid and v is the mean ion velocity, $v = \mu F$, where μ is the mobility and F the effective force on the metal ion at the saddle point. The Einstein relation connects μ with diffusivity, $\mu = D/kT$, where D follows the usual Arrhenius relationship, Equation 1.2. The force acting on the ion is in two parts. The first is due to the electric field which is generally very small because of the screening of the metal nuclei by the electron shells and because the field that a metal conductor normally sustains is very small anyway. The second is caused by the scattering of the moving electrons, the so-called 'electron wind' effect, which is dominant. For convenience these two are combined together into an 'effective charge' Z^* of an ion[260,261] and $F = Z^*qE$, where $Z^* = Z^*_{el} + Z^*_{wind}$ which is the number of electronic charges the metal atom ought to have in order to experience the same force caused exclusively by the electric field, E. The metal atom really does not have a charge Z^* but is affected by the electric current density j which is connected to E through $j = \rho E$, ρ being the resistivity of the material. On combining all these we get

$$J = Nq\rho j(D_0/kT)(Z^*_{el} + Z^*_{wind})\exp(-\Delta E/kT). \qquad (1.13)$$

Electromigration has been observed in metal single crystals and in polycrystalline wires and thin films.[261,262] The difference between the two former cases and the latter are twofold:

(i) The resistivity ρ of a sample is due to the scattering of electrons both by thermally vibrating lattice atoms and by imperfections (defects, GBs and surface): Matthiessen's rule.[263] At room temperature the electron mean free path (m.f.p.) in the metal lattice is in the range of 20 to 40 nm[263,264] while in the 200–300 °C range the m.f.p.s are half to three quarters as much.[264] In single crystal and polycrystalline wires having grain size $r \gtrsim 1$–$2\ \mu$m, lattice scattering determines ρ[57] while in thin films with $r \sim 10$–30 nm, and thickness $\sim$50–100 nm, GBs and surface scattering have the dominant effect on resistivity.

(ii) Air suspended thick (>few microns) metal single crystals and wires can carry, at the most, $j \simeq 10^3$ A cm^{-2} when the temperature reaches 400–600 °C, and for $j \simeq 10^4$ A cm^{-2} they usually melt because of Joule heating (note: power density $= \rho j^2$ in W cm^{-3}). A thin metal conductor stripe deposited on a semiconductor, with or without an intervening thin (100–300 nm) layer of oxide is extremely well heat sunk and even for $j \simeq 10^6$ A cm^2 the stripe temperature rarely exceeds $\sim$200 °C.[260–262]

Both due to the enhanced electron scattering at extended defects and the relatively lower temperatures of operation electrotransport occurs predominantly through the GBs, at a rate about 10^2 to 10^6 times that through the lattice and, in some cases, along surfaces in thin metal films on semiconductor devices.[265] When GB transport dominates, N and Z^* in Equation 1.13 are replaced respectively by N_b and Z_b^*.[266] The former is difficult to define because of atomic relaxation in GBs and the latter is often different from the Z^* in the lattice because of local changes in the electric field and the electron wave functions of both the metal atom and the electron wind. Also, a multiplication factor δ/r is added to the expression in Equation 1.13 (Section 1.3.3).

In general, failure occurs when the vacancies flowing opposite to the migrating atoms coalesce along GBs which join end to end to create a discontinuity of the stripe. This open circuit type failure is catastrophic. In some devices and ICs even a small increase in the resistance of the stripe, due to local mass depletion, can cause device or IC malfunction and may be regarded as a failure criterion.

While at low temperatures a steady electro-mass-transport occurs through GBs, causing stripe resistance to increase, catastrophic failure is often preceded by mass flow divergence caused by local heating for $j > 10^6$ A cm^2 and lattice electromigration also plays an important role. Even today there exists confusion and apparently contradictory explanations of reported facts. Stripe conductor failure is often preceded or caused by cathodic depletion but this does not necessarily imply that the metal ions move towards the anode at all temperatures, current densities, stress conditions through all the different possible paths (e.g. the lattice, GBs and the surface) in stripes deposited on diversely different substrates. Nonetheless it is now well accepted that in Au layers deposited onto an adhesion layer such as Ti, Ta, Mo, etc. the Au atoms move towards the anode.[249,266–268] In alloyed ohmic contacts on GaAs the Au atoms also move from cathode to anode causing depletion of the drain (the drain is for electrons, positive) side of the active region of power MESFETs[269,270] (see Figures 1.10 and 1.11(d)). For aluminium, the most important interconnecting metal in microelectronics, the ions move towards the anode most likely both through the lattice and GBs[260,271] in the region where it fails, that is at high temperatures. At lower temperatures and current densities the direction may be different and even the occurrence of surface backstreaming cannot entirely be ruled out.[266]

We shall explore the different possible failure mechanisms in the next section.

In general Z^* in Equation 1.13 decreases with increasing temperature[249,266,272] and its effect is to reduce ΔE.

The median time to failure (MTF) t_{50}, which depends upon the ion flux divergence, follows the general form

$$\text{MTF}, t_{50} = (KhwlT/j^n D_0)\exp(\Delta E'/kT) \qquad (1.14)$$

(a)

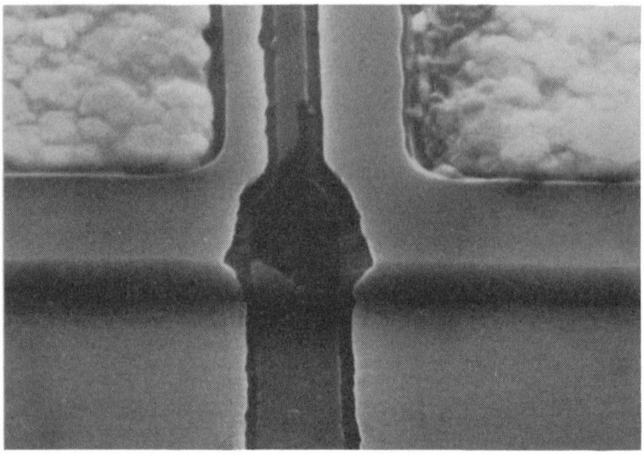

(b)

Figure 1.10. Electromigration induced degradation and failure in 1 watt GaAs microwave (X-band) power FET units having conventional interdigital layout geometry with Al gates and In–Ge–Au Ohmic contacts. (a) Gate void caused by current crowding rather than step coverage related metallization discontinuity. (b) Migration of Au in In–Ge–Au from the hotter drain regions to the cooler padded region at the top. The large plated gold source close to the active region keeps the source contact sufficiently cool and practically no Au build up occurs there. (By courtesy and permission of P. M. White of Plessey Research (Caswell) Ltd., Allen Clark Research Centre)

where K is a constant, h, w and l ($\leqslant 250\ \mu m$[266]) respectively are the height, width and length of the conductor having a low ($\leqslant 0.5$) aspect ratio (h/w). The exponent $n \simeq 1$ for a well heat-sunk conductor operating at a reasonable temperature, e.g. $\leqslant 150$ °C, $n \simeq 2$ to 3 when temperature divergence exists somewhere along the stripe (i.e. for a not-so-well heat-sunk stripe operating at high current densities) or n can be as high as 10 to 15 at very high current densities.[260,262,266] K is a function of the failure mode and this, together with a changing n (with j) implies that any extrapolation of accelerated life testing carried out at high j and T to low stress (normal operation) regions is valid only if extreme precautions are taken.

In accelerated life testing the MTF is obtained by plotting failure time against cumulative per cent failure in a log-normal plot[273,274] and then by finding the time t_{50} to 50 per cent failure. The divergence σ, given by the slope,[114] increases as the number of independent failure modes increases. In practice a high MTF with a large divergence may represent poor reliability because a large number of devices may fail at shorter times than, e.g. devices having somewhat lower MTF and a very small divergence.

The alternative to log-normal distribution is the Weibull distribution[275,276]

$$P_s(t) = \exp[-(pt)^{\alpha}] \tag{1.15}$$

where $P_s(t)$ is the percentage of surviving objects at time t, p is the failure rate and α, whose value may change with different ranges of time, indicates different failure modes.

1.6.2 Failure modes in lateral electromigration and their remedies

Local flux divergence causes film cross-sectional area to decrease. This can happen either when the film thickness becomes smaller or when a vertical void appears along a GB, which effectively reduces stripe height. Flux divergence can occur (see Figure 1.11):

(i) At a GB triple point, where two outgoing GBs flux paths against one incoming path creates local positive flux divergence and a void is produced[262] (Figure 1.11(a)).

(ii) When grain sizes change abruptly from large to small, the large number of outgoing flux paths through the small grained region creates voids where the large and small grains meet[262,266] (Figure 1.11(b)). Grain sizes may change due to unequal grain growth during post metallization sintering or when the nature of the underlying surface changes along the stripe. For example, when the stripe goes from a single crystal semiconductor surface onto the surface of an insulating film (SiO_2 or Si_3N_4) (Figure 1.11(c)) or when it goes from one over to a second polycrystalline underlayer, grain sizes, texture, or all three may change.

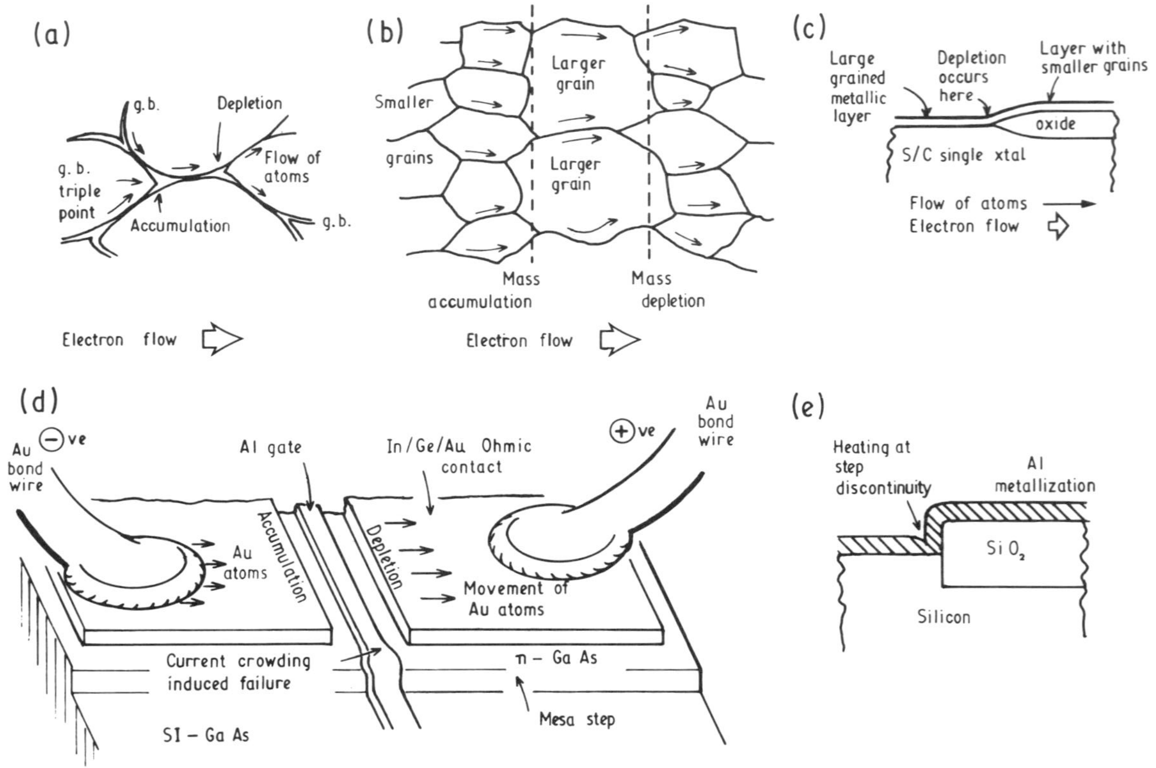

Figure 1.11. Possible failure modes caused by electromigration. (a) Accumulation and depletion at GB triple points caused by flux divergence. (b) Flux divergence due to spread in grain sizes. (c) Flux divergence due to change in film structure from single crystal to polycrystalline–amorphous underlayer. (d) Schematics of failure mechanisms in GaAs FETs. An accumulation may cause nodules to protrude out and short circuit the gate. (e) Flux divergence caused by localized heating at a metallization step discontinuity

(iii) In regions where current crowding occurs[266] either because of a gradual change in conductor width or when the uniform stripe collects charge along its length such as in the gate of a normally-off MESFET during the turning on period or of a normally-on MESFET at reverse bias due to leakage.[270,277] (See Figures 1.11 (d) and 1.10 (a).)

(iv) When any physical parameter associated with the stripe induces a local hot spot, material is removed quickly from the heated region while the colder part is unable to supply an equal atomic flux and voids are generated. Such hot spots can be caused by current crowding, (iii) above, or by local contamination and structural discontinuity in the stripe, such as a metallization step[278,279] (Figure 1.11(e)).

We shall now explore a few practical metallization systems and summarize suggested remedies.

1.6.2.1 Aluminium

Al usually has a very thin layer of surface oxide which partially inhibits vacancy generation necessary for mass transport. But internally generated vacancies migrate and create local stress which often cracks the surface oxide open and hillocks form at the expense of void production elsewhere. In many cases the discontinuities produced bear no relationship to the cathode or anode side of the stripe.

Because of surface oxides, surface transport does not play a significant part in failure. The activation energies for failure in small grain conductors vary from 0.48 to 0.60 eV;[260,262] in texturized Al in which migration has to occur from one open space to another through an intervening 'fit' region (see Figure 1.1(b)), $\Delta E' \simeq 0.73$ eV; in large grain (1–8 μm) Al $\simeq$ 0.9 eV;[280,281] and in single crystals $\simeq$ 1.3 eV;[262,282] all of these having similar surface conditions.

Surface passivation by BSG and PSG reduce electromigration failure considerably not only by reducing vacancy formation but also because of grain growth during glassing at 400–600 °C.[262] Grain growth by preannealing helps but a divergence in grain size does not.[57] Large grained Al stripes with CVD glass passivation has $\Delta E' \simeq 1.2$ eV, close to that for a single crystal layer.[260]

However, an increased $\Delta E'$ does not necessarily ensure an increased MTF, the pre-exponent in Equation 1.14 has to be higher as well. The additions of 2–4 atomic per cent Cu,[262] 6 atomic per cent Mg,[283] 2 per cent Cu–2 per cent Si, a few per cent Cu–few per cent Mg, Cu–Mg–Ni[260] have all increased the MTF of Al conductor by factors from 10 to 100. Both Mg and Cu segregate at the GBs to form Al_3Mg_2 and $CuAl_2$ respectively and hence reduce D_{gb} by two orders of magnitude in the range 100–200 °C while increasing D_l by about a factor of 10 for Al self-migration. Unfortunately, Cu itself migrates towards the anode[272] and its depletion near the cathode causes discontinuity failure.[284]

More recently the annealed, sandwiched Al(Cu)–Tr.M–Al(Cu) structures, where Tr.M are transition metals such as Cr, Ti, Pt, Ni, Cu, Hf, Zr, Ta, etc. have yielded extremely long lifetimes,[241] regardless of the T_m of the compounds formed (see Section 1.5.2). Similar increases were also achieved by the use of rare-earth–aluminium intermetallics.[285] In all these cases the intermetallics formed at the GBs inhibiting GB motion as well as hindering vacancy creation and flow.

1.6.2.2 Gold and Silver

Gold stripes fail both at random spots because of local discontinuities and by cathodic depletion (Section 1.6.1). Unlike Al it does not form surface oxides and small increases in resistance are often caused by surface ripples as opposed to hillock or void formation in Al. The ΔE values for Au drift velocity lie in the range ~0.6–1.0 eV.[249,268,286] Most life test data give $\Delta E \simeq 0.9$ eV, suggesting GB transport.[266]

Silver allows substantial surface migration in the 100–200 °C range with $\Delta E \simeq 0.3$ eV while in the 200–300 °C range transport is mostly through GBs with $\Delta E = 0.9$ eV.[266] Because of this high surface activity, vacancies can move along the surface, and become deposited at GB-surface intersections causing local thinning of the film. The phenomenon is called 'grooving'[262,265,287] and occurs in the absence of internal void formation. Surface activities occur in Au to a lesser degree.

To reduce surface generation or movement of vacancies or both in Au and Ag, glassing may be useful but only with an intervening adhesion layer such as Ti and Cr.[262] Even a thin layer of deposited Cr reduces Ag surface migration.[287] In Au the addition of Ta reduces GB self-diffusion and hence electrotransport.[245,266,288] As discussed earlier in Section 1.5.3, GB segregation is essential for a solute to have any desired effect.

Recently, electromigration failure of heavily doped poly-Si conductors have also been observed, the failure being caused by the migration of dopants, B or P, with localized increase in resistance followed by heating and thermal runaway.[289]

1.6.3 Vertical electromigration

Lateral electrotransport limits the reliable operation of conductor stripes on discrete devices and ICs. A few other mechanisms which are somewhat related but in many respects different from above, have been responsible for the direct degradation of some devices. We would like to categorize these under 'vertical electromigration' since the mechanisms are dependent upon the electric field or the associated current flow perpendicular to the metal film(s) deposited onto a semiconductor surface or both. The degradation or failure modes are diverse and have not been studied as extensively as lateral electromigration. We shall give a few examples to illustrate the complexities involved.

All the devices involved are power devices of some kind, especially the power microwave diodes such as TEDs, IMPATTs, TRAPATTs, etc. The most common factor among all these is either a high DC or RF current flow (e.g. in power Gunn diodes) or a reasonably small current flow at a high reverse breakdown voltage (IMPATTs and TRAPATTs).

At high power level operation, ohmic contacts on GaAs TEDs (Au–Ge, Au–Ge–Ni, Ag–In–Ge, etc.) melt locally, penetrate into the active epitaxial layer in the form of a liquid alloy spike, reducing the impedance of the total device and eventually shorting it.[290–292] Similar failure modes were also observed in Sn contacted, Cu heat-sunk InP TEDs where the composition of the metal spike was an alloy of In and Sn with very little P. The Sn–In eutectic temperature is 117 °C and a local melting followed by eutectic formation was suggested.[293] Such failure mechanisms are common to diodes on GaAs both with[291,294] and without[294,295] heat-sinks as well as on Si.[276] The prime requirement for these failures is a direct physical contact between the semiconductor and the heat-sinking and ohmic contact metals such as Au, Ag, In, etc. which would form low T_m eutectics with the semiconductor (Si–Au $\simeq$ 370 °C, Ga–Au $\simeq$ 340 ° C, Ga–Ag $\simeq$ 375 °C, etc.). It is believed that in most cases some form of local filamentary current flows through the semiconductor active layer operating at reverse breakdown or at forward bias causing a local hot-spot to occur. Filamentary current is generally unstable and often changes the localized impedance abruptly from a high to a low value.[296] Such a current may flow simply because of buried metal or oxide precipitates in the active region which allow local Zener or avalanche breakdowns at lower total field strengths (Section 1.3.4) or through threading dislocations even in the absence of dopants or metal preciptates.[297] Which one of these is more important is not clear but it is generally believed that crystal imperfections play a decisive role together with imperfections in the metallic layers. In spite of the intervening Ti–Pt, Ti-W–Pt or Cr–Pt barriers in, e.g. Schottky barrier on GaAs and ohmic contacts to silicon devices, the heat-sinking metal (Au) somehow migrates to reach the semiconductor–metal interface, probably in regions near the hot-spots or through metallization pin-holes. Whether vertical electromigration takes place locally or if any other mechanism (Section 1.4.4) is responsible for metal transport remains unresolved. But subsequent eutectic formation and metal penetration induced thermal runaway caused by local Joule heating is rapid and the device fails catastrophically. Similar shorting may also occur along the edges of mesa-etched diodes[276] instead of somewhere along the middle of the diode metallization. Extreme control of wafer defect concentration and metallization stress and uniformity with a minimum pin-hole density are clearly the obvious ways of inhibiting such failure mechanisms.

At avalanche breakdown conditions, both electrons and holes are generated and a very high field exists at the semiconductor–metal interface for a Schottky barrier IMPATT diode. The situation is rather different from that in lateral electromigration in metal conductor stripes where a high current flows at a low field, and so entirely new degradation mechanisms may come into play. For example, in GaAs–Pt devices the electron-hole recombination energy is believed to induce the diffusion of some defects, probably Pt ions, into the GaAs active layer from the interface.[298] Non-radiative recombination causes the energy to become converted into thermal energy (phonons) and the excited recombination centres find it comparatively easy to migrate to neighbouring equilibrium states with the associated activation energies in the range 0.2–0.4 eV. The high temperature (325–400 °C) degradation mechanisms with $\Delta E \simeq 1.6$ eV are superseded by the low temperature (250–300 °C) mechanisms with $\Delta E \simeq 0.3$ eV making accelerated life-testing at elevated temperatures erroneous when extrapolated to the usual junction temperatures of 200–250 °C.

Another example is when compensating defects (probably Cr atoms) were found to diffuse into the active *n*-GaAs layer from the GaAs–Cr interface if Schottky barrier IMPATTs at 200 °C only when stressed under avalanche breakdown.[299] Simple thermal stressing had no such effect. Whether field or recombination assisted migration or both took place is not yet known.

1.7 ANALYTICAL AND DEPTH PROFILING TECHNIQUES

Advances in the microfabrication of reliable semiconductor devices have depended primarily on device physics and fabrication technology on one hand and the analytical techniques for characterizing the various solid phase interactions occurring during processing and subsequent device operation on the other. We shall therefore discuss briefly the most important techniques commonly used in research and industry. A number of good review articles,[300–312] books[313–316] and symposia proceedings[317,318] are available on electrical, surface and sub-surface analytical techniques and so we shall concern ourselves mainly with their uses and abuses rather than the detailed descriptions of the methods themselves. Where possible, we shall mention how the techniques have helped us to arrive at the various conclusions reported earlier in this chapter.

Apart from the mechanical, optical, and electrical methods used individually or in combinations, the techniques involve some form of incident beam of waves or particles such as photons, electrons and ions (see Figure 1.12), and a very broad classification according to the nature of the incident beam is shown in Table 1.7. Unless the incident beam is well focused, as in SEM, scanning auger microscopy (SAM), scanning optical microscopy (SOM), TEM and convergent beam TEM, and STEM, only area averaged informa-

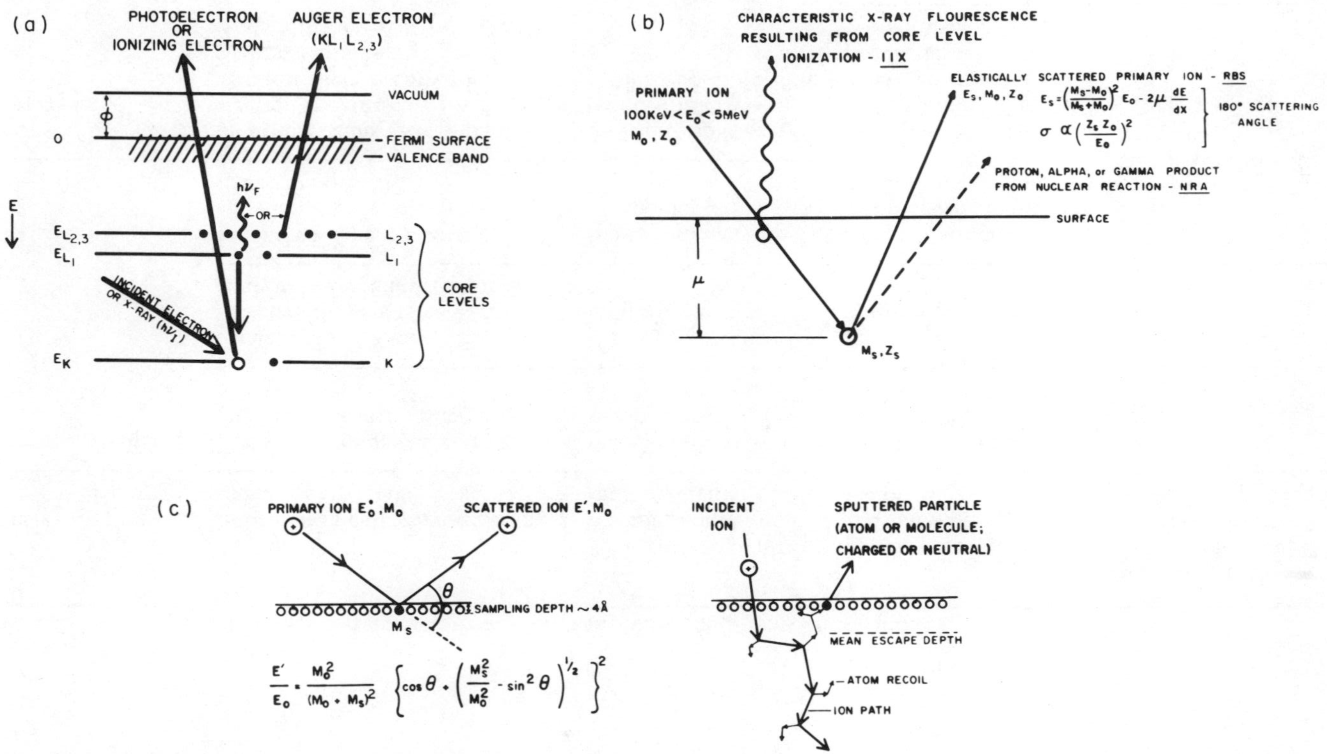

Figure 1.12. Fundamentals of materials analytical techniques, involving particle beams of various kinds (see Table 1.7) (a) Auger and photoelectron emission processes. The Auger electron has an energy $E_K - E_{L1} - E_{L2,3} - \theta$ and the photoelectron has an energy $h\upsilon_1 - E_K - \theta$. (b) Processes involving energetic ion beams: RBS, IIX, and NRA. (c) Processes with low energy light or heavy atoms as the primary beam. Left: ISS, right: SIMS, SNMS. (From Holloway and McGuire[300])

Table 1.7 Analytical techniques for characterizing/depth profiling solid-phase interactions

Primary beam	Energy	Information carrying outgoing signal	Near surface depth resolution and limiting factor	Spatial resolution and limiting factor	Relative atomic species (Z = atomic number) and material quantity detection sensitivity
Electrons	5–30 keV	Emitted secondary electrons (Scanning electron microscopy, SEM)	0.5–2 nm, electron escape depth	5–10 nm, beam diameter and escape depth	
		Characteristic X-rays (Energy or wavelength dispersive X-ray analysis, EDX and EPM or WDX)	0.5–1 μm, electron scattering volume	0.5–1 μm, electron scattering volume	Except low Z elements, 0.1–1%
		Secondary Auger electron emission (Auger electron spectroscopy, AES)	0.5–2 nm, electron escape depth	30 nm–1 μm, electron-beam energy induced disruption	Except very low Z. 0.1–0.3%
Photons (X-rays)	5–15 keV	Photo-electrons (X-ray photo-election spectroscopy, XPS or electron spectroscopy for chemical analysis, ESCA)	5–20 nm, sputtering induced intermixing during depth profiling	20 μm for scanning XPS, volume of metal where electrons are produced.	Except very low Z. 0.1–0.3%
		Bragg diffractied X-rays (XRD)		~1 mm usually	

		Anomalous diffraction and/or transmission (X-ray topography, XRT)	1–10 μm, collimation of X-ray beam	1–10 μm, collimation of X-ray beam	
Energetic light ions	1–3 MeV	Elastically scattered primary ions or atoms (Rutherford back-scattering, RBS)	20–30 nm, solid state detector 1–2 nm, magnetic analyser 2–3 nm, glancing angle RBS	0.3–3 mm usually. 5–25 μm, focused beam	0.01–1% Mass separation better at intermediate Z, quantification better at high Z
		Resonantly scattered protons, or products of nuclear reactions (NRA)			
		X-rays (ion induced X-rays, IIX)	0.1–0.5 μm, X-ray attenuation depth	Same as RBS and NRA	
Heavy ions (Ne, Ar, O_2, N_2, Cs)	0.5–5 keV	Elastically scattered inert ions (Ion scattering spectroscopy, ISS)	0.5–1 nm, subsurface scattering	1–2 μm, focused beam diameter	1 per cent
		Secondary ions (Secondary ion mass spectrometry, SIMS)	5–20 nm, beam induced roughness and intermixing	2–500 μm–ion probe. 1 μm–ion-microprobe	1 p.p.b. to 1 p.p.m.

tion is obtained. Naturally, the ratio of the dimension of relevance in the sample (e.g. the grain size, precipitate diameter, GB width, or the extent of surface wetting by the diffusant, etc.) to the incident beam diameter decides whether or not a technique should yield area-averaged information. A focused electron beam having a circle of least confusion of diameter 100 nm or an ion beam diameter of 10 μm would give us area averaged information respectively for grain sizes $\leqslant$50 nm and for non-uniform intermixing or reaction front with undulation dimentions $\leqslant$5 μm. It is therefore vitally important to use at least one each of macro- and microscopic analytical techniques to supplement each other and to avoid any misinterpretation of area-averaged information. With regard to the techniques we shall consider their trace element and compound detection capabilities and any adverse effects caused by the incident beam(s) which distort the gathered information through, e.g. the decomposition of compounds and heating or damage induced intermixing.

1.7.1 Mechanical, chemical and optical methods

Angle lapping (or bevel sectioning) together with chemical staining[319] is a traditional tool for quick estimation of silicon *p–n* junction depths. Angle lapping is also very useful for studying both planar[320] and non-planar interacted interfaces using either optical microscopy or SEM–EDX–AES,[321] for example in the studies of Si–Ti–Au and Si–Ti–Mo–Au[247] and GaAs–Au-Ge[292] contact degradation.

Selective chemical etching[6] is often used for removing unreacted metal or semiconductor layers to reveal the reaction front or recrystallized products.[192,210–212,322]

Reflectivity of metallic surfaces depends on the presence of other metals or impurities in solid solution or as compounds and its measurement has found use in the studies of interdiffusion induced alloying[323] and intermetallic formation.[324] The detection of optical phase change on reflection is the basis for phase or interference contrast microscopy[325] (e.g. the Nomarski technique) which has been used for observing carrier density changes in both Czochralski[326] and laser annealed[327] single crystal silicon. Very small phase changes are detectable and, although not very much in use in metallization induced problems, phase contrast microscopy is a powerful technique for quick assessments.

Optical transmission through most semiconductors needs visible to infrared radiation since $h\nu$ (in eV) $\simeq 1.24/\lambda$ (in μm) and the optical wavelength has to be large with $h\nu < E_g$ for transmission to occur where E_g is the band-gap of the semiconductor. IR transmission is used not only for studying complex formation in the bulk of semiconductors but also for detecting local strain fields due to metallic precipitates,[126] surface eutectic formation induced adhesion and metallization peel-off,[328] etc. Its usefulness and simplicity has caused its adaptation for routine analysis in silicon technology and it should find similar use in high reliability compound semiconductor device processing.

While both metallization and interaction induced strain are often measured by X-ray and IR transmission or X-ray diffraction techniques (Sections 1.2.3 and 1.7.7), a recent use of the Newton's ring optical interference method has claimed superior sensitivity for measuring macroscopic stress induced substrate bending.[329]

1.7.2 Electrical methods

Traditionally *I–V* (four-point and spreading resistance), *C–V* and Hall measurements are used for studying carrier density or mobility in semiconductors.[330,331] When metal–semiconductor interactions cause any change in carrier concentration or mobility or both, these techniques are used together[237,298,299] with deep level transient spectroscopy (DLTS).[331,332]

On the other hand, when a metallic layer interacts with another layer of metal or the underlying semiconductor substrate, changes in resistivity take place. Often interdiffusion has been studied using changes in resistivity[79,199,248,333,334] (see also Section 1.5.1 and Table 1.6) both in cases where intra-grain diffusion occurs[79,199,333] and when new intermetallic compounds are formed.[248,334]

1.7.3 High energy ion beam techniques

1.7.3.1 Rutherford backscattering spectrometry (RBS)

The technique has been used extensively for studying planar interactions in thin films.[335–339] It uses collimated beams of 1–3 MeV ions, mostly of helium, although protons and nitrogen ions are also used,[335] and rests on the elastic backscattering of the incident ions. The outgoing ion energy E_o equals K^2E_i for scattering from the surface (Figure 1.13(a)) where K^2 decreases with decreasing atomic mass M_2 of the target (Figure 1.13(b), top). For backscattering from the interior E_o decreases due to energy loss (Figure 1.13(b), bottom) and thus depth information is obtained. On combining these two types of information elemental quantity variation with depth (i.e. depth profiles) can be deduced for some combination of materials. When a proton beam is used, mass separation is very good for $2 < M_2 \leqslant 10$ (Figure 1.13(c)) and for helium ions it is reasonable for $10 \leqslant M_2 \leqslant 30$.[340,341] For $M_2 \geqslant 40$ mass resolution becomes so poor that atomic mass differences in the target of at least 10 to 20 a.m.u. are needed for any useful depth analysis. This is mainly because the exit ion energy depends upon the elemental composition as well as the scattering depth and confusions can and often do occur. Iterative computation is used routinely in many laboratories for depth profiling, and a program for the direct computation of backscattering spectra[342] is available which is capable of obtaining complicated depth profiles by the spectrum fitting process. All these, of course, need accurate values of ion stopping powers and ranges in the solid[343,344] and the scattering cross-sections which can be different from the calculated Rutherford (elastic) values due to minor resonances.[345] Depth resolution of RBS for normal incidence of the probing beam is typically 20 to 30 nm, limited mainly by the finite energy resolution of the solid state radiation detector which generally lies in the 10–17 keV range (f.w.h. maximum). However, for glancing angle incidence (or emergence) of the ions, depth resolution can be increased to 2–3 nm[346] at the expense of the maximum depth of analysis.

RBS gives area-averaged information since the beam diameter lies in the range 0.3 to 3 mm and non-planar interactions can cause confusion. Complementary information from SEM (EDX), chemical etching, AES, TEM or

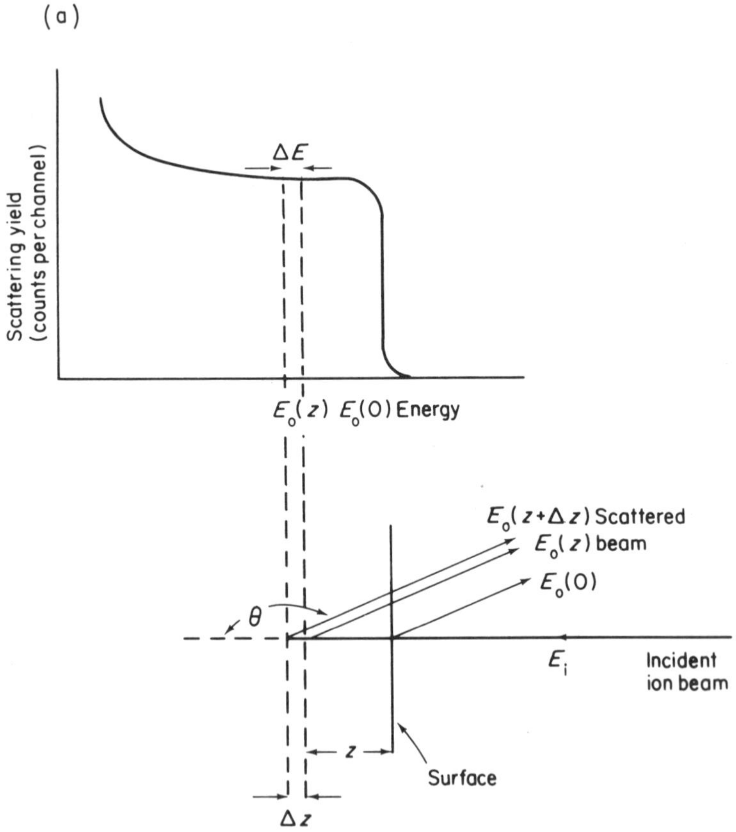

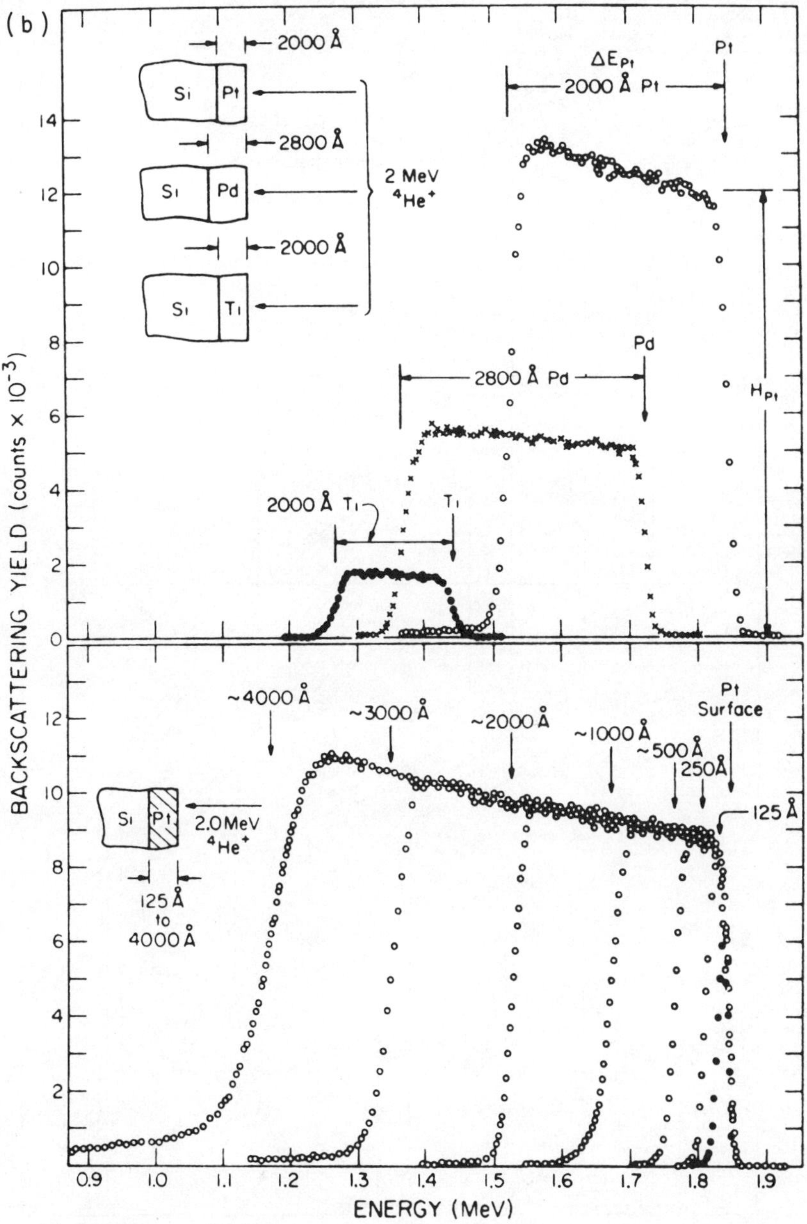

(b)
2000 Å
Si
Pt
2800 Å
Si
Pd
2 MeV
$^4He^+$
2000 Å
Si
Ti
ΔE_{Pt}
2000 Å Pt
Pt
Pd
2800 Å Pd
H_{Pt}
2000 Å Ti
Ti
BACKSCATTERING YIELD (counts × 10^{-3})
~4000 Å
~3000 Å
~2000 Å
~1000 Å
~500 Å
250Å
125 Å
Pt
Surface
Si
Pt
2.0 MeV
$^4He^+$
125 Å
to
4000 Å
ENERGY (MeV)

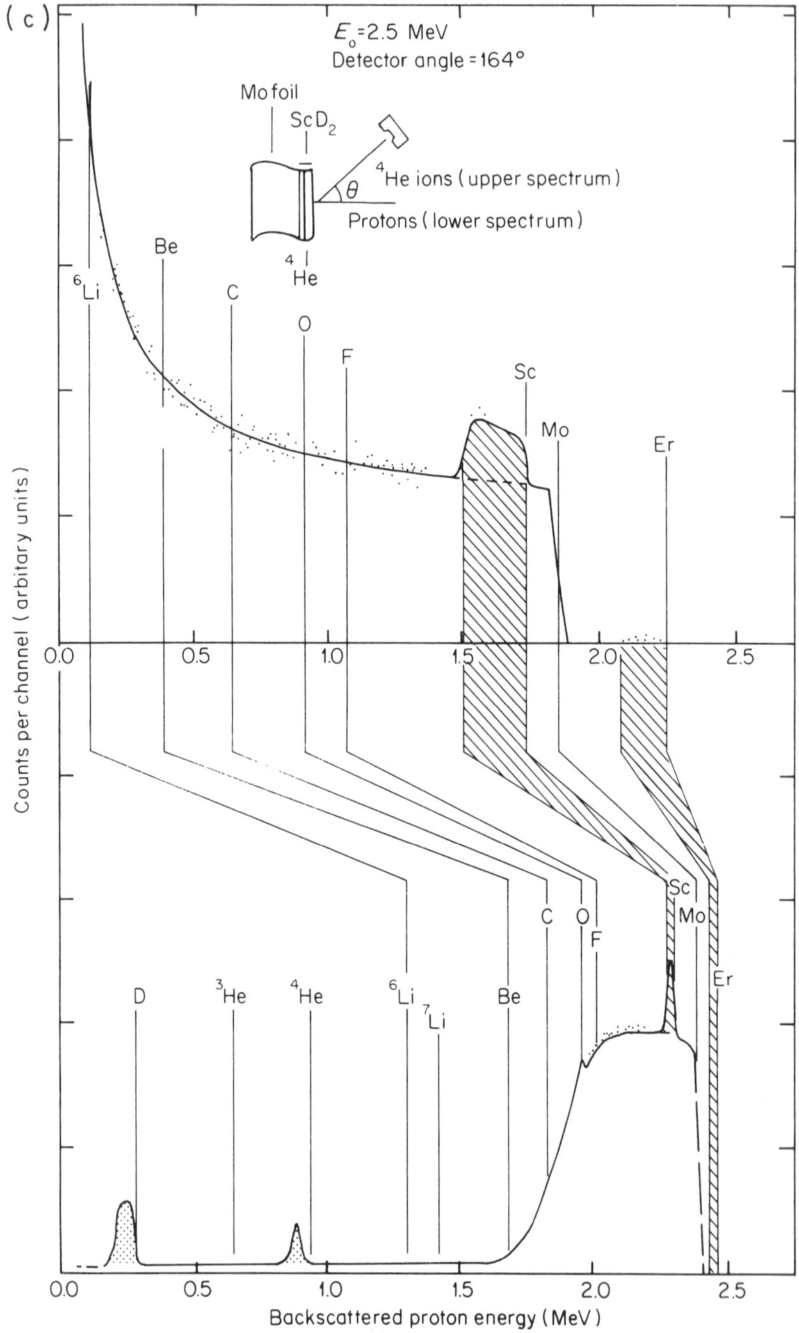
(c)
E_0=2.5 MeV
Detector angle = 164°
Mo foil
ScD_2
^{4}He ions (upper spectrum)
Protons (lower spectrum)
θ
^{4}He
^{6}Li
Be
C
O
F
Sc
Mo
Er
0.0
0.5
1.0
1.5
2.0
2.5
Counts per channel (arbitary units)
C
O
F
Sc
Mo
Er
D
^{3}He
^{4}He
^{6}Li
^{7}Li
Be
0.0
0.5
1.0
1.5
2.0
2.5
Backscattered proton energy (MeV)

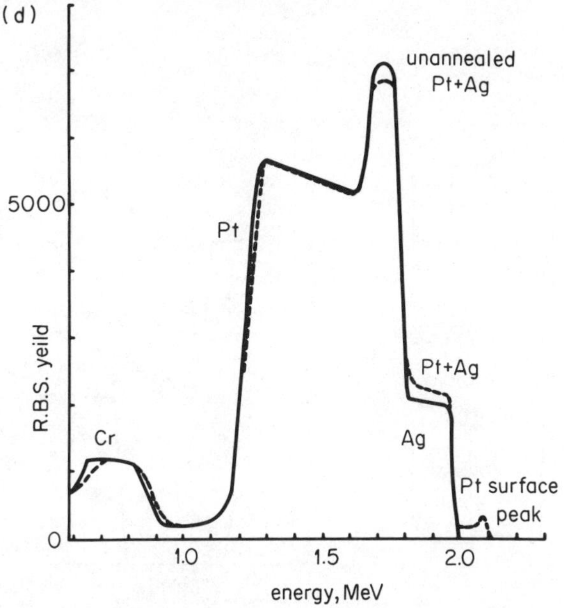
(d)
unannealed
Pt+Ag
5000
Pt
R.B.S. yeild
Pt+Ag
Cr
Ag
Pt surface
peak
0
1.0
1.5
2.0
energy, MeV

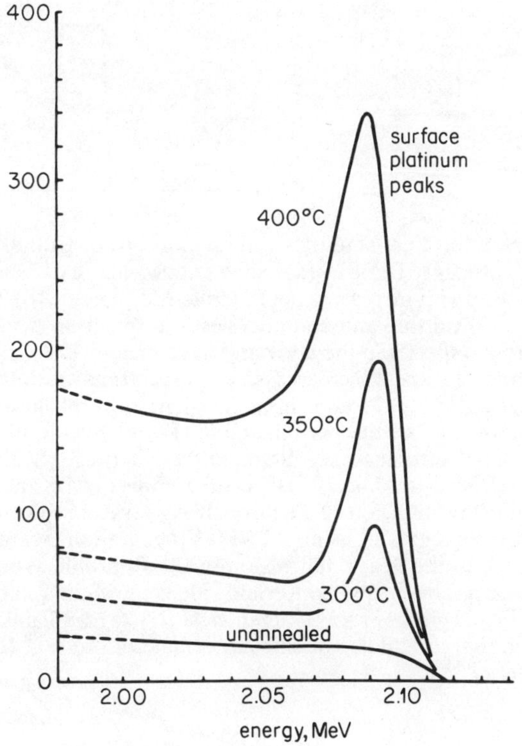
400
surface
platinum
peaks
300
400°C
200
350°C
100
300°C
unannealed
0
2.00
2.05
2.10
energy, MeV

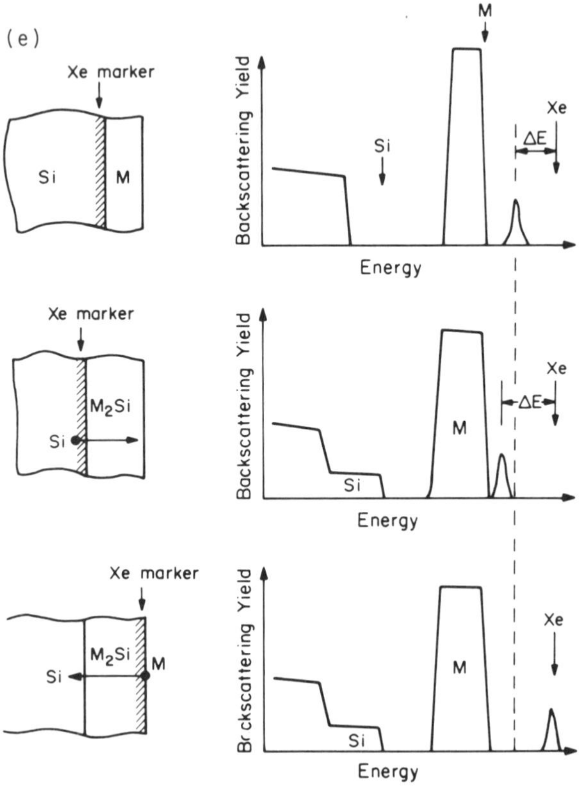

Figure 1.13. Fundamentals and applications of Rutherford backscattering (RBS). (a) Shows how ion exit energy decreases with increasing depth (from Morgan[339]). (b) Top: the yield, starting and end energies in a spectrum depends on atomic species in the layer and its thickness. Bottom: As the layer thickness increases, so does spectrum width (from Chu *et al.*[335]). (c) Protons have excellent mass resolution at the lower end of the periodic table (D to F) while helium ions show better mass resolution in the mid range (Sc to Er, here) (from Terreault *et al.*). (d) An example of RBS analysis of thin film diffusion of Pt through Ag layer and eventual surface segregation, using 2.25 MeV helium ions scattered at 135°. In the nearly flat region of the Pt profile Whipple analysis has been applied to yield values shown in Table 1.4 and Figure 1.6. (From Morgan *et al.*[100]) (e) Explains diffusion marker studies of silicide formation, see text (from Chu *et al.*[347])

even optical microscopy is essential for avoiding erroneous interpretation of data.

RBS has been used successfully in finding the diffusion species in silicide formation.[347] By incorporating an implanted layer of heavy inert gas atoms (e.g. Ar or Xe) as a marker either in the Si substrate or in the deposited metal, the displacement of the marker with relation to the surface is observed when silicides form on annealing. If the metal atoms migrate in the process then the marker moves towards the surface and *vice versa* (Figure 1.13(e)).

Generally, RBS is more suited and has been used extensively for the study of metal–metal and metal–silicon interactions where a large difference in atomic masses exists, but not as successfully in the studies of GaAs–metal interfacial reactions mainly because with a thin metallic overlayer Ga and As are not mass-separable in the RBS spectra due to their rather similar atomic weights. In the case of InP the very large difference in atomic weights between In and P almost swamps the RBS yield from the lighter P atoms (since yield $\propto Z_2^2$) within the spectrum of In.

1.7.3.2 Ion induced X-ray analysis (IIX)[348,351]

Since X-rays emitted from a depth have finite probabilities of coming out and becoming detected, it is not so useful in depth profiling the way RBS is. Nonetheless, its use in conjunction with RBS can often be used for obtaining supplementary information.

1.7.3.3 Nuclear reaction analysis (NRA)[352–353]

Because of the higher energy needed for a higher probability of nuclear penetration of the ions into the target nuclei, nuclear reactions occurring in the 1–3 MeV energy range are limited only to light atoms and has found little use in analysing metals. Most useful reactions occur in the 2–6 MeV energy range, often outside the scope of inexpensive, low energy accelerators. Low energy reactions tend to involve either an unstable isotope or deuterium and tritium ions as incident particles and the associated γ-ray production with light atoms in the scattering chamber requires shielding and remote control experimentation, which has discouraged its routine use.

All these techniques, including the detection of inner shell X-rays, are insensitive to chemical bonds, but by recording the relative atomic proportions of the constituting elements, chemical formulae of compounds can be obtained by RBS only indirectly.

1.7.4 Low energy ion beam methods

1.7.4.1 Ion scattering spectroscopy (ISS)

In this technique only the elastically scattered primary ions (such as He, Ne, Ar, Xe, etc.) in the energy range 0.2–10 keV are recorded at specular angles

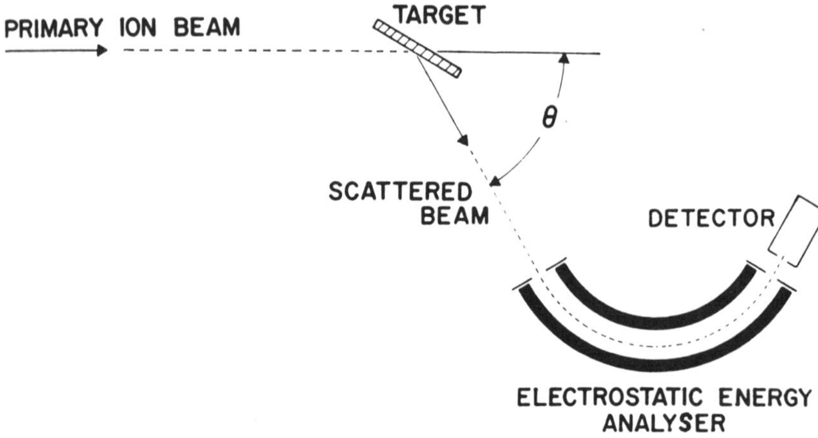
PRIMARY ION BEAM
TARGET
θ
SCATTERED BEAM
DETECTOR
ELECTROSTATIC ENERGY ANALYSER

(a)

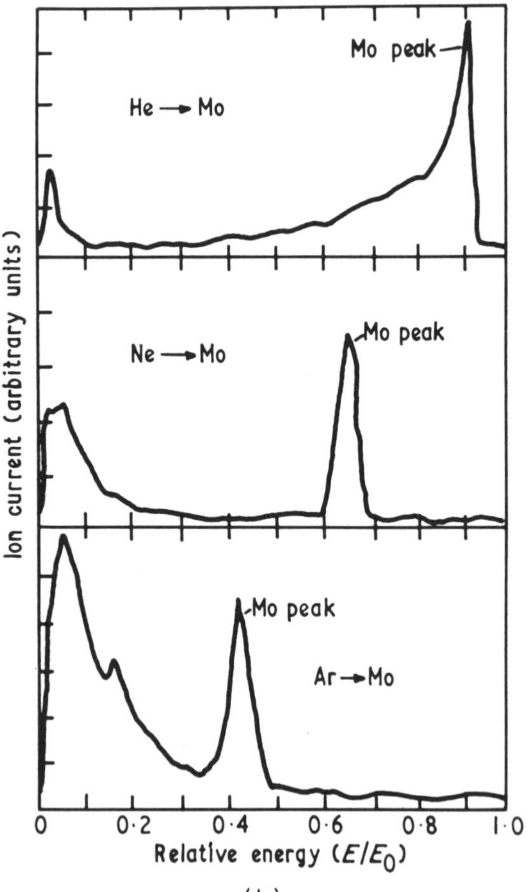
He → Mo
Mo peak
Ne → Mo
Mo peak
Ar → Mo
Mo peak
Ion current (arbitrary units)
0
0·2
0·4
0·6
0·8
1·0
Relative energy (E/E0)

(b)

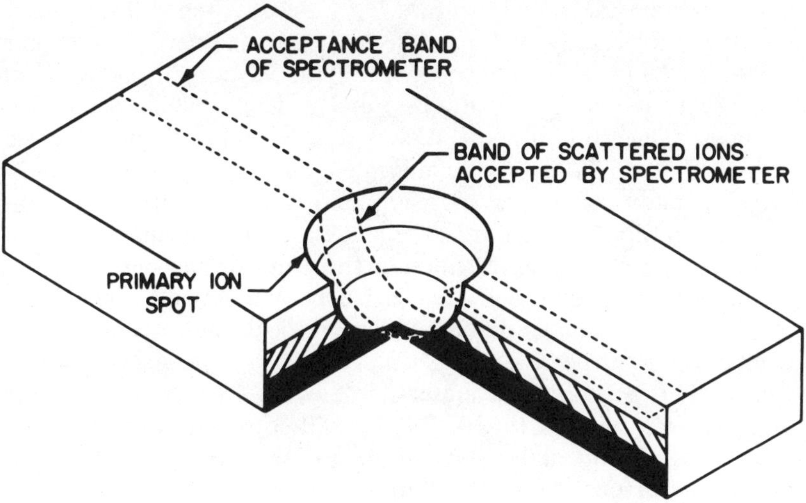

(c)

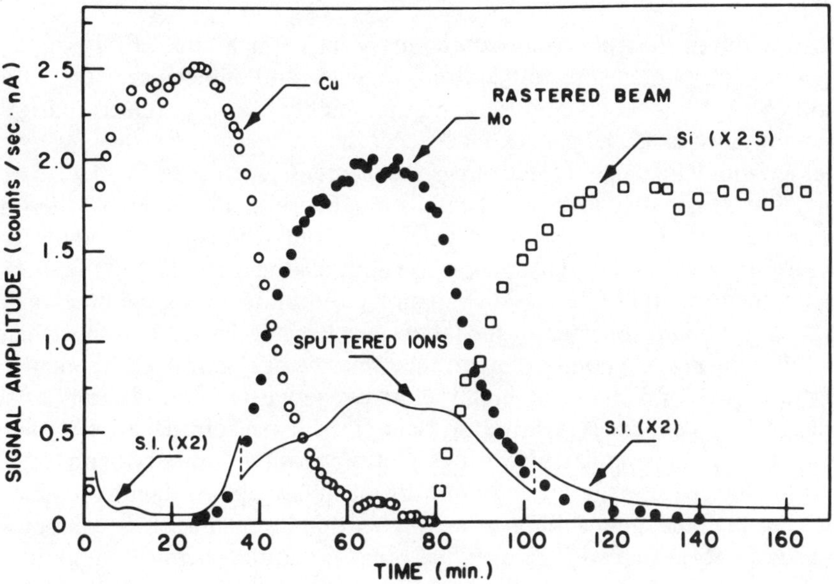

(d)

Figure 1.14. Ion scattering spectroscopy (ISS) principles. (a) Schematic set up (from Smith[354]). (b) Shows how a choice of primary particles shifts an elemental peak, and hence mass resolution around it, to the optimum energy range (from Armour and Carter[356]). (c) Exaggerated schematic of sputtering induced erosion crater formation (from Rusch *et al.*[357]). (d) Depth profile of a Si–Mo–Cu structure after crater wall induced errors are eliminated (from Rusch *et al.*[357])

of reflection from the sample surface.[354–356] Because the collision is elastic with no electronic energy exchange, the emergent ion-energy is a function of the masses of the incoming ion and target atom only; any incident ion which penetrates into the solid is neutralized and is not detected by the electrostatic analyser (Figure 1.14(a)).[354] Also chemical binding has no effect on the measurement. As in RBS, a choice of incident ion shifts the mass resolution in the atomic scale to where it is most required (Figure 1.14(b))[356] depending upon the composition of the layer to be analysed. Since the incident ion beam has a radial Gaussian flux distribution from the beam centre, the crater formed by sputter erosion has a similar shape and ions emerging from the edges of the crater makes depth profiling difficult and erroneous (Figure 1.14(c)).[357] But if the beam is rastered at least along a line and only ions from the central (flat) portion of the crater are detected, acceptable depth profiles can be obtained (Figure 1.14(d)).[357] ISS is still not in universal use but both imaging[357] and fundamental sputtering studies[358] are possible where ISS gives additional information.

1.7.4.2 Secondary ion mass spectrometry (SIMS)

SIMS involves the collection of secondary ions which could be positive or negative, elemental ions or ionized radicals and molecules.[359–368] Often 1–5 keV Ar^+ ion beams are used and then highly oxidizing (electropositive) elements such as Li, K. Na, Ti, Cr, Al, V, Zr, Ta, etc. come out as positively charged ions while highly electronegative elements, such as F, Cl, O, C, S, etc. come out as negative ions. In all cases only a very small fraction of the total number of secondary emission particles come out as ions, and the ionization process includes both the changing of the neutral atom very close to the surface during the process of its emission and direct ionization during the breaking up of chemical bonds. Because of the latter, a leakage of oxygen into the vacuum chamber during Ar^+ bombardment, the presence of electronegative elements in the matrix or the use of highly electronegative $O_2^{\pm}$ and $N_2^{\pm}$ ions as the primary beam enhances positive ion yield of the less electronegative elements dramatically. In a similar way a highly electropositive primary beam of Cs^+ ions enhances the negative ion emission of highly electronegative elements such as the noble metals. In both these cases thin layers of surface compounds or radicals form first with an eventual breakage of the chemical bond due to sputtering.

Depth profiling is performed in a way similar to ISS, by allowing the beam itself to etch through the solid to reveal underlying materials. Different schemes for avoiding the adverse effects of the sloping erosion crater walls are available (Figure 1.15(a)).[361]

Ion mass analysis is achieved by the use of either electrostatic[361,363,370] or quadrupole mass filters[362,369] and spatial resolution of 1–2 μm is achieved both by rastering a fine beam, as in an 'ion microprobe' such as the ARL-IMMA (due to Liebl[361]) and A-DIDA-3000 (due to Wittmaack[362]), or by direct ion optics, as in an 'ion microscope', such as the CAMECA IMS-300 and 3-F (due to Slodzhian[363]). Other proprietary systems[369,370] and those included in the equipment for multiple surface analysis (such as in Figure 1.17(a)) are used mainly for depth profiling only, where a large area of diameter 100–150 μm is analysed. Figure 1.15(b) shows the schematics of A-DIDA-3000,[371] the most recent addition to the sophisticated range of dedicated SIMS equipment. The schematics of CAMECA and IMMA are available elsewhere.[360,363–365]

There are a number of problems associated with sputtering induced profiling such as SIMS. Brief descriptions of these with the suggested solutions are as follows.

(i) Lower sputtering from highly packed crystal faces and the presence of low sputtering rate materials in the polycrystalline matrix causes the crater bottom to be extremely rugged, and thus depth resolution is degraded. Either the use of multiple ion-guns or rotation of the sample,

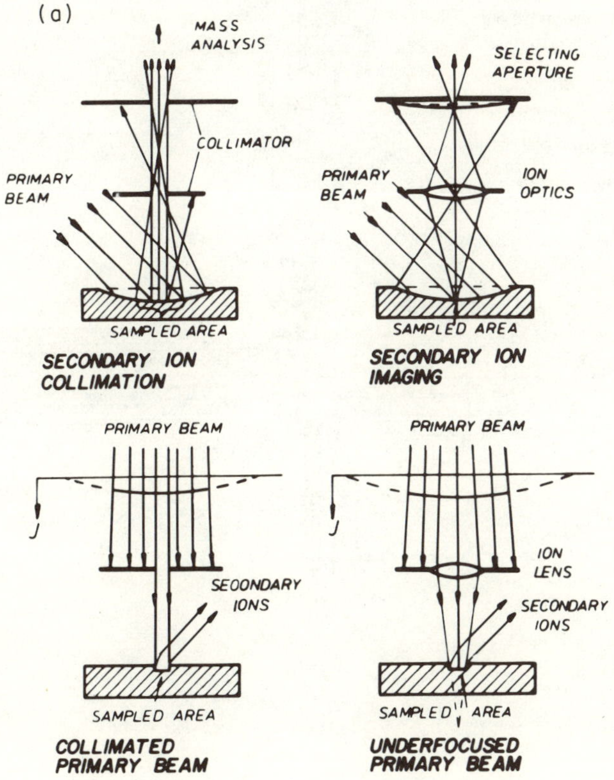

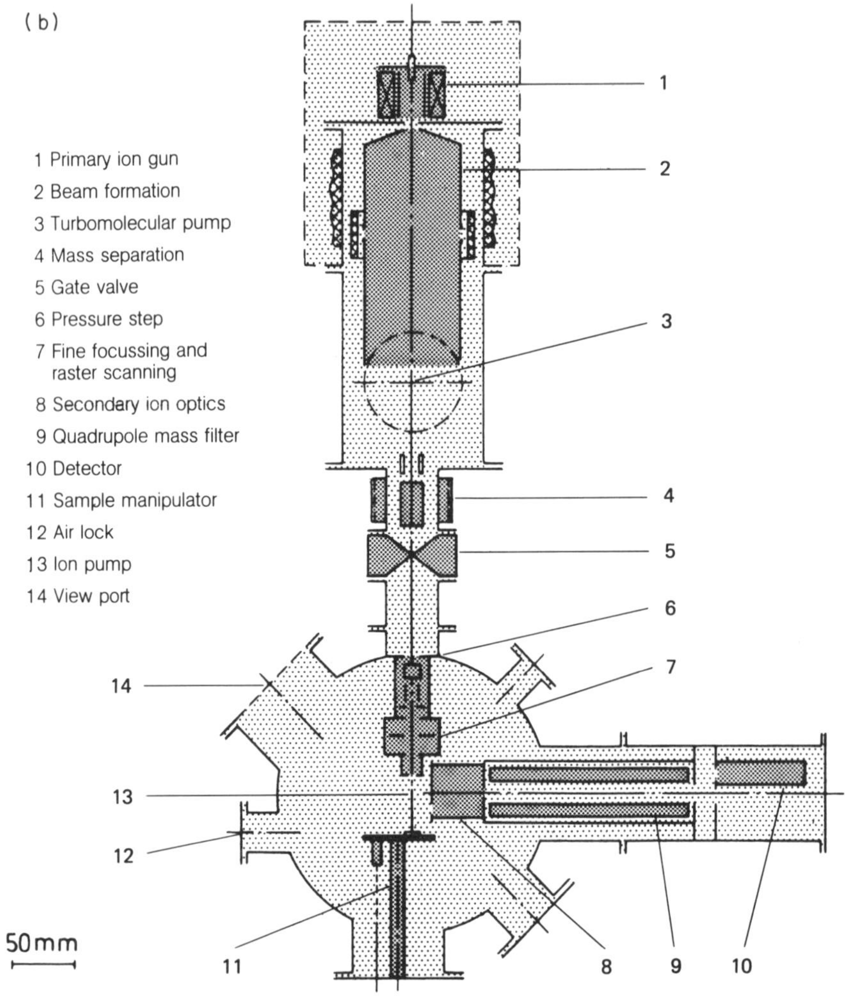
(b)
1 Primary ion gun
2 Beam formation
3 Turbomolecular pump
4 Mass separation
5 Gate valve
6 Pressure step
7 Fine focussing and raster scanning
8 Secondary ion optics
9 Quadrupole mass filter
10 Detector
11 Sample manipulator
12 Air lock
13 Ion pump
14 View port
1
2
3
4
5
6
7
14
13
12
11
8
9
10
50mm

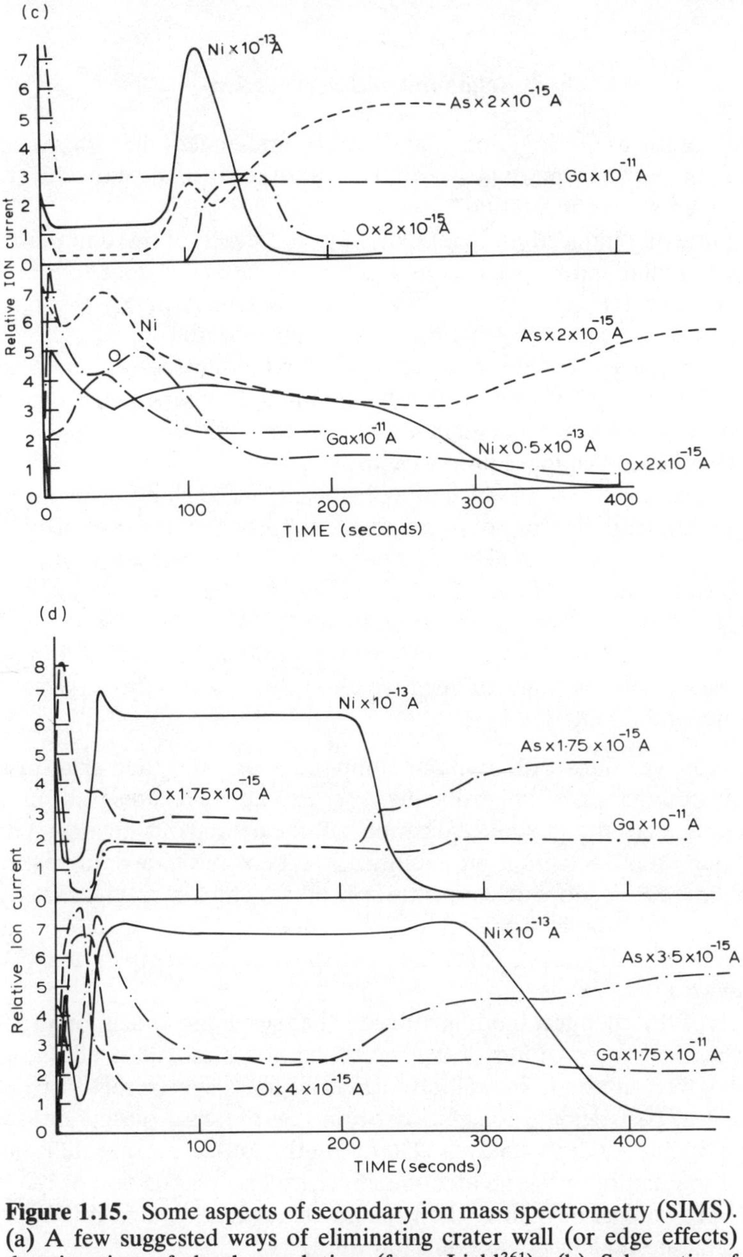

Figure 1.15. Some aspects of secondary ion mass spectrometry (SIMS). (a) A few suggested ways of eliminating crater wall (or edge effects) deterioration of depth resolution (from Liebl[361]). (b) Schematic of Atomika Ion-probe A-DIDA-3000 dedicated SIMS equipment which uses a quadrupole mass analyser (By courtesy and permission of J. L. Maul, of Atomika GmbH, W. Germany.) (c) Positive ion yield profile of as deposited (top) and 1 hour 550 °C annealed (bottom) GaAs–Ni structure using 5.5 keV O_2^{++} ions in a CAMECA IMS-300 machine. Notice that the Ga ion yield is very much higher than that of As ions and that in the unannealed sample the Ni ion emission is enhanced near the interface due to the simultaneous presence of As (from Mukherjee *et al.*[145]). (d) Similar profiles for a GaAs–Ni sample which reacted during deposition due to sample overheating, unannealed top, 1 hour 550°C annealed bottom (from Mukherjee and Morgan[312])

or the use of O_2 or N_2 ion beams are recommended, the latter is believed to cause the formation of a thin disordered or amorphous layer which prevents facet formation.[372]

(ii) Sputtering induced atomic mixing up to a depth of 5–10 nm introduces intermixing induced errors in depth profiling and reduces depth resolution considerably.[373] The only way to reduce this is to lower the incident ion energy, but then the time of analysis goes up.

(iii) Interference from ionized molecules having masses similar to the elemental isotope of interest can be eliminated by repeated plotting of an energy-bracketed region in the spectrum with time, then by visually checking the changes in ion yield.

(iv) Ion transport through insulating material due to charging changes original impurity distribution—use neutralized beams for sputtering.[374]

(v) Ion yield of a given element changes with the matrix that it is in—the so-called 'matrix effect'. A simple scaling rule based on matrix sputtering rate allows for corrections to be made.[375] This is because the degree of surface accumulation of the ionization enhancing species (oxygen for positive and caesium for negative ions) varies inversely as the sputtering rate of the matrix.

In mutlilayer films with granular compounds or intergranular diffusants or both, a combination of the problems described above creates insurmountable difficulties and often standardization of measurements against changing sputtering rates, ionization probabilities, etc. becomes almost impossible. So, except in routine dopant concentration or stoichiometry measurements in single crystal semiconductors[366–369,376] only qualitative information is obtained.[371–382] Figures 1.15(c) and (d) show such examples for GaAs–Ni interactions.[145,312]

Much of the chemical binding induced changes in the ionization probability with chemical composition of the matrix can be removed if all the emitted particles were ionized. In 'sputtered neutral mass spectrometry' (SNMS)[383] the neutral particles are ionized first in a low pressure plasma before mass separation and quantity analysis. Therefore the ion-detection rate is independent of the nature of the incident beam, chemical composition of the matrix, partial pressure of oxygen in the vacuum chamber, etc., and it is hoped that the technique finds wider use in the future.

1.7.5 Electron beam techniques

The techniques most commonly used for routine or somewhat detailed analysis of device metallization failures are:

Scanning electron microscopy (SEM)[384,385] with energy dispersive X-ray analysis[384] (EDX) or wavelength dispersive X-ray analysis (WDX) or electron probe microanalysis (EPM).[386] SEM operating in the electron beam induced current mode (EBIC),[387–389] voltage and work function contrast modes,[390–391] reflected high energy electron diffraction (RHEED),[391–393] back reflection (Kikuchi) patterns[391,394] etc.

Auger electron spectroscopy (AES),[395–397] scanning Auger microscopy (SAM),[398–401] low energy electron diffraction (LEED).[402–404]

Transmission electron microscopy (TEM), scanning TEM (STEM), transmission high energy electron diffraction (THEED) etc.[405,406]

All these techniques are topics of fundamental research and applications in themselves and it is futile to attempt to describe even their main applications in analysis here. The most important of these, however, are SEM (EDX) and AES for surface and subsurface analysis. In SEM–EDX, since X-rays generated from the entire volume containing the scattered electrons under the surface of the solid (the inverted scattering 'balloon') can be detected,[301] the sampling volume limits the spatial and depth resolution to $\sim 1\ \mu m$. Secondary and Auger electrons, however, escape only from a very thin surface layer equal to the electron mean free path which varies between 0.5 and 2 nm for electron energies in the range 50 eV to 5 keV.[407] As a result the theoretical spatial resolution is very high except in the case of AES–SAM where the need to collect sufficient number of electrons in a narrow band of energy and the destructive effect of the high intensity focused electron beam[398–401] has so far limited lateral resolution to ~ 30 nm.[398] Figures 1.16(a)–(e) show AES–SAM pictures of Ag metallized GaAs surface[398] and, for comparison Figure 1.16(f) shows an example of an SEM picture together with an EDX trace of W M_α X-rays ($h\nu \simeq 1.77$ keV) of a peeled-off region on a GaAs–W sample.[190]

While XPS (or ESCA)[409] does not involve incident electrons, it uses all the instrumentation and techniques used in electron energy analysis and detection[410–413] (Figure 1.17(a)). In AES a rather large number of scattered background electrons almost swamp the Auger electrons. But in ESCA the incident X-rays generate only a few electrons and only those very close to the surface manage to escape. As a result, while the first or the second energy derivatives of the number of emitted electrons are needed for signal separation in AES (Figure 1.17(b)) no such signal recovery process is needed in ESCA (Figure 1.17(c), (d)). Also ESCA is very useful where substrate charging is a problem. Its only deficiency is that it is not easy to focus X-rays to a small spot, and spatial resolution either does not exist (e.g. for a 1 mm sampling area) or very poor, $\sim 20\ \mu m$ in special set-ups,[414] where X-rays are generated locally using a focused electron beam incident upon a spatially separated thin metal plate.

Note that since electronic energy levels are involved, both AES and ESCA give direct chemical information provided outer shell transitions are used. Inner shell electronic energy levels are affected only minutely since these electrons do not play any prominent part in the formation of chemical bonds.

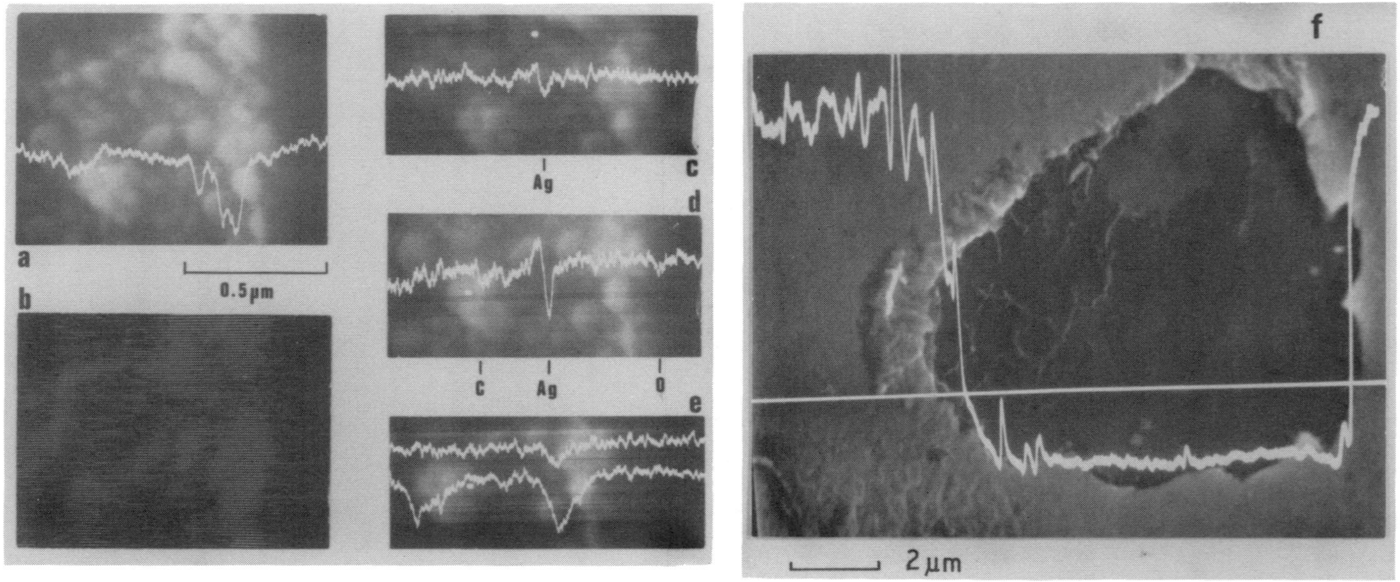

Figure 1.16. (a) to (e): Auger analysis of silver islands evaporated in situ onto GaAs. (a) SEM photograph and SAM line scan showing better than 30 nm resolution. (b) SAM map of (a) (c) and (d) d$N(E)$/dE *vs*. E plot around Ag from 150 to 650.eV at the respective spots. (e) C and Ag scans through a horizontal line passing through the spot. In (d) and (e) carbon seen is due to hydrocarbon cracking by the electron beam. (By courtesy and permission of J. A. Venables, from Venables *et al.*[398]) (f) GaAs–W sample after 250–550 °C isochronous (1 hr) stepped anneal (50 °C steps) with tungsten peel-off caused by expansion coefficient mismatch. W M_α (X-ray energy 1.77 keV) line scan using 25 keV electrons at 40° tilt. The W layer being thin (~100 nm), spatial resolution is ~0.2 μm, very much better than that dictated by scattered electrons in the bulk (from Mukherjee *et al.*[190])

1.7.6 Combinations of surface analysis and sputtering

It is only when a surface analytical technique, such as AES or ESCA, is used together with sputter etching that depth profiling becomes possible.[415–418] However, the process of sputtering, which continually disrupts the surface, brings in a number of complications (see Section 1.7.4 as well).

The first is the effects of atomic mixing, which may extend up to 10 or even 20 nm under the surface, and of sputtering induced surface roughness which increases with depth. Because of these, although the escaping electrons sample only about 1 nm under the surface, the electrons actually sample material present up to the intermixing or roughness depths respectively. The excellent depth resolutions of AES and ESCA are therefore lost immediately (Figure 1.18(a)).[373]

The second is the artifacts caused by preferential sputtering. Lighter atoms in the matrix absorb more kinetic energy during primary or secondary collisions and preferentially leave the surface. In a sample of uniform composition a steady state or a state of dynamic equilibrium is soon reached when the surface composition is so modified that the relative secondary ion or atom yield follows the matrix composition. In other words, while SIMS analysis should give us correct estimates of bulk constitution the surface composition analysed is often quite different from the starting material. This is an important artifact and painstaking standardization is needed to reduce the errors. Indeed, it is possible to collect the sputter etched material onto a clean and inert substrate placed close to the sample and analyse the deposit instead, but then the spatial and depth resolutions are severely degraded due to the collection of material removed from all the etched areas, including the crater walls.

All the other problems are similar to those listed in Section 1.7.4 excepting (ii) and similar remedies are suggested. For example, the use of N_2^+ ions give better depth resolution at larger etch depths than argon ions.[419] Also, any initial surface roughness reduces the depth resolution of any analysis involving sputter etching[420] and so the starting surface should be as flat as possible.

Since the electron beam brings in charge and excitation energy apart from the ions in AES or sputtering new phenomena come into play which have no counterpart in SIMS. The simplest is heating induced local interdiffusion.[421] Chemical breakdown of compounds by both ions[422] as well as electrons[399,400] also happens and should be avoided by reducing ion-beam and electron energies[423] and intensity either by increasing the beam diameter or by rastering, or by the use of ESCA.[424] If these basic artifacts are taken care of then the most important remaining is the so-called 'matrix effect', which is the effect of an element being present in the matrix of another.[425] Its Auger yield is different from that of its elemental state where its atoms are surrounded by atoms of the same kind. The matrix effect is due to a number of factors: (i) sputtering induced mixing and preferential sputtering, (ii) changes in peak shapes[426,427] either due to chemical bonding or changes in the mechanisms of energy loss, such as collective plasmon oscillations which is known to change with matrix composition, (iii) number density change, (iv) changes in the proportions of backscattered electrons, capable of changing the number of Auger transitions, which increases with the atomic number of the matrix, and (v) changes in the escape depth due to compositional change. Detailed theoretical ana-

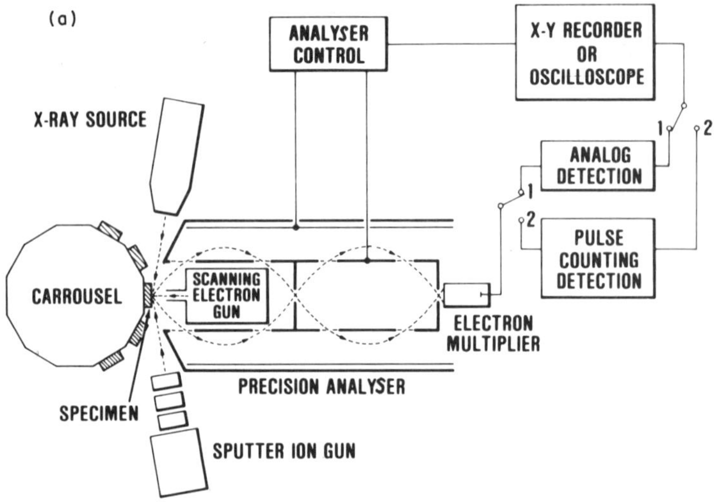

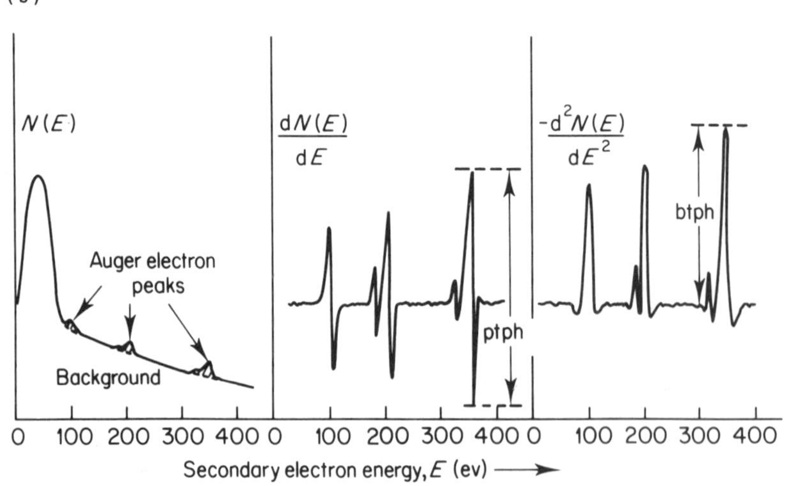

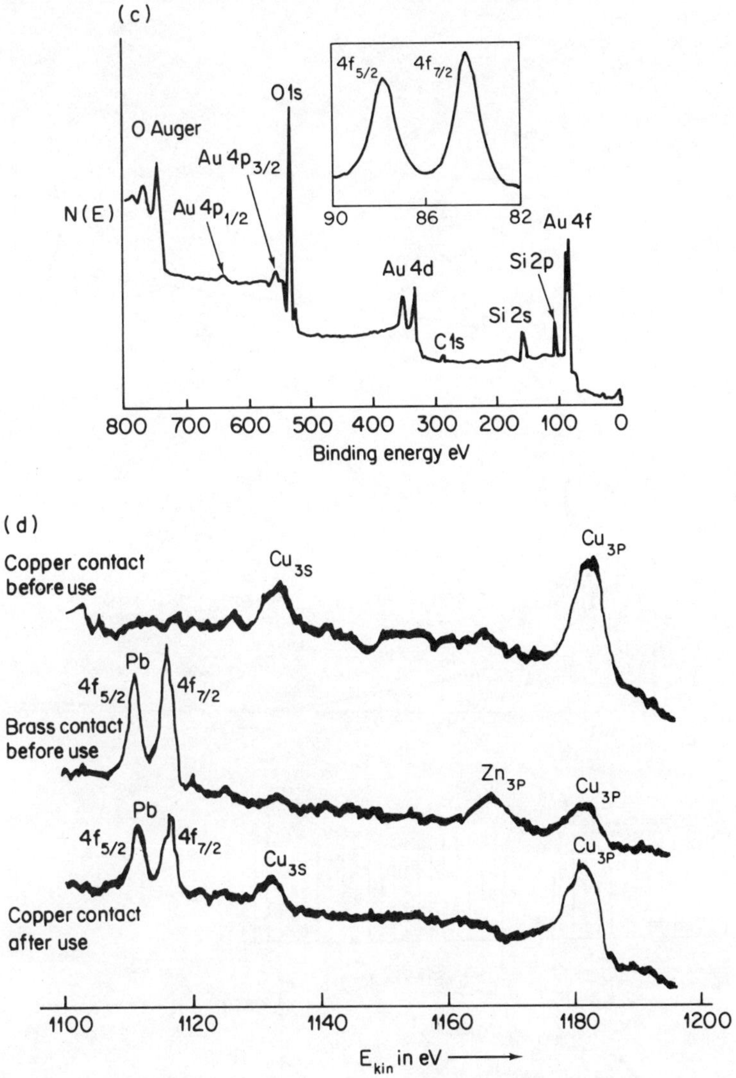

Figure 1.17. (a) Schematic of a typical AEA, ESCA–sputtering analyses set up, where a double-pass cylindrical mirror analyser (CMA) is used for electron energy analysis (by permission of Perkin-Elmer, Physical Electronics, Inc). (b) Schematics of secondary electron energy spectra where the small Auger peaks on top of high backgrounds in $N(E)$ can be enhanced by using the first and second derivatives. In quantitative analysis the area under the $N(E)$ peaks, the peak to peak height (ptph) in $N'(E)$ and the base to peak height (btph) in $-N''(E)$ are all proportional to the amount of material present, only after extensive corrections. (c) ESCA spectrum drawn in the electron binding-energy mode (by permission of Perkin-Elmer, Physical Electronics, Inc). (d) ESCA spectrum drawn in the electron kinetic energy mode (by permission of Kratos Ltd)

(a)

(b)

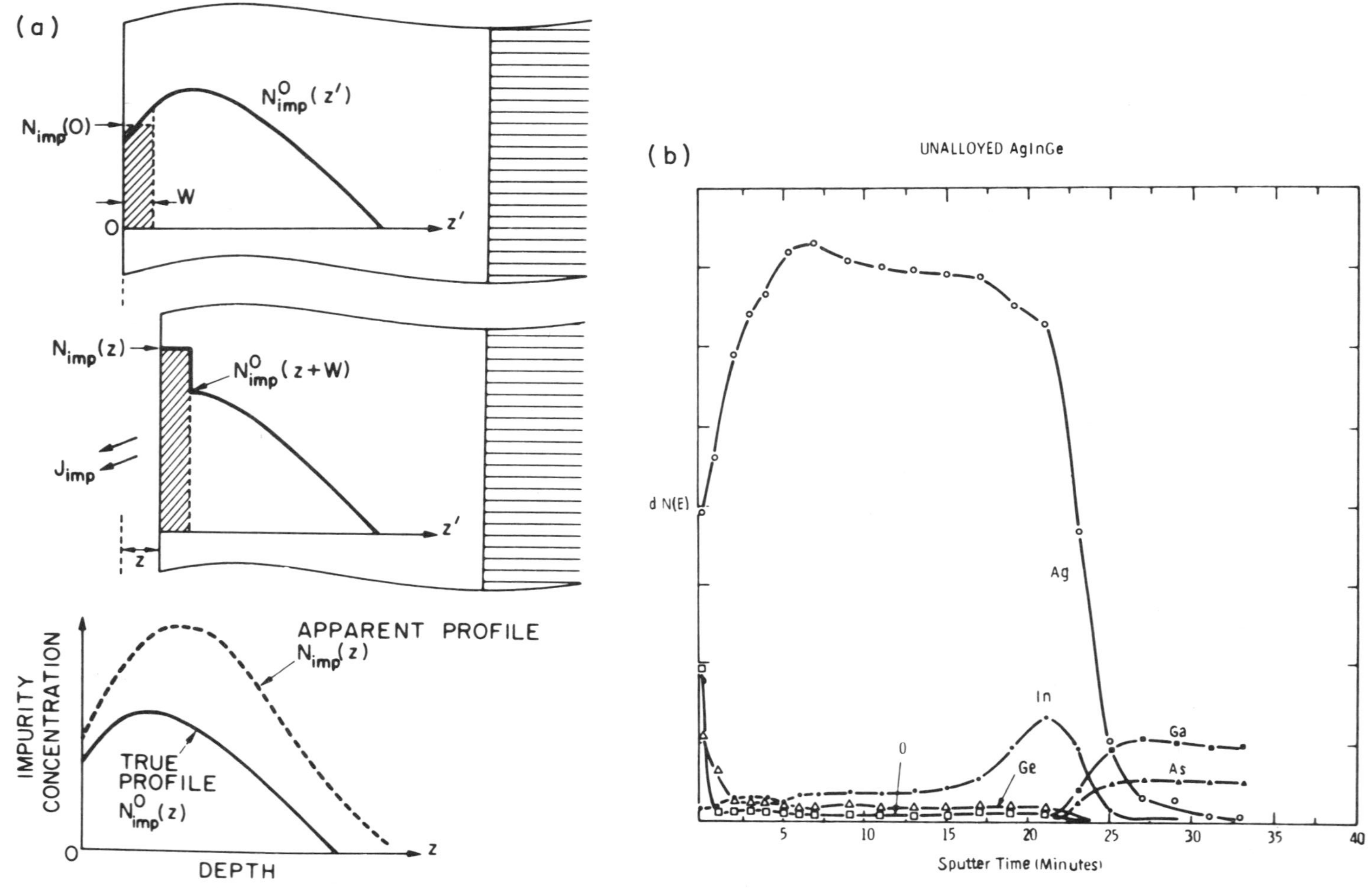

N_imp(O)
N^O_imp(z′)
W
O
z′
N_imp(z)
N^O_imp(z+W)
J_imp
z
z′
IMPURITY CONCENTRATION
APPARENT PROFILE
N_imp(z)
TRUE PROFILE
N^O_imp(z)
O
DEPTH
z
UNALLOYED AgInGe
d N(E)
Ag
In
O
Ge
Ga
As
5
10
15
20
25
30
35
40
Sputter Time (Minutes)

(c)

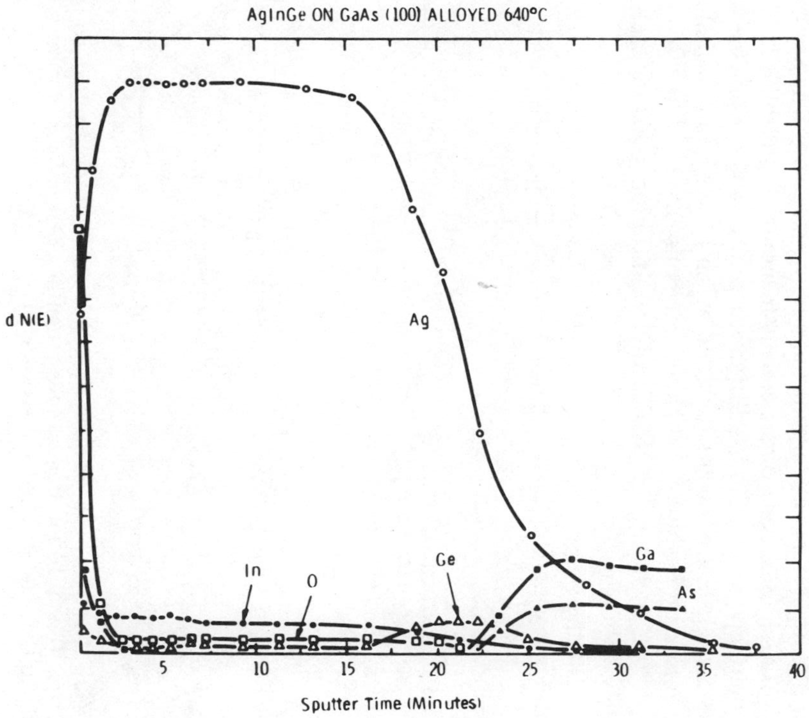

Figure 1.18 (a) Shows how sputtering induced atomic mixing can give us a wrong depth profile (from Liau *et al.*[373]). (b) and (c): Depth profiles, before and after Ohmic contact formation on GaAs by alloying Ag–In–Ge (from McGuire *et al.*[432])

lyses and computed tables exist[425] to correct for matrix effects in raw Auger profiles.

AES–sputtering has been used in a large number of semiconductor–metal interaction studies[141,170,237,238,428–432] and is being used increasingly in industry and research laboratories (Figure 1.18(b), (c)).

1.7.7 X-ray methods

Apart from XPS or ESCA, which has been discussed in Sections 1.7.5 and 1.7.6, the two most important classes of techniques are X-ray diffraction (XRD) and X-ray topography (XRT).

In XRD analysis of thin films X-rays are made incident at small angles to the film surface in order to increase the scattering probabilities at small depths and to detect small quantities of material. Two techniques are commonly employed: the Read configuration of Debye–Scherer technique[433] and the Seeman–Bohlin glancing angle diffractometer[434] especially developed for thin film applications (Figure 1.19(a) and (b)). Both are capable of revealing grain-size, strain, adhesion, texturization,[435] and intermetallic phases, but the latter is more useful in attempting quantitative studies.[436] The Seeman–Bohlin configuration uses a one-dimensionally focused (i.e. slit shaped) X-ray beam and therefore no prominent diffraction peaks appear from the underlying single crystal semiconductor substrate which makes it more suitable for thin film interaction studies.[437] These methods have been used considerably[191,438,439] including in studies of interdiffusion of completely miscible systems.[440,441]

In XRT the simplest form of contrast arises out of some distorted parts in the otherwise perfect crystal which cause the Bragg reflections to occur at

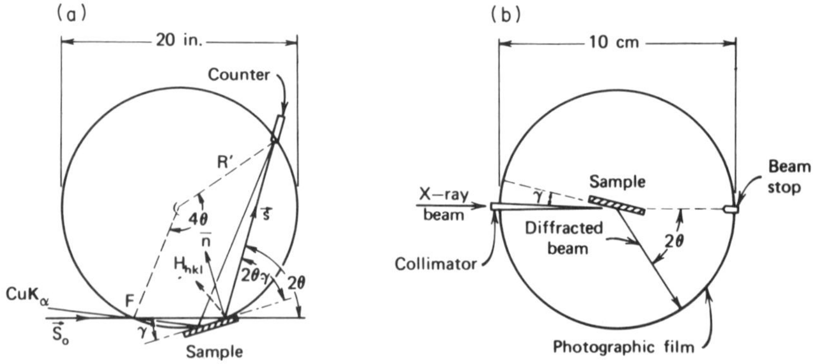

Figure 1.19. (a) The Seeman-Bohlin X-ray diffractometer (after Tu and Berry[434]) and (b) the Read configuration in a Debye-Scherer camera for thin film analyses (after Read and Hensler[433])

angles different from those from the perfectly crystalline lattice.[442–444] With ~1 to 10 μm resolution and the ability to analyse quickly 50 to 100 mm diameter slices up to ~1 mm thick in air makes XRT very useful in device fabrication studies and quality control. The three most useful configurations are shown in Figure 1.20 where (a) is for surface topography, (b) gives volume averaged information and in (c) with the crystal moving in the direction of the arrow, different sections of the crystal are observed separately and continuously. Local distortions in or on the crystal that are routinely observed are due to dislocations, stacking faults, twin and subgrain boundaries,[442] precipitate growth bands in Czochralski silicon, microvoids and large strains,[445] movements of electric and magnetic fields, and, since metallization induces interface strain, adhesion loss and local peel-offs under the metallized regions are also noticeable.[444] Metal precipitates in bulk semiconductors can also be seen, present either as impurities or added deliberately to increase strain fields near point defects[124] (Section 1.3.4). Macroscopic strain induced substrate bending caused by metallizations can also be measured by following changes in the lattice Bragg reflection angle automatically with distance traversed.[446]

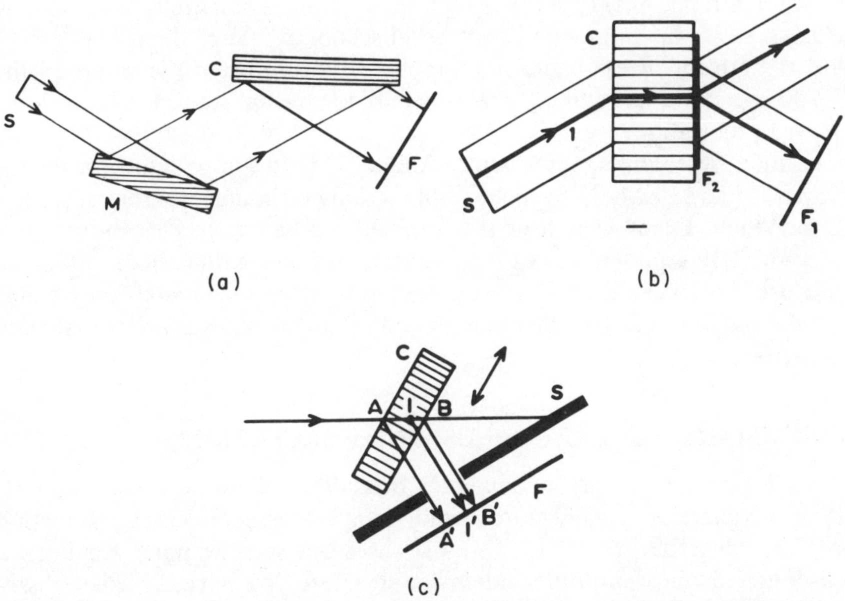

Figure 1.20. Different configurations of X-ray topography (XRT). (a) Double-crystal topographic technique. (b) Topography using anomalous transmission of X-rays. (c) Arrangement for section and projection topography (after Lang[442])

1.7.8 Convergent and divergent beam techniques

In correct configurations both convergent and divergent beams yield either higher spatial or mass resolution or the possibility of selective analysis of small volumes of material. SEM, TEM, ion-microprobe, etc. use convergent beams while Seeman–Bohlin XRD method uses a divergent beam but we shall still add a few more here which have not yet been discussed.

Both visible and IR radiation can be used in scanning optical microscopy (SOM) for example, in the optical beam induced current (OBIC) mode to study junction breakdown, charge collection, etc.[447] The method is similar to EBIC (Section 1.7.5) but the equipment set up is inexpensive and requires no vacuum. Spatial resolution is ~1 μm which is similar to that of EBIC where, in spite of the use of a finely focused electron beam, bulk scattering and diffusion of electrons determine resolution.

Convergent beam TEM–TED (D for diffraction)[448] not only selects smaller analysing volume but also gives considerably more phase information through its reciprocal lattice and has been used in the ohmic contact studies on GaAs.[192] It is a powerful technique and many TEM machines throughout the world are being converted to operate in this mode.

Focused MeV ion beams (to a diameter of 10–20 μm) have been available for selective area RBS and IIX for some time[449] but its use has not been as widespread. It has only recently been used for studies of metallization spiking in GaAs TEDs in conjunction with bevel sectioning (angle lapping).[291,292,450] With the advent of ion-beam lithography requiring beam diameters in the 10–100 nm range it is likely that focused ion beam studies will progress rapidly in the future.

The only other divergent beam technique is field ion or emission microscopy (FIM and FEM)[49,451] which is still the only technique capable of resolving individual atoms near a crystal surface and has helped us to 'see' GB structure, GB segregation, surface adatom motion and surface diffusion. Their use, however, has so far been limited to refractory metals only which have the required mechanical strength and relatively less surface activity and reconstruction.

1.8 SUMMARY, FUTURE TRENDS, AND CONCLUSION

We have considered the fundamental behaviour of the various materials interaction processes, ways of observing them and in some cases eliminating those that degrade the device. In most cases our starting point has been a well-designed and well-fabricated device structure. We have, in effect, disregarded the detailed processes used for their fabrication mainly because the technology involved is large and a very fast moving subject in itself. It is now well accepted that the rates of both miniaturization and increased reliability incorporated into the devices are basically materials limited.[452–457] Both interfacial and microstructural effects are no longer perturbations to the rather large dimension structures of the past but are parts of the functional aspects of the device itself and, with the increasing need for submicron devices[455,456] their relative importance is increasing rapidly. It is therefore hoped not only that some reasonable ground work has been established in this chapter for an appreciation of the contents of the latter works in the volume, but that the reader would find it useful to pursue a few of the many interesting aspects of the interactions and analyses which have merely been discussed briefly here because of lack of space.

1.9 ACKNOWLEDGEMENT

The author wishes to thank Professor M. W. Thompson of the University of Sussex for a Visiting Fellowship during which most of the article was written and the management of ITT—Standard Telecommunication Laboratories Ltd., Harlow, Essex for encouragement. Thanks are also due to J. A. Venables and W. Steckelmacher (Sussex), D. V. Morgan and C. J. Palmstrom (Leeds), P. M. White and A. M. Howard (Plessey—Allen Clark Research Centre), J. G. Smith and D. J. Taylor (RSRE, Baldock), P. Brook (RSRE, Malvern), J. Mun (ITT-STL Ltd.), R. M. Walser (Austin, Texas) for valuable discussions and many authors who helped by sending published materials and by permitting me to reproduce copyright materials.

REFERENCES

1. H. M. Olson, 'Device technology', Ch. 3, pp. 75–153, in *Variable Impedance Devices*, Eds. M. J. Howes and D. V. Morgan, J. Wiley and Sons, Chichester, 1978.
2. A. B. Glaser and G. E. Subak-Sharpe, *Integrated Circuit Engineering: Design Fabrication and Applications*, Addison-Wesley, Reading, M.A., 1977. Ch. 16, pp. 796–790.
3. J. M. Poate, K. N. Tu and J. W. Mayer (Eds.), *Thin films—Interdiffusion and Reactions*, J. Wiley and Sons, New York, 1978.
4. J. E. E. Baglin and J. M. Poate (Eds.), *Proc. Symp. Thin Films Phenomena—Interfaces and Interactions*, Vol. 78–2, The Electrochemical Society, Inc., Princeton, 1978.
5. L. Holland, *Vacuum Deposition of Thin Films*, Chapman and Hall, London, Sixth Printing, 1970.
6. J. L. Vossen and W. Kern, *Thin Film Processes*, Academic Press, New York, 1978.
7. K. L. Chopra, *Thin Film Phenomena*, McGraw-Hill, New York, 1969, pp. 69–75.
8. A. Y. Cho, 'Recent developments in molecular beam epitaxy (MBE)', *J. Vac. Sci. Technol.,* **16**, 275–284 (1979).

9. J. G. Erlings and H. Weerheijm, 'UHV vacuum deposition of thin solid films', *J. Vol. Sci. Technol.*, **17**, 851–853 (1980).
10. H. L. Garvin, 'Advantages of ion sputtering in microfabrications', *Electronics Packaging and Production*, **18–4**, 90–4 (1978).
11. L. I. Maissel, 'The deposition of thin films by cathode sputtering', *Physics of Thin Films*, **3**, 61–130 (1966).
12. S. Schiller, U. Heisig, and K. Goedicke, 'The role of plasmatron/magnetron systems in physical vapor deposition techniques', *Thin Solid Films*, **54**, 33–47 (1978).
13. R. W. Wilson and L. E. Terry, 'Applications of high rate ExB or magnetron sputtering in the metallization of semiconductor devices', *J. Vac. Sci. Technol.*, **13**, 157–164 (1976).
14. R. K. Waits, 'Planar magnetron sputtering', *J. Vac. Sci. Technol.*, **15**, 179–187 (1978).
15. J. A. Thornton, 'Magnetic Sputtering: basic physics and application to cylindrical magnetrons', *J. Vac. Sci. Technol.*, **15**, 171–177 (1978).
16. P. D. Reader, D. P. White, and G. C. Isaacson, 'Argon plasma bridge neutralizer operation with a 10 cm-beam-diameter ion etching source', *J. Vac. Sci. Technol.*, **15**, 1093–1095 (1978).
17. H. R. Kaufman, 'Technology of ion beam sources used in sputtering', *J. Vac. Sci. Technol.*, **15**, 272–276 (1978).
18. S. M. Kane and K. Y. Ahn, 'Characteristics of ion-beam-sputtered thin-films', *J. Vac. Sci. Technol.*, **16**, 171–174 (1979).
19. A. Brenner, *Electrodeposition of Alloys*, Vol. I and II, Academic Press, New York, 1963.
20. Session on 'Metal patterning by selective deposition', pp. 1165–1192 in the *Extended Abstracts, The Electrochemical Society Fall Meeting, Los Angeles*, Oct. 1979.
21. p. 45 in Reference 7.
22. G. A. Walker and C. C. Goldsmith, 'A structural comparison of electroless and electroplated nickel', *Thin Solid Films*, **53**, 217–222 (1978).
23. J. M. Morabito and M. J. Rand, 'Composition profile of CVD Pt and PtSi by AES and SIMS', *Thin Solid Films*, **22**, 293–303 (1974).
24. N. G. Anantha, V. Y. Doo, and D. K. Seto, 'Chromium deposition from dicumene-chromium to form metal–semiconductor devices', *J. Electrochem. Soc.*, **118**, 163–165 (1971).
25. D. P. Grimmer, K. C. Herr, and W. J. McCreary, 'Possible selective solar photothermal absorber: Ni dendrites formed on Al surfaces by the CVD and $Ni(CO)_4$', *J. Vac. Sci. Technol.*, **15**, 59–64 (1978).
26. B. O. Seraphin, 'Chemical vapour deposition of spectrally selective surfaces for high-temperature photothermal conversion', *J. Vac. Sci. Technol.*, **16**, 193–196 (1979).
27. N. E. Miller and I. Beinglass, 'Hot wall CVD tungsten for VLSI, *Sol. State Tech.*, **23–12**, 79–82 (1980).
28. S. S. Lau, R. H. Mills, and D. G. Muth, 'Temperature rise during film deposition by r.f. and d.c. sputtering', *J. Vac. Sci. Technol.*, **9**, 1196–1202 (1972).
29. A. Amith and P. Mark, 'Schottky barriers on ordered and disordered surfaces of GaAs (110)', *J. Vac. Sci. Technol.*, **15**, 1344–1352 (1978).
30. W. J. Devlin, 'R.F. sputtered Au–Mo contacts to *n*-GaAs', *Electron. Lett.*, **16**, 92–93 (1980).

31a. W. D. Westwood, 'Porosity in sputtered platinum films', *J. Vac. Sci. Technol.*, **11**, 466–471 (1974).

31b. C. Weissmantel, G. Hecht, and H. J. Hinneberg, 'Deposition of epitaxial layers by ion beam methods', *J. Vac. Sci. Technol.*, **17**, 812–816 (1980).

32. F. H. Dorbeck, 'Electrochemically deposited Schottky contacts on GaAs', *Solid-State Electron,* **9**, 1135–1137 (1966).

33. G. A. Rozonyi, J. V. DiLorenzo, and E. Heinlein, 'Fabrication of GaAs IMPATT devices using electroplated heat sinks. II Control of water curvature', *J. Electrochem. Soc.,* **121**, 426–430 (1974).

34. P. Brook, J. P. G. Farr, K. G. Sheppard, and D. Taylor, 'Silver plating in the manufacture of semiconductor devices', *Trans. Inst. Metal Finishing (G.B.),* **56**, 75–80 (1978).

35. M. H. Francombe and H. Sato (Eds.), *Single Crystal Films*, Pergamon Press, Oxford, 1964.

36. J. H. Van der Merwe, 'Interfacial misfit and bonding between oriented films and their substrates', pp. 139–163 in Reference 35.

37. K. L. Chopra, *Thin Film Phenomena*, McGraw-Hill, New York, 1969, pp. 266.

38. J. A. Venables and G. L. Price, 'Nucleation of thin films', pp. 382–436 in *Epitaxial Growth*, J. W. Matthews (Ed.), Academic Press, New York, 1975.

39. J. H. Van der Merwe and C. A. B. Ball, 'Energy of interfaces between crystals', Ch. 6, pp. 494–529 in *Epitaxial Growth*, J. W. Matthews (Ed.), Academic Press, New York, 1975.

40. N. H. Fletcher and K. W. Lodge, 'Energy of interfaces between crystals: an *ab initio* approach', Ch. 7, pp. 530–559 in *Epitaxial Growth*, J. W. Matthews (Ed.), Academic Press, New York, 1975.

41. J. P. Hirthand and K. L. Moazed, 'Nucleation processes in thin film formation', *Physics of Thin Films,* **4**, 97–136 (1967).

42. E. Grunbaum, 'List of epitaxial systems', Ch. 9, pp. 611–674 in *Epitaxial Growth* J. W. Matthews (Ed.), Academic Press, New York, 1975.

43. J. Massies, P. Etienne, and N. T. Linh, 'The interaction of silver and aluminium on GaAs (001) surfaces: a study by MBE and associated techniques', *Surf. Sci.,* **80**, 550–556 (1979).

44. J. R. Waldorp and R. W. Grant, 'Interface chemistry of metal–GaAs Schottky barrier contacts', *Appl. Phys. Lett.,* **39**, 630–632 (1979).

45. E. Bauer, 'Growth of oriented films on amorphous surfaces', pp. 43–67 in *Single Crystal Films*, M. H. Francombe and H. Sato (Eds.), Pergamon Press, Oxford, 1964.

46. G. L. Bailey, S. Fordham, and J. T. Tyson, 'The surface structure of liquid mercury', *Proc. Phys. Soc. Lond.,* **50**, 63–69 (1938).

47. W. M. Lomer and J. F. Nye, 'A dynamical model of a crystal structure. IV. Grain boundaries', *Proc. Roy. Soc.,* **A-212**, 576–584 (1952).

48. H. B. Aaron and G. F. Bolling, in *Proc. Intnl. Conf. Structure and Properties of Grain Boundaries and Interfaces*, 1971, P. Chaudhari and J. W. Matthews (Eds.), published in 'Free volume as a criterion for grain boundary models', *Surf. Sci.,* **31**, 27–57 (1972).

49. S. S. Brenner, 'Applications of field-ion microscopy techniques to metallurgical problems', *Surf. Sci.,* **70**, 427–451 (1978).

50. R. W. Hoffman, 'The mechanical properties of thin condensed films', *Physics of Thin Films,* **3**, 211–273 (1966).

51. L. Holland, *Vacuum Deposition of Thin Films*, Chapman and Hall, London, Sixth Printing, 1970, pp. 212–215.

52. D. W. Hoffman and J. A. Thornton, 'Effects of substrate orientation and rotation on internal stresses in sputtered metal films', *J. Vac. Sci. Technol.,* **16**, 134–137 (1979).
53. D. L. Allinson, 'Measurements of microstrains in thin epitaxial films', Ch. 3.6, pp. 365–379 in Reference 8.
54. R. E. Reed-Hill, *Physical Metallurgy Principles*, Van Nostrand, Princeton, 1966, p. 200.
55. R. W. Balluffi and J. M. Blakely, 'Special aspects of diffusion in thin films', *Thin Solid Films,* **25**, 363–392 (1975).
56. D. F. Stein, W. C. Johnson, and C. L. White, 'Study of grain boundary segregation using AES', Ch. 9, pp. 301–357 in *Grain Boundary Structure and Properties*, G. A. Chadwick and D. A. Smith (Eds.), Academic Press, London, 1976. (See p. 346, for example).
57. P. B. Ghate, J. C. Blair, and C. R. Fuller, 'Metallizations in microelectronics', *Thin Solid Films,* **45**, 69–84 (1977).
58. S. U. Campisano, E. Costanzo and E. Rimini, 'Grain boundary diffusion and recrystallization in Cu–Pb thin films', *Phil. Mag.,* **35**, 1333–1344 (1977).
59. W. J. DeBonte, J. M. Poate, C. M. Melliar Smith, and R. A. Levesque, 'Thin film interdiffusion, II. Ti–Rh, Ti–Pt, Ti–Rh–Au and Ti–Au–Rh', *J. Appl. Phys.,* **46**, 4284–4290 (1975).
60. J. M. Schoen, J. M. Poate, C. J. Doherty, and C. M. Melliar Smith, 'Grain boundary diffusion of Ag through Cu films', *J. Appl. Phys.*, **50**, 1910–1914 (1979).
61. K. R. Shoulders, 'Microelectrodes using electron-beam-activated machining techniques', pp. 135–293 in *Advances in Computers*, Vol. 2, F. L. Alt (Ed.), Academic Press, New York, 1961. See pp. 221–222.
62. A. Gat, L. Gerzberg, J. F. Gibbons, T. J. Magee, and J. D. Hong, 'CW laser anneal of polycrystalline silicon: crystalline structure, electrical properties', *Appl. Phys. Lett.,* **33**, 775–778 (1978).
63. J. F. Gibbons, 'Applications of scanning CW laser and electron beams in semiconductor processing', pp. 1271–1272, *Extended Abstracts*. The Electrochemical Society Fall Meeting, Los Angeles, 1979.
64. J. C. C. Fan, H. J. Zeiger, R. P. Gale, and R. L. Chapman, 'Solid phase regrowth of large aligned grains during scanned laser crystallization of amorphous Ge films on fused silica', *Appl. Phys. Lett.,* **36**, 158–161 (1980).
65. M. W. Geis, D. C. Flanders, and H. I. Smith, 'Crystallographic orientation of Si on an amorphous substrate using an artificial surface relief grating and laser crystallization', *Appl. Phys. Lett.,* **35**, 71–74 (1979).
66. M. W. Geis, D. C. Flanders, H. I. Smith, and D. A. Antoniadis, 'Graphoepitaxy of silicon on fused silica using surface micropatterns and laser crystallization', *J. Vac. Sci. Technol.,* **16**, 1640–1643 (1979).
67. R. E. Reed-Hill, *Physical Metallurgy Principles*, Van Nostrand, Princeton, 1966, pp. 330–366.
68. A. Findlay, *Phase Rule*, 9th Edition, revised by A. N. Campbell and N. O. Smith, Dover Publication, New York, 1951.
69. M. Hansen, *Constitution of Binary Alloys*, 2nd Edition, edited by K. Anderko, McGraw-Hill, New York, 1958, First Supplement, R. P. Elliott, 1965. Second Supplement, F. A. Shunk, 1969.
70. L. S. Darken and R. W. Gurry, *Physical Chemistry of Metals*, McGraw-Hill, New York, 1953.
71. J. M. Poate, K. N. Tu, and J. W. Mayer (Eds.), *Thin Films—Interdiffusion and Reactions*, J. Wiley and Sons, 1978, pp. 69 and 396.

72. P. M. Robinson and M. B. Bever, 'Thermodynamic properties', pp. 38–78 in *Intermetallic Compounds*, J. H. Westbrook (Ed.), J. Wiley, New York, 1967.
73. C. Kittel, *Introduction to Solid State Physics*, 5th Edition, John Wiley and Sons, New York, 1976, pp. 542–543.
74. B. I. Boltaks, *Diffusion in Semiconductors*, Infosearch, 1963.
75. J. Crank, *The Mathematics of Diffusion*, Clarendon Press, Oxford, 1956.
76. B. Tuck, *Introduction to Diffusion in Semiconductors*, IEE Monograph Series No. 16, IEE, London, 1974.
77. M. Abramowitz and I. A. Stegun (Eds.), *Handbook of Mathematical Functions*, Dover, New York, 1972, pp. 297, 310–311.
78. R. E. Reed-Hill, *Physical Metallurgy Principles*, Van Nostrand, Princeton, 1966, pp. 272–274.
79. J. E. E. Baglin and J. M. Poate, 'Metal–metal interdiffusion', Ch. 9, pp. 305–357 in *Thin Films—Interdiffusion and Reactions*, J. M. Poate, K. N. Tu, and J. W. Mayer (Eds.), John Wiley and Sons, New York, 1978.
80. Y. Adda, M. Beyler, and G. Brebec, 'Radiation effects in solid state diffusion', *Thin Solid Films,* **25**, 107–156 (1975).
81. N. Q. Lam, S. J. Rothman, K. L. Merkle, L. J. Nowicki, and D. J. Dever, 'Effects of proton bombardment on self-diffusion of silver: a preliminary report', *Thin Solid Films,* **25**, 157–166 (1975).
82. M. R. Mruzik and K. C. Russel, 'Equilibrium forms of small voids in metals', *Surf. Sci.,* **67**, 205–225 (1977).
83. R. E. Reed-Hill, *Physical Metallurgy Principles*, Van Nostrand, Princeton, 1966, pp. 260–264.
84. H. Schmalzried, *Solid State Reactions*, Academic Press, New York, 1974. See Ch. 5, 'Chemical diffusion in the solid state', pp. 53–78.
85. W. K. Warburton and D. Turnbull, '"Fast" diffusion in alloys', *Thin Solid Films,* **25**, 71–76 (1975).
86. W. K. Warburton and D. Turnbull, 'Fast diffusion in metals', Ch. 4, pp. 171–229 in *Diffusion in Solids—Recent Developments*, A. S. Nowick and J. J. Burton (Eds.), Academic Press, New York, 1975.
87. W. Jost, *Diffusion in Solids, Liquids, Gases*, Academic Press, New York, 1952.
88. J. O. McCaldin, 'Atom movements occurring at solid metal–semiconductor interfaces', *J. Vac. Sci. Technol.,* **11**, 990–995 (1974).
89. A. Seeger and K. P. Chik, 'Diffusion mechanisms and point defects in silicon and germanium', *Phys. Stal. Sol.,* **29**, 455–542 (1968).
90. H. C. Casey, Jr., 'Diffusion in the III-V compound semiconductors', pp. 351–430 in *Atomic Diffusion in Semiconductors*, D. Shaw (Ed.), Plenum, New York, 1973.
91. R. B. Fair, 'Recent advances in implantation and diffusion modelling for the design and process control of bipolar ICs', pp. 968–987 in *Semiconductor Silicon, 1977*, H. R. Huff and E. Sirtl, (Eds.), The Electrochemical Society, Inc., Princeton, Proceedings Volume 77-2, 1977.
92. R. L. Fogelson, Ya. A. Ugai, and I. A. Akimova, 'Diffusion of platinum in silver', *Phys. Met. and Metallogr.,* **39**, 212–214 (1975).
93. R. N. Hall and J. H. Racette, 'Diffusion and solubility of copper in extrinsic and intrinsic Ge, Si, and GaAs', *J. Appl. Phys.,* **35**, 379–397 (1964).
94. A. J. Mortlock, A. H. Rowe, and A. D. LeClaire, 'The atomic diffusion of platinum in gold', *Phil. Mag.,* **5**, 803–814 (1960).
95. W. Shockley and A. Goetzberger, 'The role of imperfections in semiconductor devices', pp. 121–125 in *Metallurgy of Semiconductor Materials*, J. B. Schroeder (Ed.), Wiley–Interscience, New York, 1962.

96. J. Mimkes, 'Pipe diffusion along isolated dislocations', *Thin Solid Films,* **25**, 221–230 (1975).
97. D. R. Campbell, K. N. Tu, and R. O. Schwenker, 'Diffusion of arsenic along dislocations in epitaxial silicon films', *Thin Solid Films,* **25**, 213–220 (1975).
98. H. P. Bonzel, 'Transport of matter at surfaces', Ch. 6, pp. 279–338 in *Surface Physics of Materials*, Vol. II, J. M. Blakely (Ed.), Academic Press, New York, 1975.
99. C. C. Chang and G. Quintana, 'Auger studies of Au diffusion through Pt films; dependence on annealing ambient', *Appl. Phys. Lett.,* **29**, 453–454 (1976).
100. S. D. Mukherjee, D. V. Morgan, and M. J. Howes, 'Interdiffusion and degradation of Si/Cr/Pt/Ag IMPATT diode metallization', *J. Electrochem. Soc.,* **126**, 1047–1053 (1979).
See also: D. V. Morgan, M. J. Howes, S. D. Mukherjee, D. J. Taylor, and P. Brook, 'Interdiffusion of Cr/Pt/Ag metallization layers on silicon IMPATT diodes', *IEE J. Solid-St. and Electron Devices,* **1**, 158–160 (1977).
101. L. G. Harrison, 'Influence of dislocations on diffusion kinetics in solids with particular reference to the alkali halides', *Trans. Faraday Soc.,* **57**, 1191–1199 (1961).
102. D. Gupta, D. R. Campbell, and P. S. Ho, 'Grain boundary diffusion', Ch. 7, pp. 161–242 in *Thin Films—Interdiffusion and Reactions*, J. M. Poate, K. N. Tu, and J. W. Mayer (Eds.), John Wiley and Sons, New York, 1978.
103. E. W. Hart, 'On the role of dislocations in bulk diffusion', *Acta Met.,* **5**, 597 (1957).
104. P. G. Shewmon, *Diffusion in Solids*, McGraw-Hill, New York, 1963.
105. J. C. Fisher, 'Calculation of diffusion penetration curves for surface and grain boundary diffusion', *J. Appl. Phys.,* **22**, 74–77 (1951).
106. R. T. P. Whipple, 'Concentration contours in grain boundary diffusion', *Phil. Mag.,* **45**, 1125–1136 (1954).
107. A. D. LeClaire, 'The analysis of grain boundary diffusion measurements', *Brit. J. Appl. Phys.,* **14**, 351–356 (1963).
108. T. Suzuoka, 'Exact solutions of two ideal cases in grain boundary diffusion problem and the application to sectioning method', *J. Phys. Soc. Japan,* **19**, 839–851 (1964).
109. J. T. Robinson and N. L. Peterson, 'Correlation effects in grain boundary diffusion', *Surf. Sci.,* **31**, 586–616 (1972).
110. P. M. Hall and J. M. Morabito, 'A formalism for determining grain boundary diffusion coefficients using surface analysis', *Surf. Sci.,* **59**, 624–630 (1976).
111. J. C. M. Hwang and R. W. Balluffi, 'Measurement of grain-boundary diffusion at low temperatures by the surface accumulation method. I. Method and analysis', *J. Appl. Phys.,* **50**, 1339–1348 (1979).
112. J. C. M. Hwang, J. D. Pan, and R. W. Balluffi, 'Measurement of grain-boundary diffusion at low temperatures by the surface accumulation method. II. Results for gold–silver systems', *J. Appl. Phys.,* **50**, 1349–1359 (1979).
113. P. M. Hall and J. M. Morabito, 'Diffusion problems in microelectronics packaging', *Thin Solid Films,* **53**, 175–182 (1978).
114. P. M. Hall and J. M. Morabito, 'A formalism for extracting diffusion coefficients from concentration profiles', *Surf. Sci.,* **54**, 79–90 (1976).
115. L. Buene, 'Interdiffusion and phase formation at room temperature in evaporated gold–tin films', *Thin Solid Films,* **47**, 159–166 (1977).
116. V. Simic and Z. Marinkovic, 'Thin film interdiffusion of Au and Ga at room temperature', *Thin Solid Films,* **34**, 179–183 (1976).

117. V. Simic and Z. Marinkovic, 'Room temperature interactions in Ag-metals thin film couples', *Thin Solid Films,* **61**, 149–160 (1979).
118. J. M. Harris, E. Lugujjo, S. U. Campisano, M.-A. Nicolet, and R. Shima, 'Studies on the Al_2O_3–Ti–Mo–Au^- metallization system', *J. Vac. Sci. Technol.,* **12**, 524–527 (1975).
119. J. F. Ziegler, J. E. E. Baglin, and A. Gangulee, 'Thermal stability of a proposed magnetic bubble metallurgy, *Appl. Phys. Lett.,* **24**, 36–39 (1974).
120. G. J. Spokel and J. M. Fairfield, 'Diffusion of gold in silicon surface', *J. Electrochem. Soc.,* **112**, 200–203 (1965).
121. M. J. Hill and P. M. Van Iseghem, 'Influence of carbon concentration on gold diffusion in silicon', pp. 715–725 in *Semiconductor Silicon, 1977*, H. R. Huff and E. Sirtl (Eds.), *Proc. Vol. 77-2*, The Electrochemical Society Inc., Princeton, 1977.
122. D. Helmreich and E. Sirtl, 'Oxygen in silicon: a modern review', pp. 626–636 in *Semiconductor Silicon 1977*, H. R. Huff and E. Sirtl (Eds.), *Proc. Vol. 77-2*, The Electrochemical Society Inc., Princeton, 1977.
123. A. J. R. de Kock, S. D. Ferris, L. C. Kimerling and H. J. Leamy, 'Investigation of defects and striations in as-grown Si crystals by SEM using Schottky diodes', *Appl. Phys. Lett.,* **27**, 313–315 (1975).
124. A. J. R. de Kock, 'Point defect condensation in dislocation-free silicon crystals', pp. 508–520 in *Semiconductor Silicon, 1977*, H. R. Huff and E. Sirtl (Eds.), *Proc. Vol. 77-2*, The Electrochemical Society Inc., Princeton, 1977.
125. A. Goetzberger and W. Shockley, 'Metal precipitates in silicon *p–n* junctions', *J. Appl. Phys.,* **31**, 1821–1824 (1960).
126. W. C. Dash, 'Copper precipitates on dislocations in silicon', *J. Appl. Phys.,* **27**, 1193–1195 (1956).
127. C. J. Varker and K. V. Ravi, 'On the origin and development of emitter "pipes" in bipolar transistors', pp. 785–794 in *Semiconductor Silicon, 1977*, H. R. Huff and E. Sirtl (Eds.), *Proc. Vol. 77-2*, The Electrochemical Society Inc., Princeton, 1977.
128. M. P. Seah, 'Surface science in metallurgy', *Surf. Sci.,* **80**, 8–23 (1979).
129. E. D. Hondros, 'Grain boundary segregation: assessment of investigative techniques', Ch. 8, pp. 265–299, in *Grain Boundary Structure and Properties*, G. A. Chadwick and D. A. Smith (Eds.), Academic Press, London, 1976.
130. M. Fortes and B. Ralph, 'A field ion microscope study of segregation to grain boundaries in iridium', *Acta Metal.,* **15**, 707–720 (1967).
131. D. Gupta, 'Grain-boundary energies and their interactions with Ta solute from self-diffusion in Au and Au- 1.2 at % Ta alloy', *Phil. Mag.,* **33**, 189–197 (1976).
132. K. Nakamura and M. Kamoshida, 'Low temperature diffusion of Al in polycrystalline Si', *J. Appl. Phys.,* **48**, 5349–5351 (1978).
133. J. C. M. Hwang, P. S. Ho, J. E. Lewis, and D. R. Campbell, 'Grain boundary diffusion of Al in poly-Si films', *J. Appl. Phys.,* **51**, 1576–1581 (1980).
134. A. Christou and W. T. Anderson, 'Material reactions and barrier height variations in sintered Al-InP Schottky diodes', *Solid State Electron,* **22**, 857–864 (1979).
135. J. M. Poate, P. A. Turner, and W. J. DeBonte, 'Thin film interdiffusion. I. Au–Pd, Pd–Au, Ti–Pd, Ti–Au, Ti–Pd–Au, Ti–Au–Pd', *J. Appl. Phys.,* **46**, 4275–4283 (1975).
136. S. Liblich and A. G. Nassibian, 'Effects of diffused nickel on the Si–SiO_2 interface', *Solid State Electron,* **16**, 1495–1499 (1973).
137. P. F. Schmidt and L. P. Adda, 'Interface-state generation by gold diffusion

through SiO_2 films on Si: MOS and neutron activation results', *J. Appl. Phys.*, **45**, 1826–1833 (1974).

138. L. P. Erikson, A. Waseem, and G. Y. Robinson, 'Characterization of Ohmic contacts to InP', *Thin Solid Films,* **64**, 421–426 (1979).
139. K. N. Tu and R. Rosenberg, 'Mass transport in layered polycrystalline thin films', *Thin Solid Films*, **13**, 163–167 (1972).
140. C. C. Chang and G. Quintana, 'Diffusion kinetics of Au through Pt films about 2000 and 6000 Å thick studied with AES', *Thin Solid Films,* **31**, 265–273 (1976).
141. A. Hiraki and M. Iwami, 'Electronic structure of thin gold film deposited on silicon substrate studied by Auger electron and X-ray photoelectron spectroscopies', *Jpn. J. Appl. Phys.,* **Suppl. 2**, 749–752 (1974).
142. J. A. Cairns, C. W. Keep, R. J. White, J. F. Ziegler, and K. N. Tu, 'A controlled atmosphere electron microscope study of the influence of oxygen on the interdiffusion of gold and iron', *Appl. Phys. Lett.,* **30**, 392–394 (1977).
143. A. Munetz and Y. Komen, 'Structural and resistivity changes in heat-treated chromium–gold films', *Thin Solid Films,* **37**, 171–179 (1976).
144. N. T. Panousis and P. M. Hall, 'Applications of grain boundary diffusion studies to soldering and thermocompression bonding', *Thin Solid Films,* **53**, 183–191 (1978).
145. S. D. Mukherjee, D. V. Morgan, M. J. Howes, J. G. Smith, and P. Brook, 'Reactions of vacuum-deposited thin Schottky barrier metallizations on gallium arsenide', *J. Vac. Sci. Technol.,* **16**, 138–140 (1979).
146. J. Gyulai, J. W. Mayer, I. V. Mitchell, and V. Rodriguez, 'Outdiffusion through silicon dioxide and silicon nitride layers on GaAs', *Appl. Phys. Lett.,* **17**, 332–334 (1970).
147. E. Kinsborn, P. K. Gallagher, and A. T. English, 'Dissociation of GaAs and $Ga_{0.7}Al_{0.3}As$ during alloying of gold contact films', *Solid-State Electron.,* **22**, 517–524 (1979).
148. T. Sebestyen, M. Menyhard, and D. Szigethy, Electron Ltd., '*In situ* measurements of arsenic losses during annealing of the usual evaporated contacts of GaAs Gunn diodes', **12**, 96–97 (1976).
149. I. Jaeger, 'The influence of cooling rate on interface segregation', *Surf. Sci.,* **74**, 656–666 (1978).
150. P. B. Hirsch, 'The role of dislocations in solid state reactions', pp. 362–386 in *Reactivity of Solids*, J. S. Anderson, M. W. Roberts, and F. S. Stone (Eds.), Chapman and Hall, London, 1972.
151. P. Wynblatt and R. C. Ku, 'Surface energy and solute strain energy effects in surface segregation', *Surf. Sci.,* **65**, 511–531 (1977).
152. D. Gupta, 'The effect of solutes on grain boundary diffusion and related phenomena and its origin', pp. 498–505 in *Proc. Symp. Thin Film Phenomena—Interfaces and Interactions, 1977*, J. E. E. Baglin and J. M. Poate (Eds.), Vol. 78-2, The Electrochemical Society, Inc., Princeton, 1978.
153. D. V. Morgan and J. Frey, 'Comments on the modification of Schottky barrier height by interfacial oxides', *Phys. State Sol.*, **a-51**, K29–K33 (1979).
154. M. L. Joshi and F. Wilhelm, 'Diffusion induced imperfections in silicon', *J. Electrochem. Soc.,* **112**, 185–188 (1965).
155. M. Nakamura, T. Kato, and N. Oi, 'A study of gettering effects of metallic impurities in silicon', *Jpn. J. Appl. Phys.,* **7**, 512–519 (1968).
156. S. P. Murarka, 'A study of the phosphorus gettering of gold in silicon by the use of neutron activation analysis', *J. Electrochem. Soc*., **123**, 765–767 (1976).
157. T. Y. Tau, E. E. Gardner, and W. K. Tice, 'Intrinsic gettering by oxide

precipitate induced dislocations in Czochralski Si', *Appl. Phys. Lett.,* **30**, 175–176 (1977).

158. C. W. Pearce, 'Defect and impurity control in silicon', pp. 1224–1226, *Extended Abstracts*, Fall Meeting, Oct. 1979, The Electrochemical Society.
159. E. J. Mets, 'Poisoning and gettering effects in silicon junctions', *J. Electrochem. Soc.,* **112**, 420–425 (1965).
160. M. Kukuchi and S. Iijima, 'A simpler method for removing copper from germanium', *J. Phys. Soc. Jpn.,* **12**, 824 (1957).
161. T. M. Buck, K. A. Pickar, J. M. Poate, and C. M. Hsieh, 'Gettering rates of various fast-diffusing metal impurities at ion damaged layers of silicon', *Appl. Phys. Lett.,* **21**, 485–487 (1972).
162. T. E. Seidel, 'Direct comparison of ion-damage gettering and phosphorus-diffusion gettering of Au in Si', *J. Appl. Phys.,* **46**, 600–609 (1975).
163. M. R. Poponiak and T. H. Yeh, 'Impurity gettering and removing gettered impurities', *IBM Tech. Discl. Bull.,* **20**, 3454–3455 (1978).
164. A. G. Cullis and T. E. Seidel, 'The annealing and gettering behaviour of inert gas ion implantation damage in silicon', pp. 1255–1256. *Extended Abstracts*, The Electrochemical Society Fall Meeting, Oct. 1979.
165. W. K. Tice and H. J. Geipel, 'Implantation gettering in semiconductors', pp. 1257–1259, *Extended Abstracts*, The Electrochemical Society Fall Meeting, Oct. 1979.
166. B. W. Wessels, 'Gettering of epitaxial gallium phosphide using phosphosilicate glass', *Electron. Lett.,* **15**, 748–749 (1979).
167. T. J. Magee, J. Peng, J. D. Hong, C. A. Evans, V. R. Deline, and R. M. Malbon, 'Back surface gettering and Cr outdiffusion in VPE GaAs layers', *Appl. Phys. Lett.,* **35**, 277–279 (1979).
168. V. R. Deline, C. A. Evans, T. W. Sigmon, and A. Lidow, 'Gettering of Cr by residual damage and encapsulation stress in ion-implanted semi-insulating GaAs', pp. 1253–1254. *Extended Abstracts*, The Electrochemical Society Fall Meeting, Oct. 1979.
169. N. Nakashima, M. Iwami, and A. Hiraki, 'Low temperature diffusion of Au into Si in the Si (substrate)—Au (film) system', *Thin Solid Films,* **25**, 423–430 (1975).
170. A. Hiraki, S. Kim, W. Kammura, and M. Iwami, 'Dynamical observations of room temperature interfacial reaction in metal–semiconductor system by Auger electron spectroscopy', *Surf. Sci.,* **86**, 706–710 (1979).
171. R. M. Walser and R. W. Bene, 'First phase nucleation in silicon–transition metal planar interfaces', *Appl. Phys. Lett.,* **28**, 624–625 (1976).
172. R. W. Bene and R. M. Walser, 'Effect of a glassy membrane on the Schottky barrier between silicon and metallic silicides', *J. Vac. Sci. Technol.,* **14**, 925–929 (1977).
173. J. C. Phillips, 'Chemistry and structure of transition metal silicide–silicon interfaces', pp. 3–20 in *Proc. Symp. Thin Film Phenomena—Interfaces and Interactions*, 1977, J. E. E. Baglin and J. M. Poate (Eds.), Vol. 78-2, The Electrochemical Society, Inc., Princeton, 1978.
174. S. U. Campisano, 'Extensive interactions between metal films at $T < T_m/2$', pp. 129–149 in *Proc. Symp. Thin Film Phenomena—Interfaces and Interactions*, J. E. E. Baglin and J. M. Poate (Eds.), Vol. 78–2, The Electrochemical Society, Inc., Princeton, 1978.
175. G. V. Kidson, 'Some aspects of the growth of diffusion layers in binary systems', *J. Nucl. Mater.,* **3**, 21–29 (1961).

176. C. Weaver, 'Diffusion in metallic films', *Physics of Thin Films,* **6**, 301–347 (1971).
177. G. J. van Gurp, 'The growth of metal silicide layers on silicon', pp. 342–358 in *Semiconductor Silicon, 1977*, H. R. Huff and E. Sirtl (Eds.), Vol. 77–2, The Electrochemical Society, Inc., Princeton, 1977.
178. F. Lukes, 'Oxidation of Si and GaAs in air at room temperature', *Surf. Sci.,* **30**, 91–100 (1972).
179. R. Pretorius, J. M. Harris, and M.-A. Nicolet, 'Reaction of thin metal films with SiO_2 substrates', *Solid State Electron.,* **21**, 667–675 (1978).
180. G. Ottaviani, 'Review of binary alloy formation by thin film interactions', *J. Vac. Sci. Technol.,* **16**, 1112–1119 (1979).
181. J. Baglin, F. d'Heurle, and S. Petersson, 'Nucleation controlled thin film interactions–silicides', pp. 952–3, *Extended Abstracts*, The Electrochemical Society Fall Meeting, Oct. 1979.
182. T. C. Tisone and J. Drobek, 'Diffusion in thin film Ti–Au, Ti–Pd and Ti–P couples', *J. Vac. Sci. Technol.,* **9**, 271–275 (1972).
183. J. K. Howard, R. F. Lever, R. J. Smith, and P. S. Ho, 'Kinetics and compound formation in thin film couples of Al and transition metals', *J. Vac. Sci. Technol.,* **13**, 68–71 (1976).
184. R. F. Lever, J. K. Howard, W. K. Chu, and P. J. Smith, 'Anomalous growth of $HfAl_3$ in thin films', *J. Vac. Sci. Technol.,* **14**, 158–161 (1977).
185. K. N. Tu and J. W. Mayer', 'Silicide formation', Ch. 10, pp. 359–405 in *Thin Films—Interdiffusion and Reactions*, J. M. Poate, K. N. Tu, and J. W. Mayer (Eds.), Wiley–Interscience, New York, 1978.
186. B. E. Deal and A. S. Grove, 'General relationship for the thermal oxidation of silicon', *J. Appl. Phys.,* **36**, 3770–3778 (1965).
187. T. G. Finstad, 'Silicide formation with nickel platinum double layers on silicon', *Thin Solid Films,* **51**, 411–418 (1978).
188. T. G. Finstad and M.-A. Nicolet, 'Silicide formation with bilayers of Pd–Pt, Pd–Ni and Pt–Ni', *J. Appl. Phys.,* **50**, 303–310 (1979).
189. A. K. Sinha and J. M. Poate, 'Metal-compound semiconductor reactions', Ch. 11, pp. 407–432 in *Thin Films—Interdiffusion and Reactions* J. M. Poate, K. N. Tu, and J. W. Mayer (Eds.), John Wiley and Sons, New York, 1978.
190. S. D. Mukherjee, D. V. Morgan, C. J. Palmstrom, and J. G. Smith, 'The thermal stability of thin layer transition and refractory metallization on GaAs', *J. Vac. Sci. Technol.*, **17**, 904–910 (1980).
191. J. O. Olowolafe, P. S. Ho, H. J. Hovel, J. E. Lewis, and J. M. Woodall, 'Contact reactions in Pd/GaAs junctions', *J. Appl. Phys.,* **50**, 955–961 (1979).
192. J. E. Loveluck, G. M. Rackham, and J. W. Steeds, 'Electron microscopy of metal–semiconductor contacts', *Inst. Phys. Conf. Ser. No.,* **36**, 297–300 (1977).
193. S. C. Kolesar, 'Principles of corrosion', pp. 155–167 in *Proc. 12th Annual Reliability Phys. Symp. Las Vegas*, 1974, IEEE, New York, 1974.
194. L. Pauling, *The Nature of the Chemical Bond*, Cornell University Press, Ithaca, New York, 1960.
195. H. B. Michaelson, 'Relation between an atomic electronegativity scale and the work function', *IBM J. Res. Develop.*, **22**, 72–80 (1978).
196. H. Koelmans, 'Metallization corrosion in silicon devices by moisture-induced electrolysis', pp. 168–171 in *Proc. 12th Annual Reliability Phys. Symp. Las Vegas*, 1974, IEEE, New York, 1974.
197. W. M. Paulson and R. W. Kirk, 'The effect of phosphorus-doped passivation glass on the corrosion of aluminium', pp. 172–179 in *Proc. 12th Annual Reliability Phys. Symp. Las Vegas*, 1974, IEEE, New York, 1974.

198. E. A. Aitken, 'Corrosion behaviour', pp. 491–516 in *Intermetallic Compounds*, J. H. Westbrook (Ed.), Wiley, New York, 1967.
199. R. Rosenberg, M. J. Sullivan, and J. K. Howard, 'Effects of thin film interactions on silicon device technology', Ch. 2, pp. 13–55 in *Thin Films—Interdiffusion and Reactions*, J. M. Poate, K. N. Tu, and J. W. Mayer (Eds.), Wiley–Interscience, New York, 1978, see p. 48 *et seq*.
200. R. Williams, 'Electrochemical reactions of semiconductors', *J. Vac. Sci. Technol.,* **13**, 12–18 (1976).
201. F. N. Fus, C. T. Hartwig, and J. M. Morabito, 'Corrosion of solder-coated Ti–Pd–Au thin film conductors in moist chlorine atmosphere', *Thin Solid Films,* **43**, 189–213 (1977).
202. D. J. Sharp, 'Corrosion inhibition in sputter-deposited thin-film systems using an intermediary layer of palladium', *J. Vac. Sci. Technol.,* **16**, 204–207 (1979).
203. A. T. English and P. A. Turner, 'Stability of conductor metallizations in corrosive environments', *J. Electron. Mater.,* **1**, 1–14 (1972).
204. J. S. Harris, Y, Nannichi, and G. L. Pearson, 'Ohmic contacts to solution-grown gallium arsenide', *J. Appl. Phys*., **40**, 4575–4581 (1969).
205. H. M. Macksey, 'Ohmic contacts for GaAs power FETs', pp. 254–261, *Inst. Phys. Conf. Ser. No.,* **33b**, (1977), 'GaAs and related compounds, 1976', L. F. Eastman (Ed.), The Institute of Physics, London, 1977.
206. M. Rolland, J. P. Nougier, D. Gasquet, and R. Aladebra, 'Low-noise ohmic contacts on *n*-type silicon', *Solid State Electron.,* **20**, 323–331 (1977).
207. A. K. Sinha, 'Catastrophic Au–Si eutectic formation in certain W/Au metallizations on oxidized silicon wafers', *Thin Solid Films.,* **20**, 115–122 (1974).
208. G. Majni, G. Ottaviani, and M. Prudenziati, 'Interdiffusion of thin Cr and Au films deposited on Si', *Thin Solid Films,* **34**, 170 (1976).
209. S. D. Mukherjee and D. V. Morgan, *Unpublished Work*, 1978.
210. K. Nakamura, M.-A. Nicolet, J. W. Mayer, R. J. Blattner, and C. A. Evans, 'Interaction of Al layers with polycrystalline Si', *J. Appl. Phys.,* **46**, 4678–4684 (1975).
211. S. S. Lau and W. F. Van der Weg, 'Solid phase epitaxy', Ch. 12, pp. 433–480 in *Thin Films—Interdiffusion and Reactions* J. M. Poate, K. N. Tu, and J. W. Mayer (Eds.), Wiley, New York, 1978.
212. J. W. Mayer, J. M. Poate, and K. N. Tu, 'Thin films and solid phase reactions—traffic in a submicron world', *Science,* **190**, 228–234 (1975).
213. J. Basterfield, J. M. Shannon, and A. Gill, 'The nature of barrier height variations in alloyed Al–Si Schottky barrier diodes', *Solid State Electron.,* **18**, 290–291 (1975).
214. J. M. Shannon, 'A majority carrier camel diode', *Appl. Phys. Lett.,* **35**, 63–65 (1979).
215. H. C. Card, 'Aluminium–silicon Schottky barriers and ohmic contacts in integrated circuits', *IEEE Tr ED-23*, 538–544 (1976).
216. E. Hökelek and G. Y. Robinson, 'Aluminium/nickel silicide contacts on silicon', *Thin Solid Films,* **53**, 135–140 (1978).
217. C. W. Horsting, 'Purple plague and gold purity', pp. 155–158 in *Proc. 10th Annual Reliab. Phys. Symp. 1972,* IEEE.
218. D. W. Bushmire, 'Resistance increase in Au–Al interconnects with time and temperature', *IEEE Tr.,* **PHP-13**, 152–156 (1977).
219. P. M. Hall and L. W. Condra, 'Aging of solder connections to TiPd Au films', *IEEE Tr., CHMT-2*, 279–283 (1979).
220. N. T. Panousis and P. M. Hall, 'Reduced gold plating on copper leads and

thermocompression bonding. Part I. Initial Characterization'. '. . . , Part II. Long term reliability'. *IEEE Tr.,* **PHP-13**, 305–309 (1977).
221. L. G. Feinstein and R. J. Pagano, 'Degradation of thermocompression bonds of Ti–Cu–Au and Ti–Cu films by thermal aging', *IEEE Tr.,* **CHMT-3**, 134–143 (1980).
222. L. G. Feinstein and J. B. Bindell, 'The failure of aged Cu–Au thin films by Kirkendall porosity', *Thin Solid Films,* **62**, 37–48 (1979).
223. R. E. Hampy, F. G. Yost, and F. P. Ganyard, 'Interfacial behaviour of Cr–Au films in the 423–573 K temperature range', *J. Vac. Sci. Technol.,* **16**, 25–30 (1979).
224. C. J. Madams, D. V. Morgan, and M. J. Howes, 'Outmigration of gallium from Au–GaAs interfaces', *Electron. Lett.,* **11**, 24–25 (1975).
225. T. J. Magee and J. Peng, 'Electron microscopy studies of the alloying behaviour of Au on GaAs', *Phys. Stat. Sol.* **a–32**, 695–700 (1975).
226. A. N. Gerritsen, 'Metallic conductivity, experimental part', pp. 137–226 in *Handbuch der Physik*, Vol. 19, S. Flugge (Ed.), Springer–Verlag, Berlin, 1956.
227. A. K. Sinha, T. E. Smith, and T. T. Sheng, 'Thin film diffusion of platinum in gold', *Thin Solid Films,* **22**, 1–10 (1974).
228. M. A. Nicolet, 'Diffusion barriers in thin films', *Thin Solid Films,* **52**, 415–443 (1978).
229. S. P. Murarka, H. J. Levinstein, I. Blech, T. T. Sheng, and M. H. Read, 'Investigations of the Ti–Pt diffusion barrier for gold beam leads on aluminium', *J. Electrochem. Soc.,* **125**, 156–162 (1978).
230. J. H. Thomas, J. M. Morabito, and N. G. Lesh, 'Ti–Cu–Ni–Au (TCNA) compatibility with resistor and bilevel crossover circuit processing', *J. Vac. Sci. Technol.,* **13**, 152–155 (1976).
231. P. S. Ho, U. Koster, J. E. Lewis, and S. Libertini, 'Material reactions in Al/Pd_2 Si/Si junction', pp. 66–73 in *Proc. Symp. Thin Film Phenomena—Interfaces and Interactions*, J. E. E. Baglin and J. M. Poate (Eds.), Vol. 78–2, The Electrochemical Society, Princeton, 1978.
232. M. P. Lepselter, 'Beam lead technology', *Bell Lab. Record*, 299–303 (1966) and *Bell System Technical J.,* **45**, 233 (1966).
233. U. C. Brown and J. R. Sim, 'Failure mechanisms in beam lead semiconductors', *IEEE Tr.,* **PHP–13**, 225–229 (1977).
234. W. C. Hagel, 'Diffusion', pp. 377–404 in *Intermetallic Compounds*, J. H. Westbrook, (Ed.), Wiley, New York, 1967.
235. A. K. Sinha, M. H. Read, and T. E. Smith, 'Silicide formation in tungsten and other refractory metallizations on PtSi on silicon', *J. Electrochem. Soc.,* **120**, 1775–1778 (1973).
236. P. B. Ghate, J. C. Blair, C. R. Fuller, and G. E. McGuire, 'Application of Ti:W barrier metallization for integrated circuits', *Thin Solid Films,* **53**, 117–128 (1978).
237. H. M. Day, A. Christou, and A. C. Macpherson, 'Interdiffusion and Schottky barrier height variation in Au–W (Ti)/*n*-GaAs contacts', *J. Vac. Sci. Technol.,* **14**, 939–942 (1977).
238. A. Nara, H. Kondo, T. Ishi, and H. Ikegawa, 'High performance Ni–Pd/GaAs Schottky barrier diode made by electroplating', *Jpn. J. Appl. Phys. Suppl.,* **17-1**, 179–184 (1978).
239. K. Nakamura, S. S. Lau, M.-A. Nicolet, and J. W. Mayer, 'Ti and V layers retard interaction between Al films and polycrystalline Si', *Appl. Phys. Lett.,* **28**, 277–280 (1976).
240. J. O. Olowolafe, M.-A. Nicolet, and J. W. Mayer, 'Chromium thin film as a

barrier to the interaction of Pd_2Si with Al', *Solid State Electron.*, **20**, 413–416 (1977).

241. J. K. Howard, J. F. White, and P. S. Ho, 'Intermetallic compounds of Al and transition metals: effect of electromigration in 1–2-μm-wide lines', *J. Appl. Phys.*, **49**, 4083–4093 (1978).
242. S. Okhawa, O. Akanuma, and H. Ishikawa, 'Atmosphere effect on migration of Co and Si in Au–Cu–Si system', *Jpn. J. Appl. Phys.*, **14**, 1589–1590 (1975).
243. J. C. M. Hwang, P. S. Ho, and R. W. Balluffi, 'Effect of surface condition on diffusion in thin films at low temperatures', *Appl. Phys. Lett.*, **33**, 458–461 (1978).
244. D. Y. Shih and P. J. Ficalora, 'The reduction of Au–Al intermetallic formation and electromigration in hydrogen environments', *IEEE Tr*, **ED-26**, 27–34 (1979).
245. D. Gupta and R. Rosenberg, 'Effect of a solute addition (Ta) on low temperature self-diffusion process in Au', *Thin Solid Films*, **25**, 171–180 (1975).
246. E. K. Storms, 'Phase relationships and electrical properties of refractory carbides and nitrides', Ch. 2, pp. 37–78, in *Solid State Chemistry*, L. E. J. Roberts, (Ed.), Butterworths, London, 1972.
247. H. Goronkin 'Bulk leakage in avalanche diodes due to silicon metallization reactions', *Solid State Electron.*, **18**, 891–893 (1975).
248. T. C. Tisone and S. S. Lau, 'Interdiffusion and compound formation in Ta–Au thin film couples', *J. Appl. Phys.*, **45**, 1667–1674 (1974).
249. K. L. Tai and M. Ohring, 'Grain boundary electromigration in thin films. II. Tracer measurements in pure Au', *J. Appl. Phys.*, **48**, 36–45 (1977).
250. J. K. Howard and P. L. Gabarino, 'Auger analysis of Ta/Ta oxide diffusion barrier to reduce Al–polycrystalline Si reaction', *J. Vac. Sci. Technol.*, **15**, 243 (1978).
251. W. J. Garceau and G. K. Herb, 'TiN as a diffusion barrier in the Ti–Pt–Au beam lead metal systems', *Thin Solid Films*, **53**, 193–194 (1978). Also: *Thin Solid Films*, **60**, 237–247 (1978), same title.
252. R. S. Nowicki and I. Wang, 'Improvement of the diffusion barrier properties of rf-sputtered molybdenum', *J. Vac. Sci. Technol.*, **15**, 235–237 (1978).
253. M. Maenpaa, H. von Seefeld, N. Cheung, M.-A. Nicolet, and A. G. Cullis, 'Characteristics of titanium nitride films prepared by r.f. sputtering', pp. 946–948, and N. Cheung, H. von-Seefeld and M.-A. Nicolet, 'Titanium nitride as a diffusion barrier in metallization schemes', pp. 949–951, *Extended Abstracts*, Fall Meeting, Oct. 1979, The Electrochemical Society Inc., Princeton.
254. M. Wittmer, 'TiN and TaN as diffusion barriers in metallizations to silicon semiconductor devices', *Appl. Phys. Lett.*, **36**, 456–458 (1980).
255. R. S. Nowicki, J. M. Harris, M.-A. Nicolet, and I. V. Mitchell, 'Studies of the Ti–W/Au metallization on Al', *Thin Solid Films*, **53**, 195–205 (1978).
256. P. M. Schaible and L. I. Maissel, p. 190 in *Trans. 9th National Vacuum Symp., 1962*, G. H. Bancroft (Ed.), Macmillan, New York, 1963.
257. A. J. Learn and R. S. Nowicki, 'Method for minimizing silicon regrowth in Al films', *Appl. Phys. Lett.*, **35**, 611–613 (1979).
258. A. Christou and H. M. Day, 'A stabilized tantalum diffusion barrier for the gold metallization system', *J. Electron. Mater.*, **3**, 25–35 (1974).
259. G. E. McGuire, W. R. Wisseman, and P. H. Holloway, 'Diffusion studies of Au through electroplated Pt films by Auger electron microscopy', *J. Vac. Sci. Technol.*, **15**, 1701–1705 (1978).

260. J. R. Black, 'Physics of electromigration', pp. 142–149 in *Proc. 12th Annual Reliab. Phys. Symp.,* Las Vegas, 1974, IEEE.
261. H. B. Huntington, 'Effect of driving forces on atom motion', *Thin Solid Films,* **25**, 265–280 (1975).
262. F. M. d'Heurle and R. Rosenberg, 'Electromigration in thin films', *Physics of Thin Films,* **7**, 257–310 (1973).
263. C. Kittel, *Introduction to Solid State Physics*, 5th Edition, J. Wiley, New York, 1976, p. 171 *et seq*.
264. J. S. Blakemore, *Solid State Physics*, W. B. Saunders, Philadelphia, 1970, p. 170 *et seq*.
265. R. Rosenberg, A. F. Mayadas, and D. Gupta, 'Grain boundary contribution to electro-transport', *Surf. Sci.,* **31**, 566–585 (1972).
266. F. M. d'Heurle and P. S. Ho, 'Electromigration in thin films', Ch. 8, pp. 243–303 in *Thin Films—Interdiffusion and Reactions*, J. M. Poate, K. N. Tu, and J. W. Mayer (Eds), J. Wiley, New York, 1978.
267. I. A. Blech and R. Rosenberg, 'On the direction of electromigration in gold thin films', *J. Appl. Phys.,* **46**, 579–583 (1975).
268. I. A. Blech and E. Kinsborn, 'Electromigration in thin gold films on molybdenum surfaces', *Thin Solid Films,* **25**, 327–334 (1975).
269. D. A. Abbott and J. A. Turner, 'Some aspects of GaAs MESFET reliability', *IEEE Tr.,* **MTT-24**, 317–321 (1976).
270. P. M. White, B. L. Hewett, and J. A. Turner, 'Reliability investigations of GaAs power FETS with aluminium gate metallization', pp. 405–409 in *Proc. European Microwave Conf.*, Paris, 1978.
271. K. Kemmochi and K. I. Hirano, 'Electromigration of grain boundaries in aluminium', *Thin Solid Films,* **25**, 353–361 (1975).
272. I. A. Blech, 'Copper electromigration in aluminium', *J. Appl. Phys.,* **48**, 473–477 (1977).
273. F. A. Reynolds, 'Thermally accelerated aging of semiconductor components', *Proc. IEEE,* **62**, 212–222 (1974).
274. J. R. Lloyd, 'On the log-normal distribution of electromigration lifetimes', *J. Appl. Phys.,* **50**, 5062–5064 (1979).
275. W. Weibull, 'A statistical distribution function of wide applicability', *J. Appl. Mech.,* **18**, 293–297 (1951).
276. F. Sellborg, P. Weissglas, and G. Andersson, 'A study of failure mechanisms in silicon IMPATT diodes', *IEEE Tr*, **ED-25**, 742–746 (1978).
277. P. M. White, 'Degradation mechanisms in FETs' presented at the *Colloquium on Metallizations for III-V Semiconductor Devices*, November, 1979, IEE, London. (Digest no. 1979/59).
278. H. S. Gurev, 'Sensitivity of current pulse burnout testing to the geometry of defects in aluminium metallization', pp. 187–195, in *Proc. 12th Annual Reliab. Phys. Symp*., Las Vegas, 1974, IEEE.
279. I. A. Blech, D. B. Fraser, and S. E. Haszko, 'Optimization of the Al step coverage through computer simulation and scanning electron microscopy', *J. Vac. Sci. Technol.,* **15**, 13–19 (1978).
280. M. J. Attardo and R. Rosenberg, 'Electromigration damage in aluminium film conductors', *J. Appl. Phys.,* **41**, 2381–2386 (1970).
281. J. C. Blair, P. B. Ghate, and C. T. Haywood, 'Electromigration induced failure in aluminium film conductors', *Appl. Phys. Lett.,* **17**, 281–283 (1970).
282. F. M. d'Heurle and I. Ames, 'Electromigration in single crystal aluminium film conductors', *Appl. Phys. Lett.,* **16**, 80–82 (1970).
283. A. Gangulee and F. M. d'Heurle, 'Mass transport during electromigration in Al–Mg thin films', *Thin Solid Films,* **25**, 317–325 (1975).

284. P. S. Ho, J. E. Lewis, and J. K. Howard, 'Kirkendall study of electromigration in thin films', *Thin Solid Films,* **25**, 301–315 (1975).
285. J. K. Howard and F. E. Turene, 'Rare earth–aluminium intermetallic compounds for electromigration imprővement of micron and submicron conductor lines', *IBM Tech. Disd. Bull.,* **20**, 2241–2242 (1977).
286. R. E. Hummel and H. J. Geier, 'Activation energy for electrotransport in thin silver and gold films', *Thin Solid Films,* **25**, 335–342 (1975).
287. R. Rosenberg, 'Inhibition of electromigration damage in thin films', *J. Vac. Sci. Technol.,* **9**, 263–270 (1972).
288. D. Gupta, 'On the direct measurement of diffusion at temperatures less than 0.5 T_m', *Thin Solid Films,* **25**, 231–244 (1975).
289. J. R. Lloyd, M. R. Polcari, and G. A. MacKenzie, 'Observation of electromigration in heavily doped polycrystalline silicon films', *Appl. Phys. Lett.,* **36**, 428–430 (1980).
290. D. J. Colliver and A. F. Fray, 'Limitations to the performance of planar Gunn effect devices', *Solid-State Electron.,* **12**, 671–674 (1969).
291. C. J. Palmstrom, 'Contact degradation in GaAs transferred electron devices', *Ph.D. Thesis*, 1979 (University of Leeds, England).
292. C. J. Palmstrom, D. V. Morgan, and M. J. Howes, 'Contact degradation of GaAs transferred electron devices', *Nucl. Instr. Meth.,* **150**, 305–311 (1978).
293. R. J. Tree, M. J. Josh, and C. T. Foxon, 'On a failure mechanism in InP microwave oscillators', *Solid-State Electron.,* **14**, 519–520 (1971).
294. J. C. Irvin, 'GaAs IMPATT diodes in perspective', pp. 287–298 in *Proc. 4th Biennial Cornell Electrical Eng. Conf.*, Cornell, U.P., Ithaca, New York, 1973.
295. S. D. Mukherjee, *Unpublished Work* (on proton-isolated and un-heatsunk GaAs/Ti/Au Schottky barrier diodes), 1980.
296. B. K. Ridley, 'Specific negative resistance in solids', *Proc. Phys. Soc. Lond.,* **82**, 954–966 (1963).
297. W. Schröter, 'Electronic states at dislocations in germanium and silicon', paper K2 at the *Conf. Microscopy of Semiconducting Mater.*, April, 1979, Oxford, Proc. to be published in *J. Microscopy*. See *Proc. Roy. Microscopic Society,* **14**, 2 (1974) for program.
298. W. C. Ballamy and L. C. Kimerling, 'Premature failure in Pt–GaAs IMPATTs—recombination assisted diffusion as a failure mechanism', *IEEE Tr.*, **ED-25**, 746–752 (1978).
299. J. G. Smith, 'Compensation degradation mechanism for high efficiency GaAs IMPATT diodes', *Electron. Lett.,* **15**, 74–75 (1979).
300. P. H. Holloway and G. E. McGuire, 'Chemical characterization of coatings by analytical techniques sensitive to the surface and near surface', *Thin Solid Films,* **53**, 3–18 (1978).
301. R. E. Honig, 'Surface and thin film analysis of semiconductor materials', *Thin Solid Films,* **31**, 89–122 (1976).
302. C. A. Evans, 'Thin film compositional analysis—a comparison of techniques', *J. Vac. Sci. Technol.,* **12**, 144–150 (1975).
303. J. P. Hobson, 'The relationship of solid surfaces to vacuum science and technology', *Jap. J. Appl. Phys., Suppl.,* **2**, 317–325 (1974).
304. J. R. Shappirio and C. F. Cook, Jr., 'Modern analytical techniques for failure analysis', in *Solid State Technology*, Sept., 1979, pp. 89–94.
305. J. W. Coburn and E. Kay, 'Techniques for elemental composition profiling in thin films', *CRC Critical Reviews in Solid State Sciences,* **4**, 561–590 (1974).
306. H. W. Werner, 'Characterization of ceramics by means of modern thin film and surface analytical techniques', *Science of Ceramics,* **8**, 55–79 (1975).
307. K. N. Tu and S. S. Lau, 'Thin film deposition and characterization', Ch. 5,

pp. 81–118, in *Thin Films—Interdiffusion and Reactions*, J. M. Poate, K. N. Tu, and J. W. Mayer (Eds.), Wiley, New York, 1978.
308. J. W. Mayer and J. M. Poate, 'Depth profiling techniques', Ch. 6, pp. 119–160, in *Thin Films—Interdiffusion and Reactions*, J. M. Poate, K. N. Tu, and J. W. Mayer (Eds.), Wiley, New York, 1978.
309. C. J. Powell, 'Surface characterization: present status and the need of standards', *Applications of Surf. Sci.,* **1**, 143–169 (1978).
310. A. E. Morgan and H. W. Werner, 'Modern methods for solid surface and thin film analysis', *Physica Scripta,* **18**, 451–463 (1978).
311. S. Hofmann, 'Surface and thin film analysis: concepts, capabilities and limitations', *Talanta,* **26**, 665–673 (1979).
312. S. D. Mukherjee and D. V. Morgan, 'Physical studies of contact formation and interdiffusion', presented at the *Colloquium on Metallizations for III-V Semiconductor Devices*, November, 1979, IEE, London (proceedings unpublished, abstracts in Digest No. 1979/59, IEE, Savoy Place, London).
313. V. A. Phillips, *Modern Metallographic Techniques and their Applications*, Wiley, New York, 1971.
314. A. W. Czanderna (Ed.), *Methods of Surface Analysis*, Elsevier, Amsterdam, 1975.
315. P. F. Kane and G. B. Larrabee (Eds.), *Characterization of Solid Surfaces*, Plenum Press, New York, 1974.
316. J. L. McCall and W. M. Mueller (Eds.), *Microstructural Analysis: Tools and Techniques*, Plenum, New York, 1973.
317. B. Schwartz and N. Schwartz (Eds.), *Measurement Techniques for Thin Films*, The Electrochemical Society Inc., New York, 1967.
318. *Conf. Quantitative Surface Analysis*, 16–17 Oct. 1979, Teddington, England, the papers presented have been published in various recent issues of *Surface and Interface Analysis*.
319. D. G. Schimmel and M. J. Elkind, 'An examination of the chemical staining of silicon', *J. Electrochem. Soc.,* **125**, 152–155 (1978).
320. M. Ogawa, D. Shinoda, N. Kawamura, T. Nozaki, and S. Asanabe, 'GaAs alloy type Schottky barrier contact', *NEC Res. and Dev.,* **22**, 1–7 (1979).
321. M. L. Tarng and D. G. Fisher, 'Auger depth profiling of thick insulating films by angle lapping', *J. Vac. Sci. Technol.,* **15**, 50–53 (1978).
322. A. Christou, 'SEM failure analysis techniques for thin films and microwave power transistor metallizations', pp. 971–978, in *Scanning Electron Microscopy/1974 (Part IV)*, April, 1974, IIT Res. Inst., Chicago, Illinois.
323. W. B. Nowack and R. N. Dyer, 'Diffusion in thin film couples of platinum–gold', *J. Vac. Sci. Technol.,* **9**, 279–283 (1972).
324. C. Weaver and L. C. Brown, 'Diffusion in evaporated films of gold–aluminium', *Phil. Mag.,* **7**, 1–16 (1962).
325. G. Nomarski and A. R. Weill, 'Application à la metallographic des methodes interferentielles à deux ordres polarisées', *Rev. Metallurgie (France)*, **50**, 121, (1955). Also: O. Bryngdahl, 'Applications of shearing interferometry', Ch. II, in *Progress in Optics*, E. Wolf (Ed.), Volume 4, 1964.
326. J. R. Corruthers, A. F. Witt, and R. E. Rensser, 'Czochralski growth of large diameter silicon crystals—convection and segregation', pp. 61–71, in *Semiconductor Silicon, 1977*, H. R. Huff and E. Sirtl (Eds.), *Proc. Vol. 77-2*, The Electrochemical Society, Inc., Princeton.
327. J. S. Williams, W. L. Brown, H. J. Leamy, J. M. Poate, J. W. Rodgers, D. Rousseau, G. A. Rozgonyi, J. A. Shelnutt, and T. T. Sheng, 'Solid phase epitaxy of implanted silicon by ew Ar ion laser irradiation', *J. Appl. Phys.,* **33**, 542–544 (1978).

328. S. Takasu, Y. Matsushita, H. Kotake, and T. Wada, 'Application of infrared photoelastic measurement in silicon processing', pp. 456–465, in *Semiconductor Silicon 1977*, H. R. Huff and E. Sirtl (Eds.), The Electrochemical Society, Inc., Princeton.
329. J. Angillelo, J. Baglin, J. Dempsey, F. d'Heurle, and S. Petersson, 'Stresses resulting from silicide formation', pp. 955–956, in *Extended Abstracts*, The Electrochemical Society Fall Meeting, Oct. 1979.
330. P. A. Barnes and G. A. Rozgonyi (Eds.), *Proc. Symp. Semiconductor Characterization Techniques, 1978*. The Electrochemical Society, Inc., Princeton.
331. W. M. Bullis and F. G. Vieweg-Gutberlet, 'Current trends in silicon characterization techniques', pp. 360–6, and C. T. Sah, 'Detection of defect and impurity centres in semiconductors by steady-state and junction capacitance and current techniques', pp. 868–893, in *Semiconductor Silicon, 1977, Proc. Vol. 77-2*, The Electrochemical Society, Inc., Princeton.
332. D. V. Lang, 'Deep level transient spectroscopy: a new method to characterize traps in semiconductors', *J. Appl. Phys.,* **45**, 3023–3033 (1974).
333. A. Wagendristel, 'Resistometric investigation of initial interdiffusion in Ag–Au thin film couples', *Phys. Stat. Sol.,* **a-13**, 131–138 (1975).
334. E. N. Mgbenu, 'Dominant diffusion mechanism in RhSi formation', *Thin Solid Films,* **50**, L13–L16 (1978).
335. W. K. Chu, J. W. Mayer, M.-A. Nicolet, T. M. Buck, G. Amsel, and F. Eisen, 'Principles and applications of ion beam techniques for the analysis of solids and thin films', *Thin Solid Films,* **17**, 1–41 (1973).
336. M.-A. Nicolet, J. W. Mayer, and I. V. Mitchell, 'Microanalysis of materials by backscattering spectroscopy', *Science,* **177**, 841–849 (1972).
337. T. M. Buck and J. M. Poate, 'Ion scattering for analysis of surfaces and surface layers', *J. Vac. Sci. Technol.,* **11**, 289–296 (1974).
338. S. T. Picraux, 'Mass transport between two metal layers as studied by ion scattering', *Japan J. Appl. Phys.,* **Suppl. 2**, 657–664 (1974).
339. D. V. Morgan, 'Rutherford scattering analysis: a tool for semiconductor–device technology', *IEE Jnl. Solid-St. Electron Devices,* **1**, 37–45 (1977).
340. R. S. Blewer, 'Some practical aspects of depth profiling gases in metals by proton backscattering: application to helium and hydrogen isotypes', pp. 185–199, in *Ion Beam Surface Layer Analysis*, Vol. I, O. Meyer, G. Linker, and F. Käppeler, (Eds.), Plenum Press, New York, 1976.
341. B. Terreault, J. G. Martel, R. G. St.-Jacques, and J. L'Ecuyer, 'Depth profiling of light elements in materials with high energy ion beams', *J. Vac. Sci. Technol.,* **14**, 492–500 (1977).
342. J. F. Ziegler, R. F. Lever, and J. K. Hirvonen, 'Computer analysis of nuclear backscattering', pp. 163–183, in *Ion Beam Surface Layer Analysis, Vol. I*, O. Meyer, G. Linker, and F. Käppeler (Eds.), Plenum Press, New York, 1976.
343. W. K. Chu, J. W. Mayer, and M.-A. Nicolet, *Backscattering Spectrometry*, Academic Press, New York, 1978.
344. J. F. Ziegler, *The Stopping and Ranges of Ions in Matter*, Vol. 3 (Hydrogen ions), Vol. 4 (Helium ions), Vol. 5 (Heavy ions), Pergamon Press, New York, 1977.
345. J. F. Ziegler and J. E. E. Baglin, 'Does RBS of α by nuclei obey Rutherford cross-section formula?' *J. Appl. Phys.,* **45**, 1888–1898 (1974).
346. J. S. Williams, 'Range and stopping power effects obtained from high resolution Rutherford backscattering analysis of implanted targets', *Rad. Effects,* **22**, 211–213 (1974). See also: 'The application of low angle Rutherford backscattering to surface layer analysis', pp. 223–234 in Reference 340.

347. W. K. Chu, S. S. Lau, J. W. Mayer, H. Mueller, and K. N. Tu, 'Implanted noble gas atoms as diffusion markers in silicide formation', *Thin Solid Films,* **25**, 393–402 (1975).
348. F. Folkmann, 'Analytical use of ion-induced X-rays', *J. Phys.,* **E-8**, 429–444 (1975).
349. S. A. E. Johansson and T. B. Johansson, 'Analytical application of particle induced X-ray emission', *Nucl. Instr. Meth.,* **137**, 473–516 (1976).
350. A. B. Campbell, 'Characterization of thin films on metals using proton induced X-rays', *Thin Solid Films,* **64**, 445–454 (1979).
351. H. Kerkow, 'Measurement of the thickness of thin layers by proton induced X-rays', *Thin Solid Films,* **62**, 73–82 (1979).
352. J. W. Mayer and E. Rimini (Eds.), *Ion Beam Handbook for Material Analysis*, Academic Press, New York, 1977.
353. V. Valkovic, *Nuclear Microanalysis*, Garland STPM Press, New York, 1977.
354. D. P. Smith, 'Analysis of surface composition with low-energy backscattered ions', *Surf. Sci.,* **25**, 171–191 (1971).
355. W. Heiland and E. Taglauer, 'Investigation of surface topography of oxygen on nickel single crystals by helium ion backscattering', *J. Vac. Sci. Technol.,* **9**, 620–623 (1972).
356. D. G. Armour and G. Carter, 'Analysis of backscattered ions as a technique for the study of surfaces', *J. Phys.,* **E-5**, 2–8 (1972).
357. T. W. Rusch, J. T. McKinney, and J. A. Leys, 'ISS imaging and depth-profile analysis using beam rastering and signal gating', *J. Vac. Sci. Technol.,* **12**, 400–402 (1975).
358. G. C. Nelson, 'Search for preferential sputtering in Ag/Au alloys', *J. Vac. Sci. Technol.*, **13**, 974–975 (1976).
359. H. W. Werner, 'Secondary ion mass spectrometry and its application in thin film and surface layer research', *Japan. J. Appl. Phys.,* **Suppl. 2,** 367–374 (1974).
360. H. W. Werner, 'Introduction to secondary ion mass spectrometry', pp. 324–441, in *Electron and Ion Spectroscopy of Solids*, L. Fiermans, J. Vennik, and W. Dekeyser (Eds.), Plenum, New York, 1978.
361. H. Liebl, 'Secondary-ion mass spectrometry and its use in depth profiling', *J. Vac. Sci. Technol.,* **12**, 385–391 (1975).
362. K. Wittmaack, 'Secondary-ion mass spectrometry as a means of surface analysis', *Surf. Sci.,* **89**, 668–700 (1979).
363. G. H. Morrison and G. Slodzhian, 'The ion microscope opens new vistas in many fields of science by its ability to provide spatially resolved mass analysis of solid surfaces', *Analytical Chem.,* **47**, 933A–943A (1975).
364. J. S. Colligon, 'Surface compositional analysis using low energy ion bombardment induced emission processes', *Vacuum,* **24**, 373–388 (1974).
365. R. E. Hurley, 'Ion beam systems for surface analysis', *J. Phys.,* **E-5**, 1025–1036 (1972).
366. C. W. Magee, 'Depth profiling of *n*-type dopants in Si and GaAs using Cs^+ bombardment negative secondary ion mass spectrometry in ultrahigh vacuum', *J. Electrochem. Soc.,* **126**, 660–663 (1979).
367. A. J. Socha, 'Analysis of surfaces utilizing sputter ion source instruments', *Surf. Sci.,* **25**, 147–170 (1971).
368. A. Benninghoven (Ed.), *Secondary Ion Mass Spectrometry—SIMS II*, Springer–Verlag, Berlin, 1979.
369. C. W. Magee, W. L. Harrington, and R. E. Honig, 'Secondary ion quadrupole mass spectrometer for depth profiling—design and performance evaluation', *Rev. Sci. Instrum.,* **49**, 477–485 (1978).

370. H. Tamura, T. Kondo, I. Kanomata, K. Nukamura, and Y. Nakajima, 'In depth analysis in selected area with ion microprobe analyser', *Japan. J. Appl. Phys.*, **Suppl. 2**, 379–382 (1974).
371. Courtesy J. L. Maul, *Atomika GmbH*, Munchen, W. Germany.
372. W. O. Hofer and H. Liebl, 'Depth profiling of Cu–Ni sandwich samples by secondary ion mass spectrometry', *Appl. Phys.*, **8**, 359–360 (1975).
373. Z. L. Liau, B. Y. Tsaur, and J. W. Mayer, 'Influence of atomic mixing and preferential sputtering on depth profiles and interfaces', *J. Vac. Sci. Technol.*, **16**, 121–127 (1979).
374. D. V. McCaughan, C. W. White, R. A. Kushner, and D. L. Simms, 'Differences in sodium transport in SiO_2 films caused by ion and neutral-particle bombardment', *Appl. Phys. Lett.*, **35**, 405–407 (1979).
375. V. R. Deline, W. Katz, C. A. Evans, and P. Williams, 'Mechanism of the SIMS matrix effect', *Appl. Phys. Lett.*, **33**, 832–835 (1978).
376. A. M. Huber, G. Morillot, N. T. Link, P. N. Favennec, B. Deveau, and B. Toulouse, 'Chromium profiles in semi-insulating GaAs after annealing with a Si_3N_4 encapsulant', *Appl. Phys. Lett.*, **34**, 858–859 (1970).
377. H. W. Pickering, 'Ion sputtering of alloys', *J. Vac. Sci. Technol.*, **13**, 618–621 (1976).
378. J. W. Coburn, 'Sputtering in the surface analysis of solids: a discussion of some problems', *J. Vac. Sci. Technol.*, **13**, 1037–1044 (1976).
379. R. L. Gerlach and L. E. Davis, 'Semiquantitative analysis of alloys with SIMS', *J. Vac. Sci. Technol.*, **14**, 339–342 (1977).
380. H. B. Kim, G. G. Sweeney, and T. M. S. Heng, 'Analysis of metal–GaAs Schottky barrier diodes by secondary-ion mass spectrometry', *Inst. Phys. Conf. Ser., No. 24*, 307–319 (1975).
381. H. B. Kim, A. F. Lovas, G. G. Sweeney, and T. M. S. Heng, 'Effects of heat treatment on metal–InP Schottky barriers characterized by secondary-ion mass spectrometry', *Inst. Phys. Conf. Ser.*, **No. 33b**, 145–153 (1977).
382. J. J. Berenz, G. J. Scilla, V. L. Wrick, L. F. Eastman, and G. H. Morrison, 'Evaluation of "barrier" metals for sintered platinum–GaAs contacts', *J. Vac. Sci. Technol.*, **13**, 1152–1157 (1976).
383. H. Oechsner and E. Stumpe, 'Sputtered neutral mass spectrometry (SNMS) as a tool for chemical surface analysis and depth profiling', *Appl. Phys.*, **14**, 43–47 (1977).
384. *Scanning Electron Microscopy/1974 (Part I), Proc. of the 7th Annual Symp., IIT Research Institute*, Chicago, April 1974. See for example: N. C. Barbi *et al.* 'Light element analysis on the scanning electron microscope using a windowless energy dispersive X-ray spectrometer', pp. 151–158; C. E. Fiori *et al.*, 'Some techniques of signal processing in scanning electron microscopy', pp. 167–174; D. M. Maher, 'Scanning electron diffraction in a TEM and SEM operating in the transmission mode'.
385. G. W. Kammlott, 'Some aspects of scanning electron microscopy', *Surf. Sci.*, **25**, 120–146 (1971).
386. W. Reuter, 'Electron probe microanalysis', *Surf. Sci.*, **25**, 80–119 (1971).
387. L. C. Kimerling, H. J. Leamy, J. L. Benton, S. D. Ferris, P. E. Freeland, and J. J. Rubin, 'Analysis of impurity distribution and defect structures in semiconductors by SEM charge collection microscopy', pp. 468–480, in *Semiconductor Silicon 1977*, PV 77–2, The Electrochemical Society, Inc., Princeton.
388. R. B. Marcus, S. E. Haszko, S. P. Murarka, and J. C. Irvin, 'Scanning electron microscopy studies of premature breakdown sites in GaAs IMPATT testers', *J. Electrochem. Soc.*, **121**, 692–699 (1974).

389. P. Ashburn and C. J. Bull, 'Observations of dislocations and junction irregularities in bipolar transistors using the EBIC mode of SEM', *Solid-St. Electron.*, **22**, 105–110 (1979).
390. R. E. Thomas and G. A. Haas, 'Diffusion measurements in thin films utilizing work function changes: Cr into Au', *J. Appl. Phys.*, **43**, 4900–4907 (1972).
391. J. A. Venables, A. P. Janssen, P. Akhter, J. Derrien, and C. J. Harland, 'Surface studies in a UHV field emission gun scanning electron microscope', *Proc. Conf. Microscopy of Semiconducting Materials*, Oxford, April, 1979.
392. S. Ino. Japan, 'Some new techniques in reflected high energy electron diffraction (RHEED) applications to surface structure studies', *Japan J. Appl. Phys.*, **16**, 891–908 (1977).
393. J. A. Venables, C. J. Harland, and A. P. Janssen, 'RHEED and THEED patterns and dark-field microscopy in a UHV SEM', *Inst. Phys. Conf. Ser.*, **No. 41**, 313–318 (1978).
394. J. A. Venables, C. J. Harland, and R. bin-Jaya, 'Crystallographic orientation determination in the SEM using electron backscattered patterns and channel plates', *Proc. Electron Microscopy and Analysis Group Meeting*, 101–104 (1975).
395. C. C. Chang, 'Auger electron spectroscopy', *Surf. Sci.*, **25**, 53–79 (1971).
396. A. Van Oostrom, 'Some aspects of Auger microanalysis', *Surf. Sci.*, **89**, 615–634 (1979).
397. S. Hofmann, 'Auger electron spectroscopy', Ch. 2, pp. 89–172, in *Wilson and Wilson's Comprehensive Analytical Chemistry*, Vol. IX (Ed. G. Svehla), Elsevier, Amsterdam, 1979.
398. J. A. Venables, A. P. Janssen, C. J. Harland, and B. A. Joyce, 'Scanning Auger electron microscopy at 30 nm resolution', *Phil. Mag.*, **34**, 495–500 (1976).
399. S. Ichimura and R. Shimizu, 'Observation of electron beam damage in thin-film SiO_2 on Si with scanning Auger electron microscope', *J. Appl. Phys.*, **50**, 6020–6022 (1979).
400. A. Mogami, 'Analytical techniques with Auger scanning electron microscope', *Thin Solid Films*, **57**, 127–140 (1979).
401. J. A. Venables and A. P. Janseen, 'Developments in scanning Auger microscopy', pp. 280–291, in *Proc. 9th Internat. Cong. Electron Microscopy*, Toronto, 1980, Vol. III.
402. M. N. Read and P. J. Jennings, 'Surface barrier structure analysis using LEED', *Surf. Sci.*, **74**, 54–68 (1978).
403. P. J. Estrup, 'Surface characterization by low energy electron diffraction', pp. 553–589, in *Diffraction and Imaging Techniques in Materials Science*, S. Amelinckx, R. Gevers, and J. van Landuyt (Eds.), North-Holland, Amsterdam, 1978.
404. C. B. Duke, 'Electron-solid interactions: their nature and consequences', *J. Vac. Sci. Technol.*, **15**, 157–169 (1978).
405. S. Amelinckx, R. Gevers, and J. van Landuyt (Eds.), *Diffraction and Imaging Techniques in Materials Science*, North-Holland, Amsterdam, 1978, Vols. I and II.
406. J. J. Hren, J. I. Goldstein, and D. C. Joy (Eds.), *Introduction to Analytical Electron Microscopy*, Plenum Press, New York, 1979.
407. M. P. Seah and W. A. Deuch, 'Quantitative electron spectroscopy of surfaces: a standard data base for electron inelastic mean free paths in solids', *Surf. and Interface Analysis*, **1**, 2–11 (1979).
408. Perkin-Elmer, Physical Electronics, Inc.
409. D. Briggs (Ed.), *Handbook of X-ray and Ultraviolet Photoelectron Spectroscopy*, Heyden, Philadelphia, 1977.

410. D. R. Penn, 'Quantitative chemical analysis by ESCA', *J. Electron. Spectroscopy and Rel. Phen.*, **9**, 29–40 (1976).
411. C. J. Powell, 'The physical basis for quantitative surface analysis by Auger electron spectroscopy and X-ray photoelectron spectroscopy', pp. 5–30, in *Quantitative Surface Analysis of Materials*, ASTM STP 643, 1978.
412. B. Feuerbacher and B. Fitton, 'Photoemission spectroscopy', Ch. 5, pp. 151 **et seq**. in *Electron Spectroscopy for Surface Analysis*, H. Ibach (Ed.), Springer–Verlag, Berlin, 1977.
413. P. W. Palmberg, 'A combined ESCA and Auger spectrometer', *J. Vac. Sci. Technol.*, **12**, 379–384 (1975).
414. C. T. Houland, 'Scanning ESCA: a new dimension for electron spectroscopy', *Appl. Phys. Lett.*, **30**, 274–275 (1977).
415. G. K. Wehner, Japan, 'Sputtering in thin film analysis methods', *J. Appl. Phys.*, **Suppl. 2**, 495–500 (1974).
416. S. Hofmann, 'Evaluation of concentration–depth profiles by sputtering in SIMS and AES', *Appl. Phys.*, **9**, 59–66 (1976).
417. D. M. Holloway, 'Applications of depth profiling by Auger/sputter techniques', *J. Vac. Sci. Technol.*, **12**, 392–399 (1975).
418. P. M. Hall and J. M. Morabito, 'Compositional depth profiling by Auger electron spectroscopy', *CRC Crit. Rev. Solid St. Mater. Sci.*, **8**, 53–67 (1978).
419. S. Hofmann and A. Zalar, 'Auger electron spectroscopy depth profiling of Ni/Cr multilayers by sputtering with N_2^+ ions', *Thin Solid Films*, **60**, 201–211 (1979).
420. S. Hofmann, J. Erlewein, and A. Zalar, 'Depth resolution and surface roughness effects in sputter profiling of NiCr multilayer sandwich samples using Auger electron spectroscopy', *Thin Solid Films,* **43**, 275–283 (1977).
421. S. Hofmann and A. Zalar, 'Electron beam effects during the sputter profiling of thin Au–Ag films analysed by Auger electron spectroscopy', *Thin Solid Films,* **56**, 337–342 (1979).
422. T. J. Chuang, C. R. Brundle, and K. Wandelt, 'An X-ray photoelectron spectroscopy study of the chemical changes in oxide and hydroxide surfaces induced by Ar^+ ion bombardment', *Thin Solid Films,* **53**, 19–27 (1978).
423. T. J. Chuang and K. Wandelt, 'Sputter profiling through Ni/Fe interfaces by Auger electron spectroscopy', *IBM J. Res-Develop.,* **22**, 277–284 (1978).
424. Y. Mizokawa, H. Iwasaki, R. Nishitani, and S. Nakamura, 'Quantitative chemical depth profiles of anodic oxide on GaAs obtained by X-ray photoelectron spectroscopy', *J. Electrochem. Soc.,* **126**, 1370–1374 (1979).
425. P. M. Hall and J. M. Morabito, 'Matrix effects in quantitative Auger analysis of dilute alloys', *Surf. Sci.,* **83**, 391–405 (1979).
426. J. T. Grant, R. G. Wolfe, M. P. Hooker, R. W. Springer, and T. W. Haas, 'Automatic correction for effects of Auger line-shape on depth profiles', *J. Vac. Sci. Technol.,* **14**, 232–235 (1977).
427. G. Moore, H. Guckel, and M. G. Lagally, 'Auger investigation of boron-doped SiO_2/Si', *J. Vac. Sci. Technol.*, **14**, 70–74 (1977).
428. K. E. Singer, 'The application of Auger depth profiling to metal–semiconductor contacts', *Thin Solid Films,* **57**, 115–126 (1979).
429. L. L. Kazmerski, 'Grain boundary and interdiffusion studies in compound semiconductor thin films and devices utilizing Auger electron spectroscopy and secondary-ion mass spectroscopy', *Thin Solid Films,* **57**, 99–106 (1979).
430. O. Wada, S. Yanagisawa, and H. Takanashi, 'Thermal reaction of Ti evaporated on GaAs', *Appl. Phys. Lett.,* **29**, 263–265 (1976).
431. A. Christou and K. Sleger, 'A comparison of Ta and Al Schottky—barrier gates for GaAs FETs using μ-spot Auger electron spectroscopy', *Inst. Phys. Conf. Ser.,* **No. 33b**, 191–200 (1977).

432. G. E. McGuire, W. R. Wisseman, R. D. Ragle, and J. H. Tregilgas, 'Investigation of the Ag–In–Ge system used for alloyed contacts to GaAs', *J. Vac. Sci. Technol.*, **16**, 141–144 (1979).
433. M. H. Read and D. H. Heusler, 'X-ray analysis of sputtered films of beta-tantalum and body-centred cubic tantalum', *Thin Solid Films,* **10**, 123–125 (1972) and references therein.
434. K. N. Tu and B. S. Berry, 'X-ray study of interdiffusion in bimetallic Cu–Au films', *J. Appl. Phys.*, **43**, 3283–3290 (1972) and references therein.
435. R. W. Vook, 'X-ray diffraction', Ch. 3.5, pp. 339–364, in *Epitaxial Growth* (Ed. J. W. Matthews), Academic Press, New York, 1975.
436. S. S. Lau, W. K. Chu, J. W. Mayer, and K. N. Tu, 'Evaluation of glancing angle X-ray diffraction and MeV^4 He backscattering analysis of silicide formation', *Thin Solid Films,* **23**, 205–213 (1974).
437. J. W. Mayer and K. N. Tu, 'Analysis of thin film structures with nuclear backscattering and X-ray diffraction', *J. Vac. Sci. Technol.*, **11**, 86–93 (1974).
438. J. O. Olowolafe, K. N. Tu, and J. Angilello, 'Contact reaction between Si and Pd–W alloy films', *J. Appl. Phys.*, **50**, 6316–6320 (1979).
439. K. N. Tu, W. N. Hammer, and J. O. Olowolafe, 'Shallow silicide contacts', *J. Appl. Phys.*, **51**, 1663–1668 (1980).
440. C. R. Houska, 'X-ray examination of diffused films', *Thin Solid Films,* **25**, 451–464 (1975).
441. M. Murakami, D. de Fontaine, and J. Foder, 'X-ray diffraction study of interdiffusion in bimetallic Ag–Pd thin films', *J. Appl. Phys.*, **47**, 2850–6 (1976). Also: pp. 2857–2861, 'X-ray diffraction study of interdiffusion in bimetallic Ag–Cu thin films'.
442. A. R. Lang, 'Techniques and interpretation in X-ray topography', in *Diffraction and Imaging Techniques in Materials Science*, S. Amelinckx, R. Gevers, and J. van Landuyt (Eds.), North-Holland, Amsterdam, 1978.
443. A. Authier, 'Section topography', Ch. 5, pp. 145–189, in *X-ray Optics: Applications to Solids*, H.-J. Queisser (Ed.), Springer–Verlag, Berlin, 1977.
444. G. H. Schwuttke, 'X-ray topography', pp. 281–308 in *Epitaxial Growth*, (Ed.), J. W. Matthews, Academic Press, New York, 1975.
445. T. Abe, K. Kikuchi, and S. Shirai, 'Oxygen striations in silicon crystals', pp. 95–104, and K. Kohra and T. Matsushita, 'Characterization of silicon with highly collimated X-rays', pp. 441–455, in *Semiconductor Silicon, 1977, Proc. Vol. 77-2*, The Electrochemical Society, Inc., Princeton.
446. G. A. Rozgonyi and D. C. Miller, 'X-ray characterization of stresses and defects in thin films and substrates', *Thin Solid Films,* **31**, 185–216 (1976).
447. T. Wilson, J. N. Gannaway, and D. Walsh, 'A scanning optical microscope for the inspection of semiconductor materials and devices', paper H5 at the *Conference on Microscopy of Semiconducting Materials*, Oxford, April, 1979.
448. J. W. Steeds, 'Convergent beam electron diffraction', pp. 387–422, in *Introduction to Analytical Microscopy*, J. J. Hren, J. I. Goldstein, and D. C. Roy (Eds.), Plenum, New York, 1979.
449. J. A. Cookson, A. T. G. Ferguson, and F. D. Pilling, 'Proton microbeams, their production and use', *J. Radioanalytical Chem.,* **12**, 39–52 (1972).
450. C. J. Palmstrom, D. V. Morgan, and M. J. Howes, 'Contact-degradation studies on GaAs transferred-electron devices using a focused backscattering technique', *Electron Lett.,* **13**, 17 (1977).
451. B. Loberg and H. Norden, 'High resolution microscopy of grain boundary structures', Ch. 1, pp. 1–43, in *Grain Boundary Structures and Properties*, G. A. Chadwick and D. A. Smith (Eds.), Academic Press, London, 1976.

452. J. E. E. Baglin and J. M. Poate (Eds.), *Proc. Symp. Thin Film Interfaces and Interactions, 1979*, The Electrochemical Society, Inc., Princeton.
453. *Physics Today*, November, 1979, Special issue: *Microscience*, Am. Inst. Phys. Inc., New York.
454. *Physics Today*, May 1980, Special issue: *Thin Films*, Am. Inst. Phys. Inc., New York.
455. Proc. NSF Workshop on *Opportunities for Microstructures Science, Engineering, and Technology*, Nov. 1978, Airlie, VA, USA. (Workshop Chairman: J. M. Ballantyne, NRRFSS, Cornell University, Ithaca, New York.)
456. C. B. Duke, 'Frontiers in microelectronics and thin film components', *J. Vac. Sci. Technol.,* **17**, 1–8 (1980).
457. L. E. Murr, 'Future microelectronic devices: materials aspects and interfacial phenomena', *Microelectronics Journal,* **10–4**, 12–19 (1979).

Reliability and Degradation
Edited by M. J. Howes and D. V. Morgan

CHAPTER 2

Mechanisms and interfacial layers in silicide formation

G. OTTAVIANI and J. W. MAYER

2.1 INTRODUCTION

In integrated circuits constraints which lie outside the semiconductor itself are the metallization layers that provide a contact between the semiconductor and the outside world. Processing techniques including diffusion, ion implantation, oxidation, and gettering are sufficiently well developed so that in themselves they often do not limit device performance or degradation. As device size reduces to micron dimensions metallization becomes a major factor.

There are at least two major divisions in metallization schemes: contact to the semiconductor and interconnections between active devices. In the early development of integrated circuits one metal and one process temperature were used to perform both contact and interconnection functions. With the increased complexity of integrated circuits the one metal approach is not adequate and, therefore, different metals, depositions and heat treatments are used for the two different functions. This is a natural development because the requirements for interconnection—good conductivity, resistance to electromigration and corrosion, adhesion—are different than the requirements for contracts. In this chapter we concentrate on the metal–semiconductor contact even though the concepts which we develop may apply to interconnects.

There are two broad classes of metal–silicon reactions: eutectics and compounds (silicides). The Al–Si eutectic system dominated contact metallization schemes in the early years when device dimensions exceeded several microns in depth and lateral extent. The primary requirement of the metal was to make contact to diffused regions.

The major limitation of Al is the relatively high solubility and diffusivity of Si in Al which leads to the dissolution of silicon into aluminium during thermal processing and the consequent erosion of the surface. The erosion is not

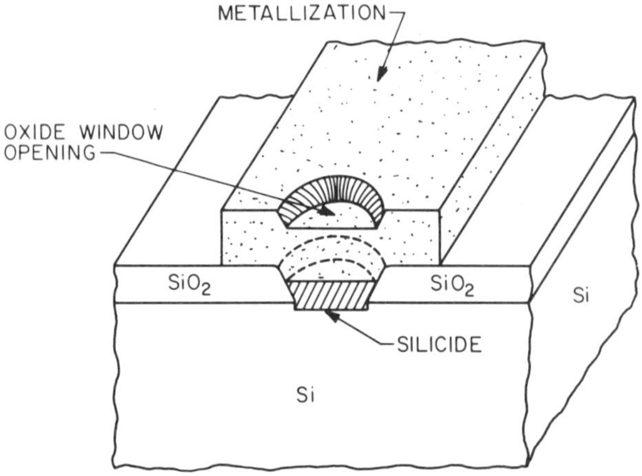

Figure 2.1. Schematic view of a metallization scheme for a silicon device structure. A silicide contact to the silicon is made in the opening etched in the oxide and the upper metallization (Al) layer provides the interconnection with other contact areas

uniform and pits can extend deeply enough in the silicon to short-circuit *p–n* junction characteristics.

Silicide forming systems offer the advantage that uniform and reproducible contacts are formed at the silicon–silicide interface. There are a wide variety of silicide forming metals and consequently there is a choice of barrier heights on *n*-type silicon between 0.5 and 0.9 eV. Silicide formation can also be achieved in the temperature range of 200–600 °C, a range that is compatible with device processing technology. A typical metallization scheme, Figure 2.1, involves a silicide and a metal.

In this chapter we are primarily concerned with silicide forming systems. The applications of such silicide forming metals require an understanding of the reactions that occur during thermal processing. For example, impurities such as oxygen can influence the rate of formation of silicides. To determine the amount of silicon consumed for a given metal thickness, one must know the identity and the thickness of the phases which are formed. Electrically active dopants in the silicon can redistribute at the silicon–silicide interface. The interconnect metal may react with the silicide layer necessitating the use of a diffusion barrier. The purpose of this chapter is to present a general view of the physical mechanisms involved in silicide formation.

2.2 OVERVIEW OF SILICIDE FORMATION

2.2.1 Sample preparation and analysis

In the fabrication of devices for integrated circuits the metal layer, typically 500–2000 Å thick, is deposited on a SiO_2 patterned wafer. The metal reacts with silicon in the narrow openings in the oxide layer. For most analytical studies of silicide formation, the metal is deposited directly on an unpatterned silicon wafer. We point out this difference in substrate preparation at an early stage so that the reader may be aware of possible differences in the formation behaviour of silicides in small area and large area structures. For example, for the silicon–Al system formation of pits have been noted at the periphery of the oxide windows.

Generally, the samples are prepared by depositing thin metal films either by electron-gun evaporation under dry vacuum conditions or by sputter-deposition in a well controlled atmosphere. The samples are then annealed either in vacuum or in a controlled atmosphere.

At every stage in sample preparation impurities have been shown to affect silicide formation. We will discuss these effects in a later section (Section 2.4.4), and summarize the general features here. The native oxide layer on the silicon surface can slow or block silicide formation at low temperature. The impurity content of the metal film is also important. Oxygen is known to affect interface sharpness, to retard growth kinetics and to change the sequence of phases that are formed. Consequently the studies of silicide formation under conditions where the films were deposited under poor vacuum conditions are suspect. The same comment can be made for samples annealed under relatively poor ambient conditions. Therefore, the control of impurities will play a role in device reliability as well as in analytical studies of silicide formation.

Various techniques can be used to analyse the samples and the choice depends upon the particular problem to be solved. Generally, it is always useful to use more than one. In compound formation the major interest has been in identifying the various phases formed and to measure their growth as a function of annealing temperature or time. Auger Electron Spectroscopy (AES) combined with ion sputtering[1] and MeV Ion Rutherford backscattering[2] are the two techniques mainly used to obtain growth kinetics and chemical composition. Ion backscattering is a fast technique, is non-destructive, and is so convenient that most silicide kinetic studies have been made with this technique.

The positive identification of the phases is provided by glancing angle X-ray diffraction.[3] The diffraction patterns can be obtained using various systems and again the one chosen depends upon the kind of sample analysed. Transmission electron microscopy (TEM) is used to investigate the microstructure of the silicide layers.

2.2.2 General features

The phase diagrams of metal–silicon couples in silicide forming systems can be represented in a general fashion by the Pt–Si phase diagram, Figure 2.2. There are well-defined phases and eutectic points. The lowest temperature at which a liquid appears is greater than 800 °C and is well above metallization process temperatures commonly used in integrated circuit fabrication. In silicon device processing, silicide formation is truly a solid phase reaction.

The Pt–Si phase diagram[4,5] shown in Figure 2.2 has three eutectic points and six stable phases and at least one metastable phase (not shown). A dramatic demonstration of the existence of all six stable phases is found in bulk diffusion couples.[6] An electron microprobe scan across such a couple is shown in Figure 2.3. The steps in the silicon concentration reflect the change in composition as one traverses from one phase to the next. The steps are relatively sharp and well defined indicating that phases are well separated one from each other.

In terms of semiconductor metallization layers, the distances observed in bulk diffusion samples are enormous. The 100 micron PtSi layer in Figure 2.3 is three orders of magnitude thicker than the 0.1 micron PtSi layer found in Si

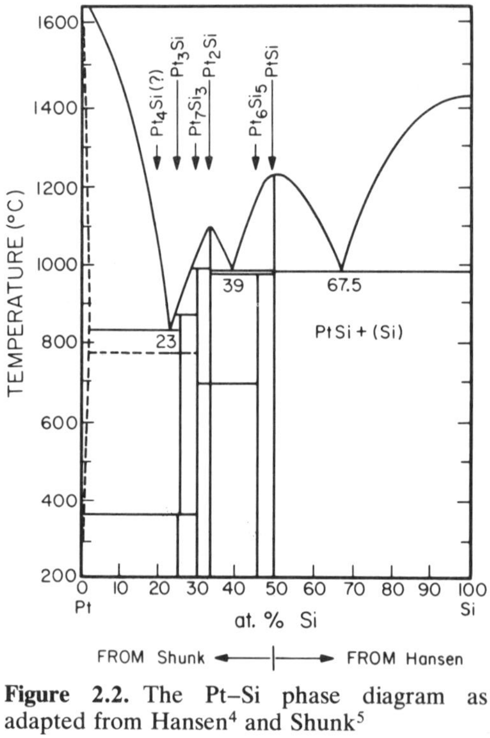

Figure 2.2. The Pt–Si phase diagram as adapted from Hansen[4] and Shunk[5]

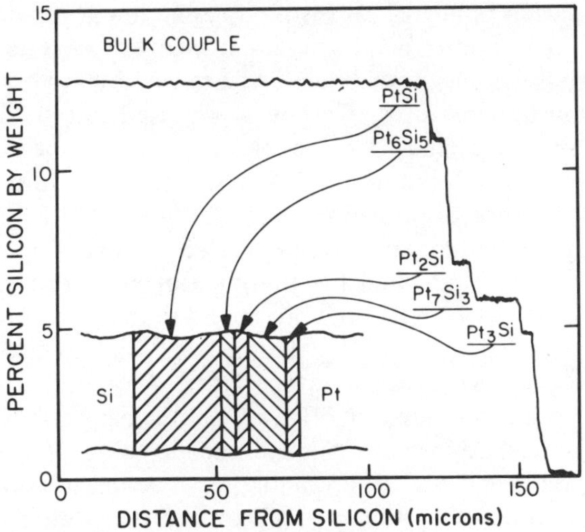

Figure 2.3. Electron microprobe scan across a silicon–platinum bulk diffusion couple annealed at 700 °C for 96 hours (from Claassen[6])

contacts. There are other differences between bulk and thin film systems. In the latter, generally no more than two silicide phases are formed at any stage in the annealing sequence. As shown schematically in Figure 2.4, one phase dominates in the initial stage of silicide formation. At longer annealing times, the width, W, of the Pt_2Si phase increases until all the platinum is consumed. A new phase PtSi develops at the Pt_2Si–Si interface.

At this point we present a model which takes into account the main features of silicon–metal interactions. We use as a reference the near-noble metals and in particular platinum. The extension to other metals is straightforward. In order to form a compound two major conditions should be fulfilled: nucleation and growth.[7] Nucleation and growth require a continuous supply of silicon and metal atoms provided by the reaction between metal and semiconductor.

The reaction is unique in that it occurs between covalently bonded silicon atoms and a metallic material. The presence of high energy phonons required

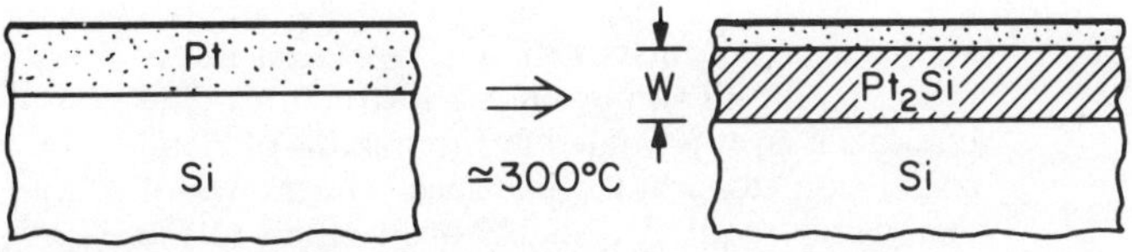

Figure 2.4. Formation of a Pt_2Si layer of thickness W at the interface between platinum and silicon

to break the silicon bonds is not justified at the low formation temperature ($\simeq$300 °C) of near-noble metal silicides. However, these temperatures are high enough to break the metal bonds. As a mechanism to explain compound formation at low temperatures, it has been proposed that the silicon bonds are softened by an intermixing between the silicon and the metal atoms. A reasonable way to get such intermixing is by interstitial diffusion of metal in silicon.[8] The existence of interstitial nickel atoms at the Ni–Si interface has been shown by channelling measurements.[9] Photo- and Auger electron spectroscopy analysis has indicated the presence of an interfacial layer having electronic properties which do not correspond to those of either the metal or the stoichiometric compound.[10–14] Channelling[15] and electron microscopy[16] also show the presence of an interface layer. A 'metallic-glass' interfacial layer was proposed to account for first-phase nucleation in silicides.[17,18]

In this intermixed layer whose thickness seems to be a few monolayers the Si bonds are weakened and are available for reaction with the metal at low temperature. A continuous supply of silicon and metal atoms is required for compound growth. The silicon atoms are supplied by the reaction at the interfacial layer and the metal atoms arrive at the Si-compound interface by

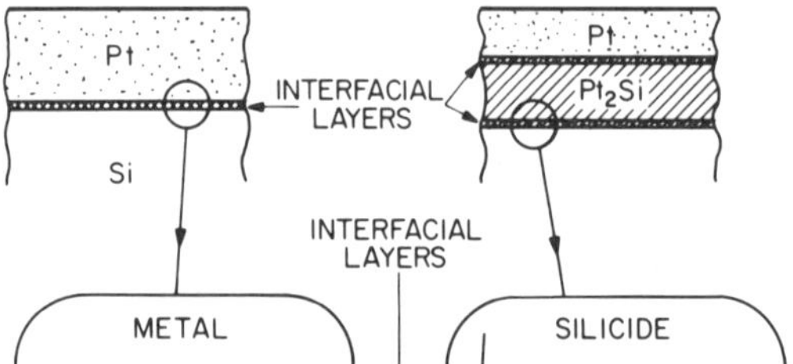

Figure 2.5. Schematics of the interfacial layers formed in Pt–Si and Pt–Pt_2Si–Si structures. The interfacial layer is formed by intermixing of metal and silicon atoms at the interface. The growth of the silicide requires a supply of Pt atoms to the interfacial layer

diffusing through the compound. The sketch in Figure 2.5 shows the main features of the model. The presence of such an interfacial layer has been shown to exist at both the metal–Si and silicide–Si interface.[14]

We believe that the interfacial layer not only is responsible for the low temperature reaction of the system but also has a composition which determines the specific compound which forms.[19] This composition can be adjusted by controlling the supply of silicon or metal atoms. In thin film couples, for example, different phases are formed when the supply of either silicon or metal atoms is reduced by consumption of the silicon or metal layer.[20,21] For the case of Pt–Si couples, the compound which first forms is Pt_2Si. As soon as the metal is consumed, the supply of metal atoms to the interfacial layer decreases and PtSi, the next phase, richer in Si than Pt_2Si, grows.[20] If the silicon layer is consumed before the metal, the supply of silicon atoms to the interfacial layer decreases and Pt_3Si, the next phase richer in metal than Pt_2Si, grows.

The model described above applies to all near-noble metals. It can also be extended to the refractory metals. A pronounced interstitial component is not now required because of the higher reaction temperature of these systems. An interfacial layer must exist and the compound formed must depend on the composition of such a layer. Hence the compound formed will also depend on the supply of either metal or silicon atoms to the interfacial layer. We also assume that the interfacial layer controls the electrical as well as metallurgical properties of the silicide/silicon system.

2.3 PLATINUM SILICIDE SYSTEM

We have chosen to illustrate the concepts of silicide formation by an overall description of the Pt-silicide system. Because of high barrier height of Pt-silicide ($\phi_B \simeq 0.87$ eV on *n*-type silicon), it was one of the first silicides mentioned in applications to Schottky barrier diodes.[22] This system continues to be in popular use in a wide variety of integrated circuit structures. The barrier height of platinum silicide to silicon can be reduced in a controllable fashion by taking advantage of dopant redistribution during silicide formation or by using an alloy combination. One of the attractive features is that it can be formed at a relatively low temperature, and its formation is not strongly sensitive to the presence of a native oxide layer at the Pt–Si interface. It is a stable, reproducible contact.

The platinum silicide contact system is not trouble-free. Impurities like oxygen contained within the film can modify the reaction kinetics and the reaction products. Moreover, aluminium, a common interconnection metal, reacts with the silicide and can degrade the electrical properties of the contact.[23–27] There is broad interest in developing diffusion barriers that will preserve the integrity of the contact.

2.3.1 Phase formation

One can pick up most of the phases indicated in the phase diagram by use of controlled thickness, thin film diffusion samples. Figure 2.6 shows the sequence of phases, a silicide family tree, that are formed in Pt–Si thin film couples.[20] The initial phase which is formed is Pt_2Si. This phase grows until all the metal or silicon is consumed. If the silicon is consumed (left side of Figure 2.6), subsequent phase formation is in the direction of increasingly metal-rich compounds. In the case where the metal is consumed and excess silicon is

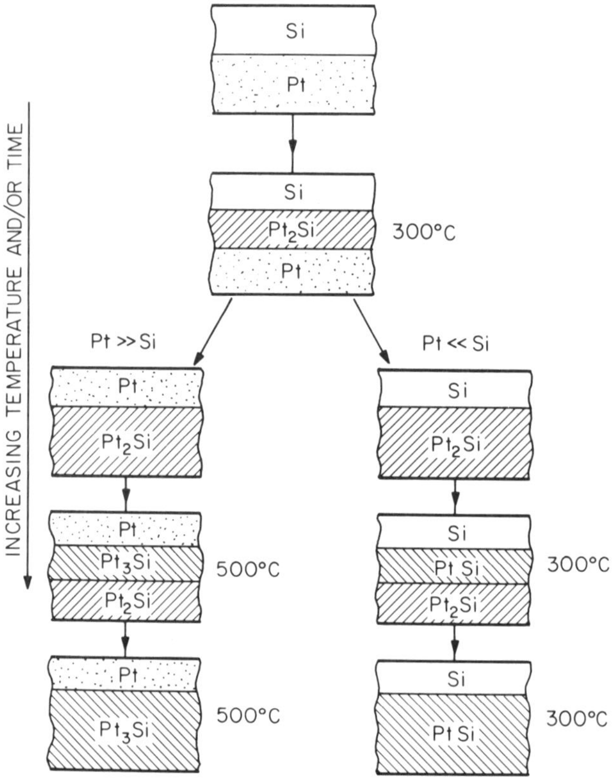

Figure 2.6. Sequence of phases formed in Pt–Si thin film couples. The initial phase formed is Pt_2Si; at longer times or higher temperatures more metal-rich (left side) or more silicon-rich silicides are formed depending on the availability of platinum on silicon. (Adapted from data given by Canali *et al.*[20])

present (right hand side of Figure 2.6), PtSi is formed. This is, of course, the situation found in standard metallization of platinum films on silicon, a progression from Pt to Pt_2Si to PtSi.

The structure of such a 'tree' cannot be explained on the basis of either thermodynamics or kinetics. According to known heat of formation values, PtSi should be formed before Pt_2Si.[7] Diffusion kinetics cannot account for the fact that the second phase forms only after the Pt or Si is consumed. We believe that the phase sequence is determined by the composition at the interface as discussed in Section 2.2.

2.3.2 Phase formation sequence

Rutherford backscattering spectrometry (RBS) is the primary technique used in studies of silicide formation kinetics to measure the thickness and composition of the growing phase.[2,28] In this section we present some RBS spectra to give an indication of the nature of the data. Figure 2.7 shows two superimposed backscattering spectra taken from a sample of platinum deposited on silicon (solid line) and from a sample annealed at 350 °C (closed points).[29] The

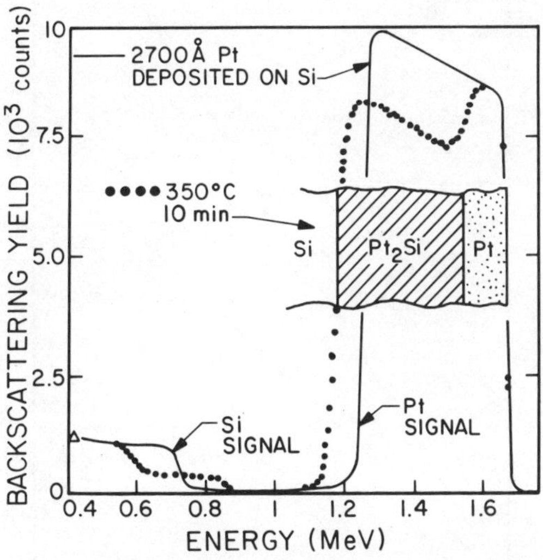

Figure 2.7. Rutherford backscattering spectra for 1.8 MeV ^4He ions incident on a 2700 Å Pt film deposited on silicon (solid line) and on the sample after annealing at 350 °C for 10 min. The backscattering signals from the Pt atoms appear at energies above 1 MeV and those from Si atoms below 1 MeV. The width and height of the steps in the signal indicate the thickness and composition of the silicide (from Canali *et al.*[29])

energy width of the signal from the platinum film indicates a film thickness of 2700 Å. After the sample was annealed, the platinum signal has a step whose height corresponds to platinum in Pt_2Si and whose width corresponds to 400 Å of Pt_2Si. The sharpness of the step indicates that the silicide layers are uniform over the 1.5 mm dimensions of the analysis beam. The location of the step indicates that growth occurs between platinum and silicon.

The first phase, Pt_2Si, exhibits $(\text{time})^{1/2}$ growth behaviour[30] as shown in Figure 2.8. This behaviour is typical of a process limited by the diffusion of one element in the growing compound; platinum is the diffusing species. On the basis of the model, it is necessary to supply platinum atoms at the Pt_2Si–Si interface in order to account for the low temperature of reaction.

When all the platinum deposited on the silicon wafer is converted into Pt_2Si, a new phase PtSi, starts to grow layer by layer at the original Pt_2Si–Si interface. The second phase, as shown in Figure 2.9, also has a $(\text{time})^{1/2}$ growth behaviour. Moreover its growth rate is not substantially different from the previous one. The onset of the second phase does not occur until after all the

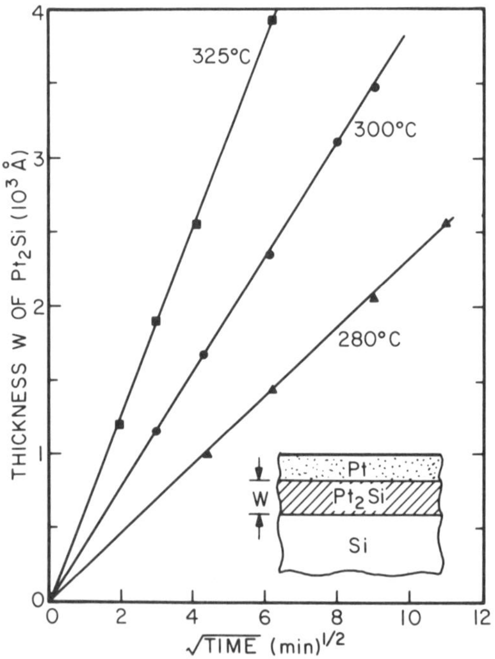

Figure 2.8. The thickness *W* of Pt_2Si as a function of $(\text{time})^{1/2}$ for samples annealed at different times. The kinetic behavior was determined from Rutherford backscattering spectra similar to those shown in Figure 2.7 (from Ottaviani *et al.*[30])

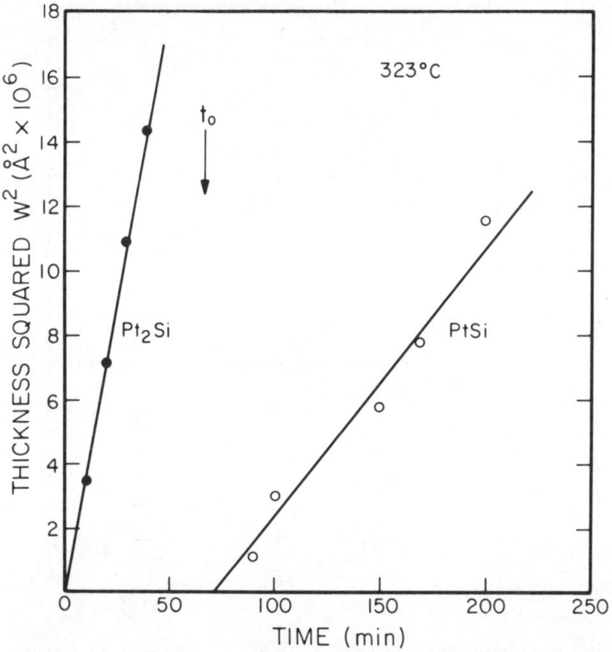

Figure 2.9. The (thickness)^2 versus annealing time behaviour of the formation of Pt_2Si and PtSi at 323 °C. The onset of the second phase PtSi occurs at time t_o when all the Pt has been consumed (from Canali *et al.*[29])

platinum has been consumed in the formation of Pt_2Si. The fact that the second phase, PtSi, grows at about the same rate as the first phase is a clear indication that kinetics alone cannot explain the delay in the formation of the second phase. We attribute the onset to a change in the availability of excess platinum atoms at the Pt_2Si–Si interface. The complete consumption of the platinum film leads to a decrease in the supply of platinum atoms and can trigger a change in composition of interface which in turn allows the formation of a more silicon rich phase. This phase in turn grows and converts Pt_2Si to PtSi.

The delay in the formation of the second phase can be tested directly by annealing samples with different platinum thicknesses together in the same annealing furnace.[31] In samples with thin platinum layers the platinum is consumed at the same rate as in thick samples. Figure 2.10 shows backscattering spectra for a thin (1150 Å) and thick (3570 Å) platinum film annealed at 325 °C. For samples annealed for 5 min, the steps in the platinum signal have the same width indicating that the same thickness of Pt_2Si is formed. For samples annealed for a longer time, the sample with the thick Pt layer continues to show the growth of Pt_2Si. In the thin sample where all the platinum

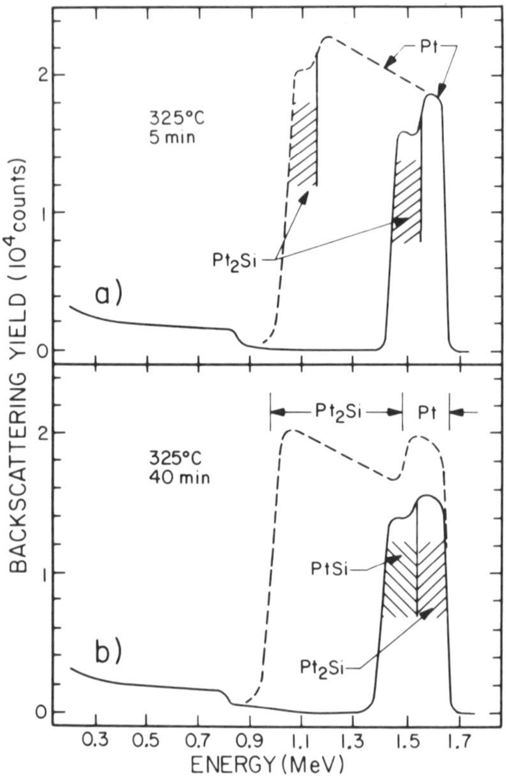

Figure 2.10. Rutherford backscattering spectra for thick (3570 Å) and thin (1150 Å) Pt films on Si annealed at 325 °C for (a) 5 minutes and (b) 40 minutes. The spectra show that the silicide phase sequence is affected by film thickness (from Canali *et al.*[31])

was consumed after a 10 minute anneal PtSi is formed. Consequently, the second phase PtSi forms in thin samples while the first phase Pt_2Si is still growing in the thicker sample. Again this is a manifestation of the influence of the supply of platinum on phase formation.

A more dramatic demonstration of the same phenomena is the decomposition of PtSi.[30] According to our model the second phase PtSi cannot exist in a Pt rich environment. Consequently a sample with platinum deposited on PtSi should form an intermediate phase of Pt_2Si after thermal annealing. As shown in Figure 2.11 for such a sample annealed at 325 °C, platinum reacts with PtSi and forms Pt_2Si at the PtSi–Pt interface. The back interface between PtSi and silicon does not shift indicating that silicon is supplied from the PtSi layer to form Pt_2Si.

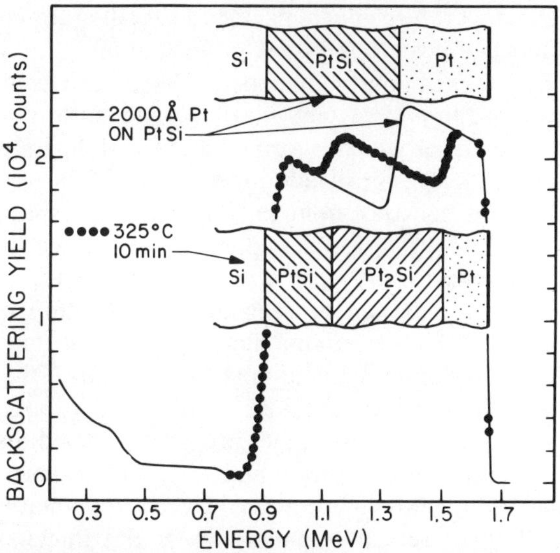

Figure 2.11. Backscattering spectra for 2000 Å Pt films deposited on PtSi (solid line) and annealed at 325 °C for 10 minutes (•). The presence of Pt causes decomposition of PtSi and formation of Pt_2Si at the Pt–PtSi interface (from Ottavani *et al.*[30])

In the experiments described in this section, phase formation was determined by the presence or absence of unreacted platinum. An alternate method of influencing the supply of platinum to the interface is to modify the transport of platinum by introducing a diffusion barrier. As we show in the next section, the incorporation of impurities, oxygen in this case, can establish such a barrier. One can also modify the supply of Pt by controlling the release of the platinum atoms from a matrix containing a mixture of platinum and a second metal. This effect occurs when an alloy of platinum and chrome is deposited on silicon and reacted to form PtSi without the formation of Pt_2Si as an intermediate phase.

2.3.3 Influence of impurities

One of the striking features of the reaction of platinum with silicon is the insensitivity of Pt–silicide formation to the presence of the native silicon oxide layer on the silicon surface. The native oxide layer contains pinholes or weak spots. Platinum can penetrate the weak areas to initiate silicide formation. It was also found[32] that the reaction kinetics were essentially unchanged when comparing silicide formation on two silicon surfaces: one prepared by the standard cleaning procedure using wet etching and other cleaned by sputter etching and high temperature annealing. Of course, the formation of platinum silicide can be blocked if there is a thick (a few hundred Å) SiO_2 layer at the interface. The presence of intermediate oxide layers such as those grown by a dip in HNO_3 can lead to an irregular interface. The major point, however, is that the native oxide layer which is formed in normal device processing does not have a major influence on silicide formation.

The same situation does not hold for impurities incorporated within the platinum film.[32a,33–36] Oxygen, for example, can slow down the reaction rate and can, at higher concentrations, stop the silicide growth.

The influence of oxygen can be seen directly by comparing the growth behaviour of silicides from platinum films deposited in ultra-high vacuum conditions (10^{-10} torr) or in conventional evaporation (10^{-7} torr) or sputtered deposition conditions. Figure 2.12 plots the $(\text{thickness})^2$ of the initial phase Pt_2Si versus annealing for samples prepared under three different conditions: ultra-high vacuum deposition and anneal,[32a] sputter deposition and vacuum anneal,[29] and sputter deposition, oxygen implantation and vacuum anneal.[33] For all three sets of samples, the initial thickness increases as

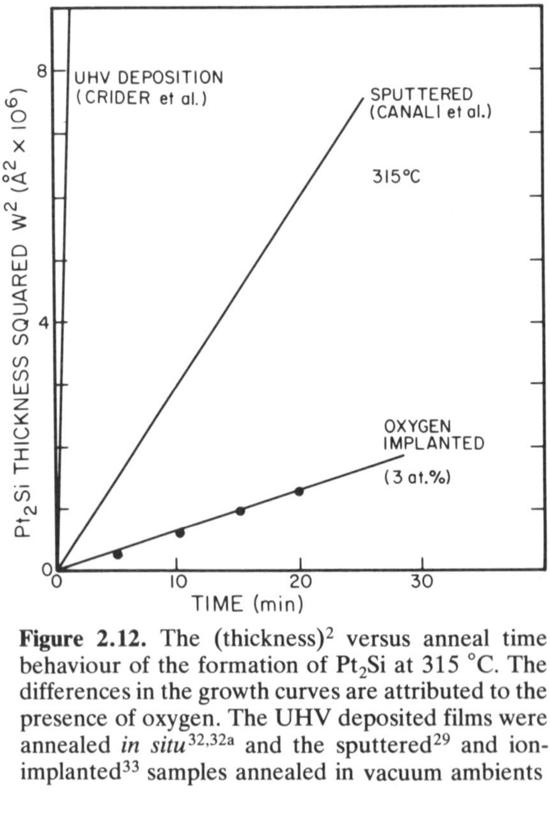

Figure 2.12. The $(\text{thickness})^2$ versus anneal time behaviour of the formation of Pt_2Si at 315 °C. The differences in the growth curves are attributed to the presence of oxygen. The UHV deposited films were annealed *in situ*[32,32a] and the sputtered[29] and ion-implanted[33] samples annealed in vacuum ambients

(time)$^{1/2}$ indicative of a diffusion limited process. The magnitude of the diffusion coefficient is a factor of 60 greater in the UHV films that in the sputter-deposited films.

Auger electron spectroscopy (AES) analysis of the sputter deposited films showed[29] that the oxygen concentration was less than 0.1 per cent. However, the formation of the compound led to an accumulation of oxygen at the Pt_2Si–Pt interface (Figure 2.13). In this case, the concentration of oxygen at the interface was about 0.8 per cent.

In the oxygen implanted sample where the implanted oxygen concentration was about 3 per cent, there was a much greater accumulation of oxygen at the Pt_2Si–Pt interface.[33] In this case the amount of oxygen was sufficient to affect phase sequence behaviour. As shown in Figure 2.14, the growth of Pt_2Si is inhibited and the second phase PtSi starts to grow before all the platinum is consumed. The growth of PtSi consumes the Pt_2Si layer and after 100 minutes at 315 °C, the reaction does not proceed further and leaves unreacted platinum at the surface. The oxygen is contained in the unreacted platinum layer.

The implanted sample represents an extreme case of oxygen contamination. Even so, samples prepared by deposition under 10^{-7} to 10^{-8} torr partial

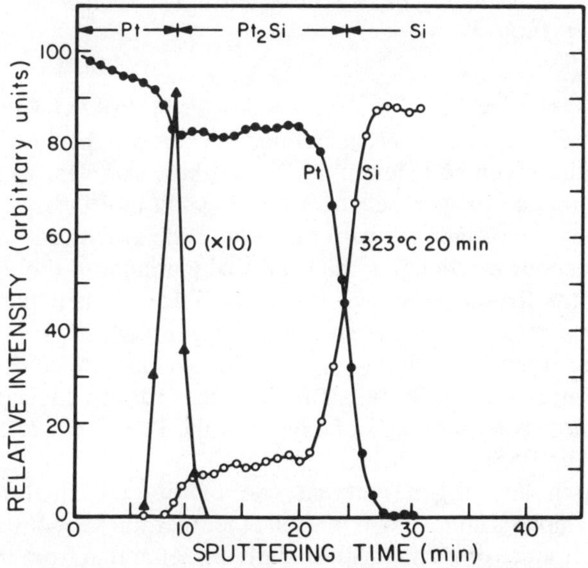

Figure 2.13. The distributions of Pt, Si, and O as determined by Auger electron spectroscopy analysis of a Pt–Si sample annealed at 323 °C for 20 minutes Oxygen accumulates at the Pt–Pt_2Si interface (from Canali *et al.*[29])

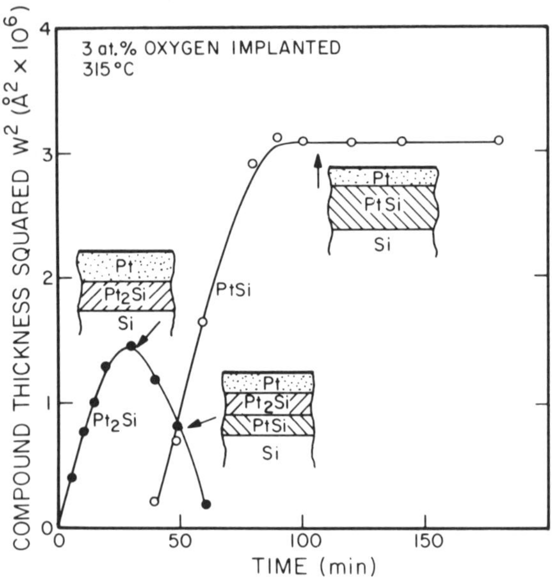

Figure 2.14. The (thickness)^2 *vs.* annealing time for the growth of Pt_2Si and PtSi in an oxygen-implanted sample annealed at 315 °C. The presence of oxygen in platinum influences the phase formation sequence (from Nava *et al.*[33])

pressure of O_2 also exhibited the presence of Pt_2Si, PtSi, and Pt at the same time.[32] The same behaviour was also noted in sputter-deposited samples contaminated with oxygen and aluminium.[34,35] In these samples, accumulation of both aluminium and oxygen occurs at the Pt_2Si–Pt interface.

The effect of the oxygen is two-fold: it slows the growth rate and modifies the phase sequence by changing the supply of platinum to the Pt_2Si–Si interface. The latter feature is shown in Figure 2.15 which depicts the phase sequence for samples with low and high oxygen concentrations. With low concentrations, the sequence observed are in agreement with the predictions of the model presented in Section 2.2. At higher oxygen concentrations, the presence of large concentrations of oxygen at the Pt_2Si–Pt interface limits the Pt supply at the Pt_2Si–Si interface.

Annealing in an oxygen ambient can modify the final structure.[37–40] Analysis of samples[38] annealed in oxygen at temperatures of 400 to 700 °C all showed the presence of a thin layer of platinum separated from the PtSi by an oxide layer. The oxide layer increases in thickness with increasing annealing temperature.

Post-annealing of PtSi films in oxygen ambients leads to the formation of a SiO_2 layer on the outer surface of the silicide film. The presence of this oxide

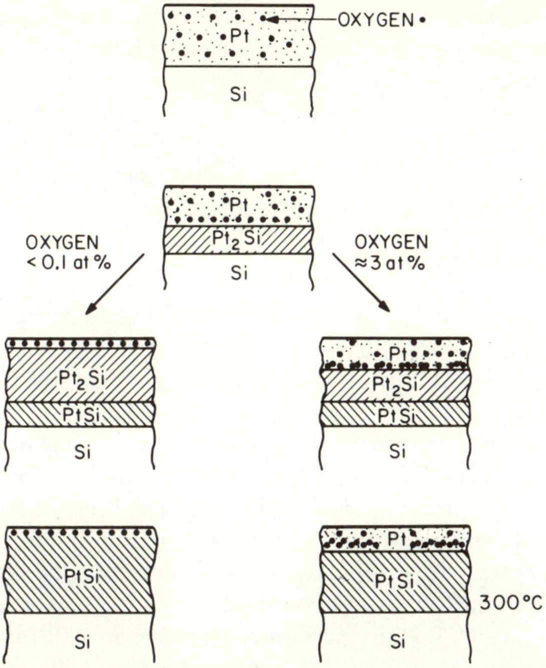

Figure 2.15. Schematic diagram of the influence of oxygen in a platinum film on silicon when the structure is annealed at 300 °C. Oxygen accumulates at the Pt–Pt_2Si interface in the initial stages of silicide growth. For long anneals of the sample containing 3 atomic per cent oxygen, the accumulation of oxygen in Pt stops silicide growth (see also Figure 2.14)

layer has been utilized in some metallization schemes in which unreacted platinum (platinum on SiO_2 adjacent to a window) is etched by aqua regia which does not react with SiO_2.

Any marked deviation from the normal sequence of phase formation is strong indication of the influence of impurities.

2.3.4 Alloys and bilayers

In the previous sections the impurities act as a diffusion barrier to the supply of platinum. Another way to change the supply of platinum is to influence the rate of release from a metal-platinum compound. This was accomplished by depositing alloys of Pt–Cr[41] and Pt–V.[42] Both chromium and vanadium react with silicon to form silicides at temperatures higher (400–600 °C) than the ones necessary to form platinum silicides. Here we use the platinum-rich

($Pt_{90}Cr_{10}$) alloy, as illustration of the behaviour. Annealing at 300 °C produces the formation of PtSi. The platinum is depleted from the alloy leaving a two layer structure with PtSi formed between silicon and the Pt–Cr alloy. There is no evidence for the formation of Pt_2Si. This follows because the supply of platinum is now governed by the depletion of the Pt–Cr alloy.[41] For the Cr-rich alloy the reaction occurs at temperatures above 600 °C and again there was no evidence of Pt_2Si formation.

2.4 PROPERTIES OF SILICIDES

There are a number of reviews which cover silicide formation.[7,43–45] In this section we concentrate on silicide forming systems in which layer by layer growth is observed. By this we mean that the growing silicide front advances uniformly with average irregularities less than a few hundred Å over mm lateral dimensions. These conditions hold for many silicide forming systems, it holds particularly for those that form in the temperature region of 200–600 °C; a temperature region of greatest interest in device fabrication.

In layer by layer growth at fixed processing temperature, the compound thickness increases with annealing time either linearly or proportional to $(\text{time})^{1/2}$. The difference in time dependence is due to the differences in the limiting mechanisms for the growth. A $(\text{time})^{1/2}$ growth rate means that the process is mainly limited by the transport through the silicide layer. The material transported can either be silicon or metal or both at the same time. The $(\text{time})^{1/2}$ rate results from the transport equation and in the increase in transport distance as the silicide becomes thicker. A linear relationship means that the growth is mainly limited by interfacial reactions.

The scheme of layer by layer growth at temperatures generally below 600 °C seems to be followed by most of silicide forming systems. There are exceptions as PdSi, $HfSi_2$, $NiSi_2$, Rh_4Si_5, Rh_3Si_4, $IrSi_3$. These compounds are formed only after annealing at high temperatures and their appearance seems to be nucleation controlled.[46]

2.4.1 Phase formation

This section is intended as a presentation, primarily in tabular form, of the formation behaviour and properties of the silicides. References are given in the earlier reviews[7,45] from which the tables were adapted. We separate silicide forming systems into the near-noble metals, the transition metals and the refractory metals. This separation is not purely arbitrary but follows the behaviour pattern of silicide formation; for example, the near-noble metals form at temperatures around 200 °C, the phase M_2Si is formed first, and the metal is the moving species. In contrast, the refractory metals form the disilicide, MSi_2. The formation temperature is higher (500–600 °C) and silicon is the moving species.

2.4.1.1 Near-noble methods

An overall picture of the reaction of near-noble metals is given in Table 2.1. This table indicates by an asterisk the first compound which is formed, in all cases, M_2Si, as well as the growth rate and activation energy for this compound. As in the case of Pt_2Si, the growth of M_2Si exhibits a $(time)^{1/2}$ dependence, and an activation energy of 1.5 eV is found. For all the metals reported in this table, the first phase forms at low temperatures and the metal is the moving species.

The second phase formed depends upon the relative quantity of material available for the reaction as indicated by the 'tree' in Figure 2.6. For the case of metal films which are deposited on Si substrate, Table 2.1 indicates that monosilicides are formed with a $(time)^{1/2}$ growth dependence and an activation energy near that of the first phase. With the exception of PdSi, the monosilicides form at temperatures near that of the first phase. The phase PdSi forms only after thermal annealing at temperatures around 825–850 °C.

Table 2.1 Silicides formed with near-noble metals.[a] The asterisks indicate the first phase formed

Silicide	Formation temperature (°C)	Activation energy (eV)	Growth rate	Moving species	Density (g cm^{-3})	Heat formation (kcal g atom^{-1})
*Co_2Si	350–500	1.5	$t^{1/2}$	Co	7.42–7.28	9.2
Cosi	375–500	1.9	$t^{1/2}$	—	6.57	12.0
$CoSi_2$	550	—	—	—	4.95–5.3	8.2
*Ni_2Si	200–350	1.5	$t^{1/2}$	Ni	7.35–7.2	11.2–10.5
NiSi	350–750	1.4	$t^{1/2}$	Ni	5.93	10.3
$NiSi_2$	≥750	—	—	—	4.93	6.9
Ni_5Si_2	400	—	—	—	—	10.3
Ni_3Si	450	—	—	—	—	8.9
*Mg_2Si	200	—	—	Mg	1.94	6.2
*Pd_2Si	100–300	1.5	$t^{1/2}$	Pd,Si	8.23	6.9
PdSi	850	—	—	—	7.69	6.9
Pd_3Si	350	—	—	—	—	—
Pd_4Si	400	—	—	—	—	—
Pd_5Si	650	—	—	—	—	—
*Pt_2Si	200–500	1.5	$t^{1/2}$	Pt	11.4	6.9
PtSi	300	1.6	$t^{1/2}$	—	12.6–7.31	7.9
Pt_3Si	400	—	—	—	—	8.1

[a]Table adapted from References 7 and 45.

For nickel and cobalt, the disilicides are listed and indeed these compounds are formed as the end phase in metal films on silicon structures. As implied by the dashes in Table 2.1 there is little known about the formation kinetics of these near-noble metal disilicides.

Thin film reactions in which metals are in excess have been carried out only for Ni, Pd, and Pt. There is progression from the M_2Si phase toward increasingly metal-rich silicides in these systems.

2.4.1.2 Intermediate metals

There are a number of silicides whose formation behaviour lies intermediate between the near-noble metal and refractory metals. The importance of these metals because of their adhesion on SiO_2 or to their barrier height warrant their inclusion. The common feature of these materials is that the monosilicide is the first phase formed. For those silicides where the kinetic behaviour has been measured in detail, the growth follows a $(\text{time})^{1/2}$ dependence, and Si is the diffusion species (Table 2.2). All the components have an activation energy somewhat less than 2 eV, with the exception of HfSi.

With metal films on a silicon substrate, the disilicide is the second phase formed. For iridium and rhodium more Si-rich phases can be formed at temperatures above 800 °C.[47,48]

Table 2.2 Silicides formed with intermediate metals[a]

Silicide	Formation temperature (°C)	Activation energy (eV)	Growth rate	Moving species	Density (g cm^{-3})	Heat formation (kcal g atom^{-1})
FeSi	450–550	1.7	$t^{1/2}$	Si	6.16	8.8
$FeSi_2$	550	—	—	—	4.54	6.2
RhSi	350–425	1.95	$t^{1/2}$	Si	8.5	8.1
Rh_2Si	400	—	—	—	—	—
Rh_4Si_5	825–850	—	—	—	—	—
Rh_3Si_4	925	—	—	—	—	—
HfSi	550–700	2.5	$t^{1/2}$	Si	16.71	—
$HfSi_2$	750	—	—	—	7.98	—
TiSi	500	—	—	—	4.24	15.5
Ti_5Si_3	550	—	—	—	—	17.3
$TiSi_2$	600	—	—	Si	4.10	10.7
MnSi	400–500	—	—	—	5.82	13.3
$MnSi_2$	800	—	—	—	5.24	2.6
IrSi	400–500	1.9	$t^{1/2}$	—	—	8.0
$IrSi_{1.75}$	500–1000	—	—	—	—	6.1
$IrSi_3$	1000	—	—	—	—	4.6

[a]Table adapted from References 7 and 45.

2.4.1.3 Refractory metals

The refractory metals exhibit a common behaviour pattern. The silicides all form at relatively high temperatures (450–650 °C) with the disilicides as the only phase; this is also the end phase which one may predict from the equilibrium phase diagram. Silicon appears to be the predominantly moving species.

These metals are all strongly influenced by the presence of impurities. The presence of oxygen influences the growth kinetics and consequently we cannot state with confidence that the growth follows a linear time dependence or that the activation energy follows the values given in Table 2.3. For example, Nb samples prepared in ultra-high vacuum conditions exhibit a $(\text{time})^{1/2}$ dependence, while a linear relationship has been found on samples prepared under poor vacuum conditions.[49]

2.4.2 Epitaxial silicides

In four cases, the crystalline structure of the silicides has a close match with the underlying silicon substrate and epitaxy has been observed.[50–59] From a technological standpoint such epitaxial silicides provide a high-conductivity layer which could act as a template for epitaxial growth of deposited silicon layers. Silicide layers between crystal silicon regions opens the possibility for device structures with buried interconnects or novel devices such as the metal-base transistor.

Epitaxial layers of Pd_2Si, PtSi, $NiSi_2$, and $CoSi_2$ have been formed. The structure of both $CoSi_2$ and $NiSi_2$ are cubic (CaF_2) with a lattice constant close to that of silicon. The mismatch of the lattice parameter compared to that of Si is 0.5 per cent for $NiSi_2$ and 1.1 per cent for $CoSi_2$. With these silicides, epitaxial growth have been found on $\langle 111 \rangle$ and $\langle 100 \rangle$ oriented Si crystals.

Table 2.3 Silicides formed with refractory metals[a]

Silicide	Formation temperature (°C)	Activation energy (eV)	Growth rate	Moving species	Density (g cm^{-3})	Heat formation (kcal g atom^{-1})
$CrSi_2$	450	1.7	t	—	4.98	7.7
$MoSi_2$	525	3.2	t	Si	6.27–6.3	8.7–10.5
$NbSi_2$	650	2.1	$t^{1/2}$	—	5.66	10.7
VSi_2	600	2.9, 1.8	$t, t^{1/2}$	Si	4.64–4.42	24.3
WSi_2	650	3.0	$t, t^{1/2}$	Si	9.28–9.4	7.3
$TaSi_2$[b]	650	3.7	t	Si	9.06	8.7–9.3
$ZrSi_2$	700	—	—	—	4.86	12.9–11.9

[a]Table adapted from References 7 and 45.
[b]Activation energy and growth rate from J. E. E. Baglin (private communication).

The hexagonal structure of Pd_2Si and the orthorhombic structure of PtSi necessitate use of ⟨111⟩ oriented silicon.

The epitaxial nature of the silicide layer has been investigated by channelling and Rutherford backscattering techniques. These measurements show that high quality epitaxial layers of $NiSi_2$ and Pd_2Si can be formed. The layers of $CoSi_2$ and PtSi have a pronounced spread ($\simeq 0.5°$ or greater) in the angular distribution of crystallite axes oriented with respect to the silicon crystal axes.

The epitaxial quality of $NiSi_2$ and Pd_2Si layers improves with increased annealing temperatures. Process temperatures of 700 to 800 °C are used for good epitaxy. The epitaxial quality also depends on layer thickness. For example, with $NiSi_2$ layers greater than 2000 Å, fractures are found.[58]

2.4.3 Stress

The processes of thermal annealing and silicide growth introduce stress.[60,61] Silicides have a different thermal expansion coefficient than silicon and there is a volume difference between the metal–Si and silicide–Si systems.

Measurements of the curvature of silicon wafers have been made using X-ray diffraction techniques.[60] During isothermal annealing of Pd–Si structures, for example, compressive stress increases with time due to volume changes during silicide formation. The stress reaches a maximum when all the palladium is consumed (Fig. 2.16). Further annealing promotes the relaxation

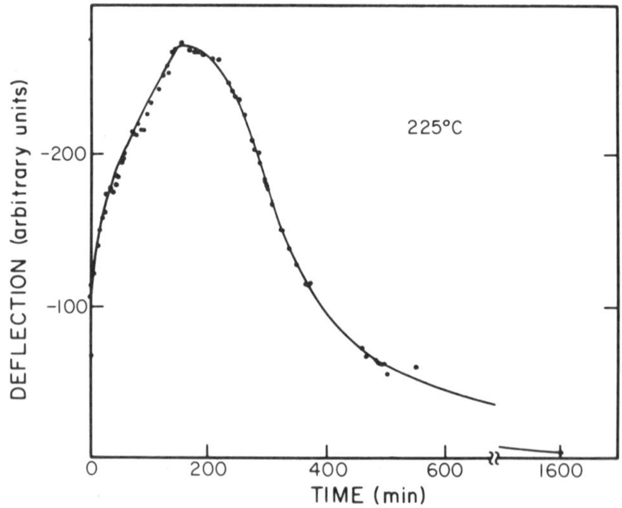

Figure 2.16. The deflection (compressive) of the substrate as a function of time for a 2000 Å Pd film on ⟨100⟩ Si annealed at 225 °C. The stress increases during Pd_2Si formation (from Angilello, *et al.*[60])

of the stress itself. In the example shown, about an order of magnitude longer time is required to relax the stress than to reach the maximum. When the sample is cooled to room temperature a tensile stress now develops. This stress is mainly due to the differences in thermal expansion coefficient of the silicon wafer and silicide film.

This example indicates that the amount and sign (tensile or compressive) of the stress depends upon the thermal history of the sample. Several silicides have been investigated. In all cases stress has been found generally tensile in nature, in the 10–20 $\times 10^{19}$ dyn cm^{-2} range.

2.4.4 Influence of impurities

Impurities modify the growth kinetics of silicide formation and the kind of compound formed. We emphasize the role of oxygen, the impurity which has a strong effect on silicide formation. Depending on the location of the impurities in the sample, a broad range of impurity phenomena can be roughly classified as: (1) impurities contained in the metallic layers, (2) impurities introduced during the annealing, (3) impurities at the Si–metal interface, and (4) impurities in silicon.

Impurities contained in the film generally slow down the kinetic behaviour. The influence of oxygen on Pt–Si was described in Section 2.3.3. Disilicide formation in refractory metals is also affected by impurities. For example, in the formation of VSi_2, the presence of oxygen in amounts of 10 per cent can significantly decelerate the reaction.[62] With higher oxygen content in the vanadium films, V_3Si rather than VSi_2 is formed.[63] In terms of the model presented in Section 2.2, we believe that the presence of oxygen modifies the properties of the interface.

Impurities can also be introduced in the film during thermal annealing and a strong effect on compound formation has been found. As an example, nickel behaves differently if the annealing is performed in vacuum or in an atmosphere with some oxygen.[31] Figure 2.17 shows the backscattering spectra of a 1700 Å film on a (100) silicon substrate before and after annealing at 325 °C in a vacuum furnace (10^{-6} torr) or in commercially pure flowing nitrogen. While vacuum annealed samples show the normal evolution (Ni_2Si forms first, then NiSi when all nickel is reacted) in samples annealed in nitrogen (Figure 2.17(b)) the coexistence of Ni_2Si, NiSi, and unreacted nickel was observed. This behaviour strongly suggests that oxygen diffuses through the nickel to the silicide–metal interface and modifies the supply of nickel to the Ni_2Si–Si interface. In a related experiment, it was found that oxygen implanted in a nickel film influences the sequence of phase formation and that oxygen accumulates at the Ni_2Si–Si interface.[64] Analysis by X-ray photoelectron spectroscopy showed the oxygen was bonded to silicon at the silicide–Ni

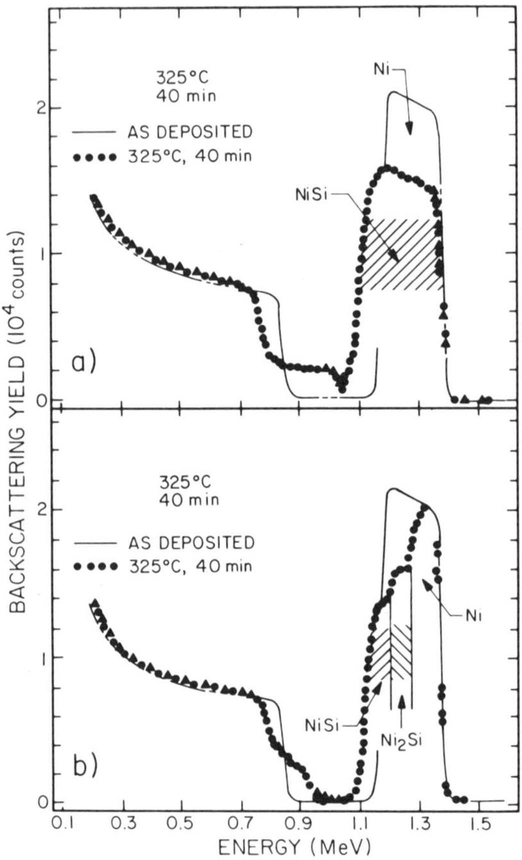

Figure 2.17. Rutherford backscattering spectra for a 1700 Å Ni film on silicon annealed for 40 min. at 325 °C in (a) vacuum ambient or (b) flowing nitrogen. Impurities introduced during annealing affect the silicide phase sequence (from Canali *et al.*[31])

interface. We believe that the build-up of the oxide layer forms a diffusion barrier.

In the case where one anneals a silicide in an oxygen ambient, SiO_2 can be formed on top of the silicide. As shown in Figure 2.18(a), SiO_2 is formed on WSi_2 after processing in steam at 1000 °C.[65,66] Radio tracer measurements of the growth of SiO_2 on $CoSi_2$ at 1000 °C show that the self-diffusion coefficient of Si in the silicide layer is very high at 1000 °C.[67] The transport of Si in SiO_2 is very low at 1000 °C so that the growth of SiO_2 occurs at the silicide–oxide interface.

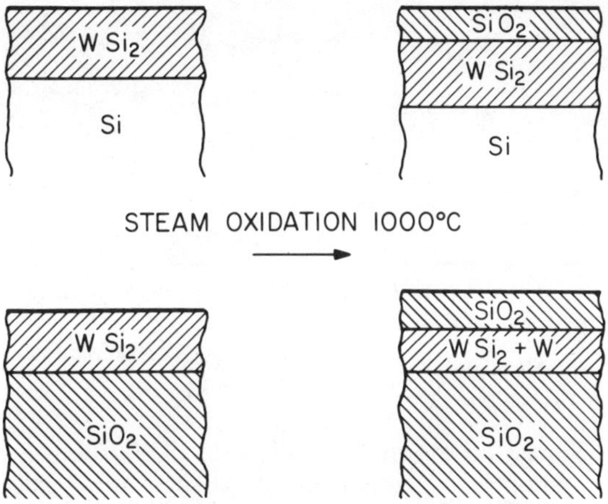

Figure 2.18. Schematic showing the formation of SiO_2 on sputtered deposited WSi_2 annealed at 1000 °C in steam. Formation of tungsten-rich silicides occurs in samples deposited on an SiO_2 substrate

The disilicide structure can be maintained as long as there is an adequate supply of silicon. Depending upon annealing conditions and sample configuration, metal-rich silicides can be formed during oxidation. If the supply of silicon from the substrate is not adequate, the incorporation of silicon in SiO_2 will deplete the amount of silicon in the silicide layer leading to the growth of more metal-rich silicides. For example, with WSi_2 deposited on SiO_2, oxidation causes the formation of W_5Si_3 (Figure 2.18(b)).

The impurities contained at the metal–silicon interface affect silicide formation by decreasing or inhibiting the silicon supply. An extreme case is when the metal is directly deposited on SiO_2.[62,68–70] After heat treatment, various cases can be observed: no reactions as in the case of Au; a metal oxide as in the case of Al, Pb, and Sn, and metal silicide and metal oxide as in the case of refractory metals. The influence of SiO_2 on vanadium silicide formation is shown in Figure 2.19. The disilicide is formed from vanadium on silicon but V_3Si and VO_x are formed if the substrate is SiO_2.[62,68] In the latter case, higher processing temperatures are required and vanadium rather than silicon is the moving species through the V_3Si layer.[71]

Impurity redistribution within the silicon substrate can occur during silicide formation. Dopants such as arsenic or antimony introduced into silicon by ion implantation before metal deposition can accumulate at the silicide–Si interface during silicide growth.[72] The accumulation of dopant atoms at the interface leads to a change in the electrical properties of silicide–silicon contacts. The concept was used as a method of lowering the barrier height of PtSi–*n*-Si Schottky barrier diodes.[73]

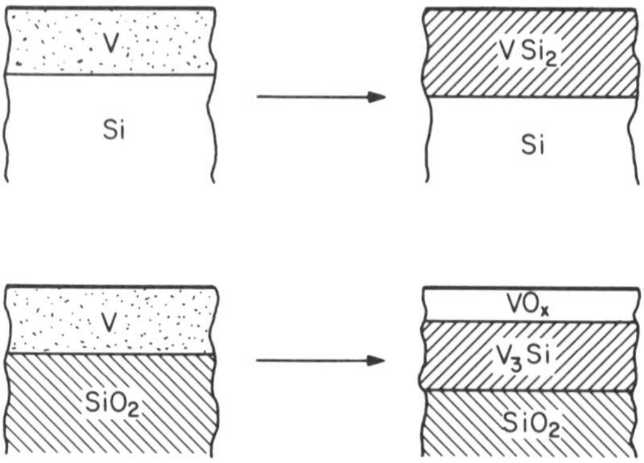

Figure 2.19. Schematic diagram of silicide formation in vanadium films deposited on Si and SiO_2. With a silicon substrate, VSi_2 is formed (Si is the moving species) and with SiO_2, V_3Si is formed (V is the moving species)

2.5 ALLOY AND BILAYER SILICIDES

It must be apparent to the reader that with the possible exception of impurity effects there is a clear understanding of the empirical behaviour of single metal silicide formation. An obvious extension of past experience is to consider the application of multilayer or alloy metal films in order to meet the more stringent requirements imposed by increasingly complex integrated circuit design. The shrinkage of dimensions toward one micron lateral and submicron depth dimensions provides a strong push for development of more sophisticated metallization schemes.[74] One requirement is to have a uniform silicide contact that requires only a limited consumption of silicon. This cannot be achieved under production conditions by the deposition of hundred Å layers. The thinness of such layers and the difficulty of obtaining uniform and reliable coverage over a wafer virtually precludes their use. Another requirement is to provide a diffusion barrier between the silicide contact and the interconnect metallic layer.

An ideal structure is shown in Figure 2.20. Two-metal silicides can be used to achieve such a structure; one approach is to deposit a sequence of layers of metal forming silicides (bilayers), and the other approach is to codeposit a mixture of metals (alloys). There are so many possible metal combinations that for the convenience of the reader we consider only two broad categories; the near-noble metal silicides, Ni, Pd, Pt, and the refractory metal silicides of the group IV, V, VI metals, Ti, V, Cr, W, etc.

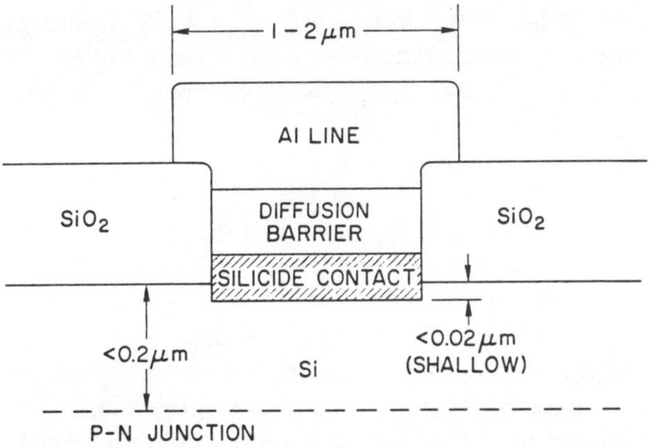

Figure 2.20. Cross-sectional view of a shallow silicide contact with a diffusion barrier between the silicide and the aluminum interconnect line (from Tu *et al.*[41])

2.5.1 Alloy silicides

The alloy systems investigated to date have been obtained by either co-deposition using dual electron-guns or sputter deposition onto a silicon wafer. The systems have been combinations of either near-noble and refractory metals,[41,42,75,76] as PtCr, PdV, and PtV or two refractory metals as TiW.[77]

Although the details of reaction temperatures and kinetics can be different, the general behaviour is the same for all the near-noble and refractory metal combinations.[42,75] We illustrate this behaviour pattern by using Pt-rich ($Pt_{90}Cr_{10}$) and Cr-rich ($Pt_{10}Cr_{90}$) alloys.[41] For the Pt-rich, $Pt_{90}Cr_{10}$, sample the formation of platinum silicide was detected at 300 °C; the amount of PtSi increases with annealing up to 600°. The platinum is depleted from the alloy leaving a two layer structure; a layer of PtSi next to the silicon and an outer layer of Cr enriched PtCr alloy. For the Cr-rich alloy, $Pt_{10}Cr_{90}$, there is no structural evidence for silicide formation for annealing at 400 °C for up to four hours. As expected, there are large reactions at annealing temperatures of 450 °C since single layers of Cr form $CrSi_2$ at 450 °C.

The results of alloy formation in the PtCr system are bounded between the reaction temperature of Pt (around 200 °C) and that of Cr (450 °C). A wider temperature range is obtained with a refractory metal, V, which reacts with silicon at higher temperature, 600 °C. In all these alloy systems investigated, annealing up to temperatures as high as 800 °C has not resulted in a ternary system, but rather in phase separation with the near-noble metal silicide next to the silicon and the refractory disilicide formed near the surface.

Ternary compounds have been obtained by reacting at 700–800 °C films of W-30 atomic per cent Ti deposited by RF sputtering on silicon substrates.[77]

The results show a layer by layer growth of a $Ti_xW_ySi_2$ compound without phase separation. Further systematic work is required in alloy systems made by refractory materials before a general picture can be presented.

2.5.2 Bilayer silicides

On the basis of the reaction temperature of the metals, the bilayers can be classified in three main categories, shown in Figure 2.21. The metals are indicated only for simplicity in the discussion and similar results have been obtained in other systems. In the systems investigated the only changes observed are in the reaction temperature, but not the general behaviour.

In the case of bilayers constituted by near-noble metals[78,79] heat treatment at low temperatures produced the formation of the metal-rich silicide compound Ni_2Si. The other metal does not react with Si even when all the nickel is completely consumed. At higher temperatures platinum penetrates by grain boundary diffusion and forms a silicide at the interface between silicon and the Ni_2Si compound. One example of such grain boundary diffusion is shown

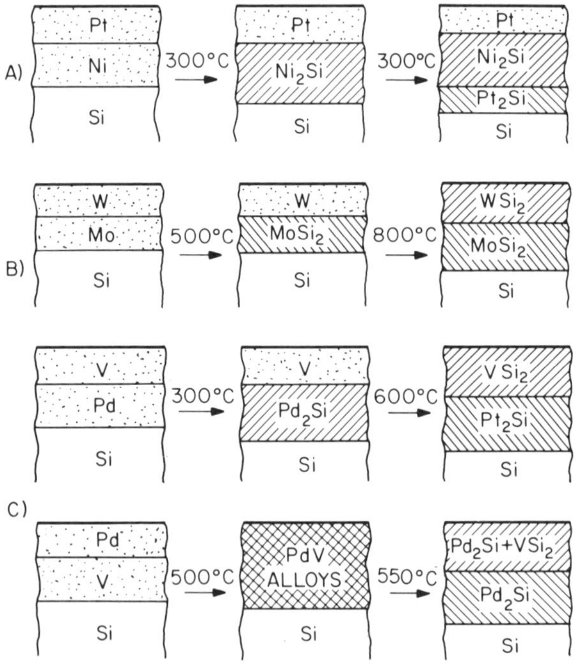

Figure 2.21. Composite diagram of silicide formation for bilayers composed of (a) near-noble metals (Pt–Ni), (b) refractory metals (W–Mo), (c) near-noble and refractory metals (V–Pd)

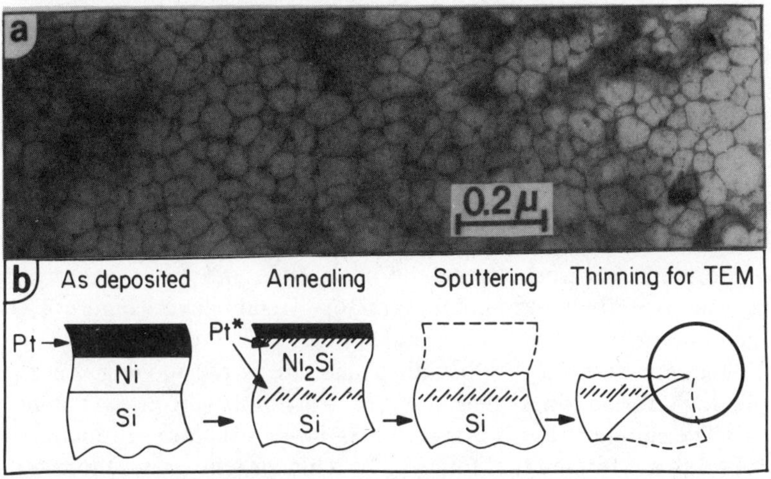

Figure 2.22. (a) TEM micrograph of sample with Pt deposited on a Ni film on Si and annealed at 350 °C for 20 minutes, (b) sample preparation procedure. The TEM view shows platinum (dark regions) decorating the grain boundaries in Ni_2Si (from Finstad[79])

in the TEM micrograph of Figure 2.22. The schematic shows the sample preparation. The cell-like structure indicates that platinum decorates the grain boundaries of Ni_2Si.

These results can be interpreted on the basis of the model for compound formation outlined in Section 2.2. It is not surprising that Ni reacts first since Ni is in contact with silicon and the formation mechanisms of Ni_2Si and Pt_2Si seem to be the same. In the near-noble systems the metal is the main moving species. After the nickel has been consumed, further reaction needs more metal atoms at the silicon surface, atoms which are provided by interstitial diffusion of platinum.

The details of the transformation at longer times and higher temperatures varies from sample to sample. Above 700 °C the distribution of Pt and Ni becomes homogeneous. This result along with X-ray diffraction data indicates the formation of ternary silicides.

A bilayer of two refractory metals behaves in a different way. For Mo deposited on Si and W deposited on Mo (Figure 2.21(b)), the formation of $MoSi_2$ is completed after 500 °C as predicted from results with Mo only.[80,81] The tungsten film remains intact without reacting with the molybdenum film. When the temperature is increased to 800 °C, the tungsten film is transformed to WSi_2. With further anneal up to 1000 °C, there was still clear separation of the $MoSi_2$ and WSi_2 phases without evidence for ternary formation.

The results with Mo–W system are not surprising because $MoSi_2$ is formed at temperature below WSi_2. More insight was found from the work with Ta-V bilayers, where vanadium reacts at higher temperatures than tantalum does.[82] The same behaviour of phase separation and kinetics of formation were found irrespective of the order of deposition. The only difference is that the reaction temperature is lower when vanadium is in contact of silicon and higher with tantalum. These results agree with the fact that silicon is now the moving species and the refractory metals tend not to interdiffuse. Consequently, no layer reversal can be expected as was found in the case of near-noble metal bilayers.

The intermediate case between the near-noble and the refractory metal bilayer, is provided by a near-noble refractory metal bilayer (Figure 2.21(c)). The reactions of structures of Pd and V with Si are governed by the low formation temperature of Pd_2Si with palladium as the moving species and with the high formation temperature of VSi_2 with silicon as the moving species.[75] The simplest case is that of the Si–Pd–V structure in which uniform layers of Pd_2Si are formed first followed by a higher temperature formation of uniform layers of VSi_2. With the bilayer structure Si–V–Pd higher temperatures are required. The first step provides intermixing between palladium and vanadium. Later Pd_2Si is formed in contact with the underlying Si and at even higher temperatures a non-uniform structure of Pd_2Si and VSi_2 is formed in the outer layer. In these structures, the tendency is to form an alloy between the V and Pd and then to release the Pd from the Pd–V alloy to form Pd_2Si.

A major problem with the use of silicides as Schottky barrier contacts is the degradation due to the reaction of the interconnect metallization, usually Al, with the silicide.[23–27,83–86] It has been shown that the barrier height of PtSi decreases 0.2 eV when the Al contact is heated to 610 °C for 30 min.[23]

In order to prevent degradation of the Schottky contact it is necessary to prevent a reaction between PtSi and Al. One method to accomplish this is to interpose a sacrificial layer between PtSi and aluminium. This concept of the use of a sacrificial layer is an extension of the near-noble refractory metal bilayer. An example of such sacrificial barrier is to use chromium as a barrier to the interaction of Pd_2Si with aluminium.[85] In such samples chromium reacts with Pd_2Si to form $CrSi_2$ and chromium reacts with aluminium to form a Cr–Al compound. As long as a layer of unreacted chromium is present the integrity of the Pd_2Si layer on the silicon substrate is preserved. In this case, the lower temperature reacting layer is in contact with the silicon and forms a silicide at relatively low temperatures. At higher temperatures, the high temperature silicide and Al–metal compounds are formed. This concept was also shown directly with a Pt–Cr alloy layer in contact with Al.[41] The outer chromium layer preserves the stability of PtSi contact for up to 1 hour at 450 °C. The contact degraded after annealing at 500 °C for 1 hour.

If the metallization system must go through metallization procedures above 450 °C, other barriers must be considered. One possibility is to use a bilayer or alloy of tungsten with platinum. For example, it has been shown that WSi_2

can be formed at 750 °C on samples formed by depositing thin film of tungsten on PtSi.[83,84]

There are other approaches to the development of diffusion barriers that can be used to prevent reaction between the Schottky barrier contact and Al.[87] We have described the use of sacrificial barriers. Alternative approaches involve the use of inert barriers such as TiW[24] or TiN.[88]

2.6 ION MIXING AND LASER ANNEALING

The previous sections were concerned with silicide formation under standard furnace annealing conditions. There are two recent developments, ion beam mixing, and laser annealing, which also have been used to promote silicide formation. Both involve the deposition of energy in the submicron region and rapid heating and cooling in the region where the energy is deposited. The advantages of such directed energy deposition techniques are that the silicide can be reacted in a localized region and that the silicides can be formed without heating the entire wafer. Beyond these generalities, the two techniques are sufficiently different so that we will discuss them separately. Although these techniques have not yet been used in integrated circuit fabrication both offer unique capabilities which may be exploited at a later time.

2.6.1 Laser annealing

We use the term laser annealing to denote both laser and electron beam annealing.[89,90] To date, there has not been any direct evidence to indicate a major difference in results achieved for the equivalent amounts of power absorbed in the sample. Both are basically short-duration, near-surface heating processes. There are two regimes: in one the energy is sufficient to melt the silicon and silicide formation occurs during the solidification of a molten, metal–Si mixture; in the other lower energy pulses are used and silicide formation occurs in the solid phase. With very short pulses of about 10^{-7} sec duration (typical of *Q*-switched lasers), the sample is heated for such a short time that silicides do not grow in the solid phase. In single-pulse *Q*-switch lasers, high energy pulses are required to form a molten layer. With longer pulses of 10^{-2} to 10^{-3} sec duration (typical of scanned CW beams), the sample is heated for a sufficiently long time so that silicides can be formed in the solid phase. Consequently, the two regimes, liquid phase and solid phase, are often associated with the type of laser, *Q*-switched or CW.

Well-defined silicide phases have been formed by scanned CW-laser and electron beams.[91,92] With electron beams both near-noble and refractory metals were formed. The results indicate that silicide formation with scanned CW beams is equivalent to that achieved by thermal annealing but at higher temperatures and shorter times than one would use in conventional furnace processing.

With *Q*-switch laser irradiation, melting produces a mixture of metal and silicon over the melt depth.[93–99] Upon termination of the pulse, cooling occurs and the liquid–solid interface moves to the surface. In all cases where crystalline layers are formed, laterally non-uniform (cellular) structures are formed with columnar regions of silicon surrounded by thin walls of silicide. The structure is shown on the right side of Figure 2.22 (the silicide regions appear black in the micrographs). The process is shown in the left side of Figure 2.22. Upon solidification the metal is pushed away from the solid–liquid boundary and accumulates in the liquid until the solidification occurs. The cellular formation follows the classical description of segregation and constitutional super-cooling.[100]

The non-uniform structure of the solidified layer is reflected in the presence of multiple silicide phases in *Q*-switch laser annealed systems.[98] Laser irradiation of Pd layers on Si led to the formation of Pd_5Si, Pd_4Si, Pd_3Si, Pd_2Si and PdSi. A similar collection of phases was found in the Pt–Si system.

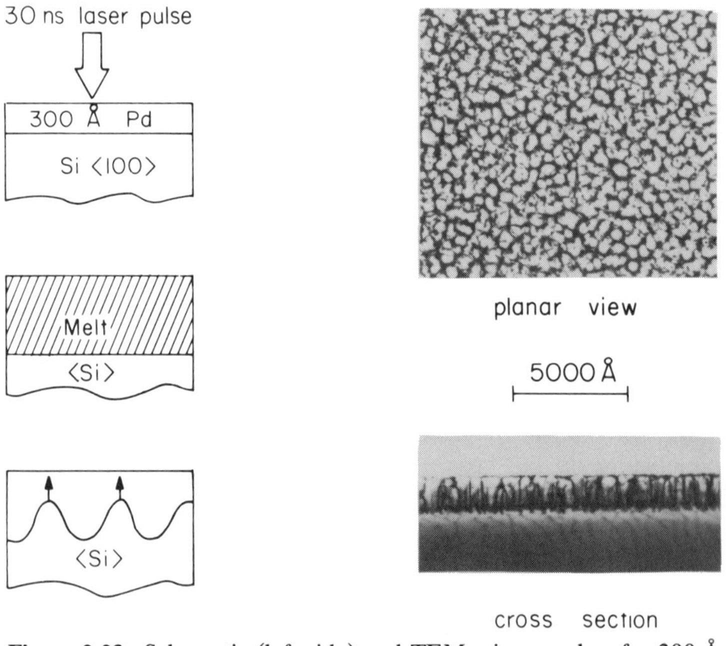

Figure 2.23. Schematic (left side) and TEM micrographs of a 300 Å Pd film on silicon irradiated with a 30 ns Nd–Yag laser. In the TEM view the dark areas are platinum. The formation of the cellular structure is attributed to constitutional supercooling of the Pt–Si molten layer

2.6.2 Ion beam induced silicide formation

The penetration of energetic ions through the metal–semiconductor interface leads to strong interfacial reactions.[101,102] In near-noble metals on Si, ion mixing leads to the same phase, M_2Si, found in thermal annealing.[102,103] For Pt–Si systems, the growth of Pt_2Si is proportional to (ion dose)$^{1/2}$ as shown in Figure 2.23. This result indicates that the mixing process is similar to radiation-enhanced-diffusion. The dependence of the silicide thickness on ion mass is correlated directly with the amount of energy deposited in atomic collisions. At higher ion doses, when all the metal film is consumed in the formation of M_2Si, ion beam mixing can lead to the formation of a more silicon-rich phase[104] or can produce an amorphous layer.[105]

With refractory metals, implantation at temperatures above room ambients is required to form well-defined, crystalline, disilicide phases.[106–108] The mixing process is temperature dependent as shown in Figure 2.24 which plots the amount, Q, of Si atoms intermixed with the Cr layer as a function of reciprocal temperature. At low temperatures, mixing is dominated by dynamic collision processes (cascade mixing) and hence is not strongly temperature sensitive. In this regime, mixing produces graded interfaces. At higher temperatures, radiation-enhanced-diffusion becomes dominant and, similar to thermal annealing, layer by layer growth of crystalline silicides are formed (Figure 2.25).

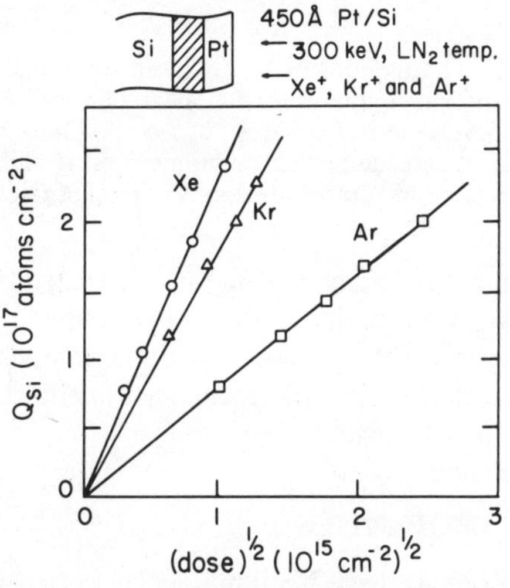

Figure 2.24 The amount of Si, Q_{Si}, in an intermixed PtSi layer versus ion (dose)$^{1/2}$ for 300 keV ions implanted through a Pt film on Si. The amount of Si increases as (dose)$^{1/2}$ (from Tsaur *et al.*[102])

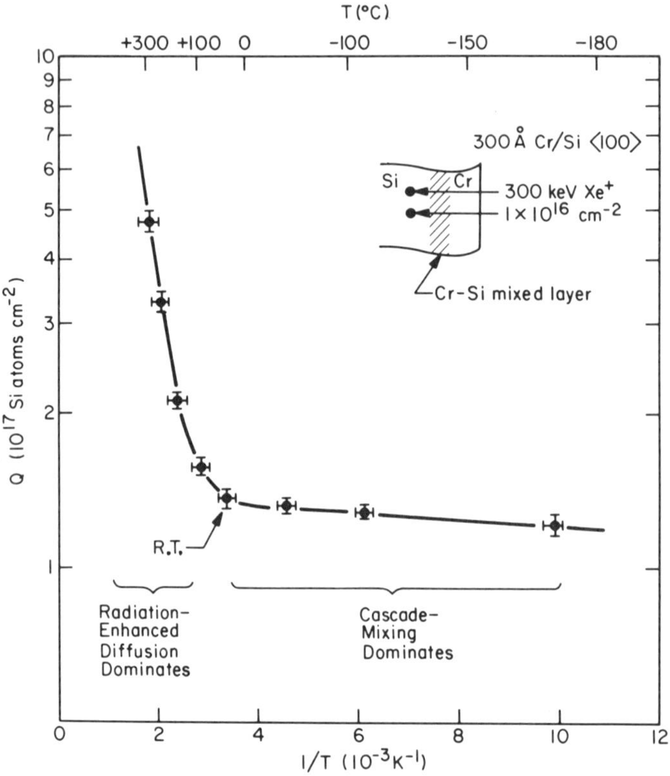

Figure 2.25 The amount of Si, Q, in an intermixed CrSi layer versus reciprocal temperature for a Cr film on Si bombarded with 300 keV Xe ions. To a dose of 1×10^{16} ions cm^{-2}. At low temperatures, cascade mixing dominates and at higher temperatures, radiation enhanced diffusion is dominant (from Tsaur[108])

It is unlikely that ion-beam mixing will be used directly in forming metal–semiconductor contacts due to the radiation damage produced in Si by the ions which penetrate the metal–semiconductor interface. However, ion mixing might be applied to the interconnect metallization to improve adhesion or to form special silicide phases.

2.7 ELECTRICAL PROPERTIES

In the previous sections we have been mainly concerned with the metallurgical properties of silicides. The importance of silicides in integrated circuits is associated with their electrical properties. The major use of silicides in device structures has been to provide a contact, either ohmic or rectifying. Due to the requirements imposed by VLSI, it has been proposed to use silicides as interconnects.[109,110] In this usage, the resistivity rather than the barrier height of the silicide is the key feature.

2.7.1 Metal–semiconductor contacts

The most important property of a metal–semiconductor interface, from the standpoint of electronic device applications, is the potential barrier or barrier height between the Fermi level and the majority carrier band edge of the semiconductor at the interface. The barrier height plays a central role in the electrical properties of the metal–semiconductor contact because it determines the current–voltage characteristics of the structures.[111–113]

A large number of investigations on barrier heights are reported in the literature covering many silicide-forming systems. On the other hand, the theoretical understanding of metal–semiconductor contacts lags well behind experimental findings.

Recently, the availability of ultra-high vacuum systems and the development of new techniques of surface analysis allows a re-examination of the physical mechanisms involved in Schottky barriers. The most striking result is the discovery that the barrier height is already established if a few monolayers of a metal is deposited on silicon. Most of these metals produce an interfacial layer having properties different from the pure compound and the pure metal. These findings shifted the interest of researchers from bulk silicide properties to interfacial layer properties.

The barrier height is determined by the density of interface states in the energy gap of silicon, in agreement with previous models, and these states are due to the interfacial layer between silicon and silicide. The position of these interface states is determined by the metal and their density can depend upon the interfacial composition.

Figure 2.26 shows the values of the barrier height measured in a SiPt system after annealing at various times and temperatures.[114] The phases quoted here have been identified by backscattering techniques and X-ray diffraction measurements. The main feature of these results is that the barrier height does not depend significantly upon the compound which is formed. Similar results were found in the Ni-silicide systems.[115]

These results indicate that (i) platinum reacts with silicon and forms an interfacial layer even if the deposition is done on substrates held at about 85 °C; and (ii) if the composition of the interfacial layer changes, the corresponding change in the density of states is not enough to produce an appreciable effect on the position of the Fermi level.

The dependence of the barrier height measured in various silicide forming systems on n-type silicon is plotted in Figure 2.27 as a function of the 'eutectic temperature' determined from phase diagrams.[116] The eutectic temperature is used as a medium to characterize the properties of the interfacial layer. The

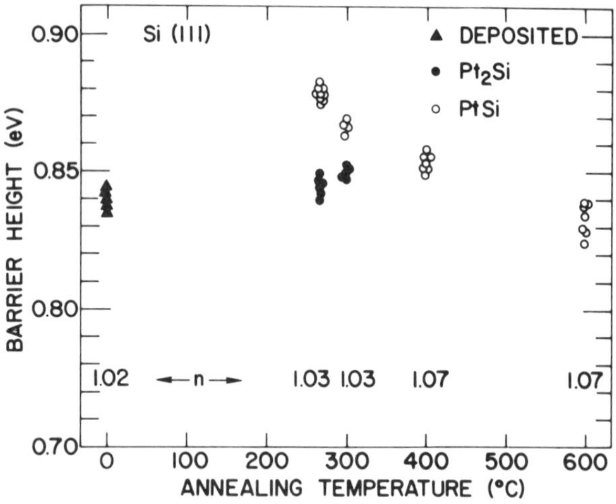

Figure 2.26. The barrier, ϕ_B, in eV versus process temperature for Pt films deposited on silicon. The phases indicated in the figure have been identified by X-ray diffraction and Rutherford backscattering (Ottaviani *et al.*[114])

correlation shown in Figure 2.26 is a further proof of the existence of the interfacial layer. The barrier height and hence the position of the surface states in the gap depends upon the kind of metal in the interfacial layer.

In Figure 2.27, we have correlated the barrier height with eutectic temperature to show the importance of the interfacial layer. Another approach, also based on the concept of an interfacial layer, is to correlate the barrier height with the geometric mean of the barrier heights of the constituents.[117] This correlation also gives a reasonable fit between measured and calculated values with the exception of PtSi and RhSi.

Actual devices are generally made in an environment which is more difficult to control than that in a laboratory where ultra-high vacuum systems and sophisticated techniques can be used to produce and characterize the device itself. Undesirable effects, as for instance non-uniform interreactions[118,119] can occur, leading to unexpected behaviour. Moreover several aspects of silicide formation, such as stress, are still under investigation. These can also affect the properties of silicon–metal contents even if the devices are prepared in the best conditions.

Other sources for non-uniform contacts come from the penetration of Al through the silicide. The non-uniform interaction produces regions in the contact area which modify the electrical properties of the whole structure. Recently the problems of non-uniform interactions have been investigated.[118,119] It has been shown that by coupling I–V and C–V characteristics it could be possible to gain more insight on the uniformity in the contact area.

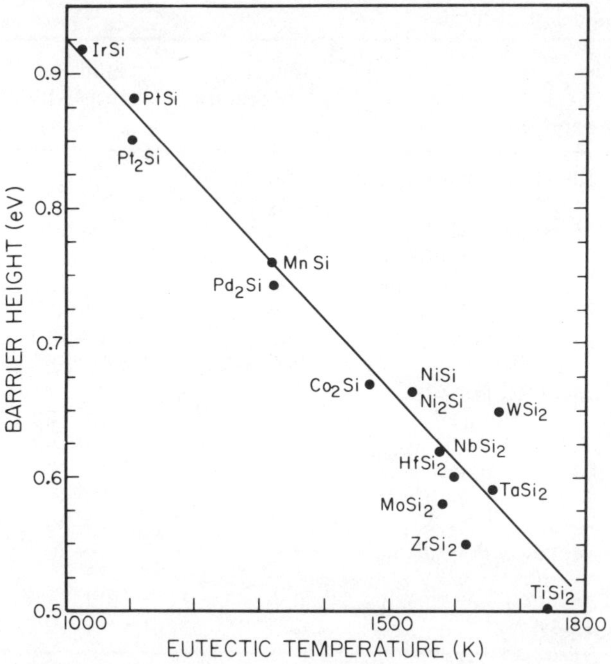

Figure 2.27. The barrier height, ϕ_B, in eV versus eutectic temperature for silicide forming systems (from Ottaviani *et al.*[116])

Moreover the results obtained can explain why barrier heights measured with C–V technique are different than the ones obtained with I–V measurements.

2.7.2 Silicide resistivity

Silicides play an increasingly important role as interconnects as device dimensions on integrated circuits are reduced. In this application the resistivity rather than barrier height becomes the important issue. The change in emphasis will lead to an increase in the number of studies dealing with electron transport properties of silicides. One may anticipate that stress and crystallite grain size as well as the silicide phase will be important parameters. It will be important to investigate the correlation between electrical and structural properties. A detailed investigation of the electrical properties of Pd_2Si has been reported.[120] Electron concentration and mobility have been measured over a wide temperature range.

Low resistivity values, about 25 $\mu\Omega$ cm, were found for $TiSi_2$ in studies of titanium on polycrystalline silicon and of cosputtered (titanium plus silicon) film on SiO_2 plus polycrystalline silicon.[121] The resistivity, stress, and surface

Table 2.4 Silicide resistivities[a,b]

Silicide	Sample preparation	Annealing temperature (°C)	Film resistivity (μΩ cm)	Bulk resistivity (μΩ cm)
$TiSi_2$	Metal on polysilicon	900	13–16	123
	Alloy	—	25	—
$ZrSi_2$	Metal on polysilicon	—	35–40	160
$HfSi_2$	Metal on polysilicon	—	45–50	—
VSi_2	Metal on polysilicon	800	50–55	9.5
$NbSi_2$	Metal on polysilicon	800	50	6.3
$TaSi_2$	Metal on polysilicon	900–1000	30–70	8.5
	Alloy	900–1000	50–55	—
$CrSi_2$	Metal on polysilicon	700	600	1500
$MoSi_2$	Alloy	1000	50–100	20–22
WSi_2	Alloy	1000	35–100	33.4
FeSi	Metal on polysilicon	500	150–200	—
$FeSi_2$	Metal on polysilicon	700	>1000	—
RhSi	—	—	70–80	—
$CoSi_2$	Metal on Polysilicon	900	18–20	—
	Alloy	—	25	—
PtSi	Metal on polysilicon	800	28–35	—
PdSi	Metal on silicon	—	18	—

[a]Adapted from the format used by S. P. Murarka.[122]
[b]Additional data from S. Zirinsky, IBM Research Center.[123]

appearance had different thermal behaviour in the cosputtered (Ti + Si) and deposited (Ti) systems in the temperature range measured. Such studies are required in other silicide-forming systems. It is also important to investigate possible correlation between electrical and structure properties.

We present here in tabular form, Table 2.4, the electrical resistivity measured in thin film silicides compared with the values obtained in bulk samples.[121] The table contains also information on sample preparation and annealing conditions.

2.8 ACKNOWLEDGEMENTS

We acknowledge, with pleasure, our many interactions and discussions with colleagues at Caltech, IBM Research Laboratories and the University of Modena. Since many of our silicide investigations involved travel between different laboratories, we thank our wives, Diana and Betty, for their patience and understanding. Financial support was provided in part by the Office of Naval Research (L. Cooper) and Consiglio Nazionale delle Ricerche (CNR).

REFERENCES

1. C. C. Chang, 'Analytical Auger electron spectroscopy'. In *Characterization of Solid Surfaces*, P. F. Kane and G. B. Larrabee (Eds.), Plenum Press, New York, 1974, Chapter 20.
2. W. K. Chu, J. W. Mayer, and M.-A. Nicolet, *Backscattering Spectrometry*, Academic Press, New York, 1978.
3. K. N. Tu and S. S. Lau, 'Thin film deposition and characterization'. In *Thin Films—Interdiffusion and Reactions*, J. M. Poate, K. N. Tu, and J. W. Mayer (Eds.), Wiley–Interscience, New York, 1978, Chapter 5.
4. M. Hansen, *Constitution of Binary Alloys*, McGraw-Hill, New York, 1958.
5. F. A. Shunk, *Constitution of Binary Alloys, Second Supplement*, McGraw-Hill, New York, 1969.
6. R. S. Claassen, 'Instrumentation for materials research', *Physics Today*, November, 1976, p. 23. Data from W. F. Chambers and E. L. Burgess (unpublished).
7. K. N. Tu and J. W. Mayer, 'Silicide formation'. In *Thin Films—Interdiffusion and Reactions*, J. M. Poate, K. N. Tu, and J. W. Mayer (Eds.), Wiley–Interscience, New York, 1978, Chapter 10.
8. K. N. Tu, 'Selective growth of metal-rich silicide of near-noble metals', *Appl. Phys. Lett.*, **27**, 4, 221–224, Aug. (1975).
9. N. W. Cheung and J. W. Mayer, 'Lattice location experiment of the Ni–Si interface by thin crystal channeling', *Phys. Rev. Lett.*, **46**, 10, 671–674 (1981).
10. T. Narusawa, S. Komiya, and A. Hiraki, 'Diffuse interface in Si (substrate) –Au (evaporated film) system', *Appl. Phys. Lett.*, **22**, 8, 389–390, April (1973).
11. A. Hiraki, S. Kim, W. Kammura, and M. Iwami, 'Dynamical observation of room temperature interfacial reaction in metal–semiconductor system by Auger electron spectroscopy', *Surface Sci.*, **86**, 706–710 (1979).
12. J. A. Roth and C. R. Crowell, 'Application of Auger electron spectroscopy to studies of the silicon/silicide interface', *J. Vac. Sci. Technol.*, **15**, 4, 1317–1324, July/Aug. (1978).
13. L. Braicovich, C. M. Garner, P. R. Skeath, C. Y. Su, P. W. Chye, I. Lindau, and W. E. Spicer, 'Photoemission studies of the silicon–gold interface', *Phys. Rev.*, **B, 20**, 12, 5131–5141, 15 December (1979).
14. P. J. Grunthaner, F. J. Grunthaner, and J. W. Mayer, 'XPS study of the chemical structure of the nickel/silicon interface', presented at Estes Park, Colorado, Physics of Compound Semiconductors Interfaces Conference, 1980, *J. Vac. Sci. Technol.*, **17**, 5, 924–929 (1980).
15. N. W. Cheung, R. J. Culbertson, L. C. Feldman, P. J. Silverman, K. W. West, and J. W. Mayer, 'Ni on Si (111): reactivity and interface structure', *Phys. Rev. Lett.*, **45**, 2, 120–124 (1980).
16. W. J. Schaffer, R. W. Bené, and R. M. Walser, 'Structural studies of thin nickel films on silicon surfaces', *J. Vac. Sci. Technol.*, **15**, 4, 1325–1331, July/Aug. (1978).
17. R. M. Walser and R. W. Bené, 'First phase nucleation in silicon–transition–metal planar interfaces', *Appl. Phys. Lett.*, **28**, 10, 624–625, May (1976).
18. R. W. Bené and R. M. Walser, 'Effect of a glassy membrane on the Schottky barrier between silicon and metallic silicides', *J. Vac. Sci. Technol.*, **14**, 4, 925–929, July/Aug. (1977).

19. G. Majni, C. Nobili, G. Ottaviani, M. Costato, and E. Galli, 'Gold–aluminium thin film interactions and compound formation', submitted for publication in *J. Appl. Phys.* (1981).
20. C. Canali, G. Majni, G. Ottaviani, and G. Celotti, 'Phase diagrams and metal-rich silicide formation', *J. Appl. Phys.*, **50**, 1, 255–258, Jan. (1979).
21. C. Canali, L. Silvestri, and G. Celotti, 'Metal-rich Pd–silicide formation in thin film interactions', *J. Appl. Phys.*, **50**, 9, 5768–5772, Sept. (1979).
22. D. Kahng and M. P. Lepselter, 'Planar epitaxial silicon Schottky barrier diodes', *Bell System Tech. J.*, **44**, 1525–1528, Sept. (1965).
23. H. H. Hosack, 'Electrical and mechanical features of the platinum silicide–aluminium reaction', *J. Appl. Phys.*, **44**, 8, 3476–3485, Aug. (1973).
24. P. C. Parekh, R. C. Sirrine, and P. Lemieux, 'Behavior of various silicon Schottky barrier diodes under heat treatment', *Solid-State Elect.*, **19**, 493–494 (1976).
25. H. Grinolds and G. Y. Robinson, 'Study of Al/Pd_2Si contacts on Si', *J. Vac. Sci. Technol.*, **14**, 1, 75–78, Jan./Feb. (1977).
26. E. Hökelek and G. Y. Robinson, 'Aluminium/nickel silicide contacts on silicon', *Thin Solid Films*, **53**, 135–140 (1978).
27. G. J. van Gurp and W. M. Reukers, 'Aluminium–silicide reactions. II. Schottky-barrier height', *J. Appl. Phys.*, **50**, 11, 6823–6926, Nov. (1979).
28. A. Hiraki, M.-A. Nicolet, and J. W. Mayer, 'Low-temperature migration of silicon in thin layers of gold and platinum', *Appl. Phys. Lett.*, **18**, 5, 178–181, March (1971).
29. C. Canali, C. Catellani, M. Prudenziati, W. H. Wadlin, and C. A. Evans, Jr. 'Pt_2Si and PtSi formation with high purity Pt thin films', *Appl. Phys. Lett.*, **31**, 1, 43–45, July (1977).
30. G. Ottaviani, G. Majni, and C. Canali, 'Dissociation of PtSi, NiSi, and PdGe in presence of Pt, Ni, and Pd films, respectively', *Appl. Phys.*, **18**, 3, 285–289, March (1979).
31. C. Canali, F. Catellani, G. Ottaviani, and M. Prudenziati, 'On the formation of Ni and Pt silicide first phase: the dominant role of reaction kinetics', *Appl. Phys. Lett.*, **33**, 2, 187–190, July (1978).
32. C. A. Crider, J. M. Poate, J. E. Rowe, T. T. Sheng, and S. D. Ferris, 'The effect of impurities on the PtSi–Si interface and the PtSi surface', *Appl. Phys. Lett.*, **36**, 7, 534 (1980).
32a. C. A. Crider and J. M. Poate, 'Growth rates for Pt_2Si and PtSi formation under UHV and controlled impurity atmospheres', *Appl. Phys. Lett.*, **36**, 6, 417–419, March (1980).
33. F. Nava, G. Majni, S. Valeri, and G. Ottaviani, 'Influence of implanted oxygen on the growth of platinum silicides', *J. Appl. Phys.* (1981) (to be published).
34. J. M. Poate and T. C. Tisone, 'Kinetics and mechanism of platinum silicide formation on silicon', *Appl. Phys. Lett.*, **24**, 8, 391–393, April (1974).
35. J. B. Bindell, J. W. Colby, D. R. Wonsidler, J. M. Poate, D. K. Conley, and T. C. Tisone, 'An analytical study of platinum silicide formation', *Thin Solid Films*, **37**, 441–452 (1976).
36. S. Danyluk and G. E. McGuire, 'Platinum silicide formation: electron spectroscopy of the platinum–platinum silicide interface', *J. Appl. Phys.*, **45**, 12, 5141–5144, December (1974).
37. T. J. Kingzett and C. A. Ladas, 'Effect of oxidizing ambients on platinum silicide formation', *J. Electrochem. Soc.*, **122**, 12, 1729–1732, Dec. (1975).
38. R. J. Blattner, C. A. Evans, Jr., S. S. Lau, J. W. Mayer, and B. M. Ullrich, 'Effect of oxidizing ambients on platinum silicide formation', *J. Electrochem. Soc.*, **122**, 12, 1732–1736 (1975).

39. M. Severi, E. Gabilli, S. Guerri, and G. Celotti, 'Structural and electrical characteristics of platinum–silicide–silicon contacts as influenced by sputter etching and annealing ambient', *J. Appl. Physics,* **48**, 5, 1998–2003, May (1977).
40. M. J. Rand and J. F. Roberts, 'Observations on the formation and etching of platinum silicide', *Appl. Phys. Lett.,* **24**, 2, 49–51, Jan. (1974).
41. K. N. Tu, W. N. Hammer, and J. O. Olowalafe, 'Shallow silicide contact', *J. Appl. Phys.,* **51**, 3, 1663–1668, March (1980).
42. G. Ottaviani, K. N. Tu, J. W. Mayer, and B. Y. Tsaur, 'Phase separation in alloy–Si interaction', *Appl. Phys. Lett.,* **36**, 4, 331–333, Feb. (1980).
43. J. M. Andrews, 'The role of the metal–semiconductor interface in silicon integrated circuit technology', *J. Vac. Sci. Technol.,* **11**, 6, 972–984, Nov./Dec. (1974).
44. P. G. Ghate, J. C. Blair, and C. R. Fuller, 'Metallization in microelectronics', *Thin Solid Films,* **45**, 69–84 (1977).
45. G. Ottaviani, 'Review of binary alloy formation by thin film interactions', *J. Vac. Sci. Technol.,* **16**, 5, 1112–1119, Sept./Oct. (1979).
46. J. Baglin, F. d'Heurle, and S. Petersson, 'Nucleation-controlled thin film interactions during silicide formation'. In *Thin Film Interfaces and Interactions*, J. E. E. Baglin and J. M. Poate (Eds.), *The Electrochemical Society Inc., Princeton*, 80–2, 341 (1980).
47. S. Petersson, J. Baglin, W. Hammer, F. d'Heurle, T. S. Kuan, I. Ohdomari, J. de Sousa Pires, and P. Tove, 'Formation of iridium silicides from Ir thin films on Si substrates', *J. Appl. Phys*., **50**, 5, 3357–3365, May (1979).
48. S. Petersson, R. Anderson, J. Baglin, J. Dempsey, W. Hammer, F. d'Heurle, and S. LaPlaca, 'The thin-film formation of rhodium silicides', *J. Appl. Phys.,* **51**, 1, 373–383, Jan. (1980).
49. R. J. Wagner, S. S. Lau, J. W. Mayer, and J. A. Roth, 'Growth kinetics of VSi_2 and $NbSi_2$ thin films'. In *Thin Film Phenomena—Interfaces and Interactions*, J. E. E. Baglin and J. M. Poate (Eds.), *The Electrochemical Society, Inc., Princeton, New Jersey*, 59–65 (1978).
50. R. W. Bower, D. Sigurd, and R. E. Scott, 'Formation kinetics and structure of Pd_2Si films on Si', *Solid State Elect.,* **16**, 1461–1471 (1973).
51. D. Sigurd, R. W. Bower, W. F. van der Weg, and J. W. Mayer, 'Characterization of polycrystalline layers by channeling measurements', *Thin Solid Films,* **19**, 319–328 (1973).
52. S. S. Lau and D. Sigurd, 'An investigation of the structure of Pd_2Si formed on Si', *J. Electrochem. Soc.,* **121**, 1538–1540 (1974).
53. K. N. Tu, E. I. Alessandrini, W. K. Chu, H. Kräutle, and J. W. Mayer, 'Epitaxial growth of nickel silicide $NiSi_2$ on silicon', *Japan. J. Appl. Phys.,* **Suppl. 2**, 1, 669–672 (1974).
54. H. Ishiwara and S. Furukawa, 'Separate estimate of crystallite orientations and scattering centers in polycrystals by backscattering technique', *J. Appl. Phys.,* **47**, 4, 1686–1689, April (1976).
55. U. Köster, K. N. Tu, and P. S. Ho, 'Effect of substrate temperature on the microstructure of thin-film silicide', *Appl. Phys. Lett.,* **31**, 9, 634–636, Nov. (1977).
56. H. Ishiwara, K. Hikosaka, and S. Furukawa, 'Channeling and backscattering studies of the crystalline perfection and the thermal stability of epitaxial PtSi films on Si', *J. Appl. Phys.,* **50**, 8, 5302–5306, Aug. (1979).
57. H. Ishiwara, K. Hikosaka, N. Nagatomo, and S. Furukawa, 'Characterization of epitaxial metal silicide films grown on silicon', *Surface Sci*., **86**, 711–717 (1979).

58. S. S. Lau and N. W. Cheung, 'Epitaxial growth of the nickel disilicide phase', *Thin Solid Films*., **71**, 117–127 (1980).
59. H. Ishiwara, 'Characterization of epitaxial silicide films'. In *Thin Films Interfaces and Interactions*, J. E. E. Baglin and J. M. Poate (Eds.), *The Electrochemical Society, Inc., Princeton*, **80–2**, 159 (1980).
60. J. Angilello, J. Baglin, F. d'Heurle, S. Petersson, and A. Segmüller, 'Stress in silicides formed by the interaction of metal films with silicon substrates'. In *Thin Film Interfaces and Interactions*, J. E. E. Baglin and J. M. Poate (Eds.), *The Electrochemical Society, Inc., Princeton*, **80–2**, 369 (1980).
61. S. P. Muraka and D. B. Fraser, 'Thin film interaction between titanium and polycrystalline silicon', *J. Appl. Phys.,* **51**, 1, 342–349, Jan. (1980).
62. H. Kräutle, M.-A. Nicolet, and J. W. Mayer, 'Kinetics of silicide formation by thin films of V on Si and SiO_2 substrates', *J. Appl. Phys.,* **45**, 8, 3304–3308, Aug. (1974).
63. R. J. Schutz and L. R. Testardi, 'The formation of vanadium silicides at thin film interfaces', *J. Appl. Phys.,* **50**, 91, 5773–5781, Sept. (1979).
64. D. M. Scott, P. J. Grunthaner, B. Y. Tsaur, M.-A. Nicolet, and J. W. Mayer, 'The effect of oxygen on the growth kinetics of nickel silicides'. In *Thin Films Interfaces and Interactions*, J. E. E. Baglin and J. M. Poate (Eds.), *The Electrochemical Society, Inc., Princeton*, **80–2**, 148 (1980).
65. S. Zirinsky, W. Hammer, F. d'Heurle, and J. Baglin, 'Oxidation mechanisms in WSi_2 thin films', *Appl. Phys. Lett.,* **33**, 1, 76–78, July (1978).
66. F. Mohammadi, K. C. Saraswat, and J. D. Meindl, 'Kinetics of the thermal oxidation of WSi_2', *Appl. Phys. Lett.,* **35**, 7, 529–531 (1979).
67. R. Pretorius, W. Strydom, C. Comrie, and J. W. Mayer, 'Silicon-31 tracer studies of the oxidation of Si, $CoSi_2$, and PtSi', *Phys. Rev. B.*, **22**, 4, 1885–1891 (1980).
68. K. N. Tu, J. F. Ziegler, and C. J. Kircher, 'Formation of vanadium silicides by the interactions of V with bare and oxidized Si wafers', *Appl. Phys. Lett.,* **23**, 9, 493–495 (1973).
69. R. Pretorius, J. M. Harris, and M.-A. Nicolet, 'Reaction of thin metal films with SiO_2 substrates', *Solid-State Electronics,* **21**, 667–675 (1978).
70. H. Kräutle, W. K. Chu, M.-A. Nicolet, J. W. Mayer, and K. N. Tu, 'Reactions of the thin metal films with Si or SiO_2 substrates'. In *Applications of Ion Beams to Metals*, S. T. Picraux, E. P. EerNisse, and F. L. Vook (Eds.), New York, Plenum Publishing Corp., pp. 193–207, 1973.
71. W. K. Chu and K. N. Tu, 'Identification of diffusion species in V–SiO_2 reactions', *Appl. Phys. Lett.,* **33**, 1, 83–85, July (1978).
72. M. Wittmer and T. E. Seidel, 'The redistribution of implanted dopants after metal–silicide formation', *J. Appl. Phys.,* **49**, 12, 5827–5834, Dec. (1978).
73. J. B. Bindell, W. M. Moller, and E. F. Labuda, 'Ion-implanted, low barrier, PtSi Schottky barrier diodes', *IEEE Trans. on Electron Devices,* **ED-27**, 2, 420–425, Feb. (1980).
74. 'Joint special issue on very large scale integration', published in *IEEE Trans. on Electron Devices,* **ED-26**, 4, April (1979).
75. J. W. Mayer, S. S. Lau, and K. N. Tu, 'Silicide formation with Pd–V alloys and bilayers', *J. Appl. Phys.,* **50**, 9, 5855–5859, Sept. (1979).
76. J. O. Olowolafe, K. N. Tu, and J. Angilello, 'Contact reaction between Si and Pd–W alloy films', *J. Appl. Phys.,* **50**, 10, 6316–6320, Oct. (1979).
77. J. M. Harris, S. S. Lau, M.-A. Nicolet, and R. S. Nowicki, 'Studies of the Ti–W metallization system on Si', *J. Electrochem. Soc.,* **123**, 1, 120–124 (1976).
78. T. G. Finstad and M.-A. Nicolet, 'Silicide formation with bilayers of Pd–Pt, Pd–Ni, and Pt–Ni', *J. Appl. Phys.,* **50**, 1, 303–307, Jan. (1979).

79. T. G. Finstad, 'Silicide formation with nickel and platinum double layers on silicon', *Thin Solid Films,* **51**, 411–424 (1978).
80. J. Baglin, F. d'Heurle, and S. Petersson, 'An alternative marker experiment in the formation of Mo and W silicides', *Appl. Phys. Lett.,* **33**, 4, 289–290, Aug. (1978).
81. J. Baglin, J. Dempsey, W. Hammer, F. d'Heurle, S. Petersson, and C. Serrano, 'The formation of silicides in Mo–W bilayer films on Si substrates: A marker experiment', *J. Elec. Materials,* **8**, 5, 641–661 (1979).
82. G. Ottaviani, 'Growth kinetics in tantalum vanadium bilayers', unpublished.
83. A. K. Sinha, M. H. Read, and T. E. Smith, 'Silicide formation in tungsten and other refractory metallization on PtSi on silicon', *J. Electrochemical Soc.,* **120**, 1775–1778 (1973).
84. A. K. Sinha and T. E. Smith, 'Kinetics of WSi_2 formation in the thin-film system W/PtSi/Si', *J. Appl. Phys.,* **44**, 8, 3465–3469, Aug. (1973).
85. J. O. Olowolafe, M.-A. Nicolet, and J. W. Mayer, 'Chromium thin film as a barrier to the interaction of Pd_2Si with Al', *Solid-State Elec.,* **20**, 413–415 (1977).
86. G. J. van Gurp, J. L. C. Daams, A. van Oostrom, L. J. M. Augustus, and Y. Tamminga, 'Aluminum–Silicide Reactions. I. Diffusion Formation, and Microstructures', *J. Appl. Phys*., **50**, 11, 6915–6922 (1979).
87. M.-A. Nicolet, 'Diffusion barriers in thin films', *Thin Solid Films,* **52**, 415–443 (1978).
88. N. W. Cheung, H. von Seefeld, and M.-A. Nicolet, 'Titanium nitride as a diffusion barrier in metallization schemes'. In *Thin Film Interfaces and Interactions*, J. E. E. Baglin and J. M. Poate (Eds.), *The Electrochemical Society, Inc., Princeton,* **80-2**, 323 (1980).
89. S. D. Ferris, H. J. Leamy, and J. M. Poate (Eds.), *Laser-Solid Interactions and Laser Processing—1978*, New York, Amer. Inst. of Physics, 1979.
90. C. L. Anderson, G. K. Celler, and G. A. Rozgonyi (Eds.), *Laser and Electron Beam Processing of Electronic Materials, The Electrochemical Society, Inc., Princeton, New Jersy*, 1980.
91. T. Shibata, T. W. Sigmon, and J. F. Gibbons, 'Formation of silicide compounds using a scanned CW–Ar laser'. In *Laser and Electron Beam Processing of Electronic Materials*, C. L. Anderson, G. K. Celler, and G. A. Rozgonyi (Eds.), *The Electrochemical Society, Inc., Princeton, New Jersey*, 520–530 (1980).
92. T. W. Sigmon, J. L. Regolini, J. F. Gibbons, S. S. Lau, and J. W. Mayer, 'Metal–silicon reactions induced by a CW-scanned electron beam'. In *Laser and Electron Beam Processing of Electronic Materials*, C. L. Anderson, G. K. Celler, and G. A. Rozgonyi (Eds.), *The Electrochemical Society, Inc., Princeton, New Jersey*, 531–535 (1980).
93. J. M. Poate, H. J. Leamy, T. T. Sheng, and G. K. Celler, 'Laser-induced reactions of platinum and other metal films with silicon', *Appl. Phys. Lett.,* **33**, 11, 918–920, Dec. (1978).
94. M. von Allmen and M. Wittmer, 'Dynamics of laser-induced formation of palladium silicide', *Appl. Phys. Lett.,* **34**, 1, 68–70, Jan. (1979).
95. M. von Allmen and M. Wittmer, 'A study of silicide formation by laser irradiation', *J. Appl. Phys.,* **50**, 7, 4786–4790, July (1979).
96. G. J. van Gurp, G. E. J. Eggermont, Y. Tamminga, W. T. Stacy, and J. R. M. Grijsbers, 'Cellular structure and silicide formation in laser-irradiated metal–silicon systems', *Appl. Phys. Lett.,* **35**, 3, 263–265 (1979).
97. M. Wittmer, W. Lüthy, and M. von Allmen, 'Laser induced reactions of magnesium with silicon', *Phys. Lett.,* **75A**, 1/2, 127–130, Dec. (1979).

98. M. Wittmer, 'Laser applications in the metallization of silicon semiconductor devices'. In *Laser and Electron Beam Processing of Electronic Materials*, C. L. Anderson, G. K. Celler, and G. A. Rozgonyi (Eds.), *The Electrochemical Society, Inc., Princeton, New Jersey*, 485–503 (1980).
99. W. T. Stacy, G. J. van Gurp, G. E. J. Eggermont, Y. Tamminga, and J. R. M. Gijsbers, 'The microstructure of laser-irradiated metal–silicon systems'. In *Laser and Electron Beam Processing of Electronic Materials*, C. L. Anderson, G. K. Celler, and G. A. Rozgonyi (Eds.), *The Electrochemical Society, Inc., Princeton, New Jersey*, 504–513 (1980).
100. B. Chalmers, *Principles of Solidification*, Huntington, New York, R. E. Krieger Publishing Company, 1977.
101. B. Y. Tsaur, S. S. Lau, Z. L. Liau, and J. W. Mayer, 'Ion-beam-induced intermixing of surface layers', *Thin Solid Films,* **63**, 31–36 (1979).
102. B. Y. Tsaur, Z. L. Liau, and J. W. Mayer, 'Ion-beam-induced silicide formation', *Appl. Phys. Lett.,* **34**, 2, 168–170, Jan. (1979).
103. G. E. Chapman, S. S. Lau, S. Matteson, and J. W. Mayer, 'Silicide formation by high-dose Si–ion implantation of Pd', *J. Appl. Phys.,* **50**, 10, 6321–6327, Oct. (1979).
104. B. Y. Tsaur, S. S. Lau, and J. W. Mayer, 'Ion-beam-induced formation of the PdSi silicide', *Appl. Phys. Lett.,* **35**, 3, 225–227, Aug. (1979).
105. B. Y. Tsaur, Z. L. Liau, and J. W. Mayer, 'Formation of Si-enriched metastable compounds in the Pt–Si system using ion bombardment and post annealing', *Phys. Lett.,* **71A**, 2, 3, 270–272, April (1979).
106. T. Kanayama, H. Tanoue, and T. Tsurushima, 'Niobium silicide formation induced by Ar–ion bombardment', *Appl. Phys. Lett.,* **35**, 3, 222–224, Aug. (1979).
107. S. Matteson, J. Roth, and M.-A. Nicolet, 'Ion induced silicide formation in niobium thin films', *Rad. Effects,* **42**, 3/4, 217–226, Sept. (1979).
108. B. Y. Tsaur, 'Ion-beam-induced interface mixing and thin-film reactions'. In *Thin Film Interfaces and Interactions*, J. E. E. Baglin and J. M. Poate (Eds.), *The Electrochemical Society, Inc., Princeton*, **80-2**, 205 (1980).
109. T. Mochizuki, K. Shibata, T. Inoue, and K. Ohuchi, 'A new MOS process using $MoSi_2$ as a gate material', *Jap. J. Appl. Phys*., **17**, 17–1, 37–42, Oct. (1977).
110. S. Zirinsky and B. L. Crowder, 'Refractory silicides for high temperature compatible IC conductor lines', *J. Electrochem. Soc.,* **124**, 388C–401C, November (1977).
111. E. H. Rhoderick, *Metal–Semiconductor Contacts*, Clarendon Press, Oxford, England, 1978.
112. V. L. Rideout, 'A review of the theory, technology, and applications of metal–semiconductor rectifiers', *Thin Solid Films,* **48**, 261–291 (1978).
113. E. H. Nicollian and A. H. Sinha, 'Effect of interfacial reactions on the electrical characteristics of metal–semiconductor contacts'. In *Thin Films—Interdiffusion and Reactions*, J. M. Poate, K. N. Tu, and J. W. Mayer (Eds.), Wiley–Interscience, New York, Chapter 13, 1978.
114. G. Ottaviani, K. N. Tu, and J. W. Mayer, 'Schottky barrier with Pt, Pd, and Ni silicides', unpublished.
115. M. Iwami, K. Ohuno, S. Kamei, T. Ito, and A. Hiraki, 'Electronic states of Ni-silicide and its relations', J. E. E. Baglin and J. M. Poate (Eds.), *The Electrochem Soc. Inc., Princeton,* **80-2**, 102 (1980).
116. G. Ottaviani, K. N. Tu, and J. W. Mayer, 'Interfacial reaction and Schottky barrier in metal–silicon systems', *Phys. Rev. Lett.,* **44**, 4, 284–287, Jan. (1980).

117. J. L. Freeouf, 'Silicide Schottky barriers: an elemental description', *Solid-State Communications,* **33**, 1059–1061 (1980).
118. I. Ohdomari and K. N. Tu, 'Parallel silicide contacts', *J. Appl. Phys.*, **51**, 7, 3735–3739 (1980).
119. I. Ohdomari, T. S. Kuan, and K. N. Tu, 'Micro-structure and Schottky barrier height of iridium silicides formed on silicon', *J. Appl. Phys.,* **50**, 11, 7020–7029, November (1979).
120. M. Wittmer, D. L. Smith, P. W. Lew, and M.-A. Nicolet, 'Electrical characteristics of palladium silicide', *Solid-State Electronics,* **21**, 573–580 (1978).
121. S. P. Muraka and D. B. Fraser, 'Silicide formation in thin cosputtered (titanium + silicon) films on polycrystalline silicon and SiO_2', *J. Appl. Phys.,* **51**, 1, 350–356, Jan. (1980).
122. S. P. Muraka, 'Refractory silicides for integrated circuits', to be published in *J. Vac. Sci. and Technol.*, **17**, 4, 775–792 (1980).
123. S. Zirinsky, private communication.

Reliability and Degradation
Edited by M. J. Howes and D. V. Morgan

CHAPTER 3

Interdiffusion of metal films on gallium arsenide and indium phosphide

D. V. MORGAN

3.1 INTRODUCTION

The most important semiconducting material in use, at the present time, is silicon. Silicon technology is highly developed[1] and forms the base material in the construction of a wide range of devices, encompassing simple *p–n* diodes (for a range of applications), bipolar, and field effect transistors. The vast experience gained in silicon technology, coupled with the fortunate occurrence of a natural, stable, insulating oxide, makes this material a formidable opponent for any new semiconductor to contend with. Consequently, in order to replace silicon in a new device structure the III-V compound semiconductors must, either offer considerable improvement in performance, or else must exhibit some new phenomenon not observed in silicon. The group III-V compound semiconductors do, however, offer such potentially valuable characteristics, which justified the extensive present day research and development programmes in this area. These include optoelectronic properties for integrated optical structures, the transferred electron effect for microwave power generation and a high mobility suited for FET structures for microwave devices and high speed logic circuits.

New materials exhibit new technological problems, which must be mastered before reliable devices of the required performance can be fabricated. One of the most important of these is the choice of metallization, which will serve both to give the desired electrical characteristics, together with a high stability in its relationship with the semiconductor. In very general terms there are two basic types of contact, the ohmic contact[2] and the Schottky barrier or rectifying contact.[3] In the former, it is required to ensure that the series resistance presented by the metal–semiconductor interface has a low value and that no charge dipole forms. In principle such contacts can be achieved by correctly matching the metal's work function to that of the semiconductor. In practice the existence of surface states precludes this possibility and in many cases the metal–semiconductor contact potential is almost independent of the metal's work function.[3] To achieve ohmic behaviour, it is necessary to dope degenerately the semiconductor immediately beneath the contact, thus ensuring that the depletion region is very narrow and electrons may tunnel through the barrier. In practice, this n^+ layer may be formed by either epitaxial growth, by ion implantation or by alloying the contact. Alloying is by far the most common process and essentially involves a controlled intermixing of the contact metal and the semiconductor to produce the desired near-contact doping. This process is in essence a controlled degradation of the metal–semiconductor interface—the intermixing, or alloying of the near surface semiconductor, giving rise to the n^+ surface region. Traditionally, the technology required for producing ohmic contacts was developed empirically—a combination of systematic experimentation and inspired guesswork.[2] It is only recently that detailed microscopic studies have emerged, which have been aimed specifically at understanding the alloying process and, in particular, the physical origin and extent of the near-surface doping.

The Schottky barrier contact represents the other extremes to the ohmic contact; here it is required to obtain highly stable metals, which are resistant to interdiffusion and must retain these properties during device operation. This can mean combinations of elevated temperatures ($\leqslant 300\ ^\circ C$), high electric fields and high current densities. Degradation of the contact during such operational stressing results in degradation of device performance. Basic interdiffusion studies are needed, in order to assess the stability of metallization systems and to identify physical features, which enhance or stabilize the interdiffusion processes.

In practical device structures the metallization systems have to be more complex, in many instances involving multilayered metallizations. The reason for this lies in the nature of the many and conflicting demands made upon the contact metal; some of the more important of these are as follows:

(i) The contact must produce the desired electrical behaviour (ohmic or Schottky).
(ii) It must be stable against interdiffusion during device operation.
(iii) It must be resistant to metallurgical reactions, and to oxidation and corrosion.
(iv) There must be the potential of selective etchability of metal films and substrate.
(v) The metal must be reasonably ductile in order to allow wire bonds to be made to the contact.
(vi) It must have good adherence to the semiconductor; thus there must be a reasonable matching of the thermal expansion coefficients, otherwise the contacts will peel away. This is particularly important in multilayer metallizations, since the number of interfaces is greatly increased.

The widespread use of multilayer systems is clearly an attempt to optimize the demands outlined above. For example, in silicon IMPATT diodes a Cr–Pt–Au system is used. The Cr is used as the ohmic contact to the silicon. Au is used as the outer contact metal, partly on account of its excellent thermal conductivity (i.e. heat sinking properties) and partly because its ductile nature allows easy bonding. However, a thin barrier layer consisting of Pt (or Pd) must be inserted between the Au and Cr to inhibit Au–Cr interdiffusion, which is known to result in Au penetration into the base silicon resulting in the degradation of device performance. Consideration will now be given to the problems of ohmic contacts and in particular to understanding the nature of the alloying process.

3.2 OHMIC CONTACTS

The literature on problems associated with the technology and fabrication of ohmic contacts to the III-V compound semiconductors is extensive. Rideout[2] has recently (1975) presented an elucidating and comprehensive account of the current state of the art, for a range of III-V and ternary systems. Table 3.1 is a summary of results for GaAs and InP, most of which involve an alloying process. Consider now, the basic interdiffusion processes occurring both during contact alloying and any subsequent intermixing taking place during device operation. The discussion will be confined to GaAs and to some extent InP, where extensive data is available. The principles identified here will provide a useful foundation for understanding other III-V materials. A point that should be stressed is that some degree of alloying appears to be a necessary ingredient, independent of any prior surface doping by growth or ion implantation. This is a result of the very high doping required for ohmic behaviour ($\leqslant 10^{20}$ atoms cm^{-3}). Ion implanted layers in GaAs are in general restricted to $\leqslant 10^{19}$ atoms cm^{-3}. Molecular beam epitaxy has been more successful in achieving the necessary doping levels.[4] Recent work on the growth of highly

Table 3.1 Ohmic contacts to GaAs and InP (after Rideout[2])

	Eg	Type	Contact		Alloy temperatures
GaAs	1.45	*p*	Au–Zn(99:1)	Electroless, evaporation	
		p	In–Au(80:20)	Preform	
		n	Au–Ge(88:12)	Evaporation	350–450
		n	In–Au(90:10)	Evaporation	550
		n	Au–Si(94:6)	Evaporation	300
		n	Au–Sn(90:10)	Evaporation	350–700
		n	Au–Te(98:2)	Evaporation	500
InP	1.26	*p*	In	Preform	350–600
		n	In, In–Te	Preform	350–600
		n	Ag–Sn	Preform, evaporation	600

Table 3.2 Ohmic Contacts to GaAs (where available N is the doping of starting material)

Metallization	Temperature (°C)	R_c (Ω cm^2) (N (cm^{-3}))	Analysis techniques	Comments	References
In	350–600	~2×10^{-4} (1×10^{17})	SEM	Holes form in GaAs—increases in size with thickness of contact. $R_c \sim 10^{13} N_D^{-1}$, $R_c \sim 10^{28} N_A^{-1.7}$. R_c increases as temperature decreases	12 13 14, 15, 16
Sn	~250–600	≤R_c(In)	SEM EPM Lapping and etching	Faceted holes in GaAs; long alloy times can produce up to 10 μm deep flat bottomed holes. Balling—can be reduced using HC1 bubbling of gas. Sn dopes GaAs *n*-type (~2×10^{19}cm^{-3}). Highly compensated layer	17, 18, 13, 15, 19, 20, 16
Pd–Ge	~500	3×10^{-4} (1.2×10^{16})	X-ray diffraction	PdGe, $PdAs_2$, PdGa phases observed. No unreacted Pd. [Sintered contact: no melting]	21
Ag–In	500	1.1×10^{-2} (5×10^{15})		Ohmic for *p*-type	22
AgNi	~550	2×10^{-5} (0.0033 Ωcm)			23, 24
AgSn	~500–650	5.5×10^{-3} (1×10^{15})	SEM	Holes ~0.2–2 μm deep in GaAs. Depth dependent upon amount of Sn. High Sn concentration results in formation of white areas (Ag–Sn) and holes filled with Ag–Sn globules. As evaporation reported during heating	25, 22, 13, 20, 16, 26
AgInGe	500–630	<10^{-4} (1.5×10^{15})	μAES EPM X-ray Topography	Ag indiffusion. As evaporation. Ag rich globules and Ge–In facet growth form. For long times Au-Ge–$Ga_{0.6}$ $In_{0.4}$ As-Ag–Ge, Ga, As, In–In-Ag-Ga-As–GaAs structure has been reported	27, 28, 18, 22, 29, 30, 31, 32
AgGeSn	500	1×10^{-3}			22

AgAuGe	500	7.6×10^{-3} (1×10^{15})			22
AgInZn	600	$<10^{-4}$ $(0.1\ \Omega cm)$		p-type GaAs	27
AuIn	350–600	3×10^{-4} (2×10^{15})	X-ray diffraction RBS SEM, EPM	Mounding—thinner films give higher R_c. In_2Au and $AuIn_2$ phases reported. In_2Au makes contact brittle resulting in poor adhesion	13, 2, 33
AuSi	~300			Not commonly used	2
AuSn	350–700	$\geqslant R_c(AuIn)$		Holes form beneath contact; increase in size for thicker films	2, 13, 16
AuTe	300–500			High contact resistance	2
AuNi	350–500		RBS SIMS EPM X-ray Diff.	As in Au. Au and Ni complex phases formed	34, 35
AuZn	400 upwards	1.2×10^{-4} (2.8×10^{17})		p-type GaAs. Zn dopes GaAs to $N_A \sim 4 \times 10^{18}$ cm^{-3} below contact	2, 36, 37
AuSbGe				Many traps induced. One associated with Au–Ge 160 meV above E_v	38
AuGe	RT 300 350–550	$\phi_b \sim 0.7–0.9$ $\phi_b \sim 0.27–0.4$ 1×10^{-6} (3×10^{16})	RBS Channeling SEM EPM	Some Au and Ge indiffusion. Ga outdiffusion into Au. Balling. Au penetration $\geqslant$300 nm in GaAs. Maximum Au concentration $\geqslant$30–40 nm below surface. Formation of Au–Ga matrix layer with Ge segregated in grains. Ge island formation beneath contact	2, 11, 5, 12, 13, 17, 18, 38, 39, 40, 41

Table 3.2 (*continued*)

Metallization	Temperature (°C)	R_c (Ω cm^2) (N (cm^{-3}))	Analysis techniques	Comments	References
AuInGe	~450		SEM TEM Electron diffraction μAES	Islands form under few hundred nm thick which are single crystals containing approximately 1Au:1Ge:1As. Au eats into GaAs in 'moats' around crystals. Ga outdiffusion into Au. Over	32, 42
	>500			alloy-results in Au:Ge:As phase disappearing. Ge migrates to interface. Forms: matrix containing -0.5Ga + 0.28In + 0.18Au + 0.2As per monolayer with Au + Ga + Ge precipitates and Ge–In facets	
NiAuGe	RT	$\phi_B \sim 0.68$	AES	Ni–Ni-Ge–Au–GaAs	
	<350	$\phi_B \sim 0.83$	μAES	Ni diffusion ending up on GaAs surface.	11, 17
	350		SEM	Some nonumiformity.	
			EPM	Ga outdiffusion.	20, 26
			RBS	Ni-Ge–Au–NiGe, Ni_2Ge–GaAs.	
			Dislocation	NiAs phase possibly forms on GaAs surface.	33, 32
	360–500	4×10^{-4}	etch	Ga outdiffusion and surface accumulation $E_A \sim 0.1$ eV.	
		(2×10^{15})		As outdiffusion and evaporation.	41, 43
		1.5×10^{-6}		Au, Ge, and Ni diffuse into GaAs.	
		(5×10^{16})		Maximum Au concentration below surface.	44, 45
				Au concentration decreases exponentially in GaAs. Contact film becomes Au–Ga matrix with Ni–Ge–As crystallites on surface. Facet growth with high Ni–Ge content at interface has also been reported. High dislocation density formed beneath contact	46

doped surface layers has resulted in specific resistances to GaAs as low as 10^{-8} ohms cm^2. Clearly, this is an important new and exciting direction in ohmic contact technology and could lead to improved highly stable contacts in future device structures.

Since the objective of this chapter is to identify the basic processes taking place during alloying, the majority of the discussion will be confined to those systems subjected to the more detailed study, that is, GaAs structures. For the GaAs system, there are a range of alloys that have evolved for fabricating ohmic contacts (see Tables 3.1, 3.2, and 3.3). The proportion of metals in these alloys and the alloy temperatures generally have evolved by empirical procedures as was mentioned earlier. In order to gain some insight into the physical processes taking place during alloying. The Au, AuGe, and Ni/Au–Ge systems will be discussed in some detail. In the latter case the role of Ni appears to be a catalyst, assisting in wetting the contact and in maintaining a uniform metal layer during the alloying process.

Consider first the simple Au/GaAs system.[5–8] Gold on gallium arsenide produces a rectifying contact with a barrier height of around 0.9 eV. When heated, however, the interface degrades and the corresponding current voltage characteristics become more leaky (Figure 3.1) as the local doping near to the interface N_D increases. In these circumstances the current transport changes from thermionic emission (TE) over the barrier to thermionic field emission (TFE).[6] Using the procedure of isochronal annealing it is possible to reduce the effective barrier height from ~0.9 for an unannealed device down to ≃0.4 eV for a 500 °C anneal, when the local N_D value becomes $\simeq 8 \times 10^{18}$ cm^{-3}. Clearly, this suggests that, in order to obtain ϕ_B values ≃0 local N_D, values in excess of 10^{20} atom cm^{-3} must be achieved.

Consider now, the problem of identifying the physical origin of this local doping. The Rutherford backscattering analysis and secondary ion mass spectrometry techniques provide evidence of two important interdiffusion processes taking place during annealing (see Figures 3.2, 3.3 (a), 3.10).

(i) Ga and to a lesser extent As all diffuse into the gold layer. The outmigration of gallium continues, until the solid solubility limit is reached.
(ii) Gold, together with any impurities it may contain, diffuses into the GaAs substrate.

Figure 3.2 shows RBS and Figure 3.3(a) CAMECA spectra for an unannealed sample and the same system after annealing at 200 °C for 20 hours. Note that the RBS technique cannot differentiate between gold moving inwards and gallium or arsenic moving out—this is a severe limitation of RBS, which can be overcome by supplementary work, with either the SIMS technique or Auger analysis. Clearly, this interdiffusion gives rise to the local doping observed experimentally. These results indicate that there are two types of gallium outmigration in the Au–GaAs system. At 180 °C the gallium

Table 3.3 Degradation of ohmic contacts with overlying metallization

Contact	Overlayer	HTS temperature	Analysis Technique	Comments	Reference
Ni–Au-Ge	Ag	327	EPM, angle lapped	Ag–Au intermixing. Ge at GaAs surface. R_c constant (40 hr)	45
	Au	T > 240°C	AES-sputter	Interdiffusion. Mainly Ga outdiffusion into Au resulting in an increase in R_c with time $E_A \sim 1\,eV$ Slight Au and Ge interdiffusion.	54 55 56
		330		Thin Ni film no degradation. (Au)–Ni-As-Ge (grain): Au-Ga (grain)-GaAs. Thick Ni: initial increase in contact resistance followed by a recovery back to initial value. Cause of initial increase-induced defects by stress, Ga vacancies due to GaAs dissociation. Slight increase in R_c with further annealing which anneals out defects and redistributes Ge and compensates Ga vacancies. (Au-Ga)–Ni-As-Ge-Au-Ga (grain layer)-GaAs.	57
	Cu	327		Forms Schottky barrier 1.2 μm below original ohmic contact. Similar reaction to Ti but faster kinetics. Rapid increases in R_c.	

	Ni	330		Initial increase in R_c then fairly constant.	
	Pt	327		Pt-Ga–Pt-Au-As-Ga–GaAs, Pt–As, Pt–Ge, and Pt–Ga compound formation.	
	Ti	327		Ti diffuses through contact and consumes GaAs layer beneath contact to form Ti–Ga–As/GaAs contact below original ohmic contact. Little Ga at GaAs surface.	
	Au/Cr	285–355		Contact resistance increases initially rapidly (first few hours) then linearly with time (after ~50 hrs) $E_A \sim 1.7\,eV$ for 50 per cent increase in R_c. Thinner contacts degrade faster initially and more degradation. Linear degradation rate independent of thickness.	
Ag–Ge–In Ag–Zn–In	Ti,Pt or Cu		EPM, angle lapped	Similar results to ones obtained with Ni/Au–Ge.	55
Pt–Au-Ge	Au	330	EPM, angle lapped	Thin Pt film no degradation (Au)–Pt-As-Ga–Au-Ga–GaAs. Thick Pt film similar reaction to thick Ni–Au-Ge. Rapid degradation followed by slower recovery (than for Ni) (Au)/Pt-Ga-Ge–Au-Ga–Pt-As-Ge–Au-Ga-GaAs becomes (Au-Ga)–Pt-As-Ge–Au-Ga–GaAs. Overlying Au film acts as a sink for Ga.	45

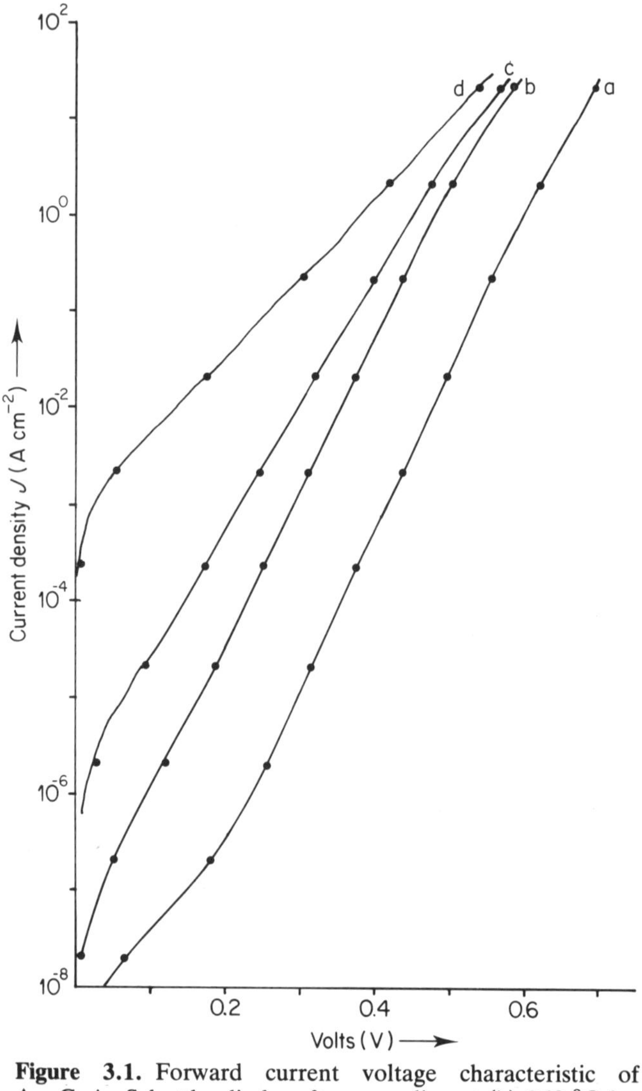

Figure 3.1. Forward current voltage characteristic of Au–GaAs Schottky diodes after annealing at (b) 260 °C for 30 minutes, (c) 300 °C for 1 hour, (d) 450 °C for 30 minutes. (a) Corresponds to the unannealed case

concentration observed in the gold, determined by Rutherford scattering, is independent of time and of about 1 atomic per cent. This suggests an initial rapid outmigration along grain boundaries, until they are are filled.

At temperatures of 250 °C the concentration of gallium is observed to be time dependent, suggesting that true diffusion is occurring in the gold, although as the concentration of gallium, as observed by SIMS or Rutherford

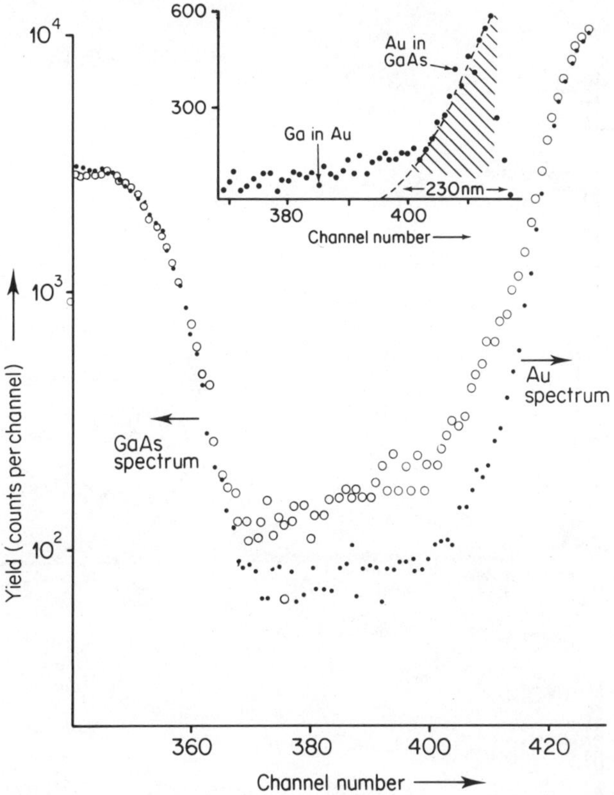

Figure 3.2. Rutherford backscattering spectra (RBS) from a gold film on a GaAs surface. These correspond to an 'as deposited' film and one annealed at 250 °C for 20 hours in an atmosphere of forming gas (see Figure 1.13 for background on RBS)

scattering, is uniform with depth, it is probably occurring laterally from filled grain boundaries. At temperatures of 350 °C the gallium concentration in the gold is observed to be almost constant, which indicates that at this temperature it takes less than 1 hour for the solid solubility limit to be reached.

The consequence of this gallium outmigration is severe and results in a marked change in the *I–V* characteristics of Au–*n*-GaAs Schottky barriers. This is not due to a reduction in the barrier height, for *I–V–T* measurements on these diodes shows that the doping density becomes progressively more *n*-type as the semiconductor becomes more gallium deficient. The precise reason for this is still unclear, for Chiang and Pearson[9] have shown that gallium vacancies are acceptors and one would therefore expect the material to become progressively more *p*-type as outmigration proceeds. There are, however, other possibilities which remain to be investigated. For example, a

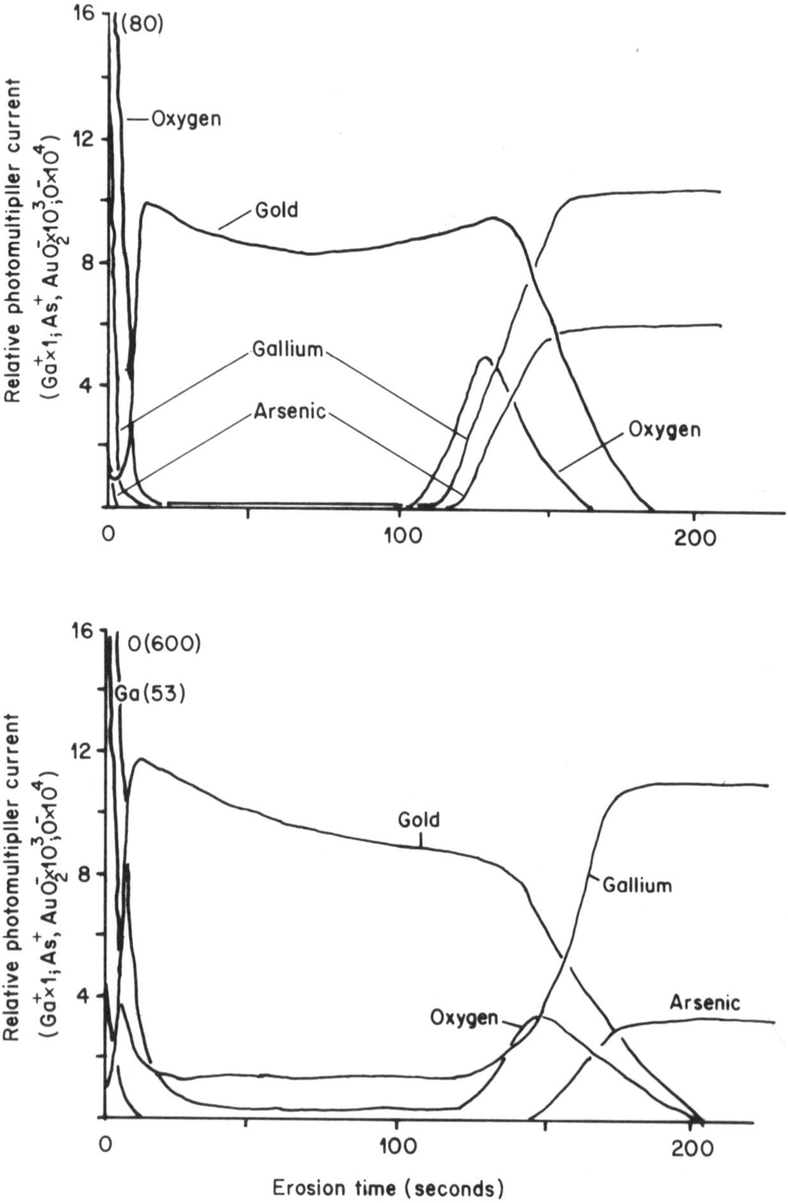

Figure 3.3. SIMS (Cameca) profiles for the Au–GaAs system as fabricated and after annealing at 250 °C for 20 hours

high degree of surface disorder may be responsible for the donor characteristics of vacancies.[10] An alternative hypothesis is that residual impurities in the gold electrode material moving into the GaAs are responsible for the n^+ doping.

Consider now, the uniformity of this annealing process. Gyulai *et al.*[5] have used SEM and optical microscopy to study the annealing of a 40 nm gold film on GaAs. Below 350 °C there was no visible change in the surface. At 400 °C and higher small localized alloyed areas begin to appear. The number of these increases up to 500 °C when liquid starts to form and all the gold is consumed in the formation of some intermetallic compound. Detailed SEM studies up to 600 °C revealed that long rectangular pits were formed due to crystallographic etching by the Ga–As–Au liquid formed during alloying and that these pits increase with increasing amounts of Au present. Since it is thought that

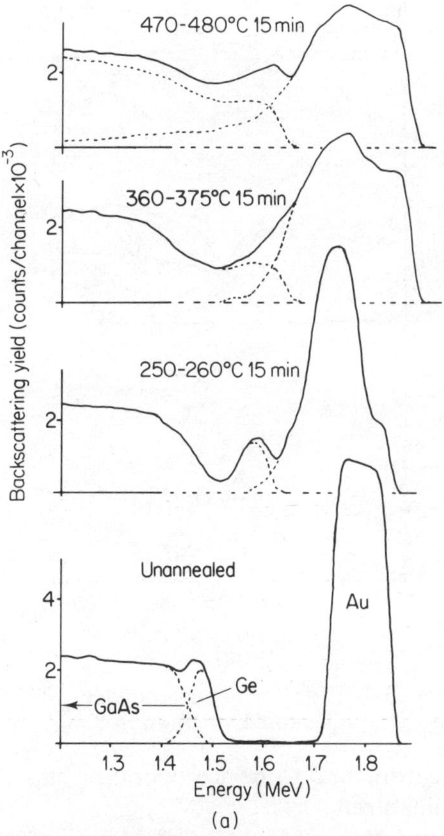

Figure 3.4. (a) RBS spectra for the Au–Ge–GaAs system after a range of annealing temperatures up to 480 °C

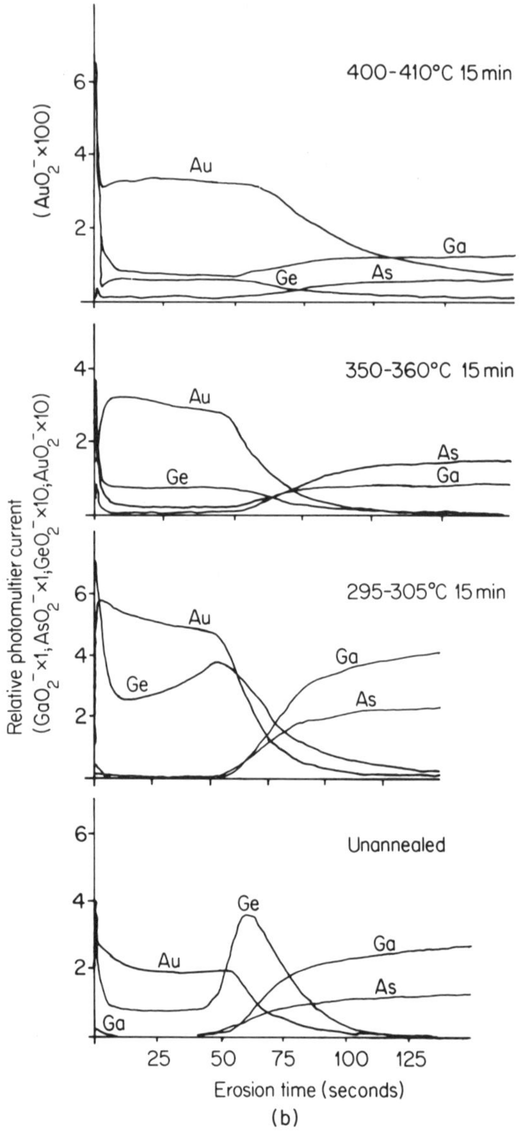

Figure 3.4. (b) The corresponding SIMS (Cameca) profiles to those shown in 3.4(a)

these etch pits are detrimental to device operation, the Au in the alloying process should be minimized.

Insofar as ohmic contacts are concerned the Au–GaAs system is in a sense of academic interest, since it is not good enough for high quality ohmic contacts and too unstable to be used for Schottky contacts. We must therefore, now consider the results of the introduction of Ge into the system. The basic alloy required for the contact is Au and Ge (88 per cent, 12 per cent). Many workers have added an overlay material, usually Ni or Pt, which is used to improve the uniformity of the alloyed contact and to reduce the tendency of the film balling up. In Table 3.2 the results of all the available data are summarized. Also shown on Table 3.2 are the results for the other important ohmic contact alloys. Clearly, from these results we deduce that the alloying process is complex. However, some general trends emerge from these studies. The Rutherford backscattering analysis and ion microprobe analysis of the Au Ge–GaAs system indicate that gallium outdiffusion to the surface occurs with the very low value of activation energy $\simeq 0.1$ eV.[11] This low value of activation energy is usually associated with a grain boundary or pin hole mechanism.

At the same time the gold and germanium intermix to form an alloy and also to penetrate into the GaAs. Figure 3.4 shows a sequence of RBS spectra and the corresponding ion microprobe depth profiles obtained during the annealing sequence. From these results we deduce that one very important consequence of the presence of Ge in the surface film is the increased interdiffusion and penetration of the system. Gold, for example, penetrates deeper into the substrate and at lower temperatures than the corresponding situation where no Ge is present. A second important characteristic of alloying is the highly non-uniform nature of the process. This is illustrated in the sequence of micrographs shown in Figure 3.5. SEM studies[7] show that there are clearly visible dark areas on the contact surface. The imaging properties of the ion microprobe showed that the dark areas contained low concentrations of Ga and high concentration of Au and Ge, compared with the rest of the contact (Figure 3.5(b)–(d)). When the contact was removed a spike formation was observed beneath the contact corresponding to the dark areas on the surface. Electron probe studies revealed that these spikes contained both gold and germanium with the gold penetrating more than 300 nm into the GaAs.[7] Gold spikes have also been reported by Gyulai *et al.*[5] but they did not report the existence of germanium in the spikes. Furthermore, this gold germanium spike formation would explain the germanium 'island' formation observed beneath the contact by Yokoyama *et al.*[12] One widely used modification to the Au–Ge system involves adding a Ni layer to the starting contacts in order to inhibit balling up during alloying. Some of the important features of the intermixing process occurring in this system are summarized in Table 3.2.

The Au–Ge and AuNi–Ge systems discussed in detail here, are perhaps, the more successful systems used up to the present day. This is true in spite of the general undesirability of having Au near to the GaAs. There are many alternative variants (Table 3.2), which have been studied and reported in the literature. In general, these systems have not received the same thoroughness of study as the Au–Ge system, particularly in terms of detailed physical

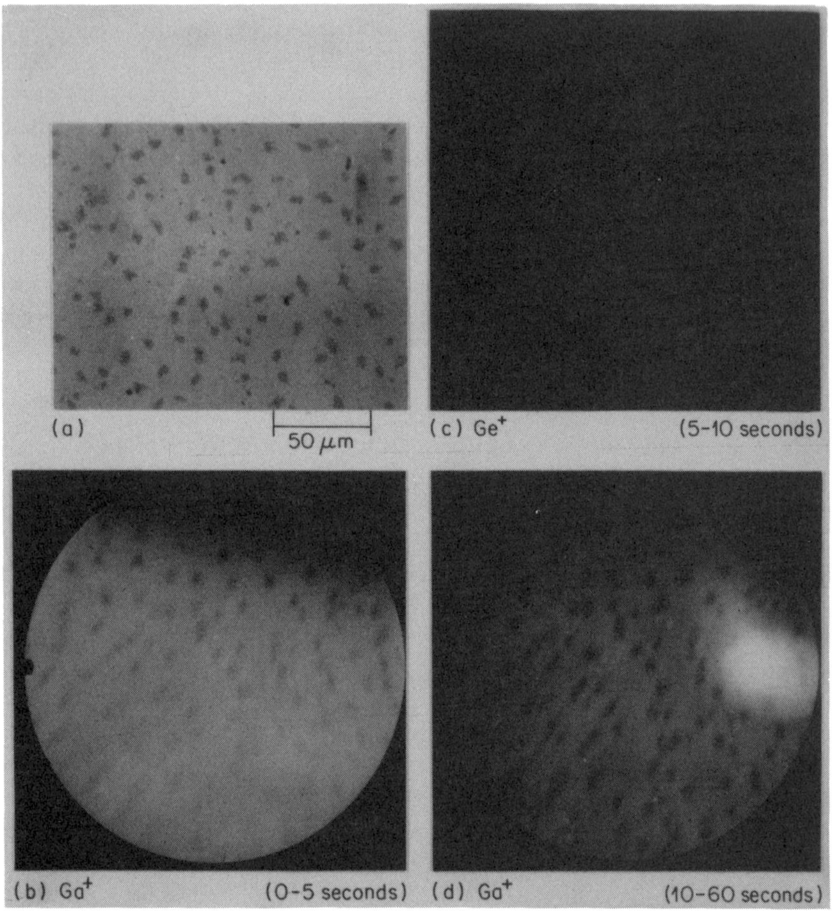

Figure 3.5. Optical micrograph of a Au–Ge film as a GaAs sample after annealing at 410 °C for 15 min. Consecutive ion microprobe images of the Au–Ge film above using (b) Ga^+, (c) Ge^+, and (d) Ga^+ ions to form the image

studies. However, the basic objective in all these is the same. A limited and controlled alloying or interdiffusion is used to produce a thin degenerately doped layer near to the interface. Thus, although many of the microscopic processes identified in the Au–Ge system may be applicable in many of these other systems, there is a need for much further work in this area.

The general technology of InP is much more primitive than GaAs and this situation is reflected in the contact technology. Some of the systems used are summarized in Table 3.1. These systems are extrapolations of those used for GaAs and most of the data is empirical. An exception to this is the work of Kim *et al.*,[88] which provides a detailed SIMS study of the alloying process.

Up to this point, the survey has been confined to the basic alloying process. In a real device structure, the ohmic contact may have to sustain relatively high current densities, while at elevated temperatures. This adds yet another dimension to the degradation process and increases the degree of degradation. Although this topic is the domain of Chapter 5, a brief mention will be made here of these effects on the basic ohmic contact.

Early workers involved with transferred electron devices used tin or indium based alloys for making ohmic contacts. Unfortunately, tin and indium readily migrate through the device, resulting in a short circuit failure.[17,43] In planar devices the initial melting of the anode was followed by migration which occurred from anode to cathode.[47,48] Kuru[49] reported a failure mode which had an activation energy, $E_A \sim 2.3$–2.6 eV and postulated that this could be due to tin diffusion, into the active layer, from a tin doped highly conducting n^+ epitaxial layer grown on top of the active layer. During this degradation process the output power decreased and the current–voltage characteristics fell with time, such that the threshold current and negative resistivity region decreased. The process is characterized by the n^+ epitaxial layer contact becoming non-ohmic and attributed to interdiffusion between the Au–Ag–Sn contact and the GaAs. Similar degradation of current–voltage characteristics was observed by Al-Moufti *et al.*[50] and Walsh.[51] The former study associated an increase in contact resistance to contact–GaAs interdiffusion.[32,45,52,53] Gold migration into the active region is also a predominant failure mode in IMPATT diodes.[53] Similarly, outdiffusion of gallium results in non-stoichiometry of GaAs at the contact–GaAs interface resulting in compensation and hence an increase in resistance. Outdiffusion of Ga was reported to be significant at temperatures above 240 °C.[56,57] Figure 3.6 shows microspot Auger analysis (μAES) of an annealed and unannealed sample: gallium outdiffusion and some redistribution of other elements has occurred; both gold and germanium can be seen to have diffused into the GaAs. Biasing the test structure accelerated the degradation.[32,56] In this example, at a bias of 5 volts the current saturated at 35 mA, corresponding to a channel current density of $\sim 10^6$ A cm^{-2}. Above 35 mA the contacts degraded rapidly with the appearance of lateral diffusion as is shown in Figure 3.6. This lateral diffusion is caused by electromigration. Electromigration causes an increase in contact resistance and has also been observed in FETs.[58] Cooke[59] reported that lateral diffusion of free gallium, after bulk diffusion to the contact surface, between the electrodes resulted in source–drain or drain–gate short circuits.

Further interdiffusion problems can arise from metal layers deposited on top or bonded to the ohmic contacts. Table 3.3 gives a summary of results from annealing studies on ohmic contacts with different overlay films. Copper, for example, diffuses rapidly through the ohmic contact layer forming a Cu–GaAs Schottky barrier beneath it.[54] Copper has been observed to diffuse through the nickel and gold plated layers used on conventional heat sinks at

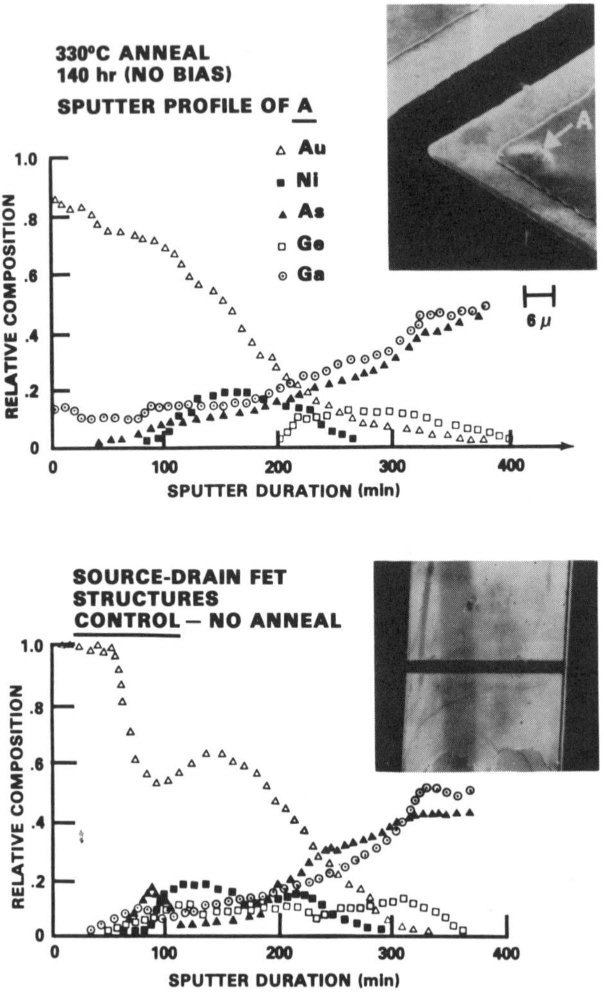

Figure 3.6. An analysis of an FET ohmic contact region before and after annealing. Marked interdiffusion of Au and Ga occurs (after Brodinsky and Davey[56])

temperatures ~300 °C[60] and therefore can be a potential reliability problem. Although the corrosion resistance of gold makes it the preferred metallization, it does unfortunately act as a sink for gallium outdiffusion. Thus a barrier layer must be used in order to improve its reliability. Hewitt *et al.*[61] for example found that a Au–Pt–Ti multilayer metallization on top of the ohmic contact results in stable contacts in FETs; the titanium acting as a barrier to gallium outdiffusion.

3.3 SCHOTTKY BARRIER METALLIZATIONS

It might be argued that the division between Schottky and oh tion systems is somewhat artificial. After all, the A Au–Ge–GaAs systems discussed earlier are only rectifying conta undergone a controlled thermal degradation process in order to ieve the desired ohmic behaviour. For this reason, the present section will concentrate on metallization systems which do not degrade as readily as the traditional ohmic contacts. There is, however, one very important difference between the Schottky barrier and ohmic contact as far as degradation during device operation is concerned; that is, relatively high fields exist at the Schottky interface. Furthermore, in the case of operation into breakdown, the contact may operate under a relatively high current density. The bulk of the reported results, however, deal only with the thermal stressing and hence only a brief mention will be made of such effects.

3.3.1 Gallium arsenide

One of the simplest rectifying metallization systems, which is still used for fundamental studies on GaAs, is the gold system discussed earlier for its importance in understanding the physical process involved in the formation of ohmic contacts. The ductility, high thermal conductivity and corrosion resistance of gold makes it attractive to the device engineer. However, due to its poor thermal stability on GaAs its use in real device structures is restricted to a component in multilayer structures.

The silver–GaAs system suffers from many of the problems of the gold contact, a relatively low annealing temperature causing interdiffusion. Other contenders for the role of Schottky contact metal include Pt, Ni, Al, Cr, Ti, Ta, Mo, and W. The results of a wide range of data on these systems are summarized in Table 3.4. In this table a list of references, analytical techniques and comments on the interdiffusion and kinetic reactions are included. The following table summarizes the main observations of a combined Rutherford backscattering and SIMS study, which were used to identify the main degradation mechanisms, together with a 'safe operating temperature' defined as that temperature below which no observable degradation could be detected. It should be emphasized at this juncture that degradation does not imply catastrophic failure of the Schottky barrier but can imply a physical change that adversely affects the operational stability of the device. In this brief survey it is not intended to present a detailed account of all the systems studied, but instead to identify some of the common physical processes occurring in these Schottky systems.

In all the results reported thus far, good stability is observed in all systems for temperatures below 250 °C. At this temperature many systems (Au, Ag,

Table 3.4 Summary of metal–GaAs–Schottky barrier interdiffusion studies

Metallization	Temperature (°C)	ϕ_b(eV)	Analysis technique	Comments on Degradation	Reference
Al	RT	~0.7	μAES	No, Ga, or As on surface. Al_2O_3 at interface surface.	66, 32
	200	~0.8	SIMS	Ga and As outdiffusion (As concentration in	
			X-ray Diffraction	Al 0.5%).	62, 74
	250	~0.9	SEM	Ga outdiffusion, Al indiffusion.	
	350	~0.87	TEM	More Ga outdiffusion. Slight As outdiffusion.	80, 81
	450–500	0.8	RBS	More Ga and As outdiffusion, Al indiffusion 20–40 nm.	
				More Al_2O_3 formed.	
				GaAs + Al precipitates near GaAs surface. Al–As crystallites at interface.	
	550			More Ga and As outdiffusion.	
				Bubbles in film 5–10 μm diameter	
Ag	RT		SEM		12, 16
	300		Contact	Poor ohmic for p^+	
	400–750		Removal	Faceted holes below contact in GaAs	22
Au	RT	~0.9	RBS Channelling	Ga outdiffuses to the surface to form Ga_2O_3. Low	6, 13
	200–300	0.7	AES	concentration of As in Au film. Au diffuses slightly	15, 5
			EPM	into GaAs (40–150 nm).	
	350–500	0.5	X-ray Diffraction	More Ga outdiffusion, As outdiffusion and evaporation.	22, 19
			SIMS	Au film nonuniform: contains a Au–Ga (21 at % Ga)	21, 34
				phase with Au segregation. Nonuniform Au diffusion	66, 83
				into GaAs; Au precipitates extend 300 nm into GaAs	84, 85
				producing a high dislocation density in GaAs. Au produces an increase in the number of effective donors to $4.5 \times 10^{18} cm^{-3}$ below the contact	
Cr	RT–300		RBS, SEM	No reaction	22
	425		SIMS	2–3 at % Ga outdiffusion into Cr.	
	550			Intermetallic formation Cr_2X(X-Ga and/or As).	72
				Cr indiffusion	

Mo	RT–450		RBS, SIMS	Stable–no reaction.	22
	>500		SEM	Bubbles on film surface due to thermal expansion mismatch	72
Ni	300		RBS, SIMS	All Ni consumed to form NiX (NiGa, NiAs)	36
	425		SEM	Formation of NiX_2 and Ni_2X_3	72
	550			Ni_2X_3 and Ga_2O_3 on surface. Ni indiffusion into GaAs	86
Pd	RT			Similar reactor kinetics to Pt	21
	500			Non-ohmic	
Pt	RT	~0.9	RBS, SIMS, AES	Slight Ga outdiffusion and surface accumulation.	9
	200–250	~0.9	X-ray Diffraction	Very small amount of As outdiffusion and surface accumulation (concentration As $< 0.2\%$ in Pt).	21, 34
	300	0.86		More Ga outdiffusion ~3–4% on Pt surface. Forms a layer structure Pt–Pt_3Ga–Pt-$PtGa_2$–$PtAs_2$–GaAs. $PtGa_2$ moves into GaAs with time.	66, 69
	350–450	0.9		Pt_3Ga moves outwards, more Ga outdiffusion and $PtGa_2$. Pt(+Ga)–Pt_3Ga, $PtAs_2$–GaAs. $GaPt_3$ has also been reported which diffuses to form PtGa and $PtAs_2$. Slight Pt indiffusion.	86, 87 72
	500	1.0		PtGa-$PtGa_2$–$PtAs_2$–GaAs forms. Pt indiffusion.	
	600			Brittle alloy forms	
Ta	RT–300		RBS, SIMS	Stable–no reaction	
	425–450		μAES	Ta indiffusion ~200–300 nm, Ga outdiffusion and TaO on surface.	32, 72
	500		EPM	More Ga and As outdiffusion. Ta indiffusion and possible TaX formation at interface (X–Ga and/or As) Ta oxidises to TaO_2	

Table 3.4 *(continued)*

Metallization	Temperature (°C)	ϕ_b(eV)	Analysis technique	Comments on Degradation	Reference
Ti	RT	~0.82	RBS, SIMS	Possibly TiAs forms. Slight oxidation of Ti (TiO_2, TiO?).	22, 25
	350–450	~0.84	X-ray diffraction EPM		
	500	0.80	AES	Oxidation of Ti. 1–2% Ga outdiffusion into Ti. Intermetallics have been reported TiAs, Ti_5As_3, Ti_2Ga_3, Ti_5Ga_4.	72, 77
	550			3–4% Ga outdiffusion. More oxydation (TiO_2)	
W	RT–500	~0.65	RBS, SIMS SEM–EDAX	Stable–no reaction. Slight increase in ϕ_b with temperature. Layers tend to peel off (T > 400 °C) due to thermal expansion mismatch	21, 64 66, 72
Au–Al				Does not form ohmic contacts	20
Au–Ga	RT	~0.75	SEM		
	250–400	~0.72		Little degradation. No 'alloyed' regions visible	84
Au–Ni	200	~0.86	SIMS	Ga outdiffusion and accumulation in Au.	
	350	~0.81	X-ray diffraction	More Ga outdiffusion. Ni diffuses into Au and GaAs. Bubbles form on surface at temperatures >300.	34
	400	~0.65		High Ga concentration in Au. As outdiffusion and more Ni diffusion into Au and GaAs; possibly Ni–As alloy forms.	
	500	Ohmic		As outdiffuses into Au. Au–Ni complex phases form. Bubbles reduce to clusters	

Au–Pt	350 500		AES	Ga and As outdiffuse; Ga edge leads As edge. Au diffuses through Pt. More Ga than Pt in Au. Au–Pt-Ga–Pt-Ga-As–GaAs	67, 87
Au–W	RT–500		X-ray diffraction	Stable–no reaction	21
Ni–Au	RT–350		RBS, EPM	No Ni accumulation at Au–GaAs interface	35
Pt–Ni	RT–200 380–480 >500	0.88 ~0.95	X-ray diffraction	No reaction. GaPt, $PtAs_2$, Ga_2Pt formed. Ga_2Pt, $PtAs_2$, GaPt Schottky barrier degraded	63
Pt–Ti	RT–350 500	~0.81 ~0.86	RBS X-ray diffraction	No reaction. Pt_3Ga–TiAs–$PtAs_2$–GaAs formed	21, 65
Pt–W	RT–500	~0.7	X-ray diffraction	No reaction	79
Ge				Does not form ohmic contact	21, 65
W–Ce				Does not form ohmic contact	21, 65
Pt–Ge				Does not form ohmic contact	21, 65
Au–Pt–Ti	RT 250			Au–Pt–Ti slight interdiffusion between Ti and GaAs Ga outdiffusion into Ti, slight As outdiffusion, Pt diffusion through Ti and into GaAs	56
W–Ti	≤600			Stable	56

Table 3.5 Metal–InP Systems

Metallization	Temperature (°C)	ϕ_b(eV)	Analysis technique	Comments on degradation	Reference
Al	⩽500 °C	0.35 to 0.6	Auger sputtering	Sintering in air caused the effective barrier height to increase when oxygen was present in Al film–growth of interfacing oxide.	89
			SIMS	Interface with semiconductor stable up to 500 °C	88
Au	⩽400 °C	0.43	SIMS	Above 350 °C Au indiffusion and In outdiffusion	88
Cr	RT–500	≃0	SIMS	Metallurgically stable—no Schottky barrier	88
Au(Ni)	RT–350	≃0	SIMS	Au, Ni indiffusion, In, P outdiffusion	88
Au(Ti)	RT–400	0.42	SIMS	Au indiffusion, P outdiffusion	88

Pt, and Ni) exhibit gallium outdiffusion and to a lesser extent some arsenic outdiffusion. This outdiffusion is balanced by an inward migration of the contact meal. Metal contacts which exhibit this effect are the ones with the lowest safe operating temperature. From the low value of activation energy ascribed to this process it would appear to be associated with a grain boundary type diffusion process, as was discussed earlier for gold contacts.

A second class of interfacial reactions are associated with materials that are able to form identifiable compounds with the base GaAs. Examples of this include platinum, palladium, nickel, and chromium; the compounds and their complexity being very dependent on the annealing temperature. A summary of some recent data on the materials that behave in this way, together with the critical temperature for the onset of compound formation, is given in Tables 3.4 and 3.5.

As an example of this process, consider first the Pt–GaAs system. This system is ideal to study with RBS technique since the Pt spectrum is quite separate from the GaAs spectrum. Figure 3.7 shows a sequence of Rutherford backscattering spectra for annealing sequences up to 500 °C and Table 3.6 is a detailed summary of the corresponding interdiffusion and solid state reactions identified from this data. In these results[62] no X-ray diffraction

Table 3.6 Platinum–GaAs reaction details[a] as a function of anneal temperature

Pt-EGaAs.	Original Pt thickness = 100 ± 7 nm (annealed in undried argon)
Anneal temperature (°C)	Interdiffusion and solid state reactions
250	No reaction
300	Ga out diffusion: 3–4 per cent to reach Pt surface
	Near Pt–GaAs interface $PtGa_2$ (or $PtAs_2$) formed, 15 nm of Pt consumed
	$PtGa_2$ interface moves into GaAs by 30 nm
	Away from the interface trace of Pt_3Ga formed at a depth of ~60 nm below Pt surface
	GaAs–$PtAs_2$, $PtGa_2$ (48 nm)—Pt, Pt_3Ga (38 nm)—Pt (60 nm)
350	More Ga out-diffusion amount undetermined
	More $PtGa_2$ formed, part of it is most likely to be $PtAs_2$
	Pt_3Ga front moves outwards
	GaAs–$PtAs_2$, $PtGa_2$ (95 nm)—Pt_3Ga (48 nm)—Pt(Ga) (40 nm)
	Here, Pt(Ga) means dissolved Ga in predominantly Pt layer
400	GaAs–$PtAs_2$, $PtGa_2$ (140 nm)—Pt_3Ga (70 nm)—Pt(Ga) (32 nm)
450	GaAs–$PtAs_2$, $PtGa_2$ (140 nm)—PtGa (120 nm)
	Very small quantity of Pt indiffusion
500	Exactly the same as 450 °C, but slightly more Pt indiffusion
550	Same as 450 °C, more Pt indifussion (to ~40 to 50 nm)

[a] A full tubulation of this kind for Pt, Ni, Al, Cr, Ti, Ta, Mo, and W is given in Reference 72

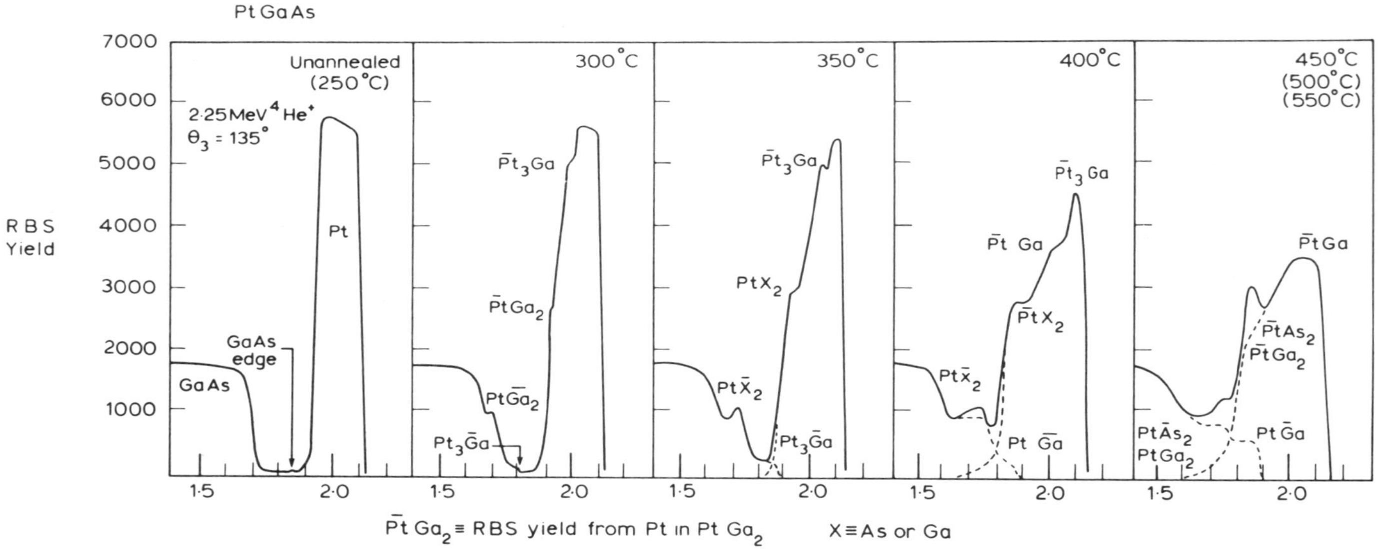

Figure 3.7. RBS spectra of Pt–GaAs samples after various stages of annealing up to 550 °C (1 hour in undried argon). A bar over one of the constituents indicates RBS yield from that element from the proposed compound

analysis was presented to identify compounds formed. Consequently, the compounds listed in Table 3.5 were ascertained by measuring relative backscattering yields from the constituent elements, using existing knowledge of GaAs compound formation obtained from X-ray diffraction studies[63–66] and other methods.[63,71] The following compounds have been identified from Figure 3.7: $PtAs_2$, $PtGa_2$, Pt_3Ga, and PtGa. Mukherjee *et al.*[62] noted that $PtGa_2$ and $PtAs_2$ cannot be separated by RBS but SIMS profiles may be used to give independent depth information on the concentrations of Ga and As. SIMS results indicate that there exists an arsenic-rich layer near to the original Pt–GaAs interface and a gallium-rich layer on the outer platinum surface. Thus from these and related studies a model for the Pt–GaAs reactions is as follows: For low temperature anneals Ga outdiffuses into Pt to go into solid solution and form compounds, such as Pt_3Ga, PtGa, and $PtGa_2$ with increasing Ga concentration, leaving an arsenic-rich layer behind. The arsenic then reacts with Pt to form the $PtAs_2$ at the Pt–GaAs interface and is responsible for the subsequent Schottky barrier characteristics of this metallization.[64,66,69] The Pt–GaAs reaction starts to occur at 300 °C and there is evidence to indicate that the reaction rate is higher when the defect concentration in the GaAs is higher,[72] possible associated with a higher doping density. A qualitative estimate of the rate of consumption of the platinum appears to be in the region of 25 nm per hour at a temperature of 370 °C.[62,68] There are indications that Ga and As are consumed in nearly equal proportions. Consequently, changes in semiconductor properties just underneath the $PtAs_2$ layer due to deviations from stoichiometry are unlikely. It has been suggested[9] that gallium vacancies act as acceptors, which would turn the relevant GaAs crystal *p*-type. If so, preferential gallium outdiffusion is certain to have a significant effect on the electrical behaviour of the Schottky barrier. In the case of Au–GaAs, discussed earlier, Ga outdiffuses more than As[6,64,66] creating gallium vacancies. Unlike platinum, gold does not combine chemically with excessive arsenic left behind to restore stoichiometry. Platinum forms $PtAs_2$ with this excessive arsenic and allows the Schottky barrier to perform even at 500 °C.[64–66] As a result of the depletion of GaAs, the PtAs front moves inwards with increasing annealing.[69] This junction 'creep' makes Pt unsuitable for example in Q or X band IMPATT applications. In this example the drift region itself is too narrow (2–10 μm) to allow further movement of one of the electrodes.

SEM inspection of the sample surfaces at all the stages of annealing shows no surface agglomeration or non-uniformity. However, even slight mechanical damage causes flaking off on the annealed Pt–GaAs samples indicating the brittleness of the compounds formed,[76] in agreement with Ogawa *et al.*[63] The effect of water vapour present in the argon atmosphere during annealing is believed not to affect compound formation significantly.

Nickel, like platinum, undergoes compound formation at annealing temperatures above 250 °C.[62] Figures 3.8(a) and (b) show Rutherford backscattering spectra and SIMS profiles of the Ni–GaAs system. (Note that in this system the RBS spectra are more difficult to analyse because of the overlap of the trailing edge of the Ni spectrum and the leading edge of the GaAs spectrum.) The compounds NiGa, Ni_3Ga_2 and Ni_3Ga are known to exist for the binary Ni–Ga system[73] and so are Ni_3As_2, NiAs, $NiAs_2$ and Ni_5As_2 for the Ni–As system. No reported evidence exists on the Ni–Ga–As tertiary system, and consequently whether the existence of an extra component severely affects compound formation of the binary kind is unknown. Note that the RBS spectra cannot separate Ga from As and hence X implies the possibility of either of these two elements. X-ray diffraction analysis is needed for a proper compound identification but this data is not available at present. From the SIMS analysis (Figure 3.8(b)) an arsenic-rich layer in which the arsenic is in chemical combination with nickel was found near the gallium arsenide side. However, gallium was identified in this layer although in small quantities. The main gallium-rich nickel compounds formed close to the surface, together with some partial oxidation of the gallium. Some evidence also exists for a nickel tail penetrating into the GaAs indicating nickel interdiffusion. A broad conclusion that may be drawn from the published results, is that nickel is unpredictable in its reactions with GaAs with a lack of reproducibility for the different starting crystals used (i.e. substrate or epitaxial). This together with its susceptibility to react with GaAs at low temperature make it unsuitable for many device applications.

Three metals with good stable thermal properties are aluminium, titanium, and tantalum; unfortunately all suffer the problem of surface oxidation. In the case of aluminium[66,74,75] Ga and As outdiffusion begins to occur at temperatures in the range 400–500 °C, the former migrating in quantity. Aluminium is also found to diffuse into GaAs, though the diffusion lengths are very small. Overall, in spite of the surface oxidation problem, aluminium is suitable, at least metallurgically, for use as a Schottky contact.

Titanium has been a traditional bonding metal in three-layer metallization systems on silicon IMPATT diodes;[62] its technology is therefore well established. Titanium–gallium arsenide reactions have been studied in considerable detail.[65,77] Both arsenic and gallium based titanium compounds such as TiAs, Ti_5As_3, Ti_2Ga_3, and Ti_5Ga_4 have been identified in reacted samples.[77] Annealing in air is found to cause electrical instabilities to the Schottky diode.[65] Figure 3.9 shows a representative RBS spectrum for an unannealed sample, and one annealed for 1 hour at 500 °C. The presence of water vapour in the undried argon causes titanium oxide (TiO_2) to form and this oxide layer allows gallium or arsenic outdiffusion. An important conclusion from these studies is that titanium is stable metallization up to about 400 °C.

Tantalum has been investigated as a Schottky barrier metal for both GaAs MESFET and IMPATT applications.[32] It has been shown that because of its

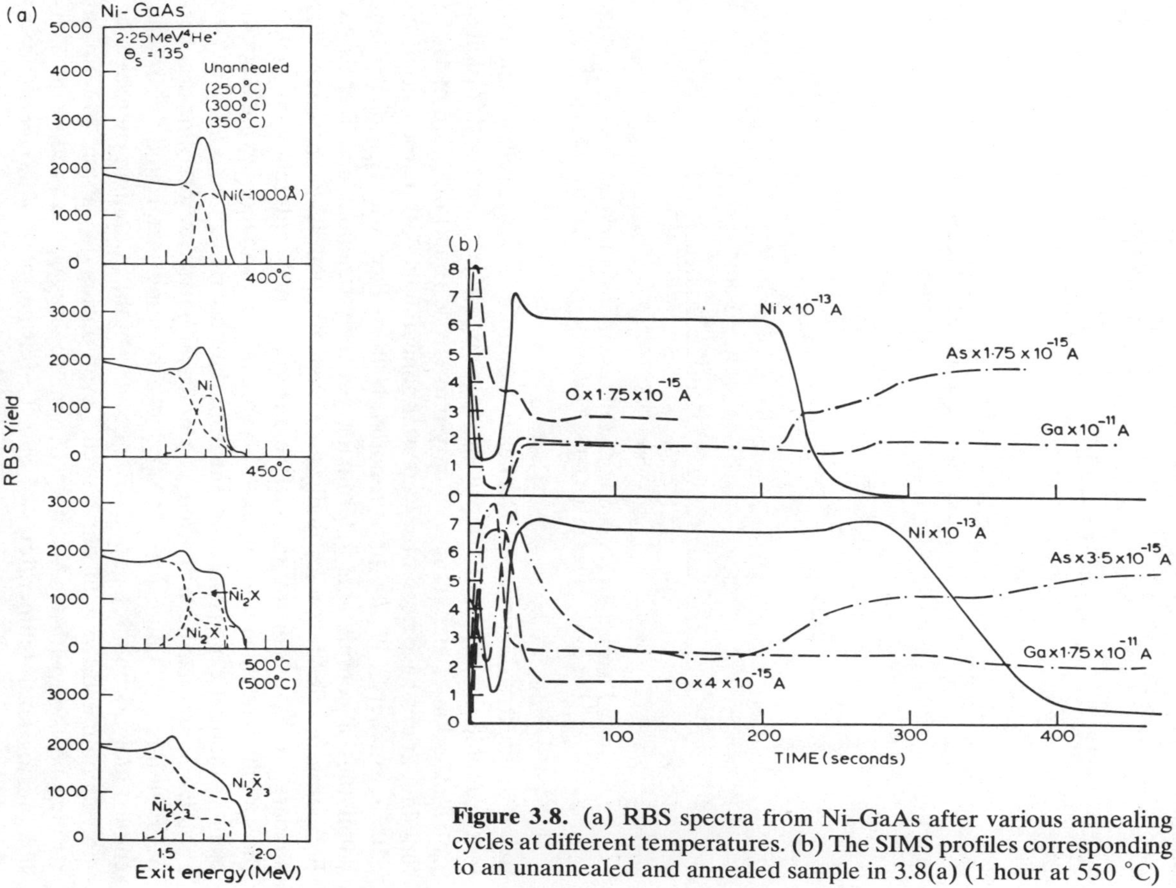

Figure 3.8. (a) RBS spectra from Ni–GaAs after various annealing cycles at different temperatures. (b) The SIMS profiles corresponding to an unannealed and annealed sample in 3.8(a) (1 hour at 550 °C)

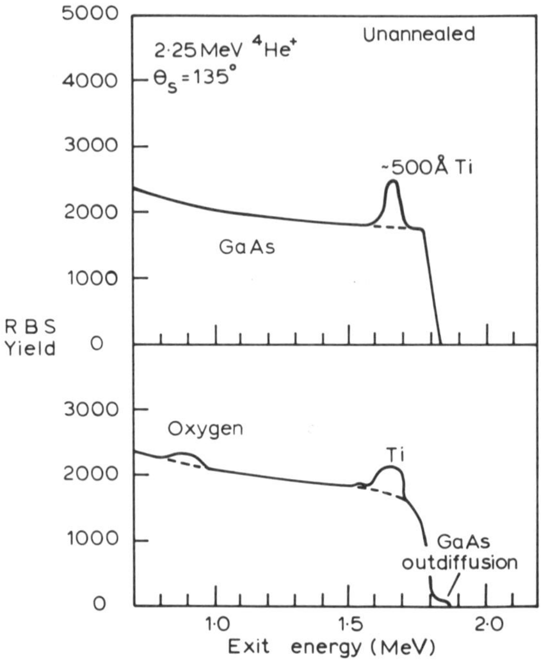

Figure 3.9. Oxidation and diffusion of Ga through the oxide layer in a Ti–GaAs sample

susceptibility to oxidation, electron beam deposition in a clean environment is more suitable than RF sputtering in an argon atmosphere. Lattice diffusion of tantalum in GaAs and grain boundary plus lattice (bulk) diffusion of Ga in tantalum grains have been observed after annealing at 450 °C for 1 hour. At 550 °C Mukherjee *et al.*[72] have reported both surface oxidation of the tantalum plus a considerable intermixing between tantalum and the GaAs. Furthermore, SEM studies indicate rather unusual layer patterns that appeared to have come off the surface due to unequal contraction on cooling the sample. The layer is thought to be an oxide of tantalum having a coefficient of thermal expansion different from the underlying GaAs.

Molybdenum has been studied as a practical barrier to interdiffusion between sintered platinum (on GaAs) and a bonding metal gold.[78,79] In this context molybdenum is a poor barrier because of the intermixing between the molybdenum and the gold which occurs even during deposition. In terms of its interdiffusion with GaAs, molybdenum appears not to produce measurable mixing up to temperatures of 550 °C. Furthermore, although the compound $MoAs_2$ is known to exist,[73] RBS and SIMS studies show no sign of such compound formation[72] or outdiffusion of gallium—the latter observation is presumably because the solid solubility of gallium in molybdenum is very small. A major problem with this system is the poor adhesion at the interface.

SEM studies[72] indicate layer 'peel off' at around 450 °C. Mukherjee *et al.*[72] suggest that this is a result of the poor match between the thermal expansion coefficients of pure molybdenum and GaAs.

A material similar in performance to molybdenum is tungsten. Electrical and metallurgical characteristics of W–*n*-GaAs have been studied by Sinha and Poate[64] for annealing temperatures in the range 350° to 500 °C. Figure 3.10 shows an 'as deposited' and an annealed spectrum and compares this with the corresponding Au case. No interdiffusion and Schottky barrier

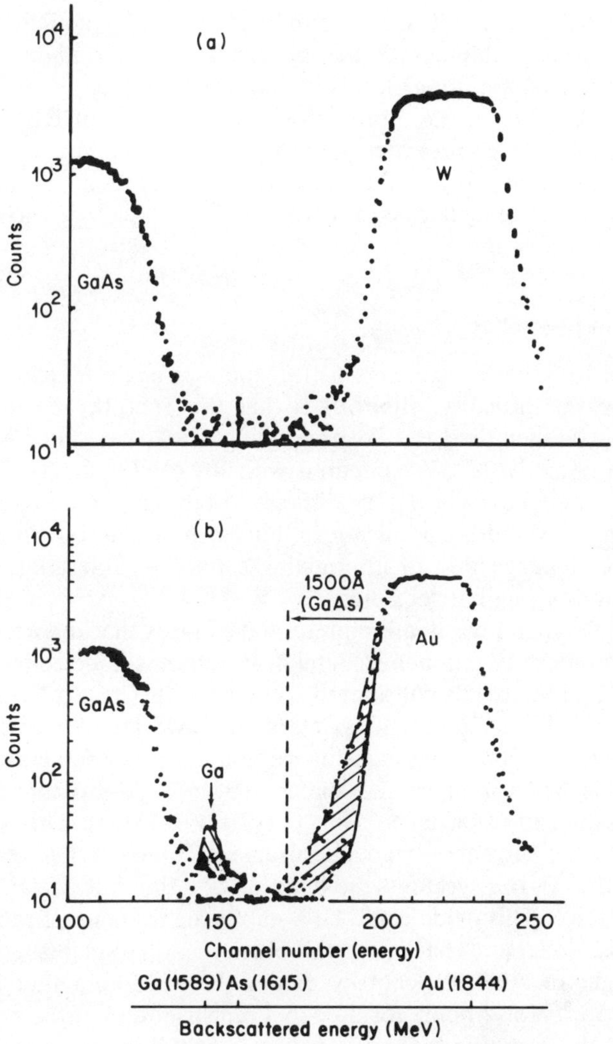

Figure 3.10. Comparison between the interdiffusion of W–GaAs and Au–GaAs for annealing up to 450 °C (after Sinha and Poate[66])

degradation was observed. Mukherjee *et al.*[67,72] confirm this pattern of behaviour but note that poor adhesion results in layer peel off at temperatures in excess of 450 °C.

Chromium is an important component as an ohmic contact in multilayer metallization schemes for power IMPATT diodes.[75] Its reactions with the other systems (platinum or palladium) have been studied and reviewed by Mukherjee *et al.*[76] In the context of a GaAs Schottky contact chromium is found to be stable up to about 400 °C, beyond which gallium or arsenic outdiffusion becomes noticeable. At 550 °C there is evidence of compound formation, the details of which could not be identified by the RBS techniques. One very serious problem with this system is that chromium is known to produce deep (compensating) levels in GaAs. Consequently, even relatively low levels of indiffusion (i.e. even below the sensitivity of RBS) can drastically alter the interface properties of the Schottky diode. Apart from developing a semi-insulating layer at the interface, deep levels at the interface are known to increase the leakage currents across Schottky diodes both in reverse bias and in low forward bias.[91]

3.3.2 Indium phosphide

With regard to Schottky contacts to InP there is again a lack of reliability data. One extra difficulty with InP is that the Schottky barrier height is considerably smaller than the corresponding GaAs case. For example the value for Au–InP is 0.5 eV compared with 0.9 eV for GaAs. These much lower barrier heights make it very difficult to fabricate low leakage contacts in, for example, MESFET applications. Furthermore, any slight degradation in the device leakage due to intermixing cannot be tolerated, due to this relatively low starting barrier height.

The most detailed investigation presented so far is that reported by Kim *et al.*,[88] who considered a number of metal–InP systems subjected to heat cycling up to 500 °C. The metals considered were Au, Al, Cr, (AuNi) and (AuTi) and the effect of heat treatment assessed by SIMS analysis correlated with terminal current voltage and capacitance voltage measurements. The SIMS analysis of the Al, In and P profiles indicates a stable Al–InP interface with no measurable interdiffusion up to 500 °C (Figure 3.11(a)). This stability was confirmed by the current voltage measurements. One possible reason for the stability of the Al–InP system is the build-up of a thin interfacial oxide at the Al–InP interface. This oxide could be a stabilizing influence on the interface. Christou and Anderson[89] have used this phenomenon to increase the effective barrier height of Al–InP Schottky diodes. It was found that heating the sample at 250 °C for 4 hours resulted in a redistribution of the oxygen in the Al film and the resulting build-up of oxygen at the interface in the form of a P_2O_5 layer (Figure 3.12) or possibly an Al_2O_3 layer; the interfacial oxide

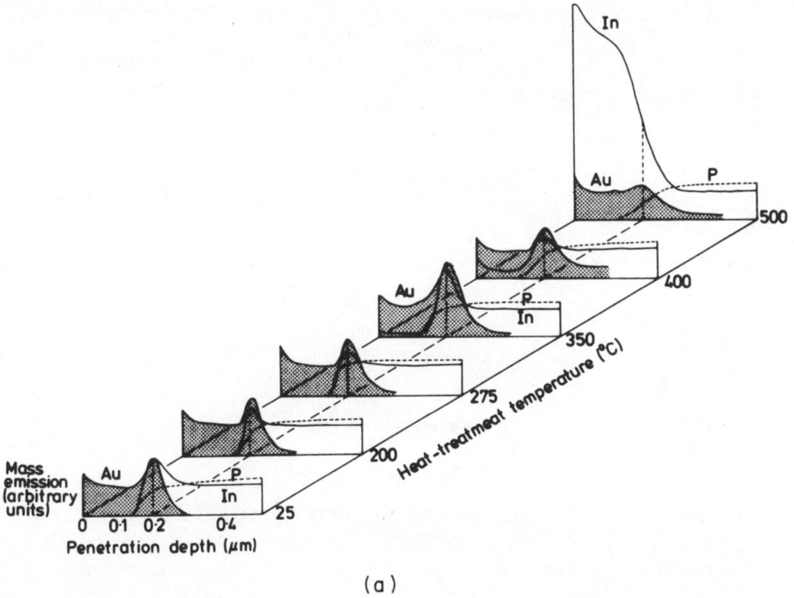

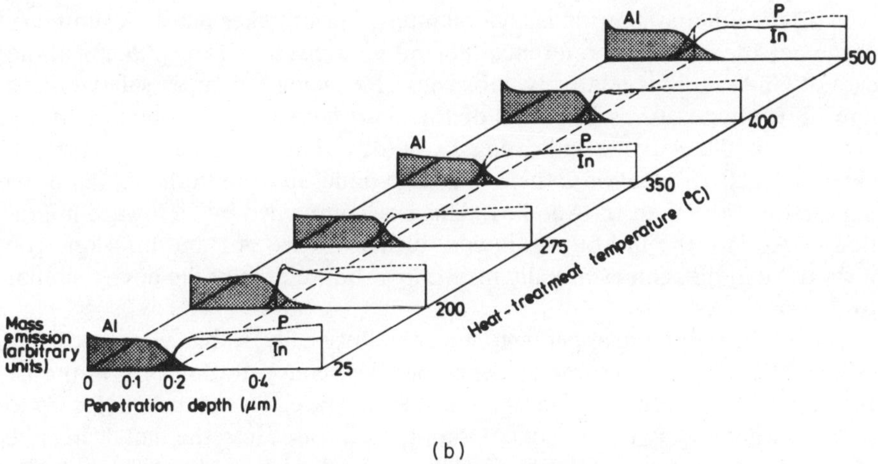

Figure 3.11. (a) Three dimensional composite of SIMS profiles for the Au–InP system. (b) The corresponding SIMS profiles for the Al–InP (after Kim *et al.*[88])

causing the reduction in the diode leakage and hence an increase in the effective barrier height.[90] The same phenomenon is seen in the data of Kim *et al.*[88] where a small but consistent increase in the effective barrier height 0.03 eV measured from the *I–V* characteristics is observed when annealing above 275 °C takes place.

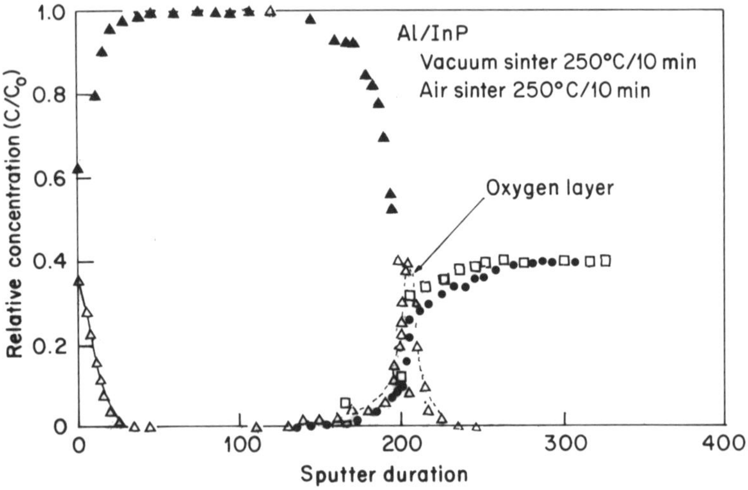

Figure 3.12. Auger electron spectra sputter profile of Al–InP vacuum sintered at 250 °C for 10 min followed by an air sinter of 250 °C (after Christou and Anderson[89])

In all the other systems investigated interdiffusion takes place. A summary of the data derived from references 88 and 89 is given in Table 3.5. As in the case of GaAs, gold is relatively poor with 350° being the upper safe working limit. For temperatures in excess of this limit increasing amounts of In are detected in the Au film indicating In outdiffusion with little detectable P (Figure 3.11(b)). At 500 °C there is a large build-up of In in the Au, indicating the function of an In–Au alloy. This is accompanied by an inward migration of Au into the InP but again very little evidence of P outdiffusion. The inward Au indiffusion eventually results in a short circuit of the device similar to that of GaAs.

The Cr–InP system appears to be metallurgically stable up to 350 °C. Above 400 °C there is some evidence of a low concentration of P in the Cr film and a slight accumulation of Cr at the interface. This picture holds up to 500 °C, when a very low In concentration outdiffuses into the metal. In spite of its apparent thermal stability, Cr will not produce a good Schottky barrier to InP. Even 'as deposited' metals will exhibit a leaky low barrier characteristic. A possible explanation for this could be a limited intermixing of the Cr into the InP to produce interface-deep trapping levels. Such levels are known to have this effect in GaAs Schottky diodes.[81] The observation of a slight concentration of Cr in the InP near to the interface would support this idea.

The two remaining InP metallizations, systems Au–Ni and Au–Ti shown in Table 3.5, were studied as barrier layers to try to avoid the degradation of Au–InP described earlier. The Ni system is of no use as a Schottky barrier, because, like Cr, it forms an ohmic contact prior to heating. With regard to interdiffusion, problems begin to occur above 350 °C. Initially the Au and Ni interdiffuse and the Au is able to penetrate through into the InP. The Ni and Au fully intermix whilst at the same time the In outdiffuses and accumulates at the surface. P also diffuses into the Ni at 350 °C and at high temperatures (>500 °C) it also penetrates the Au layer and accumulates there. As in the case of GaAs this system begins to segregate and form blisters of localized alloys at temperatures >350 °C. Clearly, the Ni functions poorly as a barrier layer and the Au–Ni interdiffusion plays a key role in the contact degradation. This follows a very similar pattern to the degradation of the Ag–Pt interface.[62]

The Au–Ti–InP system is considerably better than the Ni barrier. It does, however, have many similar characteristics. At 350 °C the Au rapidly diffuses through the Ti film and accumulates at the InP interface. Above 400 °C Ti outdiffuses and mixes with the Au film. Although the Au continues to move through the Ti, it is much less than in the Ni case. At around 500 °C the phosphorous outdiffuses and accumulates near to the surface. A summary of the main features of these results is shown in Table 3.5.

3.4. CONCLUSIONS

From the results documented in the literature and reviewed in this chapter, we can begin to draw some general conclusions on the nature of the basic alloying and degradation of metal–GaAs and metal–InP interfaces. Consider firstly the alloying required for ohmic contact formation. The following processes may be identified:

(i) Alloying results in the production of a thin near-surface n^+ (or p^+) doped layer immediately beneath the metal—current transport is then by tunnelling.
(ii) The interdiffusion causing the near-surface doping is highly non-uniform, with localized regions of deeper penetration and precipitation of the interdiffusion products.

The non-uniformity in alloying could well be the cause of the localized degradation of the contact during operation at high current densities and elevated temperatures. Alloying clearly is not the ideal process for obtaining stable uniform metallurgical interfaces. Recent work on the fabrication of ohmic contacts by MBE[4] for example, suggests an alternative technology with contacts of a potentially higher stability and lower contact resistance.

Returning to the more general problem of metal–GaAs and metal–InP interdiffusion, it is possible to identify the following physical processes taking place during thermal stressing for a range of systems:

(i) Ga or As (or both) outdiffusion into the metal film (In or P (or both) in the case of InP).

(ii) Accompanying the process above there may occur an inward migration of the contact metal into the GaAs (or InP).
(iii) Compound formation between the metal and the semiconductor components.
(iv) Oxidation of the surface metal layer.
(v) Inward migrating metal atoms may be active dopants or may produce trapping levels in the GaAs (or InP).
(vi) 'Peel off' of the metal layer.

The degree to which any of these processes may occur in any metal–GaAs or metal–InP system varies greatly for the different systems and a summary of the available data is presented in Tables 3.4 and 3.5.

One technique which would reduce contact degradation is to inhibit Ga (or As or both) outdiffusion through the metal layer. There is strong evidence which suggests that this rapid diffusion process is related to the grain boundaries and hence some of the low temperature diffusion could be inhibited by removing this diffusion path. One possibility is to incorporate some impurity into the grain boundaries which might block them. Alternatively, the growth of epitaxial single crystal metallic layers might also be a way of improving this problem. MBE is certaintly one technique which again offers a potential future solution to this problem. In the case of multilayer metallizations a problem similar in nature occurs. At low activation energy grain boundary type diffusion clearly emerges as a central component of the degradation and would demand a similar solution.

The nature and structure of metal interactions with these two semiconductors are still a long way from being fully understood. It is, however, clear that the foundation of our understanding is being laid and the physical techniques necessary for detailed analysis to be conducted are being developed to the necessary degree of sophistication.

3.5 ACKNOWLEDGEMENTS

I am very grateful to my colleagues Drs. S. D. Mukherjee, C. J. Palmström, M. J. Howes, and J. Wood for their critical comments on this chapter and for allowing me to reproduce data from our joint work.

REFERENCES

1. P. E. Gise and R. Blanchard, *Semiconductor and Integrated Circuit Fabrication Techniques*, Reston, Virginia, 1979.
2. V. L. Rideout, *Solid State Elect.,* **18**, 541 (1975).
3. E. H. Rhoderick, *Metal Semiconductor Contacts*, Clarendon Press, Oxford, 1977.
4. R. Stall, C. E. C. Wood, K. Board, and L. F. Eastman, *Electronic Letters*, 800–801 (1979).
5. J. Gyulai, J. W. Mayer, V. Rodrigues, A. Y. C. Yu, and H. J. Gopen, *J. App. Phys.,* **42**, 3578 (1971).

6. C. J. Madams, D. V. Morgan, and M. J. Howes, *Elect. Lett.,* **11**, 24 (1975).
7. C. J. Palmström, D. V. Morgan, and M. J. Howes, *Nucl. Inst. and Meth.*, **150**, 305 (1978).
8. C. J. Palmström, D. V. Morgan, and M. J. Howes, *Elect. Lett.*, **13**, 504 (1977).
9. S. Y. Chiang and G. L. Pearson, *J. Appl. Phys.,* **46**, 2980 (1975).
10. G. Y. Robinson, *Solid State Electronics,* **18**, 331 (1975).
11. N. Yokoyama, S. Onkawa, and H. Ishikawa, *Jap. J. Appl. Phys.*, **14**, 1071 (1975).
12. K. L. Klone and L. Wardinger, *J. Electrochem.,* **116**, 507 (1969).
13. S. Knight and C. Paola, *Ohmic Contacts to Semiconductors*, Ed. B. Schwartz, Electrochem. Soc., p. 2, 1969.
14. Y. A. Goldberg and B. V. Tsarenkov, *Soviet Phys. Semiconductors,* **3**, 1447 (1970).
15. C. R. Paola, *Solid State Elect.,* **13**, 1189 (1970).
16. J. Basterfield, M. J. Josh, and M. R. Burgess, *Acta Electronica,* **15**, 83 (1972).
17. Braslau, J. B. Gunn, and J. L. Staples, *Solid State Electronics,* **10**, 381 (1967).
18. R. H. Cox and T. E. Hasty, *Ohmic Contacts to Semiconductors*, Ed. B. Schwartz, Electrochem. Soc., p. 88, 1969.
19. M. McColl, M. F. Mitten, and C. A. Mead, *Solid State Electronics,* **14**, 667 (1971).
20. E. D. Edwards, W. A. Hartman, and A. R. Torrens, *Solid State Electronics,* **15**, 387 (1972).
21. A. K. Sinha, T. E. Smith, and H. J. Levinstein, *IEEE Trans. Vol. Ed-22*, **218**, (1975).
22. H. Matino and M. Tokunaga, *J. Electrochem. Soc.,* **116**, 709 (1969).
23. B. W. Hakki, *Int. Solid State Circuits Conf.,* p. 44, February 1965.
24. W. A. Schmidt, *J. Electrochem. Soc.,* **113**, 860 (1966).
25. T. B. Romachandran and R. P. Santasuosso, *Solid State Electronics,* **9**, 733 (1966).
26. T. Sebastian, M. Menyhard, and D. Szigethy, *Elec. Lett.,* **12**, 96 (1976).
27. R. H. Cox and H. Strack, *Solid State Electronics,* **10**, 1213 (1967).
28. T. Hasty, R. Stratton, and E. L. Jones, *J. App. Phys.,* **39**, 4623 (1968).
29. T. Sebastian, H. K. Hartnagel, and L. H. Herron, *Elec. Lett.*, **10**, 372 (1974).
30. T. Sebastian, H. L. Hartnagel, and L. H. Herron, *IEEE Trans.,* Ed-22, 1073 (1975).
31. B. L. Weiss and H. L. Hartnagel, *Elec. Lett.,* **11**, 263 (1975).
32. A Christou and K. Sleger, *Proc. 6th Biennial Cornell Conf.*, p. 169, 1977.
33. T. G. Finstad, T. Andreassen, and T. Olsen, *Thin Solid Films,* **29**, 145 (1975).
34. H. B. Kim, G. G. Sweeney, and T. M. S. Heng, *Inst. Physics Conf. Series*, No. 24, p. 307, 1975.
35. T. G. Finstad, *Thin Solid Films,* **47**, 279 (1977).
36. K. L. Cohen and L. Wandinger, *J. Electrochem. Soc.,* **116**, 507 (1969).
37. H. J. Gopen and A. Y. C. Yu, *Solid State Electronics,* **14**, 515 (1971).
38. A. M. Andrews and N. Holonyak, *Solid State Electronics,* **15**, 601 (1972).
39. W. J. Moroney and Y. Anand, *Inst. Phys. Conf. Ser. No. 9,* **259** (1971).
40. B. R. Pruniaux, *J. Am. Phys.,* **42**, 3575 (1971).
41. K. Heime, U. Konig, E. Kohn, and A. Wortmann, *Solid State Electronics,* **17**, 835 (1974).
42. T. E. Loveluck, G. M. Rackham, and J. W. Steed, *Inst. Phys. Conf. Series*, No. 36, 297 (1977).
43. J. S. Harris, Y. Nannichi, G. L. Pearson, and G. F. Day, *J. Appl. Phys.,* **40**, 4575 (1969).
44. G. Y. Robinson and N. L. Jarvis, *App. Phys. Lett.,* **21**, 507 (1972).
45. K. Ohata and M. Ogawa, *Proc. 12th Annual Proc. on Rel. Physics*, p. 278 (1978). Published by IEEE.

46. M. Otsube, H. Kumabe, and H. Miki, *Solid State Electronics,* **20**, 617 (1977).
47. B. Jeppson and I. Marklund, *Elect. Lett.,* **3**, 213 (1967).
48. R. J. Tree, M. J. Josh, and C. T. Foxon, *Solid State Electronics,* **14**, 519 (1971).
49. I. Kuru, *J. Jap. Soc. Appl. Phys.,* **46** (suppl.), 137 (1971).
50. M. N. Al-Moufti, S. V. Jaskolski, and T. K. Ishii, *Proc. IEEE,* **56**, 236 (1968).
51. T. E. Walsh, *Proc. 11th Annual Proc. Reliability Phys. Symp*., p. 282, 1973. Published by IEEE.
52. T. B. Ramachandram, J. L. Heaton, and E. B. Hakin, *Proc. 12th Annual Proc. Rel. Sym. Conf.,* p. 284, 1974. Published by IEEE.
53. P. Staecker, *IEDM Technical Digest*, p. 493, 1973.
54. K. Chino, Y. Wada, and K. Fukuda, *J. Jap. Soc. Appl. Phys.,* **44** (suppl.), 149 (1975).
55. K. Chino and Y. Wada, *Jap. J. Appl. Phys.,* **16**, 1823 (1977).
56. A. Brodzinsky and J. E. Davey, 'GaAs reliability programs in the U.S.', *NATO Symp. on Microwave Components*, Belgium, 1977.
57. H. M. Macksey, *Inst. Phys. Conf. Series,* **33b**, 254 (1977).
58. D. A. Abbot, J. A. Turner, *Inst. Phys. Conf. Series,* **33a**, 225 (1977).
59. H. F. Cooke, 'Small signal GaAs FET problems', *Army ECOM Contract DAA B07-76-C-1340-I*, at Avantek Inc.
60. Y. Furukawa, M. Shinoda, Y. Toyama, M. Yamamoto, S. Yoshioka, and K. Kazetani, *10th Annual Proc. Reliability Phys. Symp*., p. 64, 1972.
61. B. S. Hewitt, H. M. Cox, H. Fukui, J. V. Di Lorenzo, J. V. Schlesser, and D. E. Iglesias, *Inst. Phys. Conf. Series,* **33a**, 246 (1977).
62. S. D. Mukherjee, D. V. Morgan, and M. J. Howes, *J. Electrochem. Soc.,* **126**, No. 6, 1047 (1979).
63. M. Ogawa, D. Shinoda, N. Kowamura, T. Nozaki, and S. Asanabe, *N.E.C. Res. Dev.,* **22**, 1–7 (1971).
64. A. K. Sinha and J. M. Poate, *Japan J. App. Phys. Supplement,* **2**, 841–844 (1974).
65. A. K. Sinha, T. E. Smith, M. H. Read, and J. M. Poate, *Solid State Electronics,* **19**, 489–492 (1976).
66. A. K. Sinha and J. M. Poate, *Appl. Phys. Lett.,* **23**, 666–668 (1973).
67. C. C. Chang, S. P. Murarka, V. Kumar, and G. Quintana, *J. Appl. Phys.,* **46**, 4237–3243 (1975).
68. V. Kumar, *J. Phys. Chem. Solids,* **36**, 535–541 (1975).
69. D. J. Coleman Jr., W. R. Wisseman, and D. W. Shaw, *Appl. Phys. Lett.,* **24**, 355–357 (1974).
70. J. C. Irwin, D. J. Coleman Jr., W. A. Johnson, J. Tatsuguchi, D. R. Dekker, and C. N. Dunn, *Proc. IEEE,* **59**, 1212–1215 (1971).
71. R. B. Marcus, S. E. Haszko, S. P. Murarka, and J. C. Irvin, *J. Electrochem. Soc.,* **121**, 692–694 (1974).
72. D. V. Morgan and S. D. Mukherjee, *Internal University of Leeds Report* (for Science Research Council), 1977.
73. M. Hansen, *Constitution of Binary Alloys, 2nd Supplement* (revised by R. P. Elliott, McGraw-Hill, New York, 1965).
74. A. Christou and H. M. Day, *J. Appl. Phys.,* **4**, 4217–4219 (1976).
75. D. V. Morgan, M. J. Howes, S. D. Mukherjee, D. J. Taylor, and P. Brook, *Solid State and Electron Devices,* **1**, 158–160 (1977).
76. S. D. Mukherjee, D. V. Morgan, C. J. Palmström, and J. Smith, submitted for publication, *J. Vac. Sci. Tech*. (1980).
77. O. Wada, S. Yanagisawa, and H. Takanishi, *Appl. Phys. Lett.,* **29**, 263–265 (1976).
78. J. J. Berenz, G. J. Scilla, V. L. Wrick, L. F. Eastman, and G. H. Morrison, *J. Vac. Sci. Technol.,* **13**, 1152–1157 (1976).

79. A. K. Sinha, *Appl. Phys. Lett.,* **26**, 171–173 (1975).
80. B. J. Sealy and R. K. Surridge, *Thin Solid Films,* **26**, L19–L22 (1975).
81. N. M. Johnson, T. J. Magee, and J. Peng, *J. Vac. Sci. Tech.,* **13**, 838 (1976).
82. C. J. Todd, G. W. B. Ashwell, J. D. Speight, and R. Heckingbottom, *Inst. Physics Conf. Series No. 22*, p. 171, 1974.
93. T. J. Magee and J. Peng, *Phys. Stat. Sol.,* **32a**, 895 (1975).
84. S. Guha, B. M. Arora, and V. P. Salvi, *Solid State Electronics,* **20**, 431 (1977).
85. A. Hiraki, A. Shato, S. Kim, W. Kamura, and M. Iwami, *App. Phys. Lett.,* **31**, 61 (1977).
86. M. Ogawa, D. Shinoda, N. Kawamura, T. Nozaki, S. D. Muke, and S. Asanabe, *Nippon Electric Co. Res. and Dev. Rept.,* **22**, 1 (1971).
87. S. D. Muraka, *Solid State Electronics,* **17**, 869 (1974).
88. H. B. Kim, A. F. Lovas, G. G. Sweeney, and T. M. S. Heng, *Inst. Phys. Conf. Series,* **33b**, 145 (1977).
89. A. Christou and W. T. Anderson, *Solid State Electronics,* **22**, 857 (1979).
90. D. V. Morgan and J. Frey, *Solid State Electronics,* **22**, 865 (1979).
91. P. D. Taylor and D. V. Morgan, *Solid State Electronics,* **19**, 481 (1976).

Reliability and Degradation
Edited by M. J. Howes and D. V. Morgan

CHAPTER 4

Reliability and degradation of silicon devices and integrated circuits

J. WOOD

GLOSSARY OF ABBREVIATIONS USED IN THE TEXT

BLSJ	Beam lead sealed junction
CVD	Chemical vapour deposition
HMOS	High-performance MOS
LSI	Large scale integration: 100–1000 gates per chip
MNOS	Metal–nitride–oxide–silicon
MOS	Metal–oxide–silicon
MSI	Medium scale integration: 10–100 gates per chip
MTBF(MTTF)	Mean time before failure
PSG	Phosphosilicate glass
RAM	Random access memory device
SEM	Scanning electron microscope
SOS	Silicon-on-sapphire
SSI	Small scale integration: 1–10 gates per chip
TEM	Transmission electron microscope
TTL	Transistor–transistor logic device
VLSI	Very large scale integration: >1000 gates per chip
VMOS	V-groove MOS

4.1 INTRODUCTION

Solid state semiconductor devices are nowadays commonplace items finding many applications in complex electronic systems. The use of silicon for these devices is pre-eminent and it seems likely that its use will continue to be almost universal except for a few special applications such as optoelectronics or microwave amplifiers where gallium arsenide and other III-V semiconductor devices are in ascendency. The advantages of silicon for all other aspects are low cost and a well founded technology. Vast numbers of discrete and integrated devices are marketed and their reliability to the potential user must be assured. This is especially important, for example, in electronic telephone switching systems and computer memory where hundreds of thousands of devices are in constant operation, and so an extremely low failure rate per item must be guaranteed. Consequently, the field of reliability physics has become a very important discipline acting as a feedback loop between the servicing and quality control, and the device design and processing technologies.

The study of reliability of silicon devices and integrated circuits is well documented in relevant literature.[1–9] This review presents a survey of the failure mechanisms found in silicon devices and integrated circuits, and of the reliability problems encountered in large scale integration.

4.1.1 Accelerated stress testing

In order to investigate the degradation of (new) devices, they are artificially 'aged' using accelerated stress testing procedures to determine potential sources of failure which would curtail the operating life.[8,10–15] Generally, these tests may take such forms as, for example:

(i) Operation at rated currents and voltages at elevated temperatures e.g. 125 °C, 250 °C;
(ii) THB—temperature, humidity, bias testing: operating at rated conditions at 85 °C per 85 per cent relative humidity;
(iii) Temperature cycling, over a range perhaps as wide as −55 °C to +125 °C, to introduce mechanical stresses and strains.

These tests will be performed upon many devices over considerable periods of time to obtain statistically significant failure distributions. The most commonly found distribution is the 'bathtub' curve, Figure 4.1. The high initial failure rate—the infant mortality period—is due to 'freaks' in the population. These are devices which possess relatively coarse defects or have been damaged in manufacture, and their lifetime is consequently short. The failure rate is falling with time which is characteristic of weak samples. The use of burn-in procedures prior to shipment is designed to screen out these freaks and thereby assure the reliability of the devices in service.

The remainder of the curve indicates the failures in the main population due to intrinsic degradational processes. The flat part of the curve is characteristic of random failures, and the rising tail is the wearout phase indicating the median active life of the devices. The wearout is due to the intrinsic degradation which can be exposed by long term accelerated stress testing. The failed devices are analysed to determine the modes and mechanisms of the failures. The SEM has proved to be an invaluable analysis tool.[16,17,18] It is often backed up by more advanced techniques such as Rutherford backscattering,[18,19,20] Auger spectroscopy[18,21] and microprobe analysis[18,22,23] for chemical

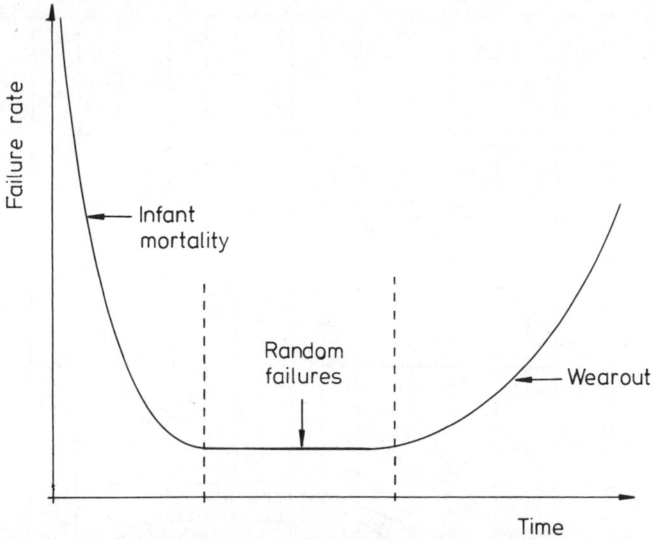

Figure 4.1. Semiconductor failure rate as a function of time. This curve shows the three regimes of failure rate due to early or freak failures, the random failures and the wearout phase

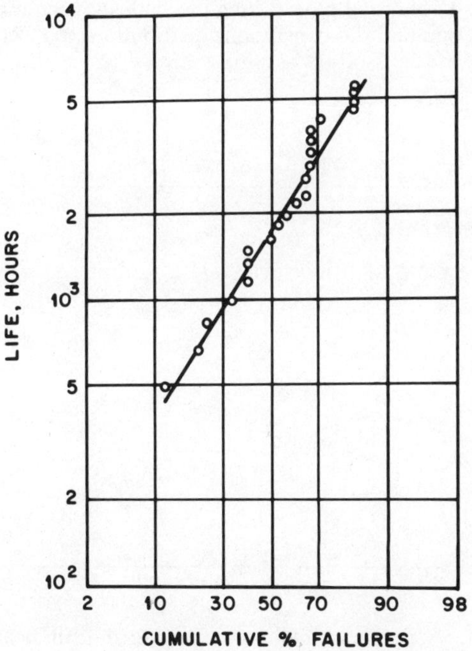

Figure 4.2. Lognormal plot of long life data showing failure distribution due to a single mode of failure. © 1971 IEEE[14]

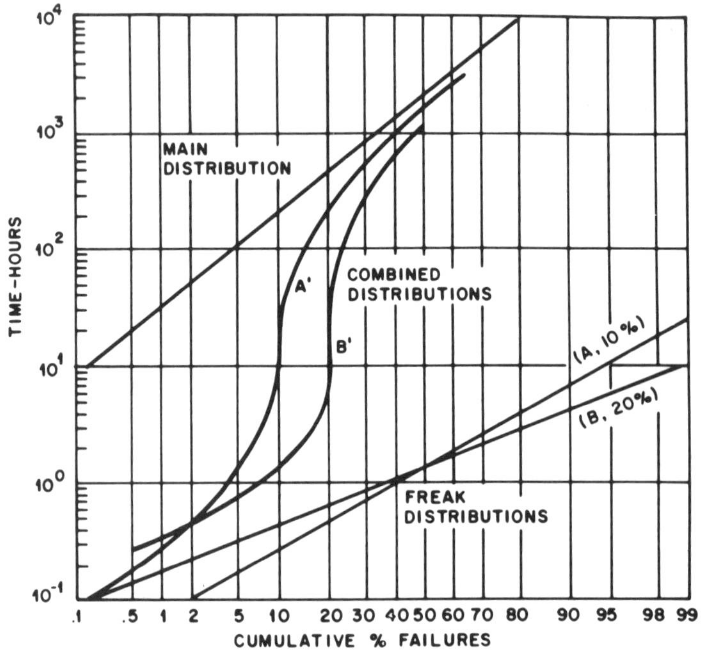

Figure 4.3. Lognormal plot of long life data showing bimodal failure distribution and the constituent distribution. © 1971 IEEE[14]

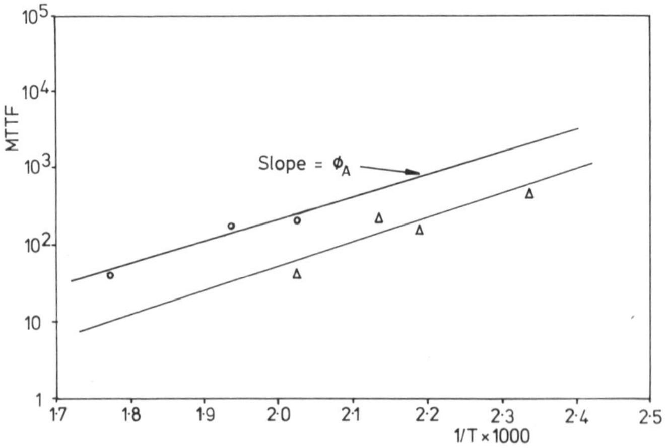

Figure 4.4. Arrhenius plot of median life data indicating the acceleration due to the activation energy ϕ_A of the failure mechanism

identification of contaminants and alloying and corrosion products. The understanding of the mechanisms of the failures enables the device design and processing to be improved, and aids the determination of which screening or burn-in procedures are effective in improving the device reliability.

The failure data may be expressed in an alternative form, Figure 4.2, which shows the cumulative failures against the logarithm of the time-to-failure: the lognormal failure distribution. In practice such a clean distribution is seldom seen, and Figure 4.3 is much more typical of accelerated stress testing and shows the bimodal distribution due to the freak and main populations. This corresponds to the bathtub curve. Having established the failure distributions for several stress levels representing a common failure mechanism, the lifetime of the device under normal operation may be predicted. This is done by extrapolation of an Arrhenius' plot of the median lifetimes at high stress level, to the operating temperature (Figure 4.4). The activation energy associated with the failure mechanism can be found from the slope of this curve. This energy indicates how susceptible the device may be to accelerated failure due to higher operating temperatures.

4.1.2 Failure mechanisms in silicon devices

A compilation of the more significant failure mechanisms is shown in Table 4.1. It shows which factors contribute to the failure process and also which factors may be increased to accelerate the stress and increase the failure rate. It also includes the activation energy of the process. These failure modes will occur in different relative proportions in devices, depending upon the type of device, its particular operating conditions, and the manufacturer.

Many of the failure processes described by Table 4.1 have been investigated and design and processing techniques subsequently implemented to prevent or reduce their occurrence. In some cases this has led to a complete change in the device processing in order to improve the reliability significantly. An example is the metallization and bonding problems associated with the aluminium tracks in silicon planar transistors. Screening has reduced the incidence of failures, and the change to a titanium–platinum–gold beam-lead system has produced a different selection of failure mechanisms, and an overall improvement in the reliability. The failure mode distributions are shown in Figure 4.5 and the absolute reliability is shown in Figure 4.6 which is the mean time-to-failure of the transistor. This figure shows how the improved packaging and progressively severe screening decreases the failure rate, and the curve for the beam-lead sealed junction transistors show the effect of new technology which is designed to reduce or eliminate specific failure modes.

Integration of many devices onto a single chip offers simultaneously increased performance and reliability, and lower cost. The reliability is improved by a reduction in the number processing steps and interconnections

Table 4.1 Time dependent failure mechanisms in semiconductor devices

Device association	Process	Relevant factors	Accelerating factors	Apparent activation energy ϕ_A: acceleration
Silicon oxide Silicon–silicon Oxide interface	Surface charges accumulation	Mobile ions $V\,T$	T	Bipolar: $\phi_A = 1.0$–1.05 eV MOS: $\phi_A = 1.2$–1.35 eV
	Dielectric breakdown	$E\,T$	E	
	Charge injection	$E\,T$	$E\,T$	$\phi_A = 1.3$ eV—slow trapping
Metallization	Electromigration	$T\,j\,A$ Gradients of T,j Grain size	$T\,j$	$\phi_A = 0.5$ to 1.2 eV (j to j^4)
	Corrosion: chemical galvanic electrolytic	Contamination Humidity $V\,T$	$H\,V\,T$	Strong humidity effect $\phi_A \approx 0.3$ to 0.6 eV (for electrolysis) V may have thresholds
	Contact degradation	T metals impurities	Varied	
Bond and other mechanical interfaces	Intermetallic growth	T impurities Bond strength	T	Al–Au = $\phi_A = 1.0$ to 1.05 eV
	Fatigue	Temperature cycling Bond strength	T extremes in cycling	
Hermeticity	Seal leaks	Pressure diffential Atmosphere	Pressure	

V Voltage; E Electric field; A Area; T Temperature; j Current Density; H Humidity

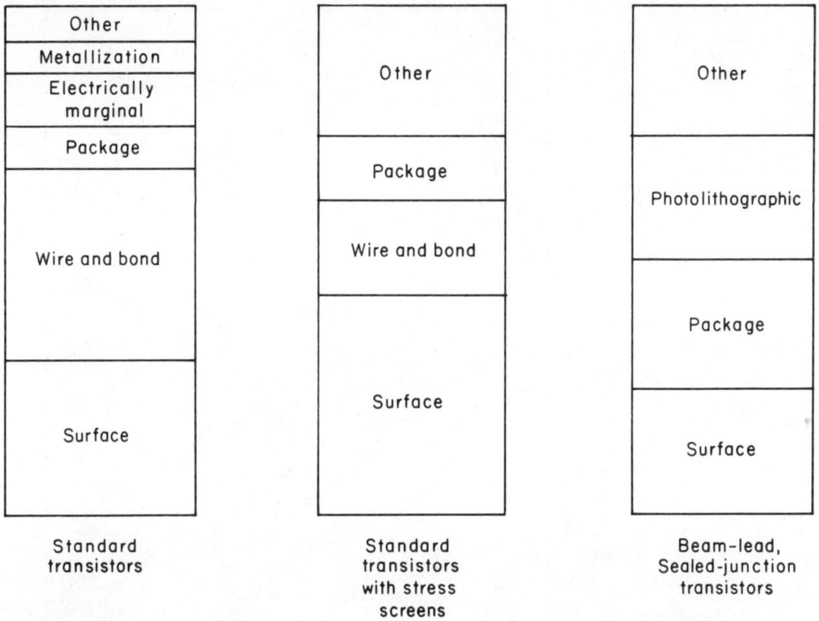

Figure 4.5. Failure mode distribution for epitaxial planar silicon transistors as modified by stress screening and a change in process technology. © 1974 IEEE[2]

in the system. However, as the trend towards greater integration continues the packing density is increased and the metallization linewidth decreased. This will introduce new failure mechanisms, e.g. parasitic transistor action between adjacent elements and may also exacerbate existing ones such as electromigration in the very fine aluminium lines.[24,25]

The failure mechanisms described below fall into the general categories of surface and bulk effects in the semiconductor and the oxide, metallization and metal–semiconductor reactions, failures arising from increasing integration, and package related failures. The measures commonly employed to combat these problems have also been included.

4.2 BULK AND SURFACE RELATED PHENOMENA

These are the failures which arise from defects or impurities within the silicon or oxide layers. The failures are observed as a drift out of specification by the device, which is still operable, and may be recoverable, or as a catastrophic breakdown.

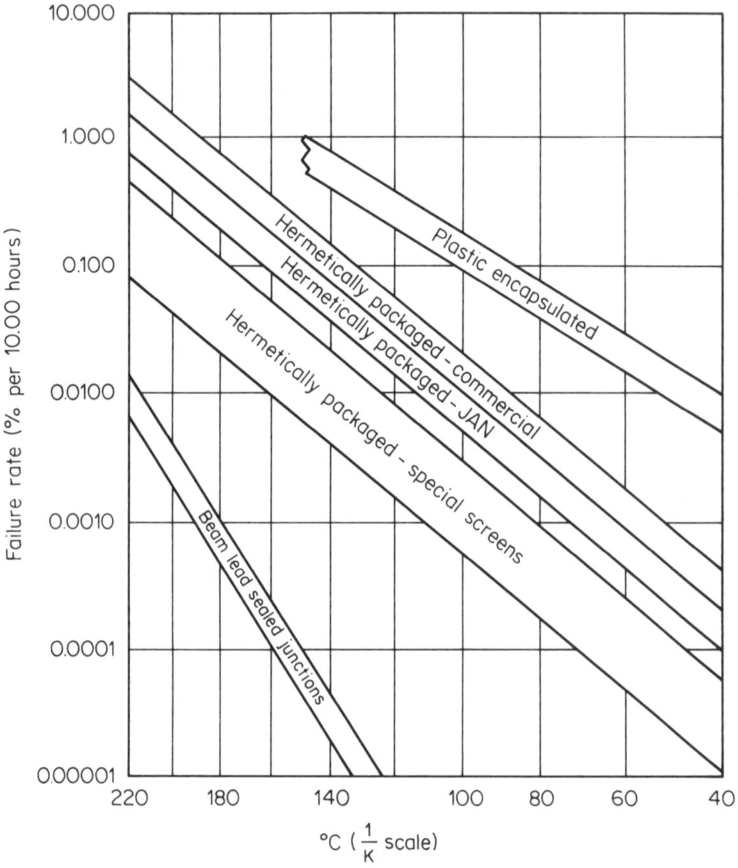

Figure 4.6. Failure rates for epitaxial planar silicon transistors. This shows decreasing failure rate in going from plastic-encapsulated transistors to hermetically packaged transistors. The effect of improved screening is also apparant. The change in process technology indicates much improved failure rate. © 1974IEEE[2]

4.2.1 Bulk defects

Gross defects such as cracks or pits in the chip of passivation come into this category. They are in general caused by mishandling of the chip, especially in the scribing and packaging processes. Mismatch between the thermal expansion coefficients of silicon and the glass passivation introduces stress and strain during subsequent thermal processing which may result in cracks or breaks. Defects of this kind tend to be yield rather than reliability problems as they are easily spotted during visual inspection of the device chips, as shown in Figure 4.7.

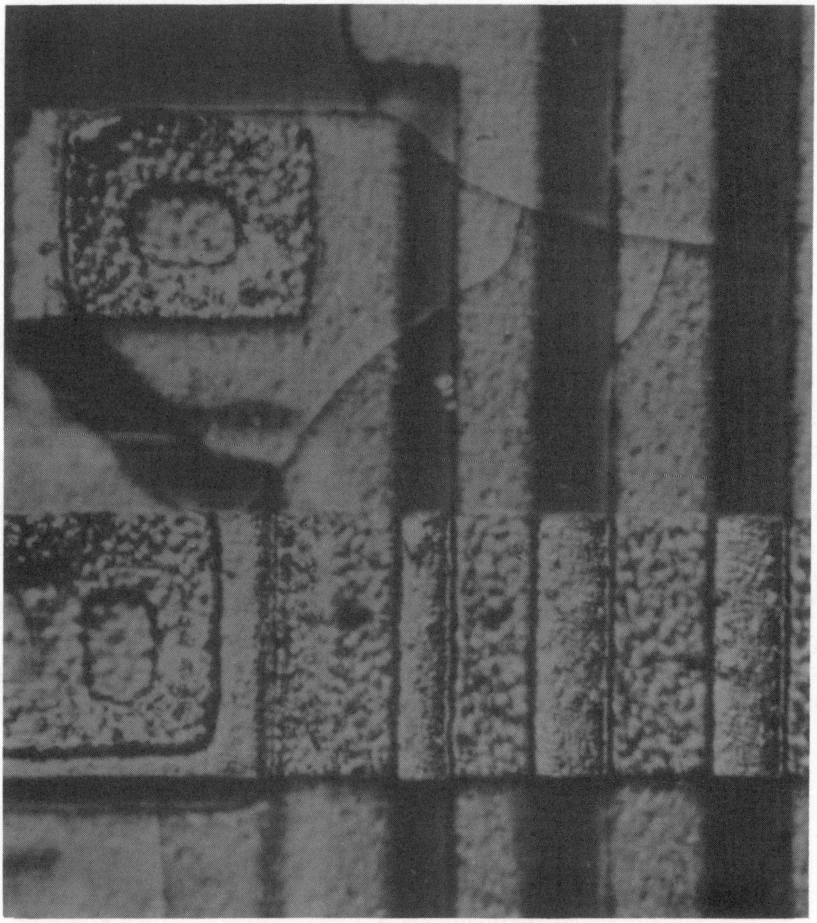

Figure 4.7. Cracked interlevel dielectric between two metallization layers[7]

Crystalline defects in the silicon tend to affect devices which operate in the avalanche mode by acting as generation–recombination centres, which enhance the diffusion of impurities. The defects may arise as a result of incomplete annealing of the damage caused by ion implantation. High temperature processes subsequent to the emitter diffusion or implantation, or operation at elevated temperatures may cause the dopants to diffuse, giving junction curvature and non-uniform current densities. These phenomena may in turn result in degradation of the device and eventual failure. The processes of dopant implantation and diffusion into silicon are well characterized however, and are not major reliability problems.

4.2.2 Gate oxide breakdown

Some failures are attributable to electrical shorts through the thermal oxide between the metallization and the silicon. The shorts may occur during operation within rated conditions, or as a result of electrical overstress. Those occurring within rated conditions are caused by localized defects such as pinholes or thin spots in the oxide, which may arise during processing. The problem is greater with MOS devices since they have a thinner oxide layer and operate at higher voltages than linear or TTL devices. The gate oxide thickness in MOS devices may be 1000 Å or less, and applied potential up to 30 V, which is a significant fraction of the breakdown voltage for the oxide. The presence of defects or impurities, or a local region of thin oxide will result in a lower dielectric strength and the device may then break down at low voltages. It is clear from Figure 4.8 that as the area of a device increases, then breakdown is more likely to occur at low voltages, due to the defects. This means that the reliability of the breakdown of the very small devices in LSI chips will be improved compared with discrete devices.[7] Electrical screening is performed at voltages higher than rated value to eliminate those devices which have regions of oxide of only marginal quality.

Because MOS devices have very high input impedance and thin gate oxides they are prone to static electricity effects and voltage transients which can exceed the maximum rated voltage and cause breakdown of the oxide. This problem must be countered by incorporating protection circuitry in the input

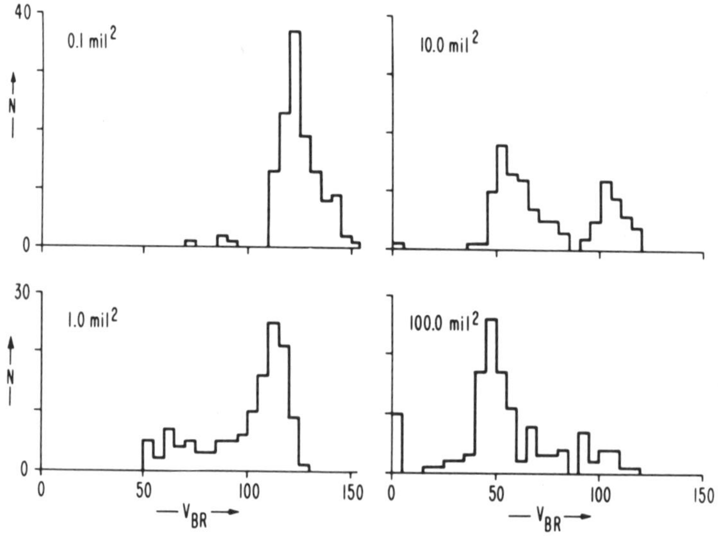

Figure 4.8. Plot of distribution of oxide breakdown voltages of different area devices[7]

of the device.[7,26] The purpose of the protection circuit is to clamp the gate to substrate potential at a safe maximum value under conditions of voltage overstress applied with either fast or slow risetime, and absorb large pulse energies. It must not affect the normal operation of the device or introduce any significant propagation delay into the circuit. Its design and fabrication must be compatible with the on-chip technology. Gate protection techniques include spark-gap devices, where the distance between two electrodes on the silicon dioxide surface are such that breakdown will occur if the safe voltage is exceeded. This technique suffers from oxide defects and surface contamination problems and more sophisticated methods are generally employed. Some typical device configurations are shown in Figure 4.9. These may include reverse biased *p–n* junctions with controlled breakdown characteristics. The breakdown must be sharp, so avalanche diodes are used. However, the

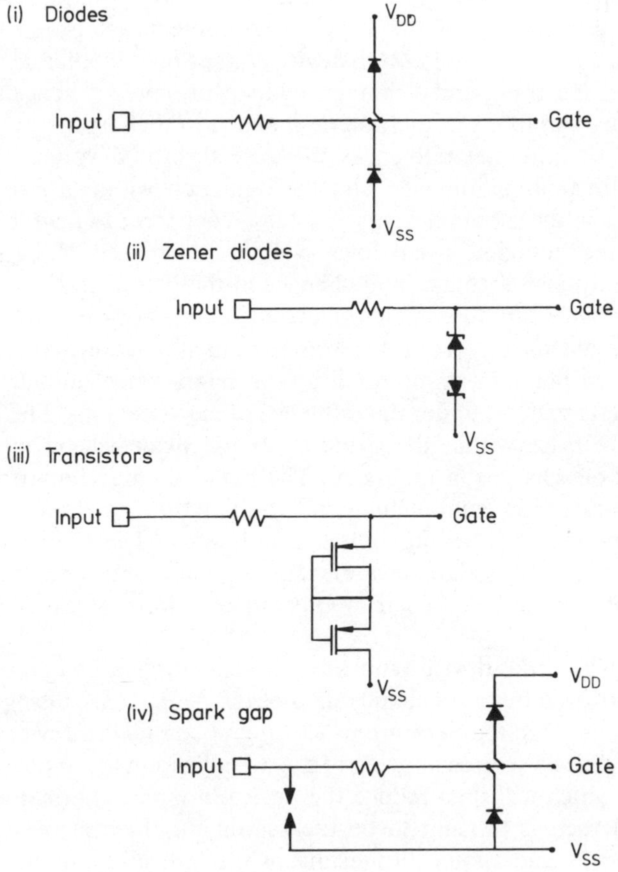

Figure 4.9. Gate protection Networks

breakdown voltage of avalanche devices is apt to 'walkout' under operation to higher voltages, and consequently the gate may be left unprotected. Other circuits include diodes and distributed (implanted) resistors, and transistor switches. These devices are so designed that the maximum acceptable voltage corresponds to the punchthrough voltage. Therefore any overstress will switch on the device which will conduct and absorb power affording protection to the gate. Low-doped devices are used so that the stored charge is low for rapid switching operation.

4.2.3 Surface instability

Many of the device failures can be attributed in some way to the instability of the potential at the silicon–silicon oxide surface: the surface potential. It is the surface potential which determines whether or not a conducting channel exists. Consequently any drift or variation in this potential will tend to degrade the performance of the device. The problems associated with surface instability are more severe in MOS devices rather than bipolar, as they have a lower doped substrate, and a thinner oxide layer, and are hence more easily influenced by variations in the potential and stored charge.

The surface potential influences the gate threshold voltage, the carrier generation–recombination rate, and the depletion width of p–n structures. The changes in these parameters will become apparent as degradation of the DC gain and junction breakdown voltage, increased leakage currents, source–drain punch-through, and changes in the gate threshold voltage and transconductance due to channel inversion. Lateral co-operative effects between adjacent emitters leading to parasitic transistor action may also be seen.

The surface potential is modified by the migration of mobile ions in the oxide or on its surface, under the influence of electrical bias. The mobile ions may activate traps within the oxide or at the oxide–silicon interface, and polarize dipole charges in the oxide. The most common impurity is sodium ions, which can arise from etching and washing, or as a result of resistance-heating deposition of the aluminium metallization. The sodium ions have a high mobility in silicon dioxide at elevated temperatures, and drift under the influence of electric field. Their mobility is enhanced by crystal defects and damage.

In cases where the device failure is not catastrophic, functional operation may be recovered by thermal annealing of the defects, dispersing the sodium ions. However, high temperature processing of completed devices may introduce other defects such as alloying of the metallization and oxide, or mechanical strain, which will then reduce the device lifetime. Thermal annealing of the ionic defects is thought to be the reason for the apparent recovery of 'failed' devices undergoing life-testing, which are allowed to cool without bias.

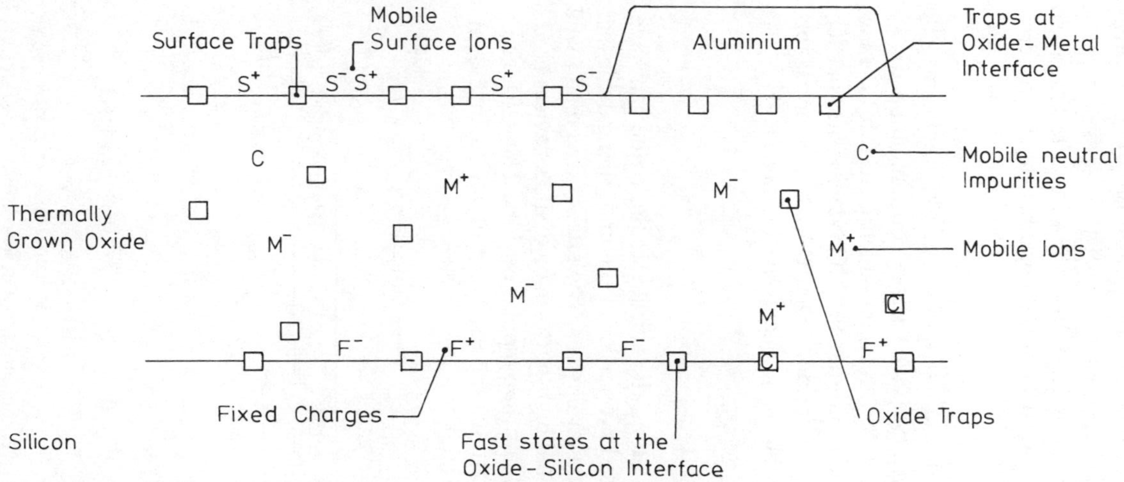

Figure 4.10. Schematic cross section of silicon–silicon dioxide interface showing locations of mobile ions, trapping centres, dipoles interface states, surface ions. A further level of dielectric can lead to more interfacial traps, but may getter the mobile ions, in the case of the PSG

The use of P_2O_5 or phosphosilicate glass (PSG) as a passivation layer over the thermal oxide is often used to getter the sodium ions: they are then fixed and unable to migrate through the oxide. Bipolar devices are also much less susceptible than MOS to ion migration effects due to the growth of PSG layers during the vapour phase emitter diffusion process, which is performed in an oxidizing atmosphere. These areas then serve to getter the sodium impurities. Failures resulting from ion migration effects may still occur even in passivated devices due to photolithographic or etching errors which give rise to misaligned or thin PSG layers and consequent incomplete gettering of the sodium ions.[27]

The use of electron-beam evaporation of aluminium may be preferred to resistance-heating which can give surface contamination. However, the electron beam causes soft X-ray emission which introduces damage in the oxide layer in the form of traps and defects. MOS devices are particularly prone to damage due to ionizing radiation.[28] Figure 4.10 summarizes the complexity of charge centres due to the oxide encapsulation layer.

4.2.4 Surface ion migration

The migration of ions along the surface of the oxide can have an enormous effect upon the electrical properties of the underlying oxide–silicon interface, in all types of silicon planar devices. The effects are greater when the oxide is thin. Therefore MOS and high-performance MOS devices tend to suffer more from surface ion problems than bipolar devices. The existence of fast states at the silicon surface promotes the existence of conducting channels, either by

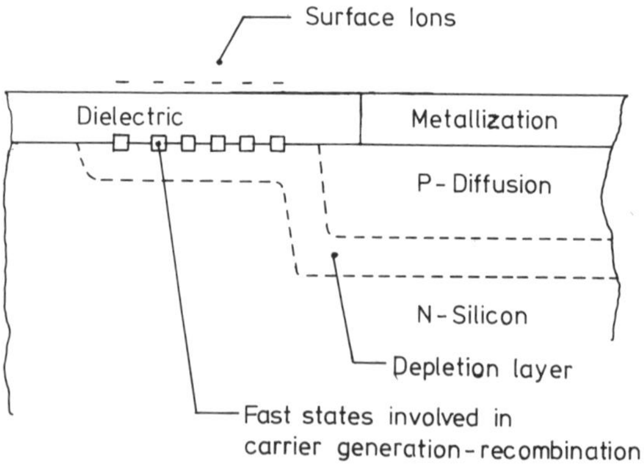

Figure 4.11. Schematic of activation of fast states at the silicon–silicon dioxide interface by mobile surface ions forming a conductive channel. The *p–n* junction width is increased and the leakage current increases

inversion of the surface layer or by trap assisted conduction. The effect is most prominent at diffused *p–n* junctions, where the channel increases the leakage current and reduces the breakdown voltage, thereby degrading the diode performance.[4,5,7,9] The formation of such a channel is shown in Figure 4.11.

Localized defects in the oxide film may be activated by surface ion migration, resulting in catastrophic breakdown of the device. The sodium ions will migrate towards the metallization in the vicinity of reverse biased *p–n* junctions and hence the depletion region will tend to widen at the surface under these unmetallized areas. Any defects in this incremental area will be activated leading to excessive leakage and breakdown of the junction (Figure 4.12).

The surface ions can also contribute to surface leakage currents by creating a conductive path between adjacent metal lines. This may take the form of an electrolytic process involving the corrosion of the metallization, which will result in the formation of voids in the metal, and hence device failure.

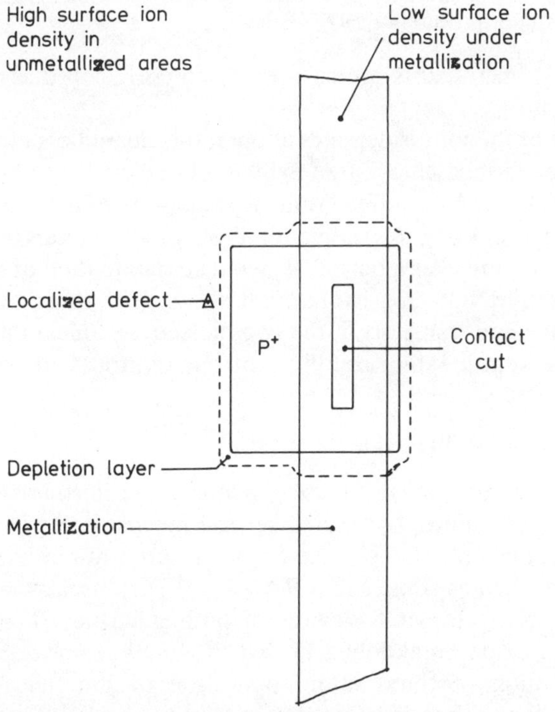

Figure 4.12. Schematic of increasing depletion width resulting in failure when it comes into contact with a local defect

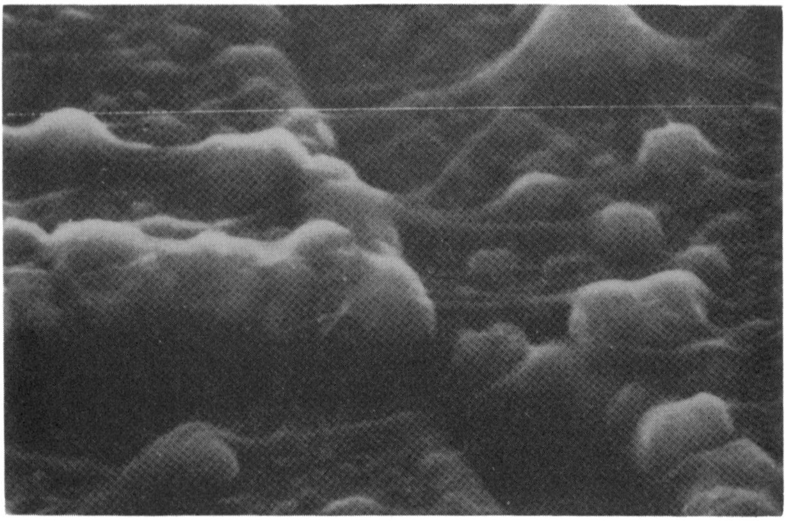

Figure 4.13. Contamination on surface due to debris which is generated by the laser scribing process. © 1977 IEEE[11]

Implanted thin film resistors are susceptible to corrosion due to surface ion migration effects.

The mobility of the ions is dependent upon the cleanliness of the surface of the oxide. Laser-scribed chips often exhibit a large amount of surface debris and have been shown to suffer from degradation induced by surface ion effects[11] (see Figure 4.13). A passivation layer of aluminium oxide or silicon nitride may be employed as a barrier to avoid contamination of the completed device. However, MNOS transistors can have unstable threshold voltage due to charge storage and transport in the two dielectrics, unless the conductivity of the nitride is kept low by carefully controlled fabrication.[7,29]

4.2.5 Ion migration in the oxide

The mobility of sodium and also hydrogen ions in the silicon oxide is high and can lead to device failure, by one of several mechanisms. Time-dependent dielectric breakdown phenomena are known to present a reliability problem in large systems such as static MOS RAMs.[30–34] Such breakdowns have been observed in silicon-gate MOS devices of both polarities. They differ from field-dependent oxide breakdowns in that the breakdowns occur with a log-normal distribution of times after application of the field. A dominant mechanism in time-dependent breakdowns is the emission of sodium ions from the metal–oxide interface under positive bias.[32] The ions then migrate under influence of the voltage towards the oxide–silicon interface where they

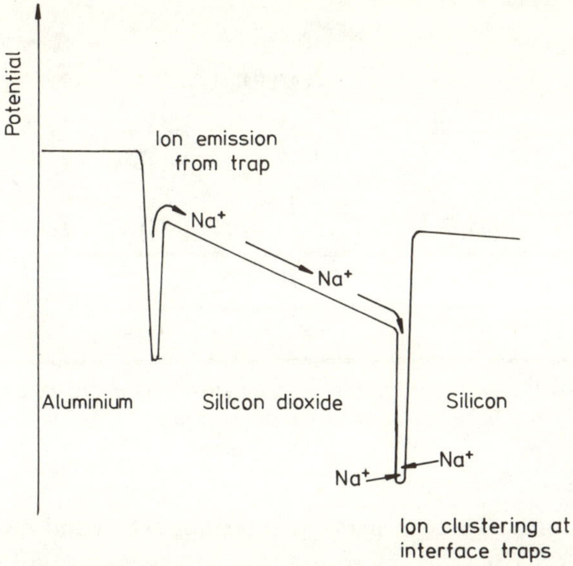

Figure 4.14. Time dependent dieletric breakdown: sodium ion emission from the gate metal drifts towards the silicon–silicon oxide interface. The build up of charge in the interface traps result in oxide breakdown

cluster around crystal defect sites. This is shown schematically in Figure 4.14. The density of ions may build up until the oxide breaks down, although if the defect or trap density at the interface is low other degradation effects may be observed as a result of the ion concentration at the interface. As a time delay occurs before the log (time) build-up of the breakdown, these potential failures can go undetected by the usual voltage overstressing tests, and hence are difficult to screen from the main population of devices.

The build-up of sodium ions at the oxide–silicon interface tends to reduce the field inversion voltage in n-channel transistors, resulting in a narrowing or even inversion of the channel giving an anomalous current flow. p-Channel devices are also susceptible to cation migration. Any positive charge build-up will increase the conducting channel width and will thus not tend to degrade device performance unless the alkali ion contamination is present in large amounts, and punch-through from source to drain occurs along the interface. If the oxide contains a large density of relatively immobile negative ions, then under high-temperature reverse biasing the anions move towards the metallization and decrease the surface charge, thus lowering the surface potential This is negative-ion instability, and its influence is dependent upon the purity of the thermally grown oxide.[7]

The build-up of charge at the interface may be great enough to cause field inversion of the underlying silicon and create a conducting channel between

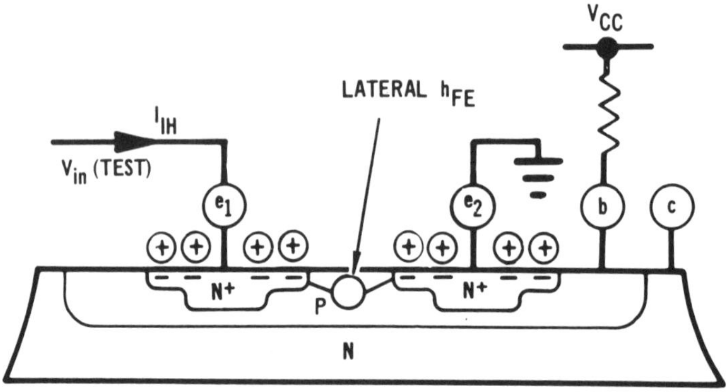

Figure 4.15. Diffusion profile of an input transistor structure showing inverted base region of a lateral parasitic transistor. © 1977 IEEE[11]

adjacent diffusions. This is a parasitic transistor action and results in device failure (Figure 4.15).[11]

4.3 METALLIZATION

Aluminium is used predominantly as the contact metal in most silicon devices. It forms good low resistance contacts to both n^+ and p type silicon when alloyed at about 550 °C for 10 minutes. It is readily deposited by either resistance heating or electron-beam evaporation.

4.3.1 Electromigration in aluminium

Electromigration is a well known phenomenon in aluminium films,[4,7,35–38] and occurs at elevated temperatures under influence of a temperature gradient, or current density in excess of about 10^5 A cm^{-2}, which is quite small considering the linewidths of VLSICs. Electromigration is due to the thermal activation of the aluminium ions which are then transported by momentum exchange with the conduction electrons. The activation energy of the process is about 1 eV and is therefore easily accelerated by thermal means. The failure modes are characteristically cracks, voids, and nodules in the metal lines, causing open-circuit failures. The cracks occur at the points of highest density such as sharp bends or constrictions in the metal lines, and grow perpendicularly to the direction of current flow. Some examples of void formation are shown in Figure 4.16.

Electromigration in aluminium is well understood and design procedures have evolved in order to eliminate its occurrence in normal circumstances in aluminium–silicon devices. However, problems may arise due to poor process

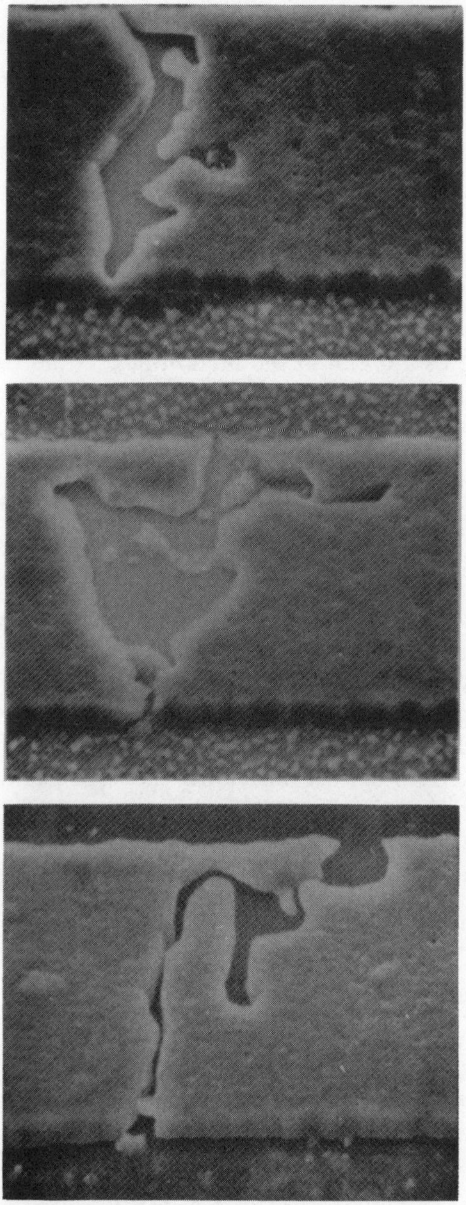

Figure 4.16. Voids formed in Aluminium lines due to electromigration at low temperatures © 1978 IEEE[38]

control, resulting in incorrect metal thickness, undercutting of the metallization during delineation, or simple scratches and smears. Such defects lead to non-uniform current densities in the metal and hence electromigration-induced failure. The processing defects can often be eliminated by screening at elevated temperatures and subsequent failure analysis.

4.3.2 Open metal at oxide steps

All devices, whether discrete or single—or multilevel integrated circuits, will have metallization lines crossing the edges of oxide or dielectric. This step is likely to be a site of poor metal coverage. Generally, the oxide is preferentially etched to provide a bevelled edge rather than a sharp step and thereby reduce the chance of a break in the metal track. If the thermal oxide has a layer of PSG upon it, cracks or breaks are much more likely to occur and this is a significant failure mechanism. The microcrack occurs at the steep upper step and is formed during the evaporation. The step is steep since the PSG etches at a much faster rate than thermally grown oxide, and produces an undercut structure (Figure 4.17). The aluminium metallization may thus be incomplete over this step. The problem is heightened by thin metal, narrow linewidths, or thick oxide or PSG layers. Failures are likely to be breaks in the metal as a result of temperature cycling, or due to electromigration in the thin high current density region creating a void.[4,7,16,35]

The microcrack may be avoided by removing all the PSG layer prior to evaporation, using thick aluminium, or rotating the substrate or using line or multiple sources during the evaporation. If all the PSG is removed, then the device may suffer from surface instability unless a passivation layer is deposited subsequently. Microcracks are difficult to detect by optical means as they are not usually normal to the surface. They are more readily observed by scanning electron microscopy (Figure 4.18).[7]

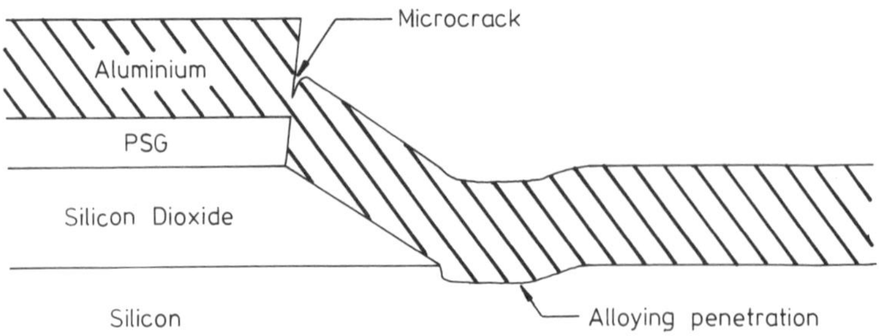

Figure 4.17. Schematic diagram of microcrack structure in aluminium metallization at an oxide step

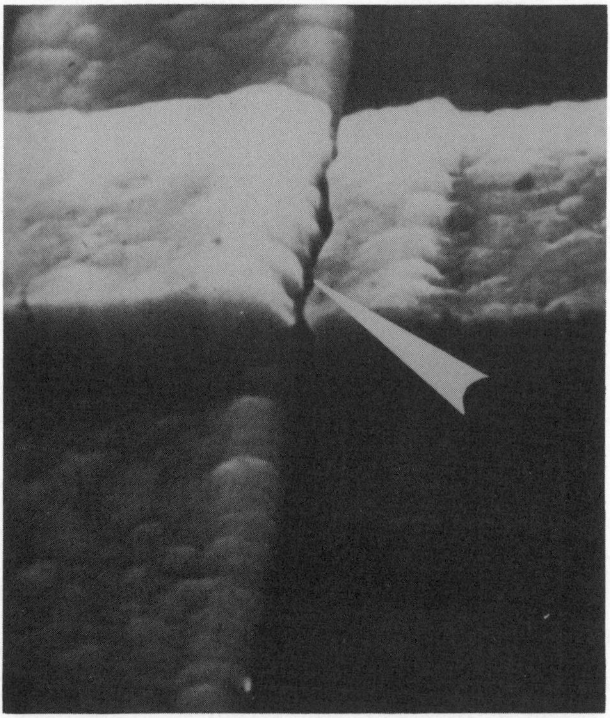

Figure 4.18. Microcrack in the upper-level aluminium metal at a step over the lower-level metallization and inter-level oxide[7]

4.3.3 Aluminium–silicon interactions

Inadequate control of conditions during alloying to form the ohmic contacts can give rise to several failure mechanisms. One problem results from over-alloying due to excessive temperature or time or both. Generally the alloying is performed at 500–550 °C for 10 minutes, below the eutectic temperature for aluminium–silicon of 577 °C, and the reaction takes place by solid state diffusion. The aluminium penetrates the silicon and the metal–semiconductor interface moves downwards. There is often appreciable dissolution of silicon in aluminium, precipitated some distance from the contact cuts. This can lead to non-uniform current flow in the metal lines. Overalloying results in pits at the contacts where the aluminium has penetrated the contact. Aluminium penetration is shown in Figure 4.19. Constrictions in the metal linewidth can lead to void formation at the edges of the contact cuts. Excessive heat treatment during subsequent processing steps can lead to this mechanism of failure.

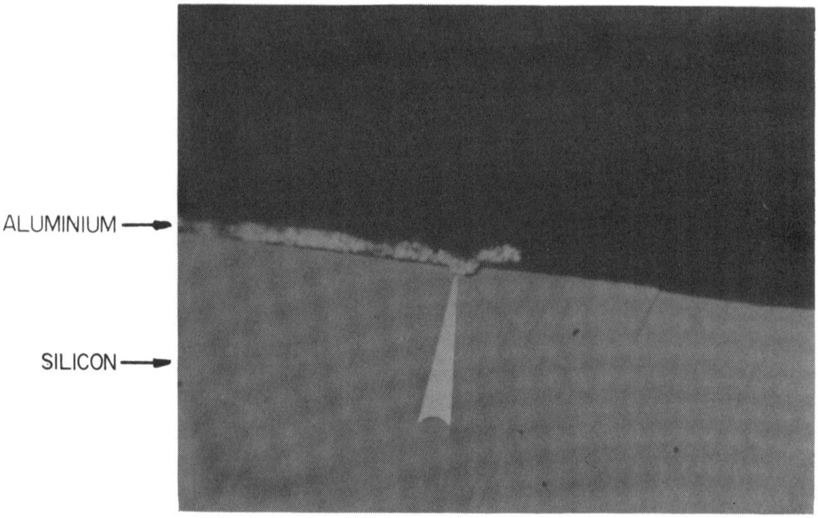

Figure 4.19. Photograph of a 90° cross-section showing the penetration of the aluminium into the silicon due to overalloying (arrowed)[7]

A failure mechanism related to overalloying arises from the tendency of aluminium to spread laterally between the silicon and the thermally grown oxide during alloying. This may lead to shorts and high local current density and field, and the associated reliability problems (see Figures 4.20 and 4.21).

At very high temperatures the silicon may form a eutectic with the aluminium, which will destroy the device. Alternatively the metal can react with the thermal oxide or PSG layer, and reduce their thickness. This increases the probability of gate oxide breakdown, at lower applied voltages, and the failure of the device. It is unlikely, however, that a device or circuit would be subjected to such a high temperature (>577 °C) under normal processing or operating conditions.

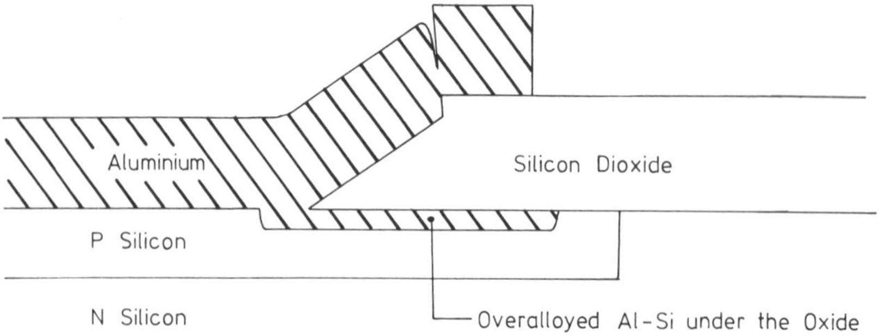

Figure 4.20. Schematic diagram of overalloying of the aluminium–silicon contact leading to lateral spread of the metal under the oxide

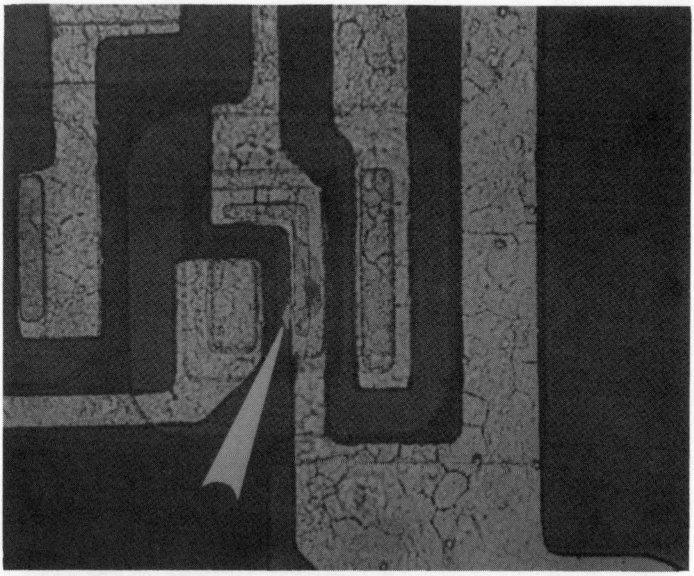

Figure 4.21. Overalloying in aluminium–silicon contacts. The aluminium has moved under the oxide layer and is visible at the emitter contact cut (arrowed)[7]

The formation of 'hillocks' on as deposited aluminium lines has been observed.[36,37,38] They are due to a discrepancy in the linear expansion between the aluminium and the oxide. The hillocks grow by self-diffusion of aluminium to relieve the compressional stress. The growth of these hillocks can cause cracks in deposited oxide or passivation layers, leading to failures due to oxide shorts or contamination and corrosion of the metal. Deposition of aluminium onto hot substrates is thought to alleviate this problem.[7] An example of a cracked passivation layer due to hillock formation is shown in Figure 4.22.

4.3.4 Intermetallic reactions

This problem arises from the need to make a connection from the chip to the outside world. The most common wire bond is made by gold thermocompression bonding achieved by a ball- or wedge-bond of fine gold wire to the aluminium contact pad. Operation or screening of the device at elevated temperatures frequently results in the formation of gold–aluminium intermetallic compounds at the bond (see Figure 4.23). This is known as 'purple plague' due to the colour of the predominant compound $AuAl_2$.[2,4,7,39] The presence of silicon acts as a catalyst to the reaction increasing the degradation of the contact. The intermetallic compound is hard and brittle but strong and

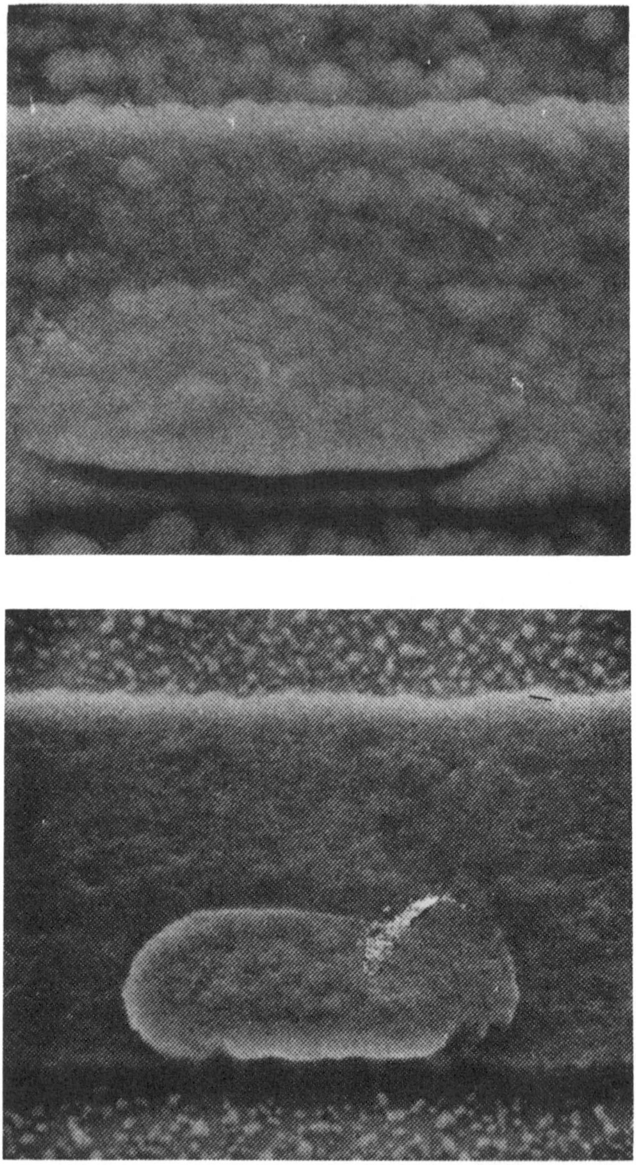

Figure 4.22. Cracked passivation layer due to growth of hillocks in the aluminium, shown before and after glass removal. © 1978 IEEE[38]

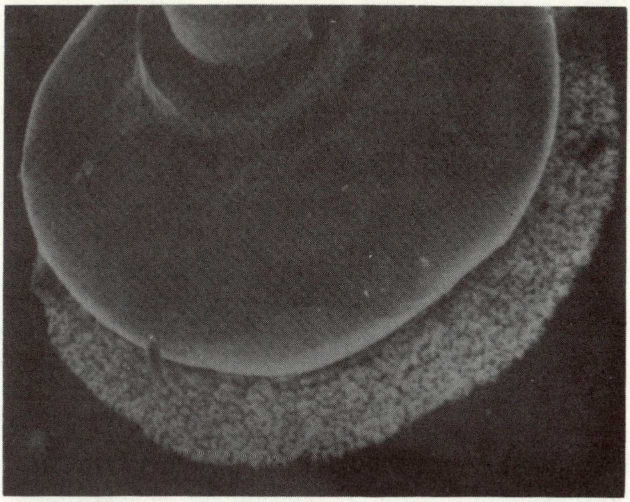

Figure 4.23. **Intermetallic** formation under gold ball bond onto aluminium. © 1979 IEEE[39]

conductive, and in itself does not degrade the performance of the bond. The presence of voids under the bond, due to the Kirkendall effect, tends to weaken the bond strength and increase the resistance. The voids tend to agglomerate and cause the bond to lift and failure of the device. The voids form due to differing rates of diffusion of the aluminium and gold, and opens can occur in the aluminium some distance from the bond, due to mass transport of ions, and occur favourably at constriction in the aluminium lines. It is generally accepted that if the device is stored or operated at temperatures below 150 °C then the gold–aluminium system is a reliable bond.[4]

An alternative is to use a single metal system: aluminium–aluminium bonds made by ultrasonic bonding technique. The aluminium wire contains 1 per cent silicon or magnesium to harden it, and no intermetallic formation is observed. The strength and reliability of ultrasonic bonding is found to be as good as gold thermocompression bonds.

The reliability of bonds in general is dependent upon the incidence of processing errors. Over or underbonding produces weak bonds which can cause failure through mishandling or thermal stress. These bonds are easily screened, either optically or by centrifuge testing which stresses the wire lead. Thus the weak samples can be eliminated from the main population.

4.3.5 Other metallization systems

Alternatives to the aluminium–gold system have been employed, in particular for operation at high chip temperature, in order to reduce the degradation problems associated with intermetallic reactions. The metallizations include tungsten– and molybdenum–gold system used in discrete and integrated circuit devices,[35] and Cr–Pt–Au system used for ohmic contacts in microwave IMPATT diodes.[40,41] The use of platinum or Pt–Ti as a barrier layer between the gold and the contact metal is becoming increasingly popular. This is the basis of beam-lead metallization on silicon devices (see Figure 4.24).[42]

The refractory metals are chosen for their adhesiveness to the oxide layer, and their unreactive nature in comparison with aluminium. They act as a barrier layer between the gold and silicon. The gold overlay provides the low resistance interconnect between devices, and presents no intrinsic bonding difficulties. The molybdenum–gold system has been widely used and the degradation mechanisms are known. One problem is the undercutting of the molybdenum layer during delineation of the system. Molybdenum, unlike aluminium, is not self-passivating and its oxide is water soluble. This leads to the possibility of solution entrapment in the undercut and thus galvanic corrosion. The molybdenum will dissolve at the anode due to self-biasing. This will lead to voids or opens in the metallization, and conducting dendrites of corrosion products have been observed to cause shorts between metal lines. Tungsten is expected to show better reliability in this respect since the oxides are far less soluble in water, being affected only by alkaline solutions.[43] The reliability of these systems is greatly improved by hermetic packaging and suitable passivation. Silicon nitride deposited by reactive ion plating is often used as the passivation layer, protecting the molybdenum from oxidation at temperatures up to 1200 °C.

Gold migration through pinholes in the contact metal is a cause of degradation. Gold diffuses very quickly in silicon, and the eutectic temperature is

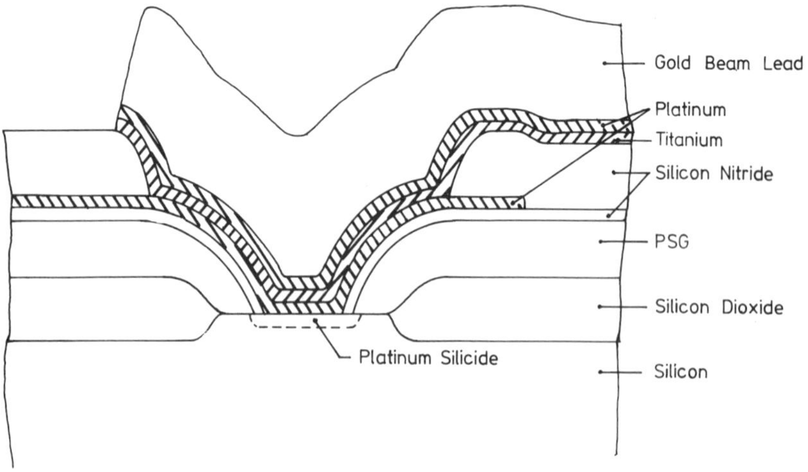

Figure 4.24. Diagram of the beam-leaded structure

quite low at 363 °C, so local hot spots can result in rapid dissolution of silicon and failure of the device. Poor adhesion of the gold to the molybdenum tracks is observed at high temperature and electrical stress. Scratches and similar processing defects also contribute to the reliability problems of the system.

Platinum is introduced as a barrier between the gold and the chromium ohmic contact in silicon IMPATT diodes. Failure of these devices occurs as a result of RF burnout in which a gold spike penetrates through the device. The gold migrates at hot spots due to high local current density at defects or pinholes in the barrier layer.[40,41]

Beam-lead metallizations are gaining acceptance due to their high reliability. A Ti–Pt barrier is employed between the plated gold beams and the contact metal. The thick beams serve as the interconnect metal, the devices are mounted in a flip-chip manner by thermocompression bonding of the beams to the gold-plated circuitry on the header. The contact metal may be aluminium, especially in multilevel systems, although metal silicides are becoming increasingly popular for the ohmic contacts. The reliability of platinum and several refractory metal silicides as contact metallizations has been investigated.[44] A passivation layer of silicon nitride is usually deposited to hermetically seal the metallizations as a protection from corrosion, and to getter any alkali ions present in the underlying oxide.

Long term tests on beam-lead sealed junction devices (BLSJ) show a very low incidence of failures.[4,27,42,45] Pinholes in the platinum barrier layer can result in gold–titanium intermetallic formation at high temperatures. This results in high contact resistance due to void formation. Poor process control can lead to contamination between the metal layers which will result in adhesion problems. The improvement in reliability is due to the elimination of wire bond failures and 'purple plague', achieved by directly bonding the beam-lead contacts to the package header. The beam-lead configuration reduces parasitic inductances commonly associated with wire bonds. Indeed, the beams' dimensions can be tailored to match their inductances and capacitances to the circuit. Beam-lead technology has been employed in fabrication of silicon microwave transistors.[45]

4.4 TRENDS IN VLSI RELIABILITY

Progress in the field of microelectronics has been extremely rapid, with larger and larger numbers of devices and functions being integrated onto single chips. All aspects of data processing have benefited from the great improvements in the cost, reliability, and performance of integrated circuit electronic devices. These improvements are brought about simultaneously by the increased gate:pin ratio which integration provides. The progress of the integration can be measured by the increasing chip size, decreasing device dimension, multilevel structure and circuit innovation. The natural limits which impose themselves upon the technological advances may be classified as:[24]

Electromigration;
Hot electron and breakdown effects;
Wiring complexity and pattern sensitivity;
Ohmic resistance and interconnections;
Power dissipation in the gates.

These will each tend to introduce reliability problems in VLSICs. The failure mechanisms which are associated with increased integration include all the

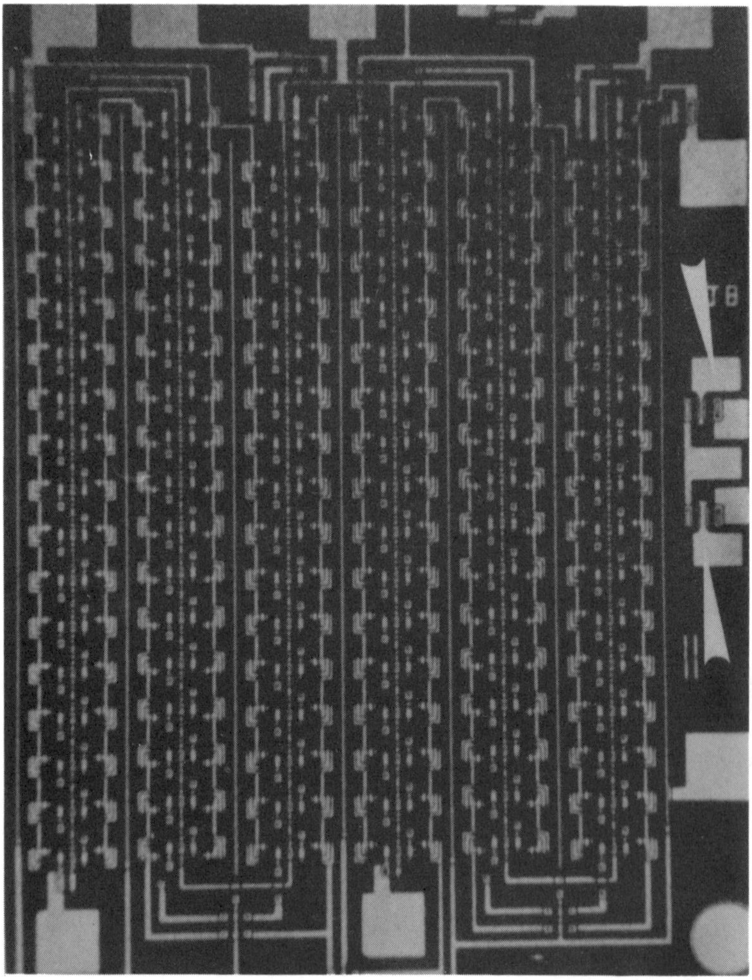

Figure 4.25. Photo of a VLSI chip with test vehicles located at the periphery of the chip[7]

mechanisms found in discrete devices, and those introduced by the device scaling. The latter degradation mechanisms are described below, classified as multilevel structure and scaling problems, hot electron effects, and defects introduced by the processing technology. The reliabilities of newer device structures and technologies, i.e. VMOS and silicon-on-sapphire are also discussed.

The screening and burn-in procedures employed for VLSICs are often found to be inadequate. This is because it is very difficult to apply consistent levels of electrical and thermal stress to each of the individual gates on the chip. Consequently some devices pass through the screening only to suffer time or stress dependent failure in service. Thus in order to make reliability predictions the chips often incorporate several test vehicles, such as MOS capacitors and transistors, so that, for example, the integrity of the oxide, or the threshold voltage of the transistors may be examined. An example of a chip-incorporating test vehicle is shown in Figure 4.25.

4.4.1 Multilevel structures and device scaling

Aluminium is used extensively as the metallization in all levels in large scale integrated circuits. As the linewidth is decreased, so the current density increased and problems related to electromigration can arise. Linewidths of about 1 μm are taken as the lowest practical limit; below this problems due to hot electron effects and non-Boltzmann conduction may also arise. This imposes restrictions upon the circuit design to avoid constrictions and current bunching. Metal silicides are used as the first level metal contact in some ICs in order to avoid the electromigration difficulties. Interconnections and higher level metallizations are made using beam-lead techniques.

In multilevel structures the metallizations are separated by sputter or chemical vapour-deposited layers of silicon oxide or PSG. The oxides are usually deposited under dry conditions, since it has been found that steam grown oxides contain impurities which considerably reduce the breakdown voltage.[46] The layers are generally about 100 nm thick. The interconnections between levels are made through windows cut into the oxide, known as vias. The slopes on the sides of the windows are angled so that the upper level of metal does not have to cross a sharp step. This eliminates the problem of microcrack formation in the metal at the oxide step, as discussed in Section 4.3.2. A similar problem occurs when second level metal and deposited oxide crosses the first-level metallization. The microcracks can be avoided by using a low-angle tapered edge in the delineated underlying metal. This technique is widely used in multilevel integrated circuits,[7] and is shown schematically in Figure 4.26. Any contamination or oxidation of the exposed aluminium in the vias before the next layer is deposited will result in a higher resistance contact due to reduced effective area. This may give rise to hot spots and high current

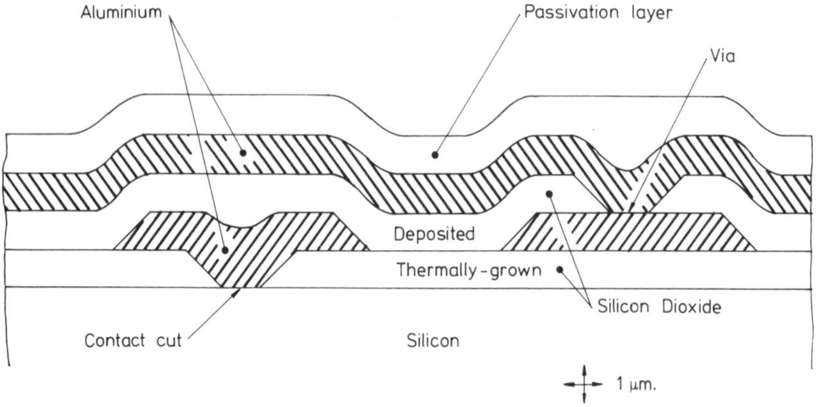

Figure 4.26. Schematic multi-level structure with two levels of dielectric showing sloping contacts cuts and vias connecting the metallization levels, to reduce the indicence of microcracks formation

densities, which may lead to electromigration related failures in the vias due to open metal.

Device-scaling techniques are frequently employed to improve the performance and density of devices. High performance MOS technology (HMOS) has been developed by reducing the channel length, gate oxide thickness and junction depth while retaining the same supply voltages to maintain compatibility with other MOS technologies. Gate oxide breakdown is thus a potential failure mechanism in such devices due to the increased field as a result of scaling of oxide thickness but not voltage. The gate oxide is typically 70 nm. High voltage screening can be used to eliminate weak oxides which could otherwise give rise to reliability problems.[47]

Polysilicon gates are often used in multilevel MOS devices in preference to an aluminium metallization in order to improve the long term reliability. Aluminium gates deposited by resistance-heating evaporation frequently contain ionic impurities which are mobile in the gate oxide, and thus the gate is subject to time-dependent breakdown wearout.[32] The mean time to failure reduces as the oxide thickness is decreased, making this an important reliability consideration for HMOS structures. A comparison of the maximum time to failure of MOS capacitors with aluminium and polysilicon electrodes has shown that the polysilicon–electrode devices to not exhibit such a time-dependent breakdown phenomenon (see Figure 4.27).[31] The use of polysilicon gates carries the performance penalty of a lower speed power product as a result of the high resistivity of the polysilicon giving a large gate time constant in comparison to metal gate devices.[25]

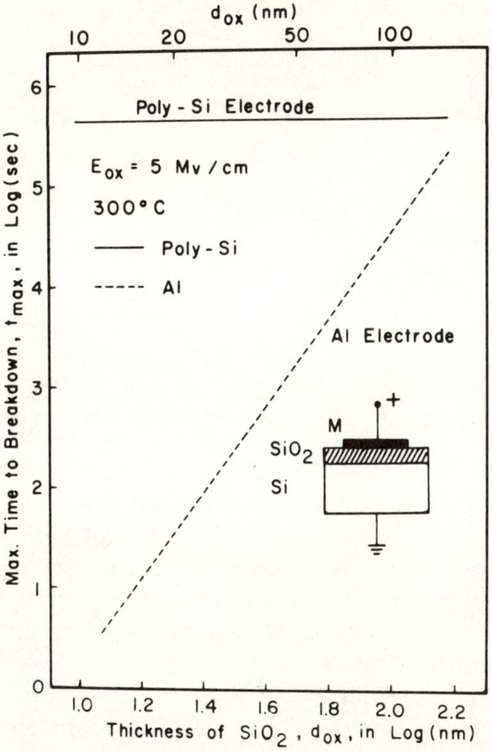

Figure 4.27. Graph of the time dependent dielectric breakdown as a function of the oxide thickness, with aluminium and poly silicon electrodes in MOS capacitor structures. © 1979 IEEE[25]

4.4.2 Hot electron effects

The reduction of device channel and gate dimension without a corresponding scaling of the voltages results in increased fields in the channel of the device, and hence the generation of hot electrons. The electrons may originate in the substrate due to thermal currents, or from the channel current. The electrons are accelerated under influence of the high field towards the drain. Impact ionization scatters the hot electrons and also multiplies the channel current. Some of the scattered electrons will have enough energy to surmount the oxide–silicon barrier (3.1 eV) and may be trapped in the oxide (see Figure 4.28). Over a long period of time the charge build-up due to the trapped electrons will cause instability in the form of threshold shift or transconductance degradation.[25,47]

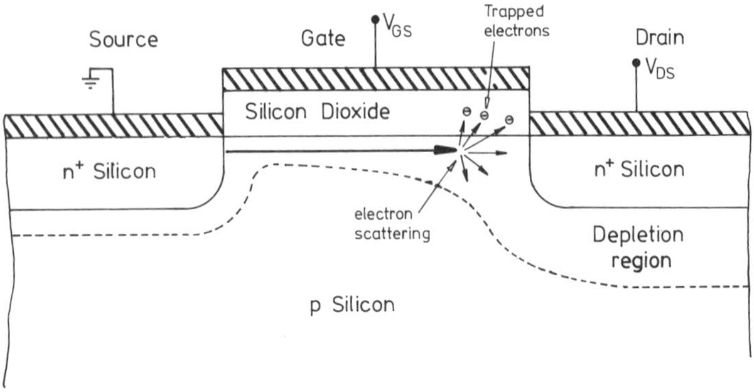

Figure 4.28. Diagram of MOSFET structure showing the scattering into the gate oxide of hot electrons which may then be trapped

The number of trapped electrons depends upon the density of available traps in the silicon dioxide. This trap density can be increased by the radiation damage as a result of the radiative processes involved in the fabrication of VLSI devices, for example electron-beam lithography and evaporation of aluminium producing soft X-rays, and reactive-ion etching.[25] Thus, VLSI is more susceptible to hot electron damage than devices employing more conventional fabrication processes.

Limits on the source–drain voltage must be imposed in order to inhibit the generation of hot electrons. This can be achieved through circuit design and device redundancy.

4.4.3 Processing faults

The complexity of the structure of multilevel devices means several photolithographic and etching operations are necessary during the device fabrication. Flaws in the devices may arise due to misalignment of the photomasks, or under- or overetching. These faults can, in some cases, pass the screening and burn-in procedures and only come to light as time-dependent failures in service. This may happen with memory devices, for instance, where it is difficult to stress every individual cells.

Failures due to poor processing observed in MOS RAMs include missing contact windows and parasitic transistors.[27] Missing metal-to-polysilicon windows in the nitride layer allows the transistor gates to 'float' (see Figure 4.29). The nitride capacitance to the floating gate is often sufficient to permit normal device operation until hot electron injection through the thin gate oxide charges the floating gate to some critical offset potential, whereupon device failure is evident.

Parasitic transistor action may occur between adjacent devices, due to over-etched or misaligned contact windows (see Figure 4.30). Failure occurs

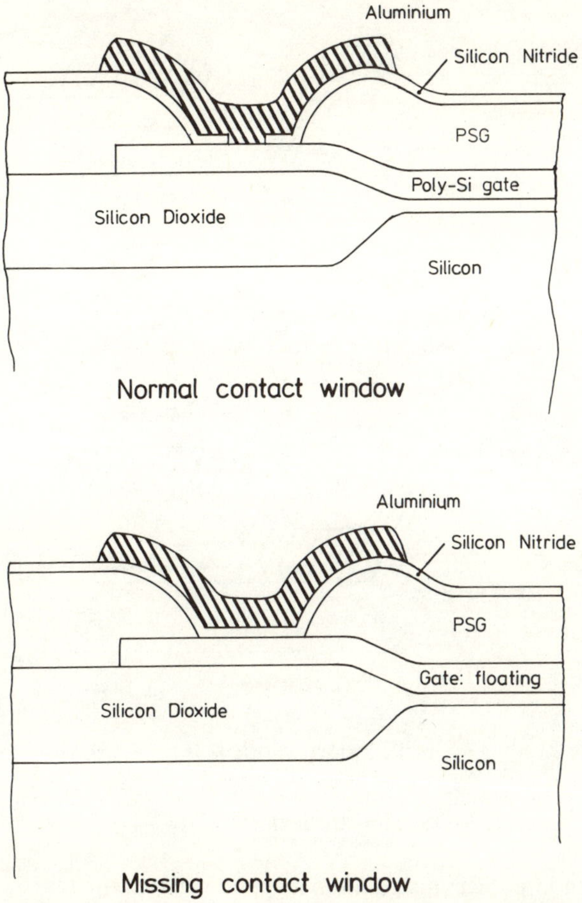

Figure 4.29. Missing contact window in the silicon nitride layer

when the charge built up on the thinned dielectric, in the field region between adjacent diffusions, is sufficient to invert the underlying silicon and create a conducting channel. The rate at which the charge builds up depends upon the local electric fields, and is thus likely to be dependent upon stored information.

A problem of oxide breakdown in multilevel polysilicon structures has been observed.[46] This is because the etch rates of the deposited oxide and polysilicon layers are different (in the usual etchants), resulting in gate oxide undercutting of the polysilicon layer(s), and consequently thinning of the oxide between the polysilicon and the substrate, or adjacent polysilicon layers, indicate Figure 4.31. The thinning can reduce the oxide to less than half its nominal thickness, with the corresponding reduction in breakdown voltage.

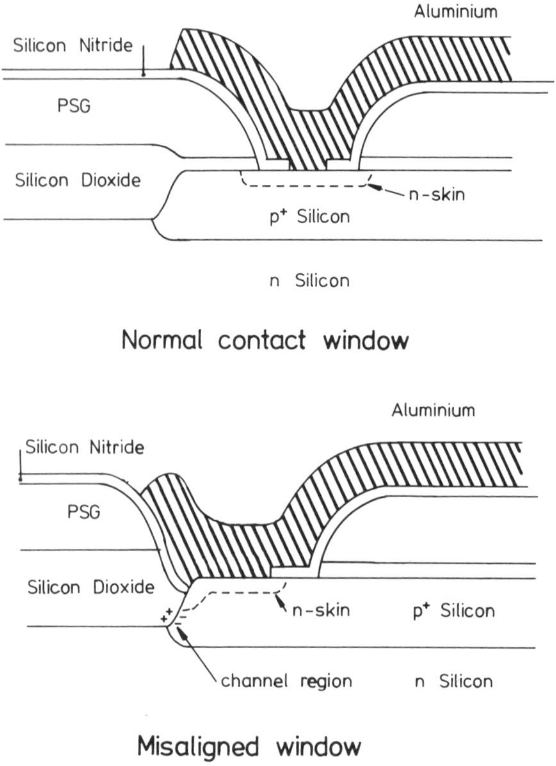

Figure 4.30. Misaligned contact window resulting in parasitic MOS transistor action

Gate oxide undercut can also lead to trapped voids in the regrown oxide again reducing the breakdown voltage. It has been found that a mixture of CrO_3 in aqueous HF will etch the polysilicon at the same rate as the oxide, and thus no undercutting will occur even with over-etching; consequently no thinning of the local oxide will be observed.

4.4.4 VMOS reliability

The VMOS technology involves V-grooved metal–oxide semiconductor transistors which utilize vertical as well as planar structures, integrated to form high speed logic circuitry. Figure 4.32 illustrates the basic components of VMOS technology. The system employs a n^+ substrate which is the source of the VMOST. The p-well is formed by the outdiffusion of controlled boron impurities in the substrate, and is important for two reasons: firstly, it controls the threshold of the VMOST; secondly, it is located near the source thereby reducing hot electron and channel shortening effects. The gate oxides for the

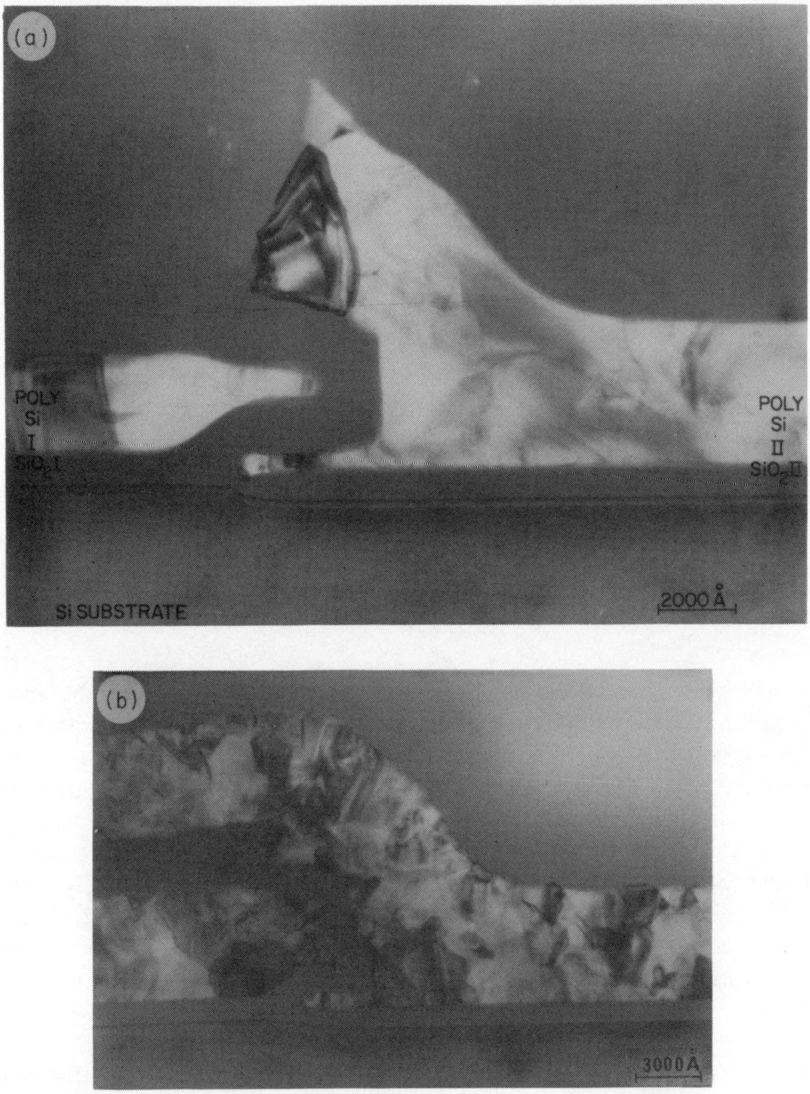

Figure 4.31. TEM cross-section through a two level polysilicon structure showing oxide thinning between the second level polysilicon and the substrate, for lateral undercut of first level oxide of (a) 3000 Å and (b) 1500 Å. © 1979 IEEE[46]

transistors are of the order 70 nm thick. The drain diffusions are on the surface, compatible with the planar high performance MOS devices.

The new technologies introduced by VMOS will be factors influencing the overall relability of the devices, in addition to the usual MOS failure mechan-

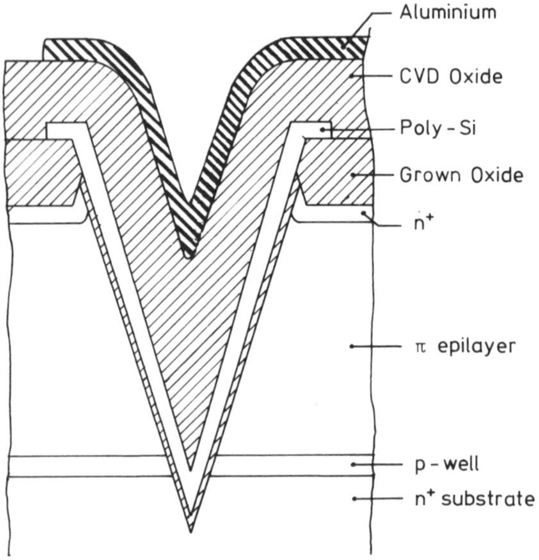

Figure 4.32. Schematic diagram of the structure of a VMOS transistor

isms—these technologies are the epitaxial *p–n* junction and the etched V-groove. The major difficulty in the formation of the *p*-well is maintaining adequate controls over the level of boron impurity in the substrate and the outdiffusion. Variations in the doping in the *p*-well will lead to changes in the leakage current of the junction and threshold shifts in the VMOS transistor. Doping variations may occur as a result of high temperature processing although at normal operating temperatures outdiffusion is unlikely to be a cause of failure. Spatial variations in the doping over the area of the chip may occur, resulting in variations in the threshold voltage from device to device. These kinds of problems are more likely to affect the yield rather than the reliability, provided adequate screening is performed. The V-groove is formed by preferential etching to expose the (111) surfaces of the silicon. The interface state density at the Si–SiO_2 interface is a function of the crystal orientation and is known to be higher for (111) silicon compared to (100) silicon used in planar devices. The VMOS structure may therefore tend to suffer more from surface instability (see Section 4.3.1) and hot electron trapping than comparable planar devices. Examination of the typical failure mechanisms in VMOS devices (see Table 4.2) shows that ionic contamination, including surface charge spreading and ionic drift and trapping in the oxide, is by far the most common.[48]

The VMOS transistor has a very sharp radius of curvature at the bottom of the groove, and the oxide breakdown is controlled by the enhanced electric

Table 4.2 Relative incidence of VMOS failure, mechanisms[48]

Ionic	75%
Oxide	20%
Mask	24%
Diode leakage	1.4%
Unknown	1.0%
Assembly	0.2%

field at this point. The breakdown voltage is typically 25–35 V compared with 60–80 V for planar devices. Thus conventional MOS input protection schemes are inadequate from VMOS. Special protection devices have been developed,[48,49] generally taking the form of *n–i–p–n* multijunction diodes which breakdown at a voltage controlled by the width of the intrinsic layer. Alternatively the geometry of the transistor can be adjusted to give a flat bottom to the groove, thus eliminating the sharp *V*. This is the U-MOS transistor, which operates in an identical manner to the VMOST, but shows better reliability with respect to oxide failures. A reliability study of VMOS devices shows that they exhibit no unique failure mechanisms, and lifetesting at high temperature predicts a reliability failure rate of less than 0.01 per cent per 1000 hours.[48]

4.5.4 Silicon-on-sapphire (SOS) reliability

Technology has been developed in order to furnish silicon integrated circuits with an insulating substrate material. A thin layer of silicon is epitaxially grown upon the sapphire, and almost ideal isolation between devices is achieved by processing the devices in individual small area islands of silicon on the insulating substrate. This feature makes SOS devices attractive for use in hostile radiation environments due to the small excitation volumes for the radiation-induced photocurrent.[50]

Large scale integrated circuits fabricated using SOS introduce new technologies which affect the reliability of the devices. The islands which form the devices are anisotropically etched from the epilayer, revealing (111) silicon surfaces at the island edges. A high density of trap states exists at the interface between these edges and the thermally grown oxide (cf. VMOS technology). This can lead to surface instability effects and trapping of mobile ions in the oxide, causing inversion of the silicon along the edges (Figure 4.33a). The conducting channel can reduce the breakdown voltage and increase the reverse leakage of exposed *p–n* junctions. However, by using a device structure in which the junction is buried within the island this problem can be alleviated. Surface leakage across devices is also enhanced by the presence of the conducting channel.[26] Metal interconnections between devices must run

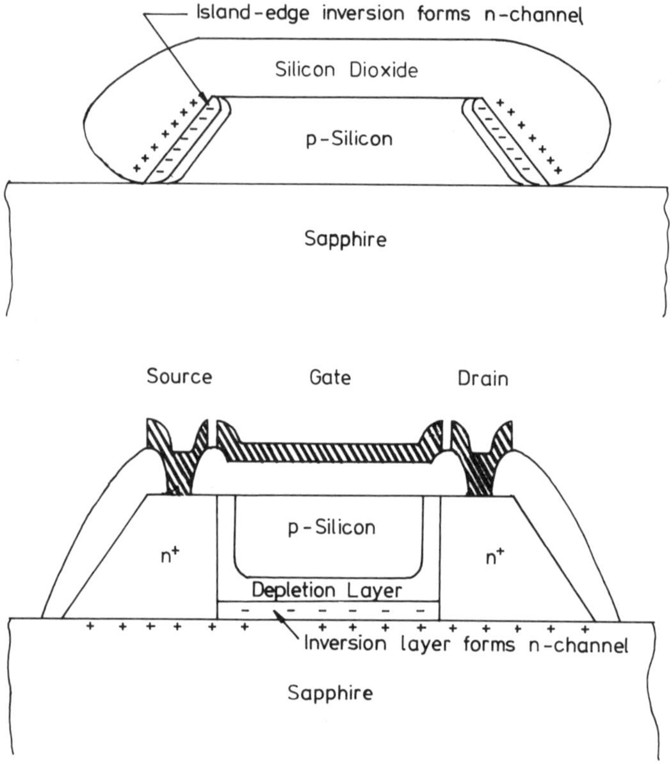

Figure 4.33. Schematic structure of silicon on sapphire device showing (a) interface traps at the island edges, and (b) the back-channel effect

over the edges of the islands. The edges are tapered by the etching process, and this reduces the incidence of microcrack formation, although the step is still quite steep. It has been found that annealing by low energy laser irradiation tends to round off the sharp edges of the islands, enabling more uniform metallization over the step (See Figure 4.34).[51]

The quality of the epitaxial silicon and especially the silicon–sapphire interface has a bearing upon the performance and reliability of SOS devices. Crystal defects and dislocations occur at the interface and act as traps for carriers and mobile ions. Positive charge trapping in the sapphire near the interface induces an n-type channel in the adjacent silicon. This would be an inversion channel in a p-type base, forming a conductive path between the source and drain: the back-channel effect. One means of preventing this effect is to increase the acceptor-density near the interface by ion-implantation. The implantation must be well controlled otherwise radiation damage at the interface will result in a higher trap density, and will conse-

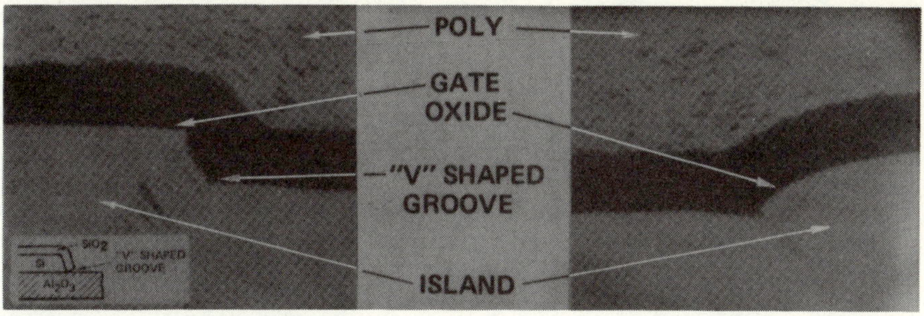

Figure 4.34. Photographs of the gate oxide coutour showing how laser annealing can soften the step. © 1980 IEEE[51]

quently worsen the effect. The back-channel effect is shown schematically in Figure 4.33(b).

Hot electron effects are also encountered in VLSI SOS devices. Impact ionization due to hot carriers can occur at the drain of MOS transistors, leading to injection of carriers into the base region and consequent current flow in the base. This induces a base voltage which in turn increases saturation current in the transistor, observed as a 'kink' characteristic. The effect of the 'kink' mechanism is a variation of the threshold voltage and hence degradation of performance. The 'kink' effect may be reduced by grounding the base to provide a low resistance leakage path.

MOS–SOS also suffer from the usual MOS degradation mechanisms, notably oxide breakdown phenomena. Gate protection circuitry is thus employed to prevent damage as a result of electrostatic or overvoltage effects.[26]

4.5 PACKAGE-RELATED EFFECTS

The package provides the physical support for the silicon device and enables electrical connection to be made to the external circuitry. The package also serves to provide electrical and mechanical protection for the device, and hence must present an inert and dry atmosphere to the surface of the semiconductor, and must give protection against mechanical and thermal shock. The package itself comprises several elements. Initially a passivation layer is often deposited onto the active surface of the silicon chip, leaving only the bonding pads uncovered. This is to isolate the surface from any contaminants which may be present in the atmosphere. The chip is then attached to a metal or ceramic header which also provides electrical connection via bond wires to the bonding pads. The packaging is then completed by encapsulation in a metal can or a ceramic or plastic package, to give a hermetically sealed enclosure.

Each element of the packaging process provides some form of protection for the device. The electrical failures are commonly due to some form of corrosion which may be related to the integrity of the surface passivation layer or the hermeticity of the external package. Mechanical failures arise from mishandling of the device, or physical shock resulting in a breakage of the device. The failure mechanisms associated with each element of the packaging are discussed below.

4.5.1 Surface passivation and corrosion

The passivation layer is frequently a glass or other dielectric layer deposited onto the device surface after the fabrication procedure has been completed. Chemical vapour deposition (CVD) of PSG is often used as the encapsulant, and acts as a getter for any mobile sodium ions which may be present in the underlying silicon dioxide. Plasma-deposited silicon nitride films are also used as passivating layers, especially in beam-lead sealed junction devices. The nitride layer is an effective barrier for alkali ions which may be left on the device surface after the final wash.[4,9] This contributes to the higher level of reliability experienced with these devices. However, silicon nitride can give rise to a change in the surface potential, which in turn affects the threshold voltage of MOSFET devices. A layer of Al_2O_3 is found to be an acceptable passivation for such devices.[4] The dielectric passivation layer can introduce further charge trapping and storage effects and give rise to interface conduction due to the activation of traps. These effects can contribute to unreliable operation and can play a part in the corrosion process.

Adhesion difficulties have been experienced with the dielectric layers: they stick well to the underlying silicon dioxide or aluminium but less well over gold metallization.[52] The CVD films are in tensile stress, and where the adhesion is poor this may cause localized lifting or cracking of the passivating layer, which can result in enhancement of the corrosion of the metallization. This is a particular problem where the conductor passes over topographical steps in the underlying oxide.

An alternative passivating material which is often used in low power devices is RTV silicone. This is a plastic coating which bonds chemically to the surface of the device and thus prevents any interface conduction. The organic plastics used for passivation purposes are often made from proprietary recipes which have a large number of ingredients, and can be a source of contamination. These impurities may attack the device surface directly or become activated under humid conditions.

Failures due to corrosion of the metallization are frequently associated with defects in the passivation, such as cracks and pinholes. These defects act as sites for the entrapment of contaminants or moisture which are instrumental in the corrosion. The failures are generally due to electrical opens caused by voiding of the metal, or shorts due to conductive particles or corrosion products between the metal tracks.

Aluminium has a stable oxide and is therefore self-passivating. In the absence of ionic contamination it is a stable metallization system. In the presence of ionic contaminants the oxide can be broken down and the aluminium attacked. The effects of sodium[53,54] and chloride[9] contamination are well known, giving electrochemical corrosion of the aluminium under the application of electrial bias. The phosphorus content of the silicon dioxide layers used in the encapsulation is an important factor in the reliability of the devices. The phosphorus is added as a getter for mobile sodium ions. However, it also affects the tensile stress of the film. The effects of the phosphorus additions to the silicon dioxide are summarized in Table 4.3. The cracking of the passivation will increase the susceptibility of the underlying aluminium to corrosion by impurities. An excess of phosphorus in the glass promotes leeching of P_2O_5 from the PSG by moisture, resulting in the formation of phosphoric acid which readily attacks aluminium.[9,52,55] In phosphorus related electrochemical corrosion the cathode is attacked,[54] distinguishing it from chloride-induced corrosion.

In contrast to the failure of aluminium metallization due to void formation, the basic failure mechanism of gold-metallized devices is the formation of conductive dendrites at the cathode causing short circuits. The presence of ionic contaminants such as chloride, other halides, or cyanide ions determines the particular mode of electrochemical reaction. The basic mechanism is the dissolution of gold at the anode in the form of a suitable ionic radical, and the subsequent reduction at the cathode from where the dendrites are observed to grow.[9,52,56,67] This corrosion is thus known as gold migration. The formation of $Au(OH)_3$ as a voluminous reaction product which has sufficient conductivity to cause failure due to leakage between conductors has been observed.[58]

Galvanic corrosion can occur in metallization systems of dissimilar metals, for example, the molybdenum–gold system, or in aluminium metallizations with gold-bond wires. The corrosion occurs at exposed regions such as bonding-pads and can proceed along the passivation–metal interface. Molybdenum oxide is soluble in water and this can lead to undercutting of the gold, solution entrapment and consequently corrosion. Aluminium–gold reactions are evident as intermetallic formation, and are discussed in detail in Section 4.3.4.

Table 4.3 Effects of phosphorus on glass properties

<table>
<tr><th></th><th>Too little phosphorus</th><th colspan="2">Too much phosphorus</th></tr>
<tr><td>Encapsulating layer</td><td><2% Cracks</td><td>>6%</td><td rowspan="2">Aluminium corrosion</td></tr>
<tr><td>Underlying oxide</td><td><6% Will not flow</td><td>>9%</td></tr>
</table>

4.5.2 Chip mount and bonding

The semiconductor chip is mounted upon a metallic or ceramic substrate by means of a low melting point solder such as gold–silicon or gold–germanium eutectic alloy, or by a glass or epoxy bond. The strength of the alloy bond may be reduced by the occurrence of solid state diffusion reactions, leading to the formation of brittle intermetallic compounds.[2] The composition of the glass must be carefully controlled in order to reduce the incidence of impurities. Small concentrations of oxide impurities can change the reaction rate and temperature range of the devitrification process which produces the bond,[2] resulting in a weak or brittle bond. In all the systems the temperature must be controlled to ensure the complete alloying of the eutectic solder or devitrification of the glass, and the thermal curing of the epoxy, otherwise a weakened bond will result. Failures of the chip-to-package bond constitute only a minor reliability hazard.[7]

The reliability problems associated with over- or under-bonding of the connecting wires to the lead frame are similar to those associated with the bonding to the bonding pads, discussed in Section 4.2.4. However, the incidence of intermetallic formation is likely to be much reduced since the bond wire and the electroplating of the header will both in general be gold. In the case of aluminium wires and gold-plated lead frame, the post temperature will be much less than the on-chip temperature, thus inhibiting the intermetallic reaction.

4.5.3 The external package

The header is enclosed in a can to complete the packaging. There are commonly three types of package: metal can, ceramic, or plastic encapsulation. The primary function of this external package is to maintain a dry and inert inner atmosphere throughout the lifetime of the device. In order to do the package seals must be made as hermetic as possible. Leak-testing of the seals is often performed by measuring the rate of escape of helium or radioactive krypton from the package.[2,4]

Metal can packages are made by welding a kovar or steel can to a gold-plated header. The seal is made in an atmosphere of nitrogen in which the moisture content is kept below about 10 p.p.m., in order to provide an inert interior atmosphere.[2] The hermeticity of the can depends upon the integrity of the welded seal, and also upon the glass–metal seals which are often used to isolate the electrical leads. These seals rely upon close matching of the coefficients of thermal expansion of the glass and metal leads to maintain the hermetic seal.

The ceramic package is often to be found on LSICs. The chip and lead frame assembly is sandwiched between two ceramic slices held together by a glass sealant. The coefficient of thermal expansion of the glass must be closely matched to that of the ceramic to prevent thermal stress cracking. The glass provides the hermetic seal, The ceramic itself is impervious to almost all gaseous contaminants and moisture, provided it has been properly sintered.[2] The ceramic also provides good electrical insulation and high thermal conductivity. The package is proven to be very strong, although brittle and especially susceptible to torsional stresses, and also thermal shock such as provided by tin-dipping of the electrical leads, for example. This can lead to cracking of the ceramic and the consequent loss of hermeticity.

The plastic package is seen most often of the the high volume-produced devices. The encapsulation is by moulding the device in the molten plastic. Originally the commercial plastics used for the encapsulation contained a high proportion of impurities, and the devices acquired a reputation for poor quality and reliability. Nowadays improved process and quality control has improved the reliability of plastic encapsulated devices to an acceptable level. The plastic must be cured and baked in order to reduce the moisture content, although it is generally accepted that water may diffuse through the plastic[9,52] or enter along the interface between the device leads and the plastic[52] due to the relative difference in the coefficients of thermal expansion of the metal and the plastic.[2] It is thought that the high reliability may be due to the plastic being chemically bonded to the device surface and thus preventing any corrosion which might take place upon a free surface.

4.6 CONCLUDING REMARKS

The trends in microelectronics have been towards an improvement in the technology resulting in better control of the device properties and lower incidence of localized material defects. The objective of device manufacturers is to design-in reliability rather than to try and achieve it by the substantial testing and screening of the completed devices. This is especially important with the trend towards VLSI. Designs are selected to ensure that the device will not be susceptible to early wear out phenomena. The designs often incorporate a large amount of redundancy of elements to guard against processing faults.

REFERENCES

1. See, for instance, *Annual Proceedings—Reliability Physics*. Conferences sponsored by IEEE Electron Devices Society and IEEE Reliability Group.
2. C. G. Peattie, J. D. Adams, S. L. Correll, T. D. George, and M. H. Valek, 'Elements of semiconductor device reliability', *Proc. IEEE,* **62**, 149–168 (1974).
3. J. Vaccaro, 'Semiconductor reliability within the US Department of Defense', *Proc. IEEE,* **62**, 169–184 (1974).
4. D. S. Peck and C. H. Zierdt, 'The reliability of semiconductor devices in the Bell system', *Proc. IEEE,* **62**, 185–211 (1974).

5. E. D. Colbourne, G. P. Coverley, and S. K. Benera, 'Reliability of MOS LSI circuits', *Proc. IEEE,* **62**, 244–259 (1974).
6. R. F. Haythornthwaite, A. R. Molozzi, and D. V. Sulway, 'Reliability assurance of individual semiconductor components', *Proc. IEEE,* **62**, 260–273 (1974).
7. G. L. Schnable and R. S. Keen, 'On failure mechanisms in large-scale integrated circuits', *Advances in Electronics and Electron Physics,* **30**, 80–138 (1971).
8. D. S. Peck, 'New concerns about integrated circuit reliability', *IEEE Trans. Electron. Devices,* **ED-26**, 38–43 (1979).
9. A. T. English and C. M. Melliar-Smith, 'Reliability and failure mechanisms of electronic materials', *Ann. Rev. Mater. Sci.,* **8**, 459–495 (1978).
10. F. H. Reynolds, 'Accelerated-test procedures for semiconductor components', *15th Ann. Proc. Rel. Phys.*, 166–178 (1977).
11. G. M. Johnson and M. Stitch, 'Microcircuit accelerated testing reveals life limiting failure modes', *15th Ann. Proc. Rel. Phys.,* 179–195 (1977).
12. L. Hirsch, 'Accelerated electrical testing for improved device reliability', *15th Ann. Proc. Rel. Phys.*, 225–226 (1977).
13. D. S. Peck, 'Practical applications of accelerated testing', *13th Ann. Proc. Rel. Phys.*, 253–254 (1975).
14. D. S. Peck, 'The analysis of data from accelerated stress tests', *9th Ann. Proc. Rel. Phys.*, 69–78 (1971).
15. B. Reich, 'A study of accelerated storage test conditions applicable to semiconductor devices and microcircuits', *IEEE Trans. Reliability,* **R-27**, 178–180 (1978).
16. R. J. Anstead, 'Failure analysis using a scanning electron microscope', *6th Ann. Proc. Rel. Phys.*, 127–137 (1967).
17. A. J. Gonzales, 'Failure analysis applications of the scanning electron microscope', *11th Ann. Proc. Rel. Phys.*, 179–184 (1973).
18. R. Kossowsky, 'Analytical techniques for electronic materials—a comparative evaluation', *16th Ann. Proc. Rel. Phys.*, 112–120 (1978).
19. D. V. Morgan, 'Rutherford scattering analysis: a tool for semiconductor device technology', *Solid State and Electronic Devices,* **1**, 38–45 (1977).
20. J. M. Poate, 'The use of high energy ion beams for surface and thin film analysis', *14th Ann. Proc. Rel. Phys.*, 297 (1976).
21. A. J. Gonzales, 'Failure analysis applications of Auger electron spectroscopy', *11th Ann. Proc. Rel. Phys.*, 185–188 (1973).
22. J. W. Colby, 'Failure analysis using the electron microprobe', *11th Ann. Proc. Rel. Phys.*, 189–193 (1973).
23. J. W. Colby, 'Failure analysis using the ion microprobe', *11th Ann. Proc. Rel. Phys.*, 194–197 (1973).
24. R. W. Keyes, 'The evolution of digital electronics towards VLSI', *IEEE Trans. Electron. Devices,* **ED-26**, 271–279 (1979).
25. H.-N. Yu *et al*. '1 μm MOSFET technology Parts I–VIII', *IEEE Trans. Electron. Devices,* **ED-26**, 318–379 (1979).
26. R. K. Pancholy, 'Gate protection for CMOS/SOS', *15th Ann. Proc. Rel. Phys.*, 132–137 (1977).
27. C. W. Green, 'PMOS dynamic RAM reliability—a case study', *17 Ann. Proc. Rel. Phys.*, 213–219 (1979).
28. H. S. Lee, 'MOS instability produced by electron beam evaporation of aluminium gates', *IEEE Trans. Electron. Devices,* **ED-25**, 795–799 (1978).
29. P. K. Chaudhari, 'Threshold voltage degradation of MNOS FET devices', *J. Electrochem. Soc.,* **125**, 1657–1660 (1978).

30. C. M. Osburn and N. J. Chou, 'Accelerated dielectric breakdown of SiO_2 film', *J. Electrochem. Soc.,* **120**, 1377–1384 (1973).
31. C. M. Osburn and E. Bassous, 'Improved dielectric reliability of SiO_2 films with polycrystalline silicon electrodes', *J. Electrochem. Soc.,* **122**, 89–92 (1975).
32. S. P. Li, S. Prussin, and J. Maserjian, 'Model for MOS field-time dependent breakdown', *16th Ann. Proc. Rel. Phys.*, 132–136 (1978).
33. D. Crooke, 'Method for determining reliability screens for time dependent dielectric breakdown', *17th Ann. Proc. Rel. Phys.*, 1–7 (1979).
34. E. S. Andick and G. R. Nelson, 'Low field time dependent dielectric integrity', *17th Ann. Proc. Rel. Phys.*, 8–12 (1979).
35. G. L. Schnable and R. S. Keen, 'Metallization and bonds—a review of failure mechanisms', *RCA Review,* **40**, 416–445 (1979).
36. J. R. Black, 'Electromigration modes in aluminium metallization for semiconductor devices', *Proc. IEEE,* **57**, 1587–1594 (1969).
37. J. R. Black, 'Mass transport of aluminium by monentum exchange with conducting electrons', *5th Ann. Proc. Rel. Phys.*, 148–159 (1967).
38. J. R. Black, 'Electromigration of Al–Si alloy films', *16th Ann. Proc. Rel. Phys.*, 233–240 (1978).
39. Da-Yuan Shih and P. J. Ficalora, 'The reduction of Au–Al intermetallic formation and electromigration in hydrogen environments', *IEEE Trans. Electron Devices,* **ED-26**, 27–34 (1979).
40. P. Staeker, 'K_a-band IMPATT diode reliability', *Proc. International Electron. Devices Meeting (Washington)*, 493–496 (1973).
41. F. Sellberg, P. Weissglass, and G. Andersson, 'A study of failure mechanisms in silicon IMPATT diodes', *IEEE Trans. Electron. Devices*, **ED-25**, 742–752 (1978).
42. M. P. Lepselter, 'Beam-lead technology', *Bell System Technical Journal,* **45**, 233–253 (1966).
43. J. A. Cunningham, C. R. Fuller, and C. T. Haywood, 'Corrosion resistance of several integrated circuit metallization systems', *J. Electrochem. Soc.,* **121**, 298–303 (1974).
44. G. Ottaviani and J. W. Mayer, 'Mechanisms and interfacial layers in silicide formation', *This volume*, Chapter 2.
45. J.-P. Pestie and J.-Y. Fourrier, 'Design and manufacture of a microwave low noise transistor having beam-leads', *IEEE Trans. Electron. Devices,* **ED-24**, 73–79 (1977).
46. A. K. Sinha, T. T. Sheng, T. A. Sharkoff, W. S. Lindenberger, E. N. Fuls, and C. C. Chang, 'Certain failure mechanisms in two-level poly-Si gate structures—a transmission electron microscope study', *17th Ann. Proc. Rel Phys.,* 35–38 (1979).
47. S. J. Rosenberg, D. L. Crook, and B. L. Euzent, 'H-MOS reliability', *IEEE Trans. Electron. Devices,* **ED-26**, 48–51 (1979).
48. J. R. Edwards, I. S. Bhatti, and E. Fuller, 'VMOS reliability', *IEEE Trans. Electron. Devices,* **ED-26**, 43–48 (1979).
49. I. S. Bhatti, E. Fuller, and B. Jenne, 'VMOS electrostatic protection', *16th Ann. Proc. Rel. Phys.*, 140–145 (1978).
50. K. G. Aubuchon and E. Harari, 'Radiation hardened CMOS/SOS', *IEEE Trans. Nuclear Science,* **NS-22**, 2181–2184 (1975).
51. G. Yaron and L. D. Hess, 'LASOS—laser annealed silicon on sapphire', *IEEE Trans. Electron. Devices,* **ED-27**, 573–578 (1980).
52. G. L. Schnable, R. B. Comizzoli, W. Kern, and L. K. White, 'A survey of

corrosion failure mechanisms in microelectronic devices', *RCA Review,* **40**, 416–445 (1979).

53. R. B. Comizzoli, L. K. White, W. Kern, G. L. Schnable, D. A. Peters, C. E. Tracy, and R. D. Vibronek, 'Corrosion of aluminium IC metallization with defective surface passivation layer', *18th Ann. Proc. Rel. Phys.,* 282–292 (1980).
54. H. Koelmans, 'Metallization corrosion in silicon devices by moisture-induced electrolysis', *12th Ann. Proc. Rel. Phys.*, 168–171 (1974).
55. R. B. Comizzoli, 'Aluminium corrosion in the presence of phosphosilicate glass and moisture', *RCA Review,* **37**, 473 (1976).
56. F. J. Grunthamer, T. W. Griswold, and P. J. Clendening, 'Migratory gold resistive shorts: chemical aspects of a failure mechanism', *13th Ann. Proc. Rel. Phys*., 99–106 (1975).
57. A. Shumka and R. R. Peity, 'Migrated gold resistive shorts in microcircuits', *13th Ann. Proc. Rel. Phys.*, 93–98 (1975).
58. R. P. Frankenthal and H. W. Becker, 'Corrosion failure mechanisms for gold metalllllizzzeations in electronic circuits', *J. Electrochem. Soc.,* **126**, 1718 (1979).

Reliability and Degradation
Edited by M. J. Howes and D. V. Morgan

CHAPTER 5

Reliability and degradation of active III–V semiconductor devices

J. E. DAVEY and A. CHRISTOU

5.1 INTRODUCTION

From a materials point of view, GaAs provides a number of significant advantages over silicon as those advantages relate to device and integrated circuit application: (a) a significantly higher low field electron mobility; (b) a band structure which supports negative differential mobility and efficient optical transitions; (c) a semi-insulating form which at once allows for homo-epitaxy and dielectric isolation; (d) a compound nature which allows combining with other III–V elements for a wide range of materials tailoring with attendant changes in band structure, crystalline, electrical, and optical properties. These advantages have resulted in a steady progression of GaAs device advancements since the early 1960s which include the following devices: Gunn diodes; light emitting diodes; heterojunction and homojunction injection lasers; solar cells; IMPATT diodes; Schottky barrier diodes; and both the junction and Schottky barrier field effect transistors. The potential for use of these devices in both military and commercial applications means that their reliability becomes a key issue. However, a survey of ongoing reliability investigations clearly shows that the assessment of device reliability has not kept pace with performance gains. Schottky barrier field effect transistor research currently dominates solid state technology developments at the higher microwave frequencies for low noise, power, and digital applications. Therefore, in this chapter, the emphasis will be on the reliability assessment of these devices for the applications cited. The detailed data provided will be, for the most part, chronological in order and the topical content presented will follow the historical evaluation of the salient reliability problems. We will, in addition, provide a brief assessment of the reliability of IMPATTs, transferred electron devices, and Schottky barrier mixer diodes. A discussion will be presented of the important materials related phenomena affecting the reliability of ohmic and Schottky contacts. Finally, a short summary of the most important factors affecting Schottky FET reliability will be provided.

5.2 SMALL SIGNAL GaAs FET RELIABILITY ASSESSMENT

In the first reported reliability studies of small signal FETs two principal failure modes were found.[1] The first failure was gate burn-out under transient voltage spiking. The proposed solution to this failure was to utilize refractory gate metallization systems and Ni was initially recommended. The second failure mechanism reported was ohmic contact migration at elevated junction temperatures. The authors recommended both thicker contact metallization (plated Au) or improved heat sinking or both. At the time of their report using improved fabrication techniques, Abbott and Turner had accumulated 500,000 device hours (room temperature) with no failures. On the basis of this preliminary data, they predicted a room temperature MTTF of 10^7 hours. In a contemporary paper,[2] results were reported on both stress and high temperature storage tests at 227, 259, and 295 °C for 1500 hours. The authors found three reliability problems: (a) gradual fall-off of I_{DS}, which was attributed to degradation of the ohmic contacts; (b) pulse burn-out of the Al gate, assigned to high electric fields at the edge of the MESA; (c) instability of drain current, noise figure and gain as a function of time, attributed to interface traps between the active layer and the SI substrate. In an expanded version of this work, Irie *et al.*[3] carried out a computer calculation on a small signal equivalent circuit for the GaAs FET in order to carry out a correlation between RF and DC parameters corresponding to the failure mode associated with the degradation of the ohmic contact resistance. In principle then, under the assumption of the validity of this correlation, one can make the simpler DC measurements to determine RF failure conditions. With contact resistance degradation as the failure mode, Irie *et al.* estimated an MTBF of 3×10^8 hours for a 100 °C channel temperature.

In 1975, Avantek began internally funded life test studies on small signal FETs and reported no failures in a two-year period for tests up to 100 °C. In a follow-on US Army funded program,[4] 50 packaged 1 μm FETs were subjected to accelerated step stress tests up to 275 °C. At that time, analysis of the cumulative failure data as a function of temperature showed that at 302 °C the average life of a device would be 168 hours or one week. They found two failures: (a) gate degradation (gate short or high leakage); (b) increase in ohmic contact resistance. These failures were associated with Ga migration through both GaAs and the surface metallization. These observations were verified by Auger analysis. Using published data on the activation energy of Ga diffusion through GaAs (0.8 to 1.0 eV) and assuming this activation energy to be correct for the Arrhenius relationship, and MTBF *vs.* temperature plot was constructed using the 302 °C, 168 hour data as a base point. In this manner an MTBF of 5×10^6 hours at 100 °C channel temperature was estimated. This value is about 100 times shorter than the Irie value and approximately 100 times longer than would be predictable from the estimate of Abbott and Turner.

One of the first detailed studies of small signal FET reliability reported was a COMSAT program at Hughes Research Laboratory.[5] From the point of view of developing a reliable device technology, a number of gate metallization systems were studied and included: Al; CrPtAu; TiPtAu; and MoPtAu. Furthermore, both epitaxial and ion implanted FETs were compared although MTBF data were reported only for ion implanted FETs with an aluminium gate and AuCr contact pad. The latter data are shown in Figure 5.1 and are compared with the activation energy obtained by Irie whose data was obtained on unbiased devices. It is to be noted that an activation energy of 1.52 eV is not in agreement with the early reports of either Abbott or Avantek. Use of the 1.8 eV activation energy to estimate MTTF yields a value in excess of 10^9 hours at 100 °C and is considered to be an optimistic estimate of device reliability. The associated 100 °C extrapolated value for the 1.52 eV curve is $\geqslant 10^8$ hours, a value considerably greater than found by Avantek or Abbott. The Hughes–COMSAT study was not sanguine about the utilization of refractory metallizations for high reliability requirements due to processing and uniformity limitations. Although interdiffusion across the Al–CrAu boundaries of the gate pad was reported, this study concluded that for small signal devices, it was the deterioration of the ohmic contacts, rather than the Al gate pad, which was the limiting failure mechanism for the device; this is in contrast with a follow-on study which found opposite results.

In a Rome Air Development Center reliability program at Hughes Research Laboratories,[6] the objectives were to determine operating life, failure modes and to establish a reliability base for small signal 1 μm × 300 μm gate ion implanted GaAs FETs. In their first set of experiments, three sets of 15 devices were monitored using DC characteristics and visual appearance and were initially life tested at 260 °C. These initial tests revealed the dominant failure mode to be gold migration in the source and drain metallization (depletion at source edge, accumulation at drain) with shorts across the source–gate, gate–drain in some cases. The failures were preceded by a decreasing I_{DS} associated with an increasing contact resistance. Based on these preliminary results, Hughes subjected additional in-house FETs to life tests at three temperatures, 240, 255, and 270 °C. After 20 hours, the 270 °C tests showed an intermetallic forming on the Al gate pad; this was *not* observed for the 260 °C tests even after 429 hours. Further testing at 270 °C resulted in additional failures due to this mechanism, while none were observed at 240 °C. It should be noted here that intermetallic formation has been found at and below 200 °C[7,8] and $AuAl_2$ formation was observed[9] after an 8–10 hour anneal at 62 °C with excessive $AuAl_2$ formation at 125 °C.

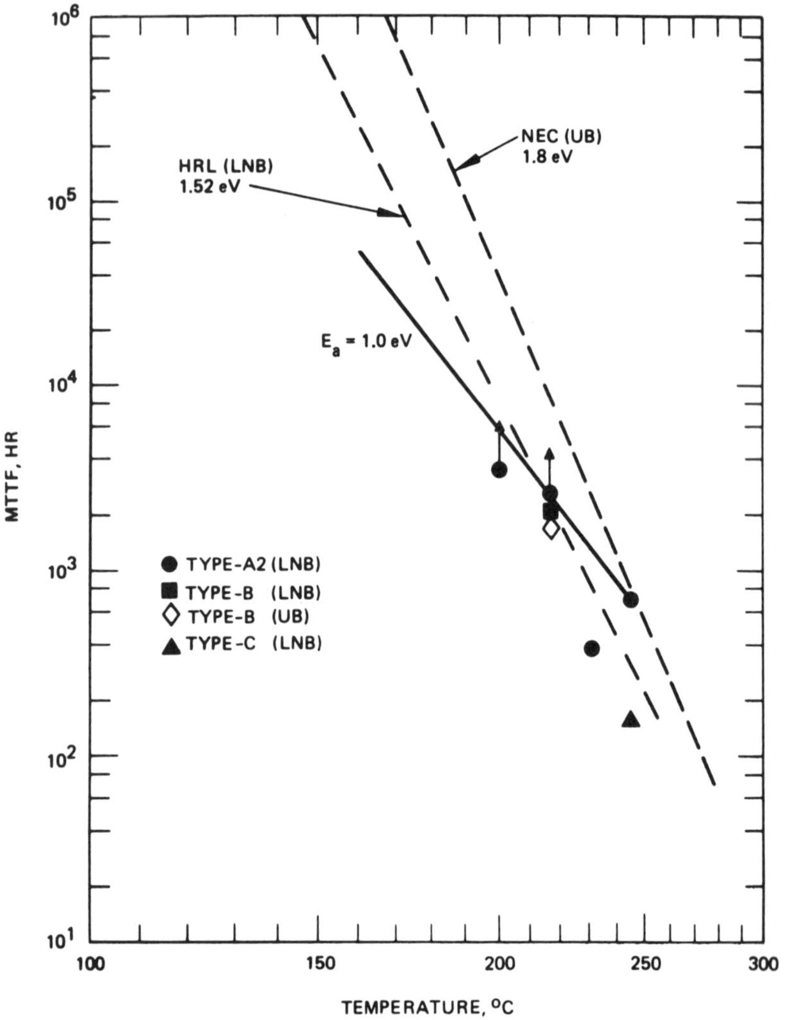

Figure 5.1. MTTF *vs.* ambient temperature for DC biased (5 V, 10 mA) Hughes GaAs FETs and unbiased NEC FETs

Using a 10 per cent change in I_{DSS} as a failure criteria HRL found no failures after 617 hours at 240 °C. They contrasted this with studies of LPE FETs for which substantial failures occurred and concluded that ion implanted FETs promise an improved MTBF; no further explanation for this conjecture was offered. *Subsequent* to these tests, (HRL) used a SiO_2 passivation over the gate pad and exposed GaAs surfaces to prevent metal migration. They also undertook to compare in-house device performance with 40 domestic and 40 non-domestic purchased chips under similar stress conditions and again using

Table 5.1 Construction details of GaAs FETs tested by HRL

FET Element	Type of FET: (A) (HRL)	(B) Purchased non-domestic	(C) Purchased domestic
Substrate	Purchased after qualification tests	Purchased after qualification tests	Purchased
Buffer layer	None	High-resistance VPE 3 μm	None
Active *n*-layer (channel)	Ion implant (Si) 2×10^{17} cm^{-3} 2500 Å	*S*-doped VPE 1.5×10^{17} cm^{-3} 2500 Å	Sn-doped LPE (750 °C)
Source–drain ohmic contact	Au-Ge–Ni–Au (1500–400–500 Å) Alloyed 30 sec at 450 °C	Au-Ge–Pt (1500–400 Å) Alloyed at 400 °C	Ge-Au–Ni–Au (1400–200–500 Å) Alloyed 45 sec at 475 °C in H_2
Overlayer	Cr–Pd–Au (1000 1000 8000 Å)	Ti–Pt–Au (2000–2000–2000 Å)	
Gate:			
Gate	Al (6000 Å) 1×300 μm	Al (4500 Å) 1×300 μm	Ti–Cr–Pt–Au (100–100–400–1700 Å)
Bond pad	Cr–Cr–Pd–Au (3000–1000–1000–8000 Å)	Ti–Pt–Au (2000–2000–2000 Å)	Ti–Cr–Pt–Au 1×250 μm gate
Glassivation	SiO_2 (1500 Å)	SiO_2 (1500 Å)	SiO_2 (2000 Å)

a 10 per cent change in I_{DSS} as a failure criterion. Table 5.1 contrasts the fabrication details of these devices; they will be referred to hereafter as A, B, and C. Both device sets were carefully characterized prior to stress; life tests were begun in May 1977. Initial tests of 19 type C devices at 245 °C had to be terminated after 69 hours because an average of 36 per cent change in V_g was necessary to maintain an I_{DS} of 10 mA. Initial tests on 20 type B devices at 245 °C showed unacceptable gate leakage current (114 microamperes, sufficient to cause Al electromigration), thus limiting tests to 220 °C. Tests of the biased type B devices at 216 °C had to be terminated after 142 hours due to large increases in I_G (up to 40 mA). Even *unbiased* type B devices similarly stressed at 216 °C showed significant gate degradation as shown in Figure 5.2. The pitting and metallic growth *were not* attributed to purple plague.

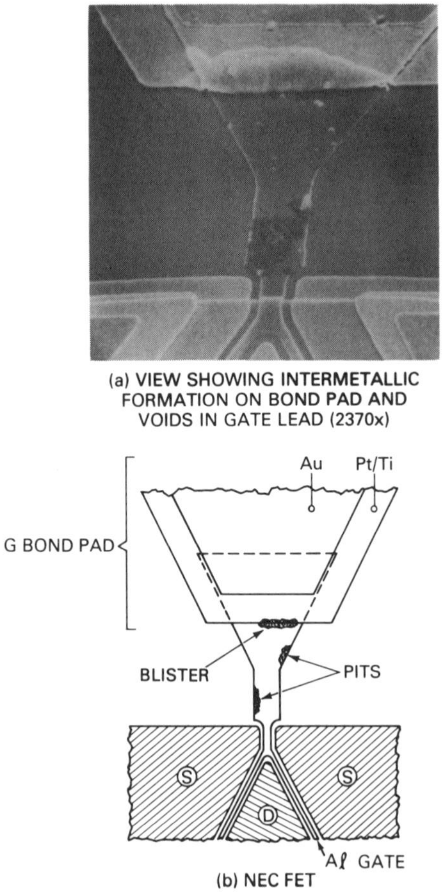

Figure 5.2. Gate degradation and void formation for unbiased type B (see Table 5.1) Al gate devices

After three stress cycles at 245 °C (261 hours total) a significant number of failures had occurred for type C devices (10 out of a total of 19). The decrease in g_m was associated with gate degradation; the mechanism of degradation was unexplained but speculated to occur through reduction of channel thickness or mobility or both. Gate–source and gate–drain shorts were observed (optically). As of the final report on the (HRL) (RADC) program, the following conclusions could be drawn: (a) type A2 (HRL) GaAs FETs generally fail because of degrading gate characteristics. This was interpreted as related to a poor SiO_2 glassification. Neither electromigration in, nor degradation of, the ohmic contacts was observed. Degradation of channel resistance was not seen. Type B FETs failed because of increasing parasitic source and drain resistance. The Au–Ge–Pt contact was not found to be reliable. Many of the type B devices failed catastrophically through intermetallic formation in the gate bond pad. A Pt–Al interdiffusion through (or along) the Ti barrier was postulated. Type C FETs failed early through rapid gate deterioration. It was suspected that the Au top layer on the gate stripe migrated down to the GaAs and degraded the Ti Schottky barrier. An MTTF curve representing HRL data on these devices is shown in Figure 5.1 by the solid line; the activation energy of 1.0 eV agrees with other observations.[10,11]

A separate and internally funded reliability program on GaAs FETs was begun in early 1976 at Hughes Aircraft Space and Communications Group.[12] Multistage RF amplifiers (7 to 8 GHz) were constructed with NE 24400, NE 38800, (HRL) and Avantec FETs (with polygallium arsenide overcoat) of 1976 vintage. Baseline measurements included g_m, open channel resistance, V_{PO}, gate leakage, I_{DSS}, and RF gain. I_{DS}, V_{DS}, V_{GS}, I_{GS} were monitored on a daily basis. Accelerated life tests were carried out at 150, 175, and 200 °C. Twenty-eight amplifiers were tested over a period up to 20,000 hours. Both the NEC devices and the (HRL) devices survived the entire test period. The devices tested at 200 °C showed both a gain degradation and a noise figure degradation at about 5,000 hours. Avantek devices tested at 200 °C showed a drastic change at about 6,000 hours not only in RF parameters but also in a decreased I_{DSS}. Visual examination gave evidence of gold diffusion through the polycrystalline GaAs passivation layer. The Avantek devices tested at 175 °C showed the same change at about 11,000 hours. Those tested at 150 °C or at room temperature showed no change during the course of testing. A rough estimate of the failure rate was that both NEC and HRL parts would be less than 0.1 FIT at 55 °C channel temperature using either a 1 dB gain decrease or 0.5 dB NF increase as a failure criterion. In this same program, Hughes also assessed the short term gain stability of NE 24400 FETs in 4-stage RF amplifiers designed for 8 GHz operation. Measurements out to 1000 hours showed no discernible changes in gain values from that observed 30 sec after turn-on.

Recently, a detailed study of the degradation mechanism of Hitachi X-band FETs has been carried out.[11] The devices were fabricated with *p*-buffers, *n*- and n^+-layers grown by VPE, AuGeNi source and drain contacts and Mo–Au Schottky gates. The devices were nominal 0.5 μm long by 300 μm wide and were protected against ambient effects by an unspecified organic coating. A total of 400 devices were tested at 175, 200, 225, and 265 °C. The parameters measured during temperature stress were: I_{DSS}; g_m; V_{PO}; R_s; R_G; gate breakdown voltage; MAG (MSG) at 8 GHz; NF_{min} (minimum noise figure). Variations in many of these parameters were plotted *vs.* time with temperature as a parameter. The significant finding was that MAG (MSG) did not show appreciable change as a function of stress while NF_{min} did show considerable degradation in every case. This is shown in Figure 5.3. The authors also

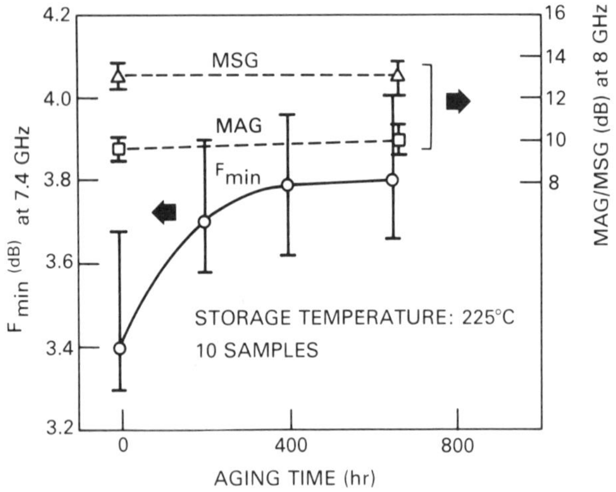

Figure 5.3. Change in minimum noise figure, F_{min}, at 7.4 GHz and maximum available and stable gain MAG, MSG at 8.0 GHz *vs*. ageing at 225 °C

correlated the NF changes with changes in zero bias drain conductance as shown in Figure 5.4. It is to be emphasized that no degradation in small signal power gain was observed as associated with the changes in the other parameters. Using a 10 per cent decrease in drain conduction as a failure criteria

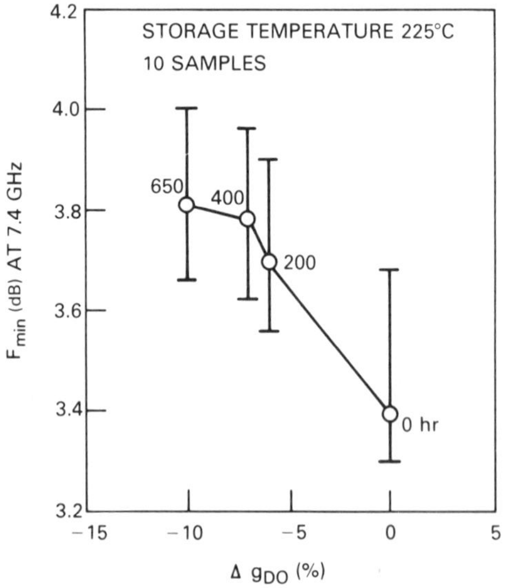

Figure 5.4. Increase in F_{min} with decrease in zero bias drain conductance Δg_{DO} with hours at 225 °C as a parameter

MTTF *vs*. temperature was determined and is plotted in Figure 5.5. The activation energy corresponds to 1.0 eV and an MTTF of 10^7 to 10^8 hours is estimated for a channel temperature of 80 °C. Mizuishi *et al.*[11] also performed detailed physical failure analysis of the devices by micro-spot Auger and ion-microscopy and showed that the primary failure mode was degradation of the Au–Ni–Au-Ge ohmic contacts through interdiffusion of Ga into the Au contact and Ni into the surface of the active layer. The result is the development of a high resistance interface layer between the ohmic contact and the body of the crystal.

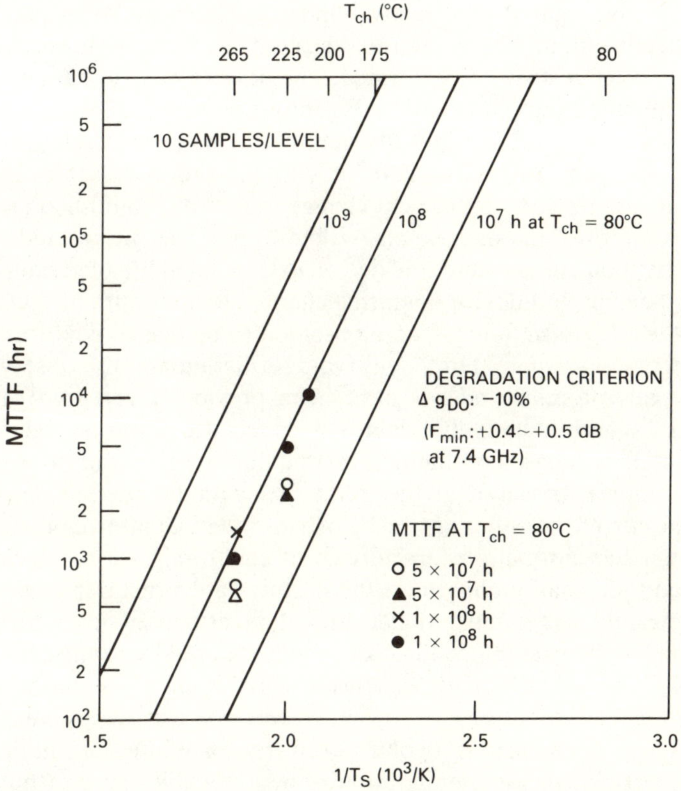

Figure 5.5. MTTF *vs*. storage temperature. The solid lines are for a 1.0 eV activation energy

In August of 1978, the US Rome Air Development Center initiated a new small signal GaAs FET reliability study at Hughes Aircraft Corporation.[13,14] It is the intent of this program to evaluate a significant number of state-of-the-art commercial devices. Results from this program are as yet preliminary in nature and not available for public dissemination. It is to be noted that this program includes a variety of other military standard tests according to

MIL-S-883B including hermeticity, temperature cycling, acceleration, random vibration, and mechanical shock; all tests were interposed with DC and hermeticity evaluation.

5.3 POWER GaAs FET RELIABILITY ASSESSMENT

In September of 1978, the first reliability investigation of GaAs power FETs was published.[15] The devices were four cell structures with 1 μm gate length and 150 μm gate width per cell. The gates were aluminum and the source and drain contacts sintered In–Ge–Au plated up with Au. Both Al and Au bond wires were investigated as directly bonded to the gate bond pad with no barrier metallization. The accelerated stress tests were performed with DC bias only; no RF tests were performed. The devices were to have application as space qualified parts with 500 mW output power, 5 dB gain and 20 per cent efficiency at 11 GHz. Life tests were performed over a range of channel temperatures from 160 °C to 220 °C with the DC parameters monitored throughout the tests; a 10 per cent change in g_m was established as a failure criteria. Analysis of the data on an Arrhenius plot was interpreted as having a 2 eV activation energy which yielded an extrapolated life of greater than 10^8 hours at the expected device operating channel temperature of 70 °C. In the case of Au–Al bonds, failures were observed to be due to void formation in the gate stripe just where it broadens (void formation was also observed in the bonding pads). Similar failures have been previously reported[10] for small signal FETs and tentatively identified as electromigration related. This interpretation was based on the bias related nature of the failure and the fact that the failures appeared at the point of maximum current density even though the current densities of 1×10^4 A cm^{-2} were about two orders of magnitude less than normally characteristic of electromigration.[16] Irvin further speculated that small grain size in the narrow gate structures might account for significantly larger current densities than estimated from the physical dimensions of the gate. Since the bias conditions used by White *et al.*[15] on the power FETs would also yield densities of $\gtrsim 10^4$ A cm^{-2}, where the gate narrows, and since the failures also occur at this point, it seems clear that the same physical mechanism is involved even though White *et al.* in their initial work did not believe electromigration to be a possible cause of failure. In a very recent study[17] a comparison was made of devices with both Al and Au wire bonds directly to the Al gate pad in an attempt to elucidate the failure mechanism: void formation was found to be the dominant failure mechanism for both Al and Au but with different activation energies. Figure 5.6 shows the Arrhenius data. Microprobe analysis showed no Au contamination of the Al bond wires. White concluded from this study that gate metal electromigration *is* the dominant failure mode for Al gate power FETs even in the absence of the Au–Al couple.

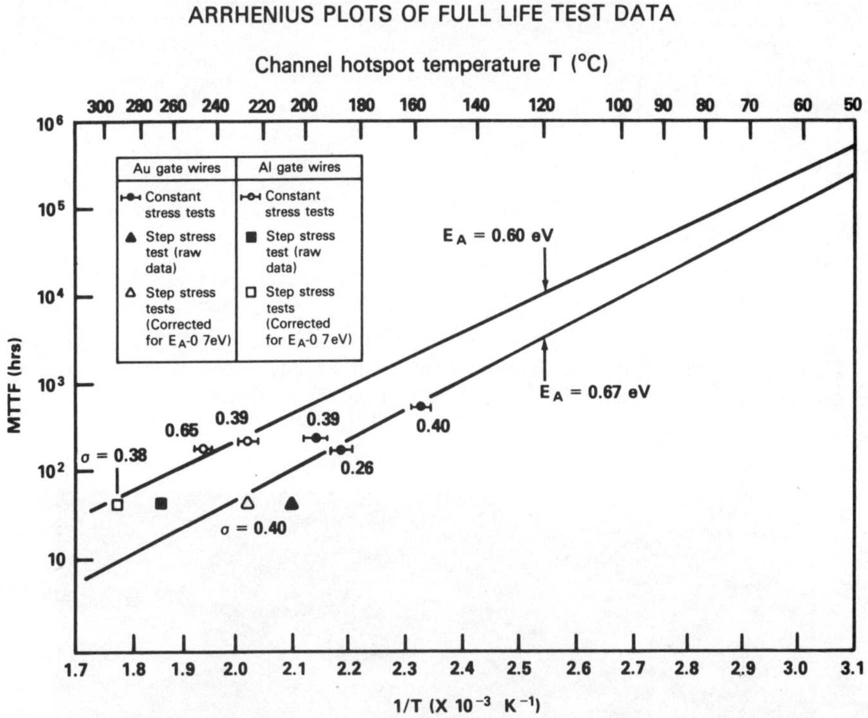

Figure 5.6. Arrhenius plots for both Al and Au bond wires to Al gate power FETs

Just one month after the initial work of White was reported on DC bias testing of power FETs, the first reliability evaluation of commercial power FETs under full microwave operation was described.[18] The FETs were Texas Instruments one watt *X*-band devices with eight Al gate stripes each of $1 \times 200\ \mu m$ width and with AuGeNi source and drain contacts. Au wire bonds were used to directly contact the Al gate bond pads with no barrier metallization. During life tests most of the devices were operated at microwave saturation with optimized slug tuner matching, optimized gate voltage and an 8 volt drain bias. Although tests were made over a range of input power, P_i, from 100 to 500 mW, normally devices were tested with $P_i = 320$ mW, for which the channel temperature was estimated to be between 120 and 130 °C. Of the 15 devices tested, 10 exhibited one or more sharp drops in power output as shown in the strip recording in Figure 5.7. In addition to these sudden drops in power during operation, the devices failed to pinch off after return to DC bias testing. All of these failures were categorized as catastrophic. Due to the construction of the devices neither high magnification optical nor SEM examination could be unambiguously utilized to locate or identify the causes of failure or both. However, the electron beam

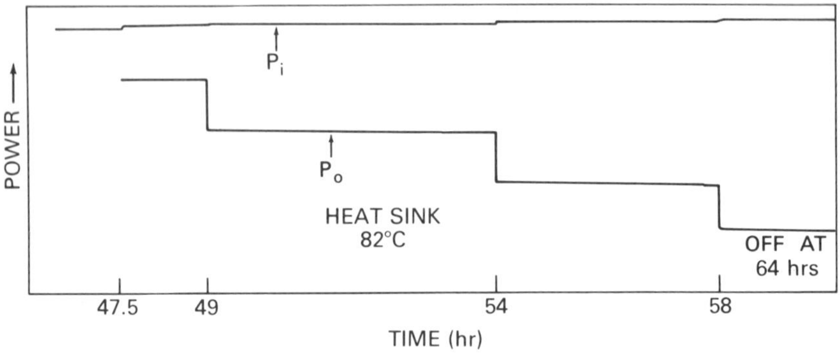

Figure 5.7. Strip chart recording of power input, P_i, and power output, P_o, of Al gate power FTEs typical of gate void occurrence. $P_i = 320$ mW

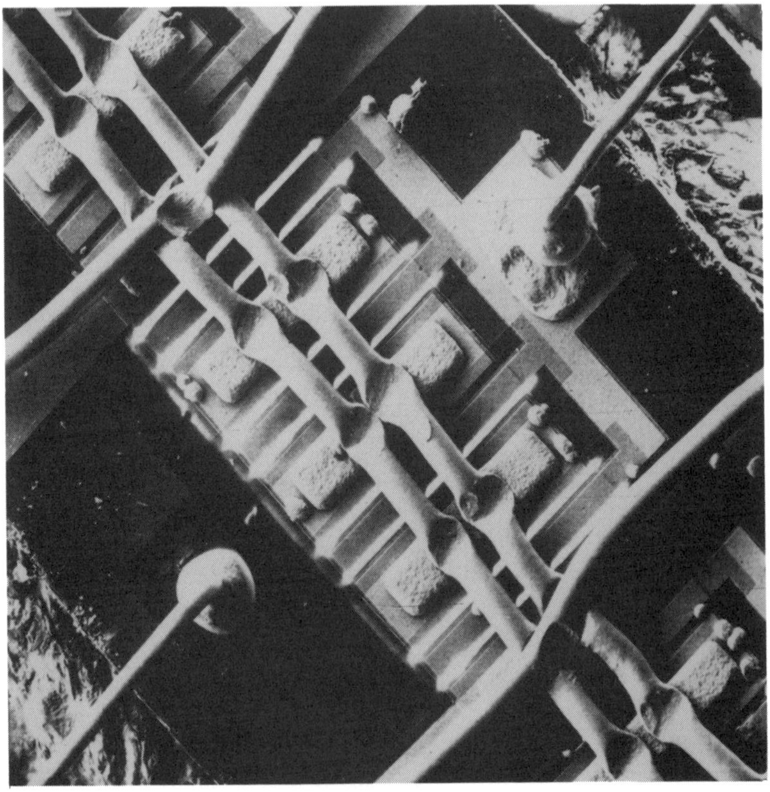

Figure 5.8. Superposition of a SEM and EBIC image; active gates show up as white stripes

induced current mode (EBIC) in the SEM clearly located void formation in the gate electrodes even at low magnification. Figure 5.8 shows a superimposed EBIC and SEM photograph on a failed device. The white strips are the good gates as revealed by EBIC. It can be clearly seen that only half of the fifth gate from the left is operative and that a void has occurred just under one of the source bond wires. Figure 5.9 is a highly magnified photograph of a void located by EBIC. All of these devices failed by gate void formation and were subsequently analysed by micro-spot Auger electron spectroscopy (AES) and energy dispersive X-ray (EDAX) techniques. The analysis of the gates revealed the presence of a surface compound of AuAl extending from the gate pad to the edge of the void, as well as the presence of Ga in the same region. Furthermore, micro-spot AES indicated a change in the concentration of Au between the gate region and the bond pad. Macpherson *et al.* interpreted these failures as directly related to gate current flow during RF operation. For a crude estimate of the magnitude of this current, they used a simple model[19] to estimate the RF current:

$$I_{\mathrm{RMS}} \approx \frac{(P_{\mathrm{I}})^{1/2}}{R_{\mathrm{I}}}$$

with R_{I} the device input resistance. For the devices tested, R_{I} is estimated to be a few ohms yielding RMS gate currents of hundreds of milliamps for the P_{I} range employed in the tests. Average gate current under normal microwave operation was monitored for a few devices and typical data is shown in Figure

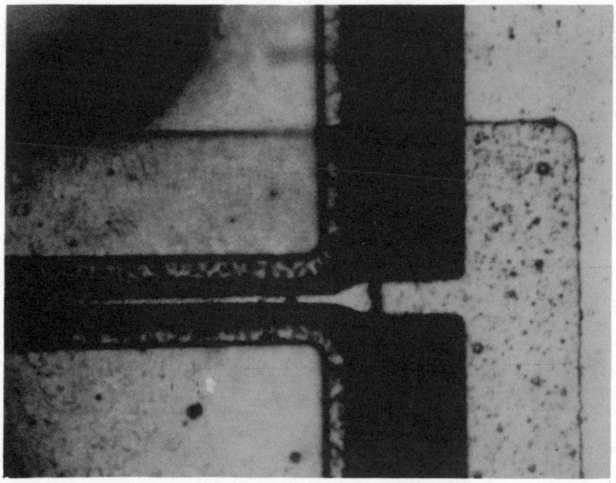

Figure 5.9. Optical photograph of a gate void located by EBIC

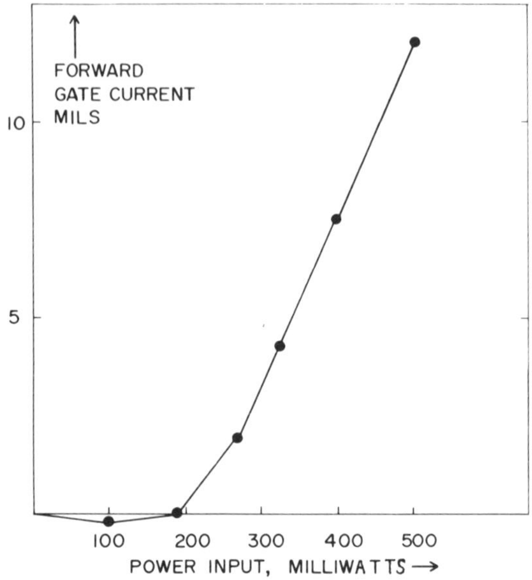

Figure 5.10. Gate current at 8 GHz. Gate voltage and tuning were adjusted for maximum power output at each data point

5.10. For the gate geometry of the devices, this average current can amount to as much as 3×10^5 amps cm^{-2} at a P_I of 500 mW while the RMS microwave gate current density can be as high as 10^7 amps cm^{-2}. The physical analysis of the failed devices indicated that in these studies the void formation took place via an Au–Al reaction in the presence of a concentration gradient driven surface diffusion of Au catalysed by the presence of Ga. Macpherson *et al.* interpreted this reaction and the subsequent void formation to be directly related to the high DC and microwave current densities present in the gates of the devices under test. Thus, the work of Irvin on small signal FETs, the studies of White on both Al and Au bonded power FETs and that of Macpherson, all indicate current related failure modes in Al gates. These results point to the inapplicability of the standard electromigration data[16] for Al on Si to GaAs devices and to the need for further research on this phenomenon.

Recent work on RF accelerated stress testing[20] includes the investigation of gold refractory gate devices made by Microwave Semiconductor Corporation (MSC)[21] as an effective comparison to the Texas Instrument Al gate devices and to help elucidate the void failure mode mechanisms. The MSC FETs (88002) were nominal 0.5 watt, flip-chip mounted X-band, devices with eight 1×150 μm gate stripes of TiWAu with Au-Ge–Ni source and drain contacts.

The DC gate–source current–voltage characteristics were always measured prior to and subsequent to step-stress testing and in addition many of the devices were characterized by gate–source capacitance using a standard CV profiler. The purpose of this test was to assess *changes* in capacitance after stressing. No attempt was made to determine absolute values of device capacitance. In these experiments, pinch-off voltage was arbitrarily defined at the voltage at which the drain current was reduced to 1 milliamp.

During the stress testing, the following parameters were continuously monitored: RF power input, RF power output, gate current, drain current, drain–source voltage, gate–source voltage and base plate temperature. For these devices non-catastrophic failure was abritrarily defined as a 1 dB decrease in power output at room temperature. Channel temperature was not measured for each device, but instead was estimated from measurements of thermal resistance as a function of temperature on typical devices using the electrical method.[22] Unlike the infrared or liquid crystal methods, the electrical method can be used when the surface of the device is not visible. The results agree well with those obtained by the device manufacturer using an infrared microscope.

A total of twenty-two MSC devices were tested from four different lots. Of these, nine failed catastrophically, three did not fail before testing was terminated and ten degraded gracefully. Cumulative failures followed a straight line on a log-normal plot and are attributed to a single failure mode. Mean-time-to-failure as a function of device temperature for devices which failed catastrophically and for devices which failed in any manner are shown in Figure 5.11. In the figure, the data point at 190 °C represents the average MTTF of four devices; the data points as a function of increasing temperature represent average MTTFs of 3, 5, 4, 2, 2, and 1 device, respectively.

Although the data shown in Figure 5.11 represents a relatively small sample of devices, it is helpful in establishing an estimate of the MTTF which can be expected at a reasonable operating temperature. From Figure 5.11, a MTTF of approximately 2×10^8 hours is projected for a channel temperature of 100 °C. The activation energy is 2.3 eV.

After failure at elevated temperatures, devices were examined with a scanning electron microscope (SEM) using both conventional and electron beam induced current (EBIC) modes of operation. In no instance was any evidence found for gate void formation for the refractory gate system. From these data void formation appeared to be a singular characteristic of Al gate devices.[17] Additionally, no physical degradation of the source or drain contacts was observed. The DC parameters of devices which failed non-catastrophically were then remeasured. The only consistent change was in the gate–source reverse characteristics. In every case of non-catastrophic failure, the amount of reverse current at low reverse bias voltage increased dramatically. Similar changes were observed in the drain–gate characteristic. The increase in

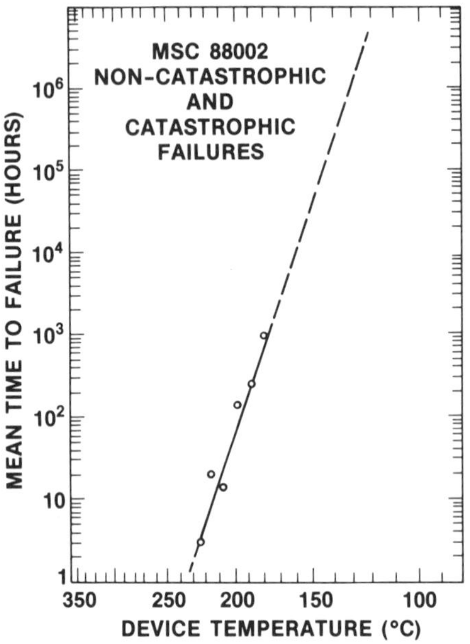

Figure 5.11. MTTF *vs.* channel temperature for both non-catastrophic and catastrophic failures under RF bias. The activation energy is 2.30 eV

reverse current may be due to a decrease in breakdown voltage. Forward characteristics also changed but not in a consistent manner.

Attempts have been made to determine the physical mechanism for the non-catastrophic device failures. Plots of device gate–source capacitance (including package) revealed that virtually no change in capacitance occurred for low values of V_{gs} after stressing. This indicates that it is unlikely that the physical cause of failure is related to the Schottky metal–GaAs interface. In addition, some of the devices which exhibited large increases in reverse current were baked in vacuum for 24 hours. No change in either gate–source or gate–drain characteristics was evident after baking, which appears to rule out surface leakage as the cause of failure. This failure mechanism is still under investigation.

MSC has reported[23] on the reliability of refractory gate power FETs tested under DC bias. The FETs were the MSC 88004 nominal 1 watt 8 GHz devices which are single cell, one micron gate with 2.4 mm total source periphery. The active layer was vapour phase grown on a similarly grown undoped buffer layer of 5 μm thickness. The source and drain contacts were Au–Ge–Ni with Au plating and no refractory barrier. The gates were Ti–W–Au. The devices are flip-chip mounted and can be hermetically pack-

Table 5.2 High Temperature with bias (from Drukier[23])

Test condition:	Constant drain current at fixed drain voltage
Temperature:	140, 150, 175, 200, 225, 250 °C
Device hours:	~2.7×10^5
Devices:	>100
Measurement times:	(25) (2^N) Hours $N = 0, 1, 2, \ldots, T \leq 200$ °C (5) (2^N) Hours $N = 0, 1, 2, \ldots T \leq 200$ °C
DC dissipation:	4–5 Watts

aged. However, all tests were performed in dry nitrogen. The failure criteria was a 1 dB change in power output at 6 GHz in a circuit with fixed bias, tuning and RF drive. DC parameters were measured throughout the tests.

A minimum of ten devices were stressed at any one time in the DC bias tests. Although extensive precautions were taken to eliminate oscillations, occasionally a failure occurred which could be related to oscillation; these devices were included in the samples used to evaluate the data. The bias test conditions used by Drukier are shown in Table 5.2. Two failure modes were observed: (a) electromigration in drain fingers and (b) gold void formation in the gates at positions of high current density. Although two failure modes were observed, Drukier did not distinguish these in the Arrhenius data as plotted in Figure 5.12, although the gate failure was described as dominant. The electromigration failure in the drain contacts of these devices would not be surprising if a flip-chip mount had not been used in view of the fact that no barrier metallization was used. The use of barrier metals has been strongly advocated[24,25] for use with the Au–Ge–Ni ohmic contacts for reliable device performance and will be discussed in Section 5.8. The void failure mode for the Ti–W–Au gates is also surprising since Cohen and Macpherson did not observe such failure in similar MSC devices under RF accelerated stress tests at similar temperatures. Although Drukier did not measure gate leakage current under test, it is implied that electromigration is the cause of these failures. This is not in disagreement with the postulate by Irvin, White, and Macpherson for Al gates or the possibility of such failures for small grain size films in general. This work further indicates the need for such a definitive study and also the need to compare DC and RF failures under the same temperature stress. The DC stress reliability data of Drukier shown in Figure 5.12 extrapolate to an MTTF > 10^7 hours at 100 °C temperature; he calculated the activation energy to be 1.85 eV.

In a very recent study[26] a comparison was made of relaxed geometry power FETs under both DC and RF bias. The devices were 2 μm gate length, 6 mm wide and were VPE grown on Cr compensated buffer layers. The gates were 1 μm thick Al with a TiPtAu barrier layer to the bond pad. The source and drain contacts were Ge–Au-Ag-Au with a TiPtAu barrier to the plated Au

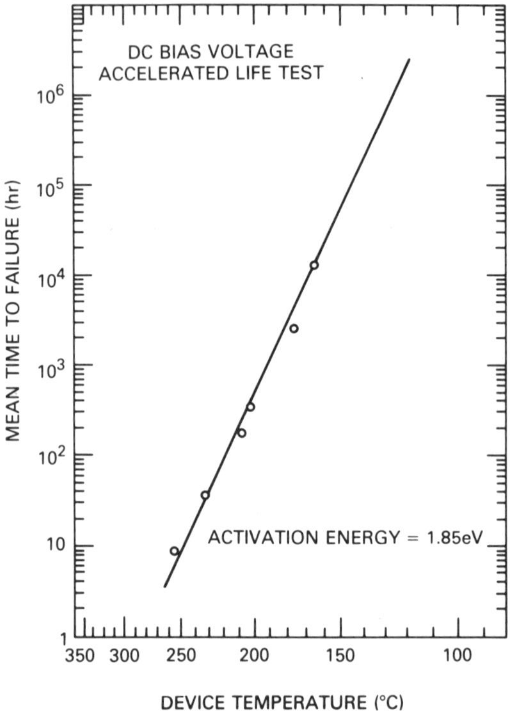

Figure 5.12. Arrhenius plot for MSC 88004 devices for DC bias life test

contact. A proprietary silicon nitride passivation system was also utilized. The failure criteria were a 5 per cent degradation in I_{DSS} for the DC tests and a 5 per cent degradation in saturated output power for the RF case. The life tests were performed at 250, 210, and 175 °C with respective sample sizes shown in Table 5.3. (According to Fukui[26] these are channel temperatures and the channel temperature was estimated from thermal resistance measurements.) No catastrophic failures have been observed in 955,000 device hours for these tests as is shown in Table 5.3; only gradual degradation has occurred. Furthermore, the authors observed no difference between the DC and RF tests. Interestingly, gate currents of 2.5×10^5 A cm^{-2} were reported with no void formation observed. This is to be contrasted with Irvin, White, Macpherson, and Drukier. Fukui's interpretation of his data when plotted on Arrhenius plot as shown in Figure 5.13 yielded a best case activation energy of 1.8 eV with an MTF of 5×10^8 hours at 110 °C and a worst case activation energy of 1.2 eV with an MTF of 2×10^8 hours at 110 °C. No change in contact resistance was measured and no changes in either the electrical or visual properties of the Al gates was seen. The degradation reported was attributed to a change in v_{sat} and the low field mobility at high temperatures.

Table 5.3 Progress of life tests

Drive	Channel temperature (°C)	Slice	Number of samples	Hours of aging (as of 31/10/79) Average	Total	Device-hours	Number of burn-out samples
DC-bias only	250	A	8	2,980	23,840	47,680	0
		B	8	2,980	23,840		0
	210	A	40	4,120	164,800	272,800	0
		B	16	3,970	63,520		0
	175	A	109	4,090	445,800	555,900	0
		B	30	3,670	110,100		0
DC + RF drive	250	A	8	3,060	24,480	34,640	0
		B	8	1,270	10,160		0
	210	A	6	1,840	11,040	43,920	0
		B	8	1,250	10,000		0
						888,000	0
DC only	210	C	16	2,780	44,480	67,400	0
DC + RF			8	2,860	22,880		0
						955,000	0

Table 5.4 Results of CW and pulsed RF stress tests on Type-A2 FETs (f = 9.7 GHz)

RF Stress Test	Input Power, W	Effect on FET
CW (N = 5)	<1	None
	1–2	Small changes (~0.2 dB) in F_{min} and G_a
	≥3	F_{min}/G_a failure
		Burnout
Pulsed (N = 2)	<1	None
	1–3	Small changes
	3–10	Burnout

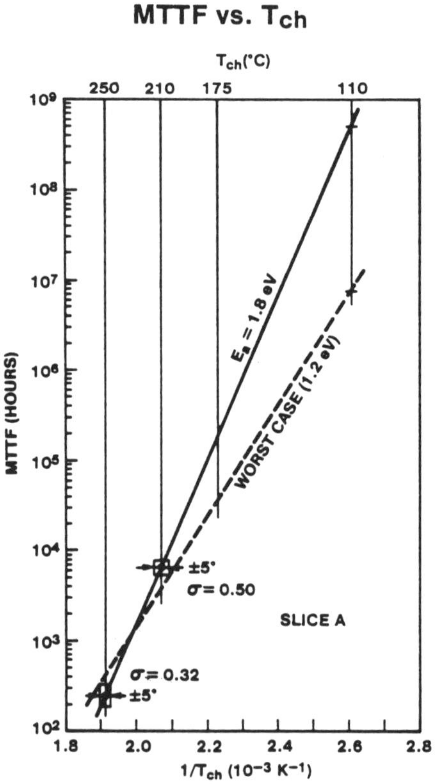

Figure 5.13. MTTF *vs.* channel temperature for Al gate (2 μm $\times$ 6 mm) power FETs with 28 dB m input power at 4 GHz

5.3.1 Other power FET reliability programs

Other recently initiated reliability programs in the United States are an internal effort at RCA Laboratories, under the sponsorship of Intelsat, Rome Air Development Center funded programs at Hughes Aircraft and Texas Instruments[14] and an internally funded program at Hughes Aircraft. These programs will be described in that order.

RCA At the present time RCA[27] has an ongoing reliability assessment program with application of power amplifiers as alternatives to TWTAs in a variety of space communications satellite systems with a design criterion of 10 years life. Since RCA requirements involve both commercially available devices and selection from large production runs, the study has been restricted to the Fujitsu FLC-30 3.0 watt and FLC-50 5.0 watt transistors which qualify in both of these categories. These devices will, in principle, meet the system design requirements of 44 per cent power added efficiency at 2.5 watts power output; the question of satisfying the 10 year life requirement is the basis for the testing program. Both of the Fujitsu devices use AuGe–Au source and drain contacts and Al gates. In the first unscreened temperature bias stress tests, under microwave power, RCA found two failure modes: gate electromigration and interdiffusion and electromigration in the source and drain contacts. The Fujitsu organization itself is presently examining the gate failure mode which dominates at low temperature (<180 °C). RCA is continuing testing of these devices subject to the following screen:

100 per cent visual
DC burn-in at 175 °C channel temperature for 168 hours
RF burn-in at 175 °C channel temperature for 168 hours.

Gate leakage is measured before and after tests. A change of >15 per cent in electrical parameters is the failure criterion. The screen is an overstress in terms of DC bias with $V_{SD} = 11$ volts.

For RF testing the following electrical parameters are employed:

$I_{DS} = \frac{1}{2}I_{DSS}$
RF input = 1 watt
$V_{DS} = 9$ volts
V_{GS} is adjusted to provide $I_{DS} = 500$ mA at testing temperature.

To date, 8 FLC-30s and 16 FLS-50s have been RF stressed up to channel temperatures of 225 °C. The FLC-30 devices degraded gracefully over 1150 hours exhibiting increases in average gate current from a few μA at the start of the tests to ~1 mA at the end of the test. (The latter value corresponds to $\sim 2 \times 10^5$ A cm^{-2} for those devices.) All of the FLC-50 devices tested at 225 °C failed within a few minutes. Devices tested at 200 °C resulted in an

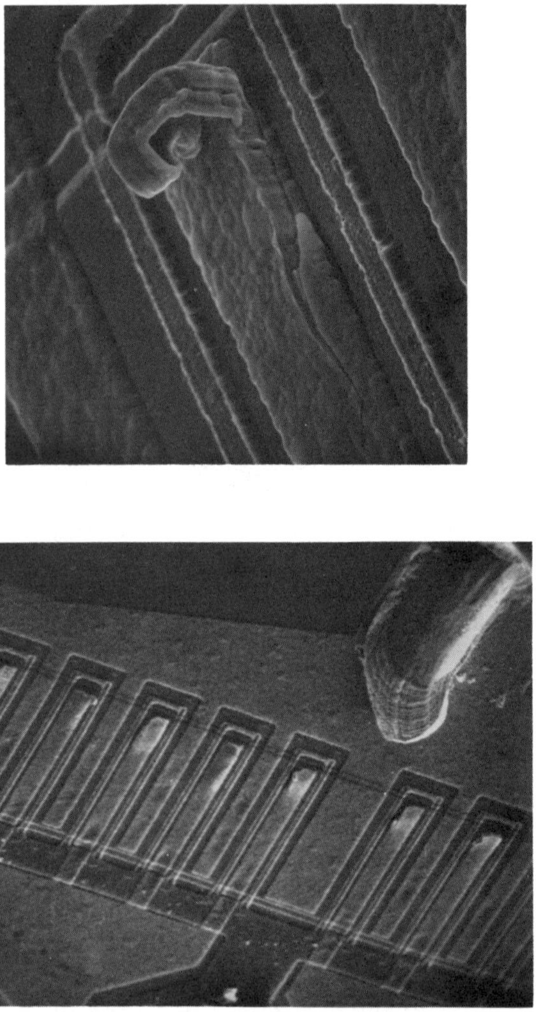

Figure 5.14. Typical electromigration effects in GaAs power FETs showing Ga whisker growth

MTTF of 2200 hours with source–drain electromigration as the primary failure mode. Based on a 1.1 eV activation energy, the extrapolated MTTF at 100 °C is 3×10^6 hours. Failed FLC-30 devices have been analysed at NRL. Figures 5.14–5.16 show electromigration, Ga interdiffusion in the contact system with subsequent Ga whisker growth and gate void formation due to the AuAl couple formation in the gate. Sputter profiling experiments at NRL indicated that the interdiffusion in the source and drain contacts proceeded into the active channel region under the bias–stress tests.

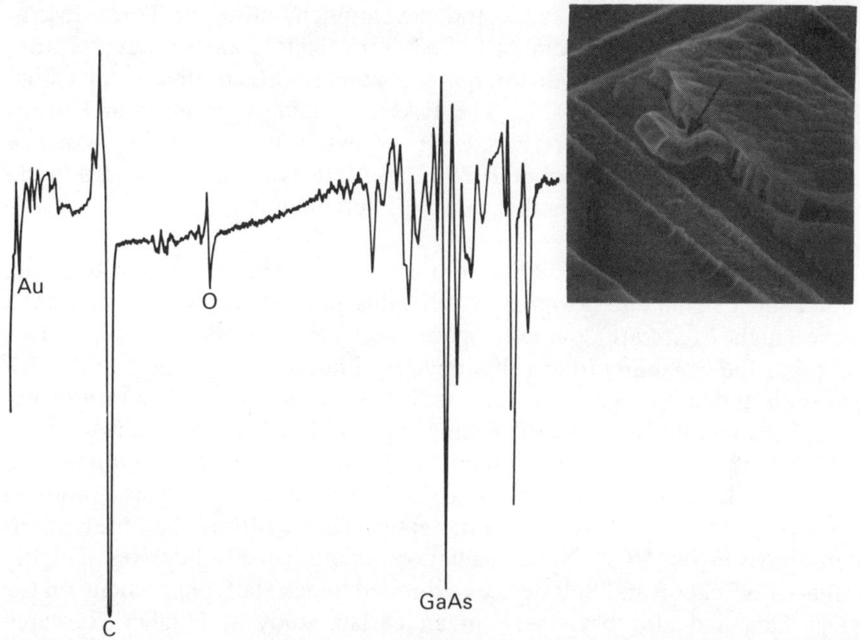

Figure 5.15. Micro-spot Auger electron spectroscopy analysis of electromigration region

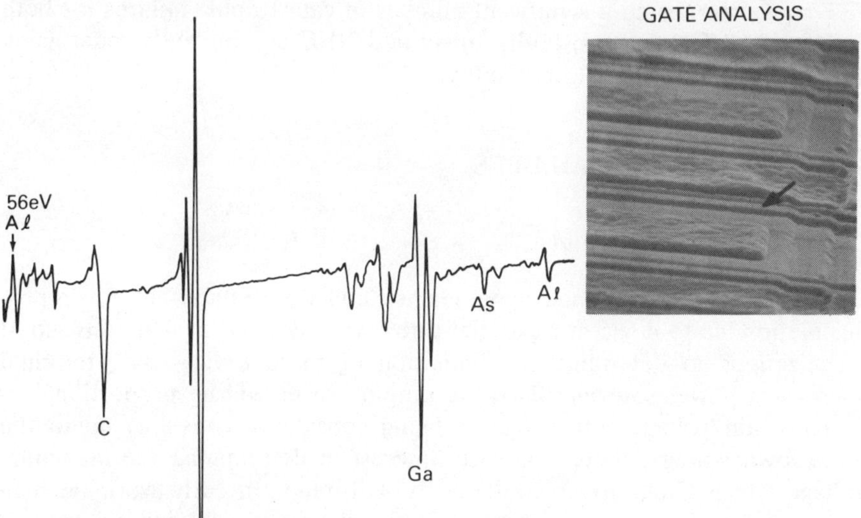

Figure 5.16. Micro-spot Auger electron spectroscopy analysis of aluminium gate region

RADC programs The research and development effort at Texas Instruments was initiated in November 1978[14] in order to compare devices from major industrial houses, both foreign and domestic. Definitive results from this study are not yet available. The RADC reliability program at Hughes Aircraft began in September 1978, however, data was restricted to studies of small signal FETs as of October 1979. These latter data are discussed in the section of this chapter on small signal FET reliability.

Hughes Aircraft (internal) One of the most comprehensive industrially funded small signal and power FET reliability programs is being undertaken by the Hughes Aircraft Company Space and Communications Group.[12] This group has the capability to test fifty devices either at DC or 8 and 4 GHz RF operation under a range of power conditions and channel temperatures between 180 and 240 °C. Since the testing began in 1979, it is too early to assess an MTTF or associated activation energy for failure. However, Fujitsu FLC parts tested showed a large percentage of failures due to electromigration in both the source fingers as well as the gates. This confirms the observations cited above in the RCA–NRL evaluations of Fujitsu FLC devices. Hughes testing of NEC 868 and 869 devices exhibited 'black spot' phenomena on the gates. This had also been seen in an earlier study at Hughes Research Laboratory on low noise FETs[28] and the appearance of spots was usually a precursor to void formation and failure (cf. Figure 5.2). The Hughes power FET study also found a significant number of catastrophic failures for both Fujitsu and NEC devices; both Fujitsu and NEC are currently redesigning these devices for improved reliability.

5.4 CATASTROPHIC FAILURES

5.4.1 Source–drain burnout

Once the properties of small signal GaAs FETs were established[29] as reproducible low noise devices, a parallel thrust was initiated in many industrial organizations to determine the limitation of these devices as 3-terminal microwave power sources. Since the output power will be proportional to source–drain voltage (other factors being considered constant) up to the breakdown voltage, there was early interest in determining the maximum voltage which GaAs FETs could safely withstand. In early work on high power devices, it was found[30] that without special processing of the source and drain contacts the safe applied voltage was less than 10 V for the geometry of the devices. However, Fukuta was able to increase the breakdown voltage to better than 17 V by etching back the active channel in the source–drain contact areas followed by regrowth of the n^+ epilayer and deposition of the standard contact metallization. In 1977, research workers from the same laboratory reported that, with increases of drain bias beyond 15 V visible emission of light occurred along the drain contacts. This light emission was attributed to impact ionization associated with the high field at this contact. However, no device failures were discussed in that report. At approximately the same time,[31] a detailed investigation of source–drain burnout, as it pertained to power FET applications, was published. The following principle results were found: (a) burnout was found to be primarily a source–drain phenomenon not directly involving the gate; (b) visual degradation effects (by direct observation of the topography of the device) were not necessarily present in failed devices in which case burnout was *via* the substrate or buffer layer; (c) visible burnout revealed the burnout path to be between source and drain, originating at the drain; (d) the burnout occurred in the substrate first; (e) the process was related to high electric fields at the drain contact causing premature avalanche (microplasmas) with concomitant electron-hole pair generation. Wemple observed white light generation along the drain contact at high source drain bias (in agreement with the Fujitsu workers) as a precursor to burnout. Following Fukuta, Wemple used n^+-layers as a means of

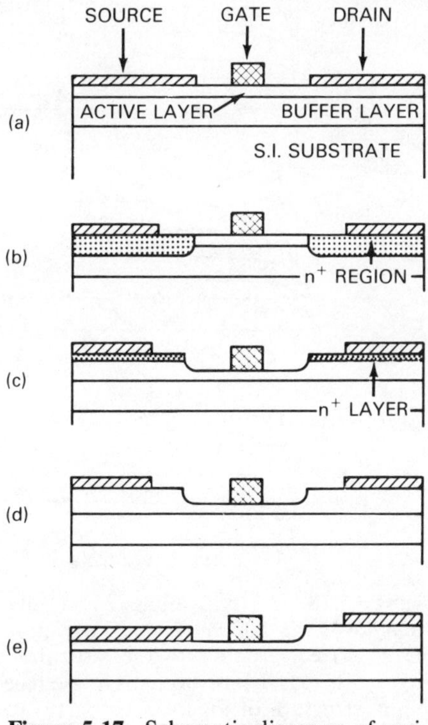

Figure 5.17. Schematic diagrams of various GaAs MESFET structures

improving breakdown voltage. In the latter work, the n^+-layer was grown over the entire active region. The n^+-layer was etched off in the gate region prior to deposition of their standard metallization. In this manner, the source–drain burnout voltage was increased to 20–30 V in slice form.

A much simpler recessed gate structure was developed[32] in order to achieve the high source–drain voltages without utilization of the n^+-contact. Figure 5.17[33] contrasts five structures. Figure 5.17(a) is the standard configuration with no recessed gate and no explicit n^+-contact region. Figure 5.17(b) shows the structure described by Fukuta.[30] Figure 5.17(c) shows the case where the n^+-contact layer is either grown by epitaxy as with Wemple or ion-implanted;[34] in this case a recessed gate is shown. Figure 5.17(d) shows the case of a thick active layer with recessed gate and no n^+-contact region. Figure 5.17(e) is a variation of 5.17(d) for simpler processing. Furutsuka compared

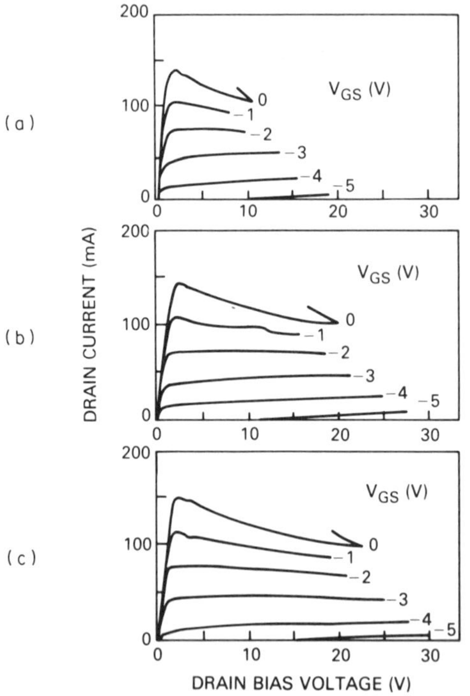

Figure 5.18. Current–voltage and drain breakdown for 500 μm wide GaAs power FETs. (a) corresponds to the structure of Figure 5.17(a); (b) corresponds to the recessed/n^+ structure of Figure 5.17(c); (c) corresponds to the simple recess of Figure 5.17(d)

the breakdown voltage for devices typified by Figure 5.17(a), (c), and (d) and the data are shown in Figure 5.18. The data indicate that the structure with recessed gate and no n^+-contacts (Figure 5.17(d)) provides equally good breakdown voltages as the structure of Figure 5.17(c) with the n^+-contacts. Microwave performance of these two structures was also compared with no significant advantage observed for the n^+-contact layer structure over its simpler counterpart.

In a related investigation[35] there was an attempt to relate light emission and burnout utilizing devices with the equivalent source, drain, and gate configurations shown in Figure 5.17. Shown in Figure 5.19 are the light intensities observed for three of these structures as measured by a multiplier with S-20 response) over the range of voltages shown. These data demonstrate that light emission is observed far below source–drain breakdown, well within the standard device source–drain voltage operating regions, but that emission intensity decreases with the more optimal structure for high drain breakdown voltage. In contrast to the results of Wemple, Yamamoto found the emission phenomenon to be sensitive to gate bias. He reported that the light intensity first decreases toward pinch-off and then increases as the reverse bias is increased; this latter phenomenon is shown in Figure 5.20. Yamamoto also

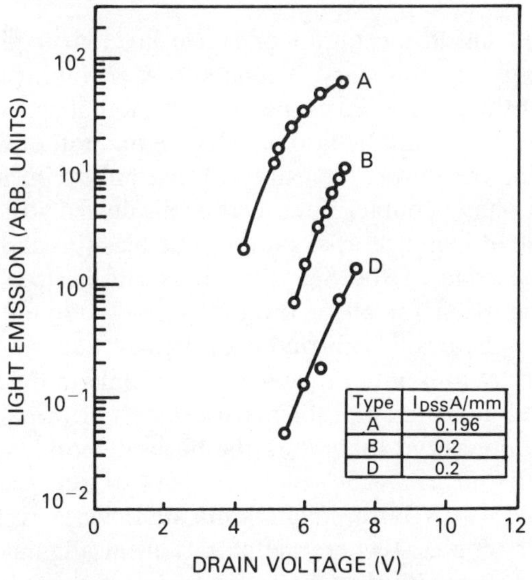

Figure 5.19. The dependence of light emission intensity *vs*. drain voltage (at zero gate bias) for different FET structures. A corresponds to Figure 5.17(a); B corresponds to Figure 5.17(b); and D corresponds to Figure 5.17(d)

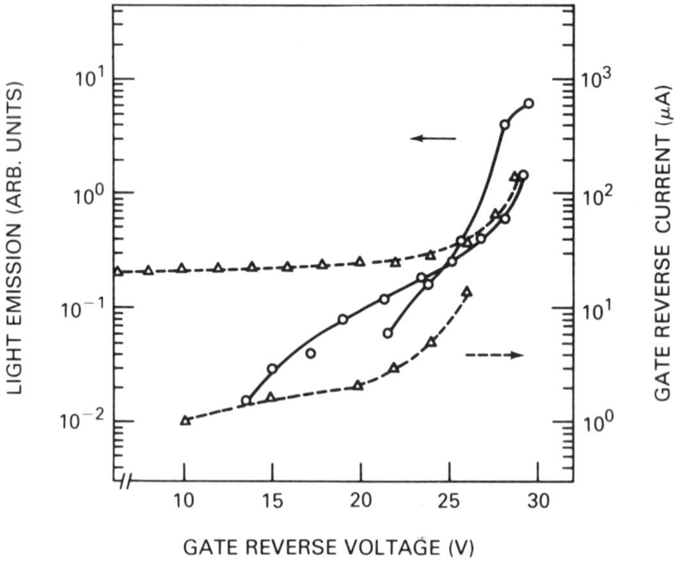

Figure 5.20. Comparison of reverse gate current and light emission from the edge of the gate *vs.* reverse gate voltage at zero drain bias

observed burnout inside the buffer or active layer near pinch-off, again in contrast to Wemple. In summary, Yamamoto *et al*. find: (a) that light emission originates at the drain edge of the active region at zero or near zero gate bias; (b) that all devices emit light to one degree or another with recessed gate structures emitting the lowest intensity; (c) that pulsed burnout originates at the GaAs–drain ohmic contact interface for all device structures at or near zero bias; (d) that at high Schottky gate reverse bias (breakdown) light is also emitted from the edge of the Schottky. It is emphasized that Yamamoto observed light emission for all of the bias ranges studied with the highest intensity found at the drain semiconductor contact under normal bias conditions. It is important to point out, however, that only in the reliability studies of Drukier have drain contact problems arisen as a principal cause of failure.

It should be emphasized that while the burnout problem has been nominally solved with the use of n^+-contact layers or recessed gates or both, Wemple observed 4–8 V burnouts in multigate devices in the regions at the edge of mesas or on mesas where metallization misalignment had occurred. This problem is not restricted to the active layer and must be kept in mind in device and circuit design.

Another potential failure mode in the case of contacts with appreciable light emission and elevated temperatures is that of non-radiative recombination assisted diffusion. While this type of failure has not been reported in either small signal or power GaAs FETs, a recent paper[36] on GaAs IMPATTs has definitively shown the role of this physical process in Pt–GaAs IMPATT failures. This phenomenon will be described in Section 5.7.

5.4.2 Pulse burnout

In one of the first reliability studies of small signal FETs for satellite application[37] significant numbers of FETs were found to fail under sudden electrical overstress during various assembly and testing operations. Believing that these failures were due to static discharge, Bellier *et al.* utilized a simple capacitance charge–discharge circuit to simulate discharge of human body capacitance. A 100 pF capacitor was charged to various positive and negative pulses and discharged from gate to ground. The devices were commercial 2 μm Al gate devices with Au wire bonded directly to the Al gate pad. Pulses were applied in stress steps doubling the successive pulse energy from 5×10^{-9} joules to failure. In many instances observable damage occurred on the surface of the devices with no discernible change in gate leakage current. Positive pulses resulted in two dominant effects: (a) a transient increase in drain to source current as a result of decreased depletion layer width and; (b) a simultaneous transient current ($\gtrsim$1 ampere for 10^{-9} sec) from gate to ground for a 200 V charging voltage. Failures were found to occur at the lowest resistance paths between gate and source and at high resistance points on the gate; that is, at physical defects in the gate. For negative voltages, failures were found to occur at points of lowest breakdown voltage. The studies showed that devices could be subject to numerous extraneous positive or negative pulses and still operate within specification but with significantly reduced potential for reliable operation. This was an important finding and mitigates in favour of extreme care in the screening and handling of these parts.

Lundgren[6] repeated the essential experiment of Bellier *et al.* but included a more definitive test protocol and then contrasted the three device types described earlier (Section 5.2, Table 5.1) which included Al and Au refractory gates. For positive voltages, the Al gate FETs failures required 70–78 V on the 100 pF capacitors with corresponding energies of from 2.5 to 3.0 ergs. The refractory gate FETs failed at lower voltages 42–65 V and more significantly their electrical characteristics indicated failure (changes of ~10 per cent) prior to catastrophic burnout. For negative voltages, the Al gate FETs failed at about 40 V (0.8 ergs) using a 100 pF capacitor. All of the Al gate FETs failed by gate–drain shorts or gate–source shorts associated with high field regions and minimum gate–drain resistance or high fields and minimum gate resistance. The refractory gate FETs exhibited random failures between source and gate. It was concluded that these studies did not support earlier findings[1,4] that gold refractory gates are more rugged than Al gates.

Lundgren also carried out the first burnout assessment of small signal FETs under high power RF pulse and CW stress. These studies are of particular importance in regard to application of these devices in radar receiver front ends where the standard limiters do not provide protection against short pulse power (also Section 5.7). The study was restricted to Al gate FETs which were tested as simple, single-stage, low noise amplifiers at 9.7 GHz. The amplifiers were biased for low noise operation and measurements were made after each exposure. CW exposure was for 1 min. Pulsed RF was with 1 μsec pulses at 100 kHz repetition rate (10 per cent duty) for 10 min at each stress level. Both CW and pulsed results are shown in Table 5.3 above. Small changes in noise figure (NF) and associated gain (G_A) occur in the range of 1–3 watts input power and burnout occurs in the range of 3–10 watts.

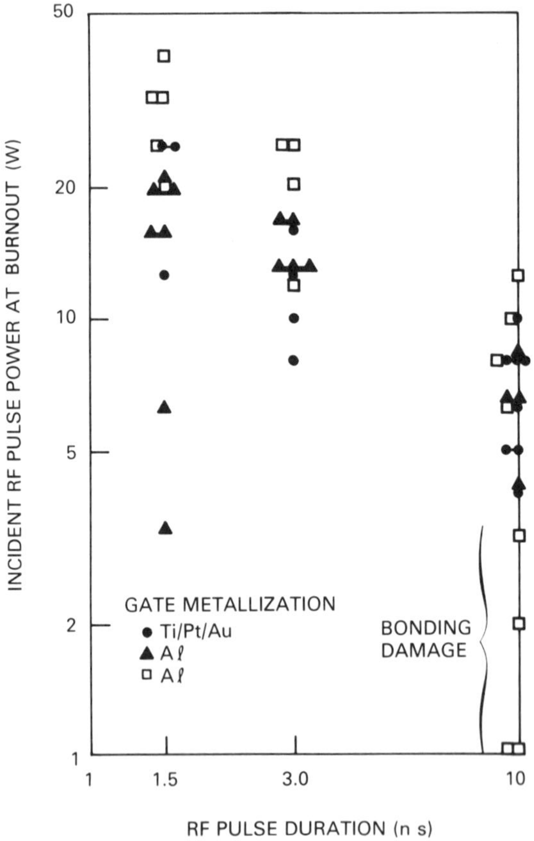

Figure 5.21. Incident microwave pulse power, P_i, required to cause burnout *vs*. pulse duration in nanoseconds. The P_i values are believed to be accurate to ± 0.5dB

Following Lundgren, further burnout studies on both 1 μm Al and Au-refractory gate FETs at 9.3 GHz were carried out[38] using nanosecond pulses. The test procedure was to use 1.5 to 10 nsec pulses with a repetition frequency of between 10 and 100 kHz for an exposure from 1 to 2 minutes. Noise figure and associated gain were measured prior to and after each exposure. Initial peak pulse power was 0.5 W and was increased in 1 dB steps. In nearly every instance, no change in noise figure was observed until catastrophic burnout occurred. Figure 5.21 shows a plot of incident RF pulse power at burnout *vs*. pulse duration. The data strongly centre in the range from 4 to 10 W for 10 nsec pulses, 12 to 25 W for 3 nsec pulses and 15 to 30 W for 1.5 nsec pulses. Two catastrophic failure modes were observed. The first was an incipient source–gate short correlated with Au migration from the source to gate, and was observed for the long pulse case. The second type of failure was a drain–source short resulting from massive channel damage observed predominantly for the short pulse case. Whalen[38] found Al-gate and Au-refractory gate FETs to have similar burnout characteristics in support of the observations of Lundgren.

5.5 RADIATION EFFECTS

At the present time, the most promising semiconductor technology for application in space radiation environments is silicon CMOS–SOS. This is so for the following reasons: (a) CMOS provide low standby power consumption; (b) the use of sapphire provides for automatic dielectric isolation, reduction of parasitic capacitance and therefore higher speed than bulk Si devices; (c) the use of sapphire minimizes radiation induced photocurrent volume thereby reducing sensitivity to transient radiation. However, GaAs MESFETs in digital (and analog) format predict, in principle, many performance advantages over CMOS–SOS for space application. Some of these advantages are: (a) 3 to 4 times higher speed; (b) homoepitaxy with no equivalent interfacial degradation in v_{sat} or mobility; (c) automatic dielectric isolation and small parasitic capacitance effects; (d) low power consumption; (e) reduced radiation sensitivity to total dose due to the absence of a gate oxide; (f) the potential for adequate transient and neutron dose hardness.

The absence of an MIS technology in GaAs FETs with both enhancement and depletion mode formats is not considered a major drawback by GaAs device and circuit proponents when considered from the point of view of radiation effects. This is so since the MOS technology in silicon has been plagued with radiation effects problems since the first adverse sensitivity problems were reported.[39] Since radiation induced effects in the MOS oxide is one of the principal causes of adverse device performance in transient and total dose environments, a technology which at once promises higher speeds, no oxide, and process simplicity is truly a strong contender for alternative use in space; such a contender is available in GaAs devices and ICs in analog and digital formats.

The promises of these advantages are uniformly advertized by proponents of the GaAs technology much as in the case of the promise of the reliability of small signal and power GaAs FETs. Unfortunately, assessment of GaAs device and circuit performance in appropriate radiation scenarios has also been inadequate.

GaAs JFET high speed planar logic circuits were initially postulated[40] to offer improved radiation hardness in comparison to equivalent silicon devices and circuits and in a later study,[41] detailed measurements of the transient response of these circuits to ionizing radiation were reported. A theoretical dose rate of $\gamma_{\text{UP-SET}} = 1.88 \times 10^{10}$ rad (GaAs) sec^{-1} was calculated for enhancement mode JFETs for a channel length of 2.5 μm, threshold voltage of 0.5 V, noise margin of 0.5 V and ion implanted channels. Measurements of transient effects were carried out on ring oscillators, buffer circuits and several memory cells. In the case of the ring oscillators suppression of oscillations was found at about 10^{10} rads (GaAs) sec^{-1} with a recovery to normal oscillations in 1 to 10 μs. The recovery time constant was postulated to be related to photoluminescent decay. For the buffer circuits, upset was observed in the majority of circuits at 1×10^{10} rads (GaAs) sec^{-1}. Approximately this same level of transient dose was required to cause loss of information in the memory cells. Irreversible failures were not observed until dose levels of about 10^{11} rads (GaAs) sec^{-1}, at which time gate burbout was observed.

In a recent study[42] a comparison was made of the transient radiation sensitivity of GaAs and SOS circuits exposed to a flash X-ray source. Devices in both materials had 1 μm gate lengths and were characterized by small signal *s*-parameter measurements which indicated pre-irradiation state-of-the-art RF performance. Significantly higher gate photo currents, as a function of dose rate, were found for the GaAs MESFETs as compared to either GaAs or SOS drain currents. This is to be expected since the Schottky gate is directly connected to the active channel (in comparison to MOS devices where the gate is isolated) and the gate sees contributions from both source and drain. Phillips[42] reported transient upset levels of $\geq 10^{10}$ rads (Si) sec^{-1} for Lockheed digital ICs, which is in agreement with Zuleeg.[41]

A broad comparison of transient effects on a variety of device or materials configurations has recently been reported;[43] the remainder of this section will comprise a summary of most significant results from this study. These authors compared devices and circuits fabricated by ion implantation directly into Cr-doped substrates and by epitaxial growth either directly on a Cr-doped substrate or on buffer layers. The implanted samples consisted of discrete power MESFETs and high-speed 2-input NOR gates prepared by and purchased from the Hughes Research Laboratories. The active regions of these experimental devices were formed by implanting 100 keV Si^+ ions (MES-

FET: 4×10^{12} cm^{-2}; NOR gate: 10^{13} cm^{-2}) into the Cr-doped GaAs substrate to obtain an electron concentration of approximately 1.5×10^{17} cm^{-3}. The power MESFETs had gate dimensions of 1×600 μm and a source–drain spacing of 5 μm, and were characterized by a pinchoff voltage (V_P) of approximately -6 V. Aluminum gate metallization was used on the discrete devices while the logic circuits employed a triple layer of Cr–Pd–Cr.

Epilayer structures consisted of experimental Ge–GaAs heterojunction gate (HJFETs)[44] and commercial GaAs power FETs prepared by MSC (88002)[23] described under Section 6.3 above. The germanium HJFET devices were also fabricated on Cr-doped substrates using an undoped buffer layer formed by VPE. The active layer, approximately 0.28 μm thick, was grown by liquid phase epitaxy (LPE) and doped with Si to obtain a carrier concentration of $\sim 1 \times 10^{17}$ cm^{-3}. In these devices a *p*-type Ge gate formed a *p–n* junction rather than a Schottky barrier gate as with the other samples. Pinchoff voltages ranged from about -5 to -7 volts.

Radiation tests were conducted using a Febetron 706 FXR and the NRL LINAC. The Febetron experiments were carried out with the machine operating in the X-ray mode and delivering a dose of approximately 100 rads per 3 ns, 600 keV pulse. LINAC testing was conducted over an exposure range of 70 to 1000 rads using 40 MeV electrons and a 60 ns pulse width. FET test samples were biased in a quiescent state during exposure with source–drain currents ranging from 500 μA up to 100 mA. Device transients were characterized as a function of temperature from room temperature (or slightly below) up to 200 °C.

In addition to the radiation experiments, FET test samples were screened for electrically-induced slow transient effects. Such effects as long-term drifts in operating parameters (I_{DS}, G_m, S_{21}) have been reported in GaAs FETs following the application of bias voltages,[45] and have been attributed to deep level traps within the material. The electrical tests consisted of superimposing voltage pulses on various applied DC gate biases and characterizing any long-term I_{DS} perturbations as a function of temperature. A modified DLTS technique was utilized in conjunction with the pulsed voltage experiments to examine trap structure. This technique, which is an adapted version of the method originally described by Lang,[46] samples I_{DS} (rather than capacitance) at selected intervals following application of a gate voltage pulse.

Typical radiation response characteristics for an implanted MESFET operating at various quiescent current levels are shown in Figure 5.22. These I_{DS} transients, generated by 3 ns X-ray exposures of only 100 rads, are characterized by peak amplitudes ranging from 90 per cent of the quiescent current at 0.5 mA to less than 5 per cent of the 10 mA level. However, other samples biased at 0.5 mA were driven into cutoff at this exposure level. The recovery time constant (time for 63 per cent transient decay) associated with these 25–27 °C data is between 10 and 14 seconds. Similar responses were

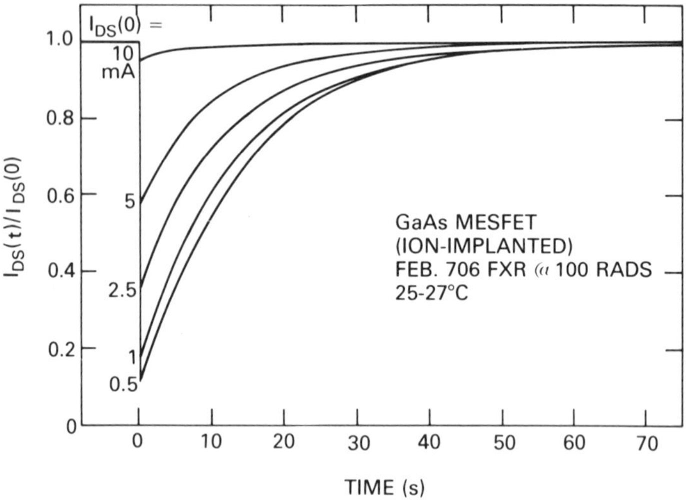

Figure 5.22. Pulsed radiation response at various I_{DS} levels

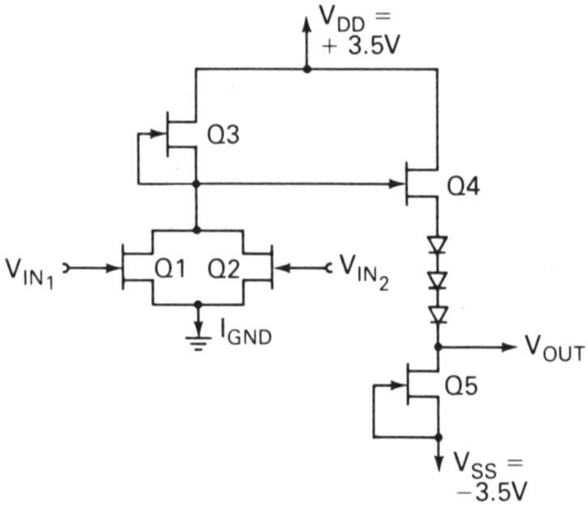

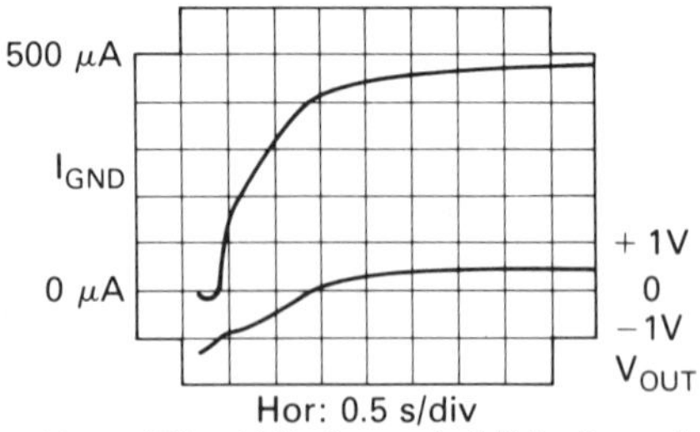

Hor: 0.5 s/div
Vert: 100 μA/div (upper); IV/div (lower)

Figure 5.23. Transient radiation response of an implanted 2-Input NOR gate when biased in the linear input range. (V_{IN_1} = 0 V, V_{IN_2} = −3.5 V)

observed on the discrete epilayer FETs (MESFET and HJFET) samples, although none of these samples were driven into cutoff from the 0.5 mA level by a 100 rad exposure.

Figure 5.23 depicts response of a 2-input GaAs NOR gate response to a 100 rad X-ray pulse when biased within the linear input voltage range. Here, in order to maximize transient response to low radiation levels, one input transistor was turned off ($V_{IN2} = -3.5$ V) while the other was biased at I_{DS} $= 500\ \mu$A ($V_{IN1} = 0$ V). Although, as seen from the upper trace, Q1 is driven into cutoff by the radiation pulse, the output voltage response is apparently governed by the turnoff transient induced Q4. Unlike the discrete MESFETs (both implanted and epilayer samples) with recovery time constants of 10 seconds or more at 25 °C, NOR gate recovery was dominated by a 1 second time constant at this temperature. Output logic level changes (high or low) of approximately 1 second duration have also been observed in

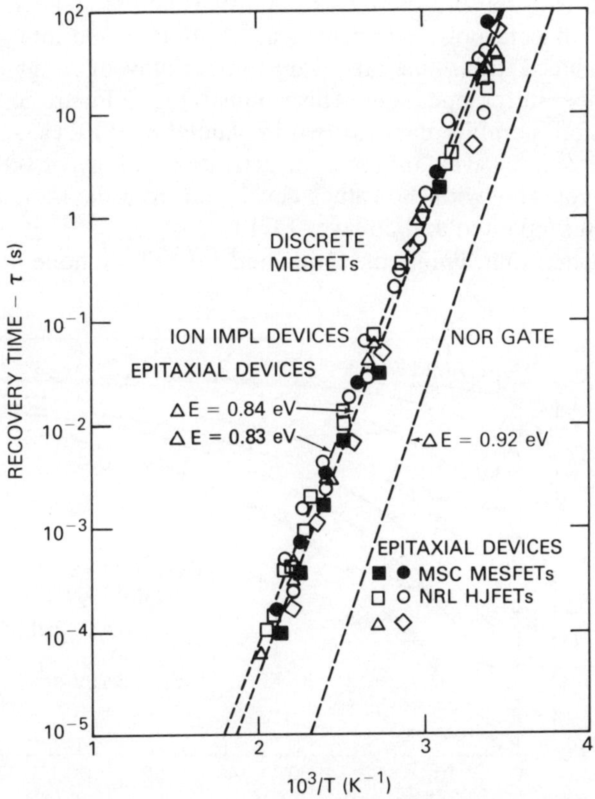

Figure 5.24. Recovery time constant *vs.* reciprocal temperature for implanted and epitaxial GaAs MESFETs exposed to transient radiation

these gates when treated at higher exposure levels under normal logic input voltage (+1 and −2.5 V).[47]

Recovery time constants plotted as a function of reciprocal temperature for implanted and epilayer devices are shown in Figure 5.24. These data are indicative of a thermal activation process dominated by single electron trapping level. Response of both implanted NOR gate and epilayer FETs is governed by an apparent ΔE of 0.83 to 0.84 eV, whereas NOR gate response is associated with an apparent ΔE of 0.92 eV.

Transient responses almost identical to those observed following pulsed irradiation could be induced in the implanted MESFETs by superimposing positive voltage pulses on the negative gate bias. A family of I_{DS} transients generated in one of these devices by 20 μsec, 1 to 6 V pulses is depicted in Figure 5.25. These data indicate that electron trapping is taking place throughout the active layer of the implanted structure. As the trapped negative charge density increases with pulse amplitude, the I_{DS} turnoff transients at the end of a gate pulse grow larger. Transient amplitudes were essentially independent of gate pulse width over a 0.1 to 100 μsec interval. Like the radiation induced transients, recovery time following a gate pulse was strongly temperature-dependent. This is illustrated in Figure 5.26 by voltage response *vs*. temperature data for two implanted MESFETs. A straight line fitted to these data gives an apparent activation energy of 0.82 eV, which compares favourably with the values developed from the radiation response data for both implanted and epilayer FETs.

It is significant that, unlike the implanted MESFETs, none of the epilayer

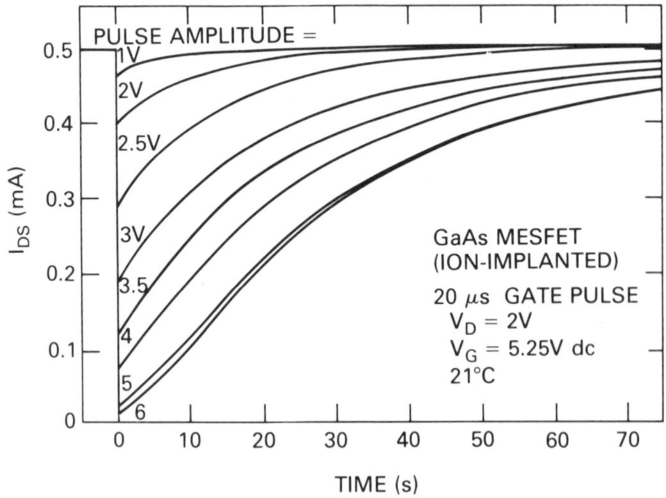

Figure 5.25. Transient I_{DS} response of implanted MESFETs to positive gate voltage pulses of various amplitudes

MESFET or HJFET samples exhibited long-term I_{DS} transients in response to applied gate voltage pulses. This indicates that the trapping centres responsible for long-term radiation-induced transients in the epilayer structures are located within the GaAs buffer layer or substrate material or both. The experiments suggested that a common centre was responsible for the phenomenon observed with ion-implanted and epitaxial FETs.

The results of Simons and King may be summarized as follows: (a) GaAs FETs in discrete or IC format, either ion-implanted or epitaxial, are susceptible to pulsed radiation with significant long-term (tens of seconds) transient recovery; (b) the same effect is seen with ion-implanted FETs in a non-radiation environment when the gate is pulsed positive; (c) a single trapping centre with capture cross section of $\sim 10^{-14}$ cm^2 and activation energy in the range 0.73 to 0.84 eV dominated FET response; (d) the trapping centre is located in the active channel for ion-implanted FETs and in the buffer layer or substrate for epitaxial FETs; (e) significant materials improvement in

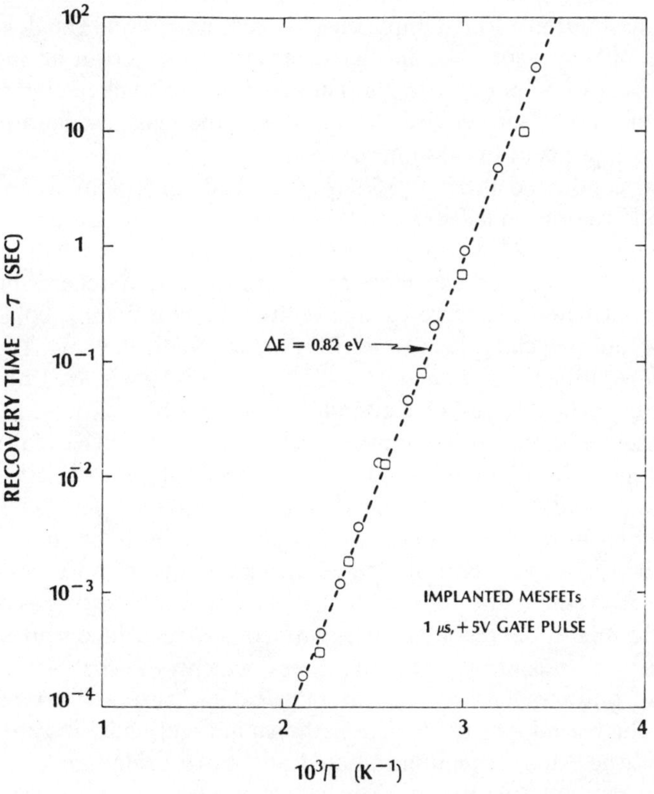

Figure 5.26. Recovery time constant *vs*. reciprocal temperature for implanted MESFETs following gate voltage pulse

GaAs will be required to harden devices and circuits to both transient radiation or positive gate pulsed effects or both.

5.6 ADVERSE AMBIENT EFFECTS

In many anticipated applications, GaAs FETs and integrated circuits will be subjected to adverse environmental conditions such as high relative humidity, ionic contamination, and temperature cycling. Reliability requirements demand that these components not degrade under conditions which result in condensed moisture and contaminants. The first investigations reported[10] were conducted at 85 °C and 85 per cent relative humidity (RH) without the presence of ionic contamination. Gate corrosion was produced by a 2-hour exposure with a 6 volt negative bias on the gate of an uncapped device. This corrosion was shown to be clearly electrolytic since in the absence of gate bias no significant corrosion was observed. Irvin also reported the observation of corroded gates of hermetically sealed devices which failed after 240 hours under bias in an 85/85 environment. The presence of unremoved photoresist may have harboured enough impurities to cause corrosion. The devices which had failed optical inspection and aged at 85/85 resulted in an incidence of corroded gates of 50 per cent in 2000 hours. The incidence of electrolytic gate corrosion among 'clean' devices which passed the leak test before and after ageing was 1 per cent in 2000 hours.

The tests conducted at 85 °C/85 per cent RH did not include conditions for water condensation on substrate surfaces. However, the presence of a high–low temperature cycle will allow H_2O to condense on the substrate in the liquid state. Recent studies[48] have considered failure mechanisms resulting from the concurrent exposure to high relative humidity, ionic contamination and temperature cycling. Low noise FETs from NEC, Plessey, Avantek, and Varian were utilized including Al gates, Au gates, passivated and unpassivated devices. The devices were cycled −10 °C to +100 °C at 95 per cent RH in the presence of Na and Cl ions at levels of 10^{14} cm^{-2}. The bias conditions were: (1) gate biased at −1 V (drain floating); (2) gate biased at −0.5 to −1 V with the presence of drain current; or (3) zero bias. The results of these investigations have shown that Al-gate FETs in both hermetic and non-hermetic configurations typically failed by a gate–source short within the first 80 hours of cycling. The failure mechanism was Al gate corrosion and Au electrolytic migration resulting in a gate to source bridge and subsequent AuAl reaction. In contrast, Al gate devices, which were cycled but unbiased, resulted in only a small increase in gate leakage current. A typical failure mode for the biased Al gate devices is shown in Figure 5.27 indicating the Au bridge and the Au–Al reaction initiated at the Au bridge area.

In this same investigation, devices with Au–refractory gates were found to be less susceptible to failure when exposure cycled in the Na and Cl contami-

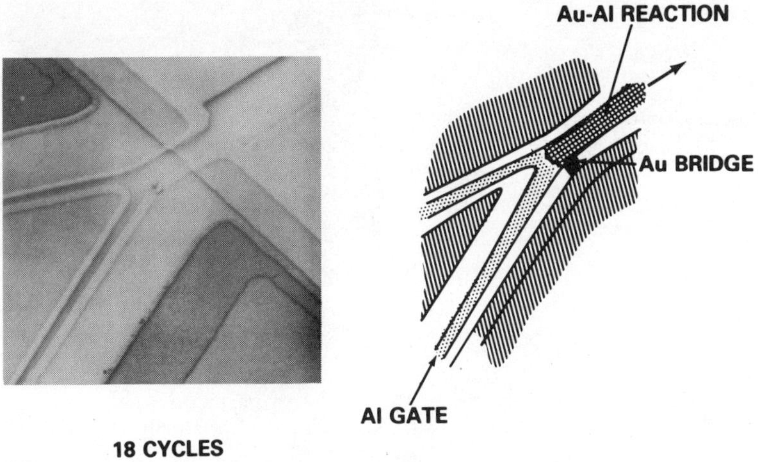

Figure 5.27. A typical failure mode for the biased Al–gate devices indicating an Au bridge and an Au–Al reaction

nated environment. Even in the non-hermetic configuration these devices typically did not exhibit catastrophic gate shorting. Little change in DC electrical characteristics were observed when the devices were exposure cycled unbiased, whether the devices were hermetic or non-hermetic. With gate biased or gate and drain biased as shown in Figure 5.28, graceful degradation was observed. Increases in gate leakage, due to migrative gold resistive short formation was accompanied by decreases in I_{DSS} and g_m due to channel corrosion.

Unpassivated Au–refractory FETs have also been tested by Hughes Aircraft Corporation[12] in order to determine the effect of moisture condensation. Avantek A8061 FET chips were used in a near cryogenic amplifier application. These FETs use a TiW–Au gate structure with the gate shaped like an inverted truncated pyramid. Four of these FETs, powered by separate gate and drain supplies were placed in an 85 °C/50 per cent RH environment for 85 days. The metallization pattern remained unchanged (i.e., no visual indication of corrosion) and gate and drain current changes remained less than 10 per cent. A second set of FETs were tested at 85 °C/85 per cent RH and again no changes in metallization appearance nor electrical properties were observed. These results are in agreement with earlier observations[49] in that TiW–Au gate FETs (Avantek) can withstand exposure to high relative humidity and ionic contamination for prolonged periods of time without degradation. The temperature cycling test, however, is the critical test since it simulates conditions for water condensation on the substrate surface. For instance, a high–low temperature cycle will include the 0 °C or −10 °C phase which requires less moisture for liquid water condensation to take place. At

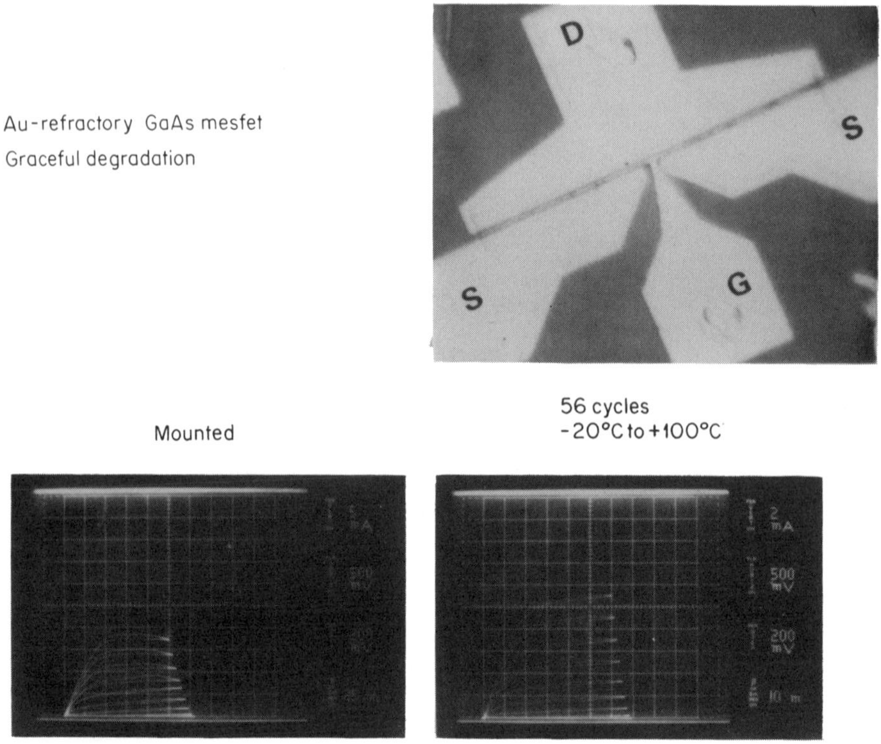

Figure 5.28. Gradual degradation of Au–refractory gate MESFETS. This device was characterized by increases in gate leakage current and decreases in I_{DSS} and g_m

100 per cent relative humidity and 25 °C, 3.1 per cent H_2O will result in liquid water on the substrate. At −10 °C, the necessary level of moisture for liquid water on the substrate will decrease to approximately 0.25 per cent.

5.7 RELIABILITY OF OTHER GaAs DEVICES

5.7.1 GaAs IMPATTs

The initial development of *n*-GaAs IMPATT diodes employed platinum Schottky barriers with plated gold which were bonded to a diamond heat sink.[50,51,52] The reliability assessment and failure analysis of these IMPATTs conducted at Bell Laboratories from 1971–74 indicated that a major mode of IMPATT degradation to be the interdiffusion of Au through Pt forming metallic spikes which extend into the GaAs. The metallic spikes so formed short-circuit the junction either just in the bulk or at the metal–GaAs interface. In addition since more than 90 per cent of the DC input power can be dissipated in the high field region, the attendant rise in junction temperature can result in concomitant increases in leakage current.[53,54] The Au filament problem was minimized by making the Pt layer thicker than 6000 Å. However, this solution introduced process related problems in the manufacture of high-efficiency IMPATTs using tailored doping profiles (e.g., hi–lo, lo–hi–lo). For such profiles, the use of a thick Pt film imposes limitations on device efficiency.[55] The thick Pt diodes were initially stabilized by forming $PtAs_2$ at the interface. However, at high operating temperatures (175–300 °C) Ga outdiffuses through the $PtAs_2$ resulting in the consumption of the GaAs epitaxial layer. The rate of GaAs epitaxial thickness decrease has been estimated to be 200 Å per 10^4 hour.[56] Even these small changes are detrimental to high reliability IMPATTs with a required MTBF of greater than 10^5 hours, since the performance of these high-efficiency diodes depends critically on the exact doping profile of the epitaxial layer. Attempts to directly replace Pt with refractory metals such as W ot Ti resulted in Schottky barriers too small for effective IMPATT operation at 175 °C.

Mahoney improved the reliability of IMPATTs to an MTBF of 10^6 hours at 175 °C by fully reacting a thin film of Pt on GaAs to form a layered structure of PtGa–PtAs–GaAs. This was followed by a W–Pt–Au metallization system. Although reliability improvement was achieved with this system, IMPATT efficiency degraded to less than 15 per cent after testing at 175 °C for longer than 1000 hours. The degradation indicates time–temperature-dependent changes in the PtGa–PtAs layers. Ideally the Pt being fully reacted with GaAs should not migrate toward the Au layer in order for the GaAs interface to remain stationary during device operation. However, it is possible that the PtGa–PtAs layer acts as a transfer layer for Ga atoms similar to the phenomenon of Si outdiffusion through the PtSi–Si structure. To date the (PtGa–PtAs)–GaAs system has not achieved the combination of high reliability and high performance mainly as a result of the instability of PtGa–PtAs compounds at temperatures of 175 °C–200 °C. In order to show the effectiveness of the reacted Pt-W–Pt–Au metallization scheme, Mahoney monitored reverse I–V characteristics at 175–200 °C. Although these studies showed no degradation in I–V characteristics, such small changes in doping profile could not be detected by his measurement technique. Additional failure analysis and accelerated stress testing at operating RF power levels must be carried out in order to assess the reacted Pt–GaAs approach for IMPATT devices.

Other reliability investigations[57,58] have been carried out using a number of high temperature metallizations. In order to determine the relative stability of metallizations used in device fabrication, Schottky barriers were formed on GaAs lo–hi–lo IMPATT structures. This structure was chosen since sequential doping profiles on packaged devices could be conveniently used to monitor the migration of the electrical junction into the material.[58] Stress temperatures of 250, 225, and 300 °C were used. Four separate systems were analysed consisting of: Al–Ti–Pt–Au; Pt–Ti–Au; Pt–TiW–Au; and TiW–Au. At 275 °C, the Pt–Ti–Au and Al–Ti–Pt–Au diodes failed within 5 hours by shorting or by high leakage currents. The Pt–TiW–Au and TiW–Au diodes were clearly superior in that no failures were observed at 275 °C up to 10 hours. Varian diodes made with the Pt–TiW–Au system clearly show a higher MTTF compared to other diodes as shown in Figure 5.29. Figure 5.30 also shows the rate of junction consumption for Pt–TiW–Au IMPATTs as observed by these authors. Clearly, the data shows that the Pt–GaAs reaction occurs by a diffusion controlled mechanism and degradation of IMPATT efficiency cannot be prevented if the Pt–GaAs is utilized.

A premature failure mode in GaAs IMPATTs was identified[36] during temperature stress ageing of diodes operating in avalanche. The failures reported prior to Ballamy and Kimerling's investigations were observed under temperature stress alone or a DC–bias condition. However, during temperature stressing with the diodes in avalanche, Ballamy and Kimerling showed that recombination enhanced diffusion dominates at temperatures below 300 °C. As a result of this stress, an activation energy of 0.2 to 0.4 eV was observed, while above 300 °C, the devices follow a log-normal failure distribution with an activation energy of 1.6 eV. Therefore, the high temperature reliability testing of IMPATTs has indicated that the junction migrates by the classical diffusion phenomenon while low temperature reliability problems in high efficiency IMPATTs are due to recombination enhanced diffusion. Since the efficiency of Read-type GaAs IMPATTs has been shown to be critically dependent on the width of the avalanche region, junction motion

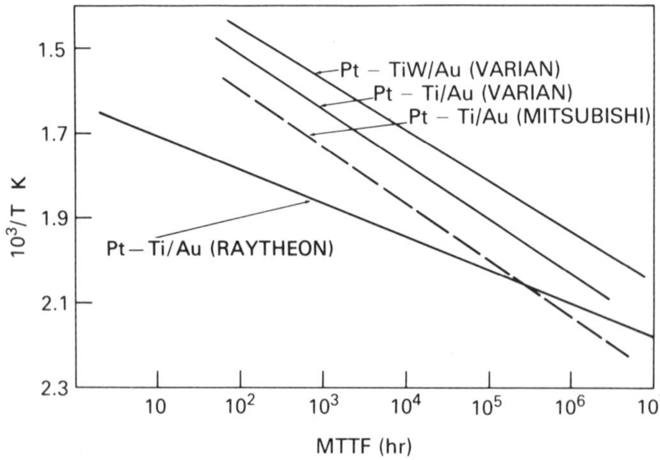

Figure 5.29. MTTF curves for GaAs IMPATTs showing the various refractory metallizations

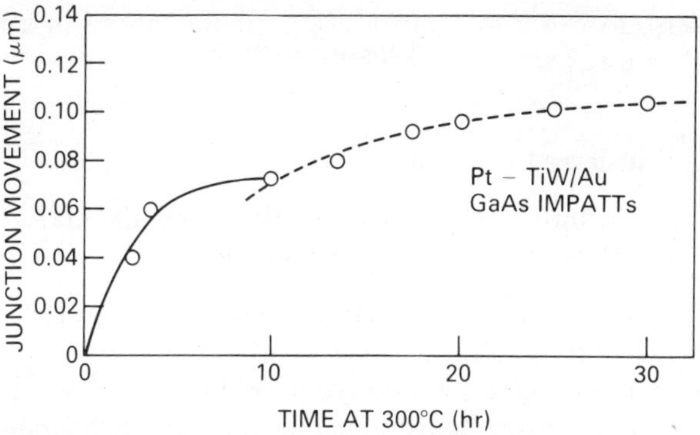

Figure 5.30. Junction movement versus time for lo–hi–lo Pt-TiW–Au IMPATTs

seriously impairs both efficiency and power output. At present, an effective approach of terminating junction motion has not evolved.

Recently, a new insight[59] was provided on the problem of the metal–GaAs junction for Read-type GaAs IMPATTs. By analysing life test data, the change in junction position was found to be proportional to the product of current density and time. Nomardi subjected ten microstrip oscillator test circuits to an RF bias–temperature stress test. The oscillators were subjected to five stress steps of 500 hours at heat sink tempertures of 25, 50, 75, 100, and 125 °C. A gradual change in IMPATT characteristics were observed at each temperature such as the monotonically increasing behaviour of the zero-current-intercept breakdown voltage. The observed change in IMPATT characteristics suggested that a slow metal–semiconductor reaction (slower than the Pt–GaAs reaction) or defect may be responsible. Hot carriers, both electrons and holes, are created by the avalanche process in each RF cycle within the field avalanche zone. A small fraction of hot carriers possess sufficient energy to create conditions for interstitial diffusion. Although these experimental results are based on low–high–low GaAs IMPATTs, the same physical mechanism would be present in high–low structures. A life time of only 5×10^4 hours can be predicted from the RF bias-temperature tests.

The utilization of refractories between the Pt and Au plating has been adopted by most investigators in order to minimize spike burnout. However, a new failure mode has been reported[60] as associated with refractory metallizations. During RF stress–temperature testing of (Pt–Ta–Au)–GaAs lo–hi–lo IMPATTs loss of adhesion was observed between the Au and Ta and between Ta and Pt. The loss of adhesion was initiated by a mismatch in thermal expansion coefficient which exists between the thin film layers and by the thermal cycling effects during RF testing. This observed failure mode must be carefully assessed since it may introduce basic limitations in utilizing various refractory films for IMPATTs.

5.7.2 GaAs transferred electron devices

Failure modes attributed to the Gunn effect in bulk GaAs diodes were initially attributed[61] to the following phenomena:

(a) Transient changes in material conductivity when a diode containing a high-field domain is subjected to an additional negative voltage spike.
(b) Recombination radiation associated with a switching of the current.
(c) Migration of tin (from the contact) in the form of filaments due to the effect of the high-field domain.

An in-depth study of the contact degradation problem has been reported.[62,63] Diffusion spikes in Gunn diodes were previously reported.[35,64] Palmstrom *et al.* in their 1977 investigation, using Rutherford backscattering techniques, showed that these spikes contained gold and germanium which penetrated more than 300 nm into the GaAs. This filament formation is thought to be due to the result of localized areas of high current density (or high fields). Since gold diffuses rapidly in GaAs, Gunn diodes with gold as an ohmic contact constituent will be highly susceptible to filament formation. The newly introduced refractory ohmic contacts may minimize this problem in Gunn diodes, although such a reliability assessment of new contacts has not yet been initiated. Clearly the standard AuGe, AuGeNi contacts are all susceptible to filament formation.

Recombination radiation associated with a switching of the current has been found to result in the current oscillation becoming noisy or disappearing altogether. Domain passages of $12–87 \times 10^6$ have been reported[65] before deterioration of the current oscillation was observed. Physically, a degradation of the anode was observed as a result of this failure mode. Again, the data reported to date indicates that standard AuGe, AuGeNi anode metallizations are not reliable for Gunn operation and can degrade due to a recombination enhanced diffusion mechanism.

All reliability tests conducted to date have shown that anode metal migration induced by the domain's electric field causes deterioration in bulk Gunn diodes. To prevent deterioration, Shoji proposed to increase the cross-sectional area sufficiently such as to force the domain to disappear before reaching the anode. Reliability tests have been performed with the result that the protected anode structure can withstand 10 times more the number of domain passages than the conventional diodes. This observation indicates the effectiveness of the protected anode structure in improving the reliability of bulk effect devices. To date, the major commercial suppliers of bulk Gunn diodes have not attempted to implement the protected anode structure.

Although bulk Gunn GaAs devices have achieved a high degree of commercialization,[66] their reliability under RF stress–temperature conditions has not been adequately assessed. An MTBF under RF conditions has not been published. Life tests at the diodes' maximum specified DC operating voltage have been reported.[67] A total of 138,000 unit hours of operation at heat sink temperatures of 75 °C were collected, resulting in a failure rate, λ, less than 16, where λ is in units of failures per million hours. The diodes, however, did show power degradation of less than 10 per cent when they were RF evaluated after the life test. Ramachandran defined 'failure' as a ten per cent degradation in power output which is indeed a very liberal criteria. Again, even these life tests are inadequate in assessing the reliability of GaAs Gunn diodes under RF operating conditions since the DC bias life test and definition of device failure may be questioned.

Planar GaAs transferred electron devices have become increasingly important for monolithic millimetre wave applications, ultra-high speed logic applications, sub-ns signal processing, analog–digital converters and for use in delay lines. These planar devices have previously been fabricated on *n*-type epitaxial layers[68] and more recently by ion-implantation into SI GaAs substrate.[69] The failure mode for the planar epitaxial devices has been reported to be a conducting channel of Sn or In depending on the ohmic contact. Excess heat is generated at the anode, where the breakdown is initiated. The mechanism starting the breakdown has been speculated to be hole injection and a combination of ion drift and diffusion of impurities.

Ion-implanted TEDs offer a number of advantages over TEDs fabricated on epitaxial material: (1) neither an active epitaxial layer nor a mesa step to form the active channel are required, (2) the devices may be completely planar which may allow gate lengths below 1 μm for operation above 20 GHz and improved Gunn domain trigger sensitivity, (3) TEDs and field effect transistors (FET), which require different active layer thicknesses, can conveniently be fabricated on the same substrate for integrated circuit applications, and (4) it may also be possible to realize improved gate reliability with a completely planar device.

A recent study[70] reported on the failure analysis of ion-implanted GaAs TEDs which failed either catastrophically by burnout or degraded gracefully during high temperature accelerated reliability testing. Devices which failed catastrophically were characterized by (1) burnout from the cathode to the anode at a critical power dissipation level, depending on device design and doping concentration, and by (2) melting in the centre of the channel near the bottom of the doping profile. The gate on the top of the channel typically remained unmelted. X-ray analysis also revealed Au and Ge in the active channel in the failed region indicating that field enhanced ion migration also contributed to the failure. In the devices which degraded gracefully, field enhanced ion migration appeared to be the dominant failure mechanism as shown by the Auger spectra of Figure 5.31. The failure mode observed in planar devices was in contrast to the failure mode reported for bulk devices or even epitaxial planar TEDs. Recombination enhanced diffusion can be considered as the primary failure mode for the ion-implanted planar TEDs which resulted in GaAs melting in the centre of the channel. When field enhanced metal migration was observed, Au was detected at the cathode side only, and Ge or Ni from the ohmic contact metallization was not present. Since the high

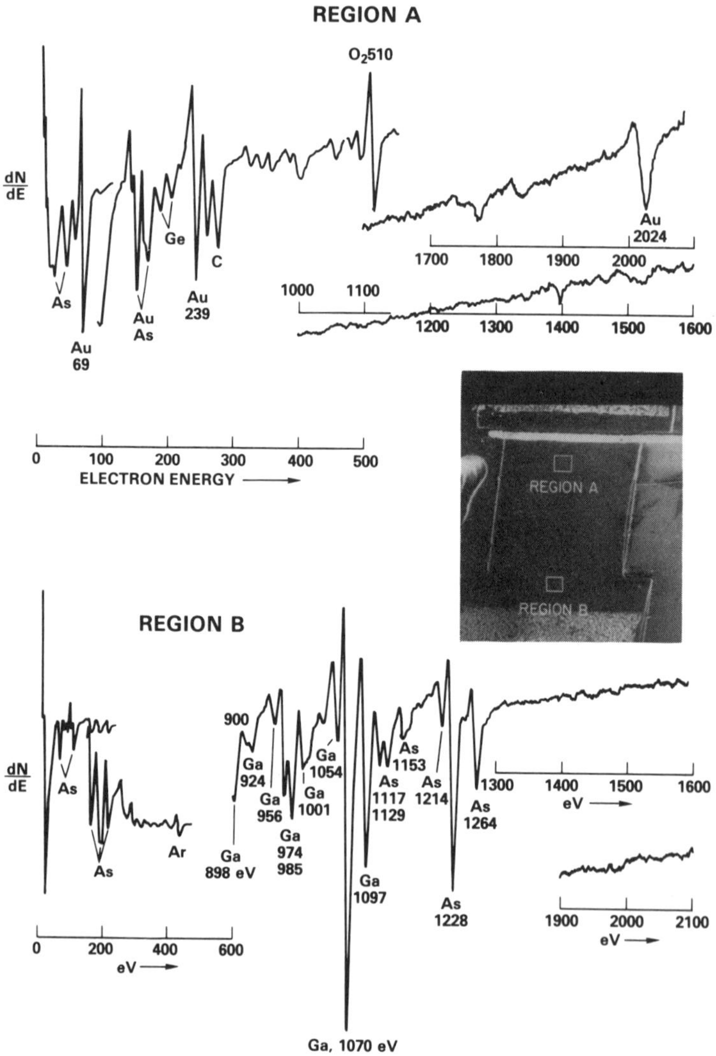

Figure 5.31. Auger analysis of an ion implanted GaAs TED showing Au in the active channel

field regions are different between ion-implanted planar devices and mesa devices, additional work should be undertaken in order to determine the following effects: (a) the role of channel temperature; (b) the effect of residual ion-implantation damage and; (c) the difference between ohmic contact metallizations, as affecting the catastrophic burnout.

5.7.3 GaAs mixer diode reliability and RF burnout

Schottky barrier mixer diodes designed for Ka-band higher frequencies have an extremely small active area (10 μm^2 or less) and are, therefore, very susceptible to RF burnout. The Schottky diodes initially developed in 1970 consisted of a AuGe–GaAs low barrier system[71] which rapidly degraded upon thermocompression bonding. These diodes were utilized for Ka-band frequencies and only when low local oscillator power was required at the expense of burnout resistance.

As the frequency is increased above 36 GHz, the RF burnout resistance has been found to decrease rapidly as shown in Figure 5.32. The burnout resistance, however, has also been shown to be highly dependent on the Schottky

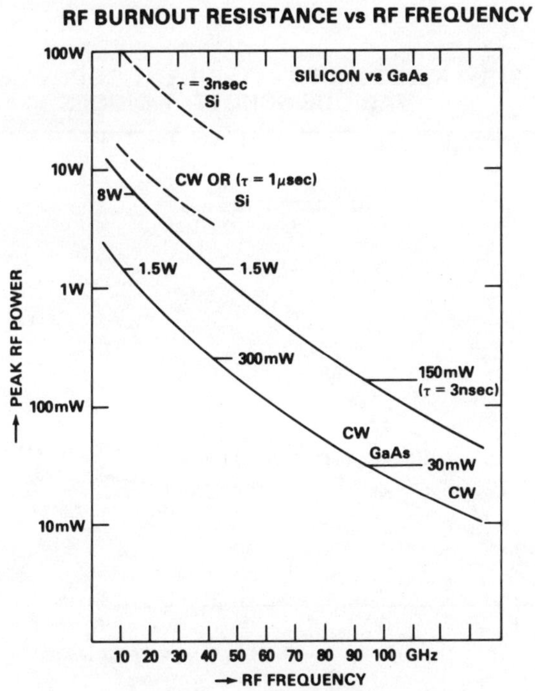

Figure 5.32. RF burnout resistance *vs*. RF frequency for GaAs millimetre mixer diodes

barrier metallization.[72] The earlier Pd–GaAs and plated Ni–GaAs mixer diodies typically failed at peak power levels of less than 300 mW (CW) and 1.5 watts pulsed (τ = 3 nsec). Improvements in burnout resistance have been achieved with (Ti–Mo–Au)–GaAs Schottky diodes and with (TiW–Au)–GaAs diodes.[72,73] The results shown in Figure 5.33 indicate that (Ti–Mo–Au)–GaAs mixer exhibits a 2 watt CW improvement over Pd–GaAs diodes. Figure 5.33 also indicates that these diodes absorb less power compared to the Pd–GaAs Schottky at 36 GHz. Finally, the (TiW–Au)–GaAs mixer at 36 GHz has exhibited a 2.5 W burnout level at CW and a 12–15 watt burnout for two 5 nsec pulses.

The SEM and micro-spot Auger electron spectroscopy analysis of these gold–refractory GaAs mixer diodes[72] indicates the presence of two different failure modes. The CW-burnout mode has been shown to be interdiffusion between the refractory metal and the GaAs. Since this is a time-dependent phenomenon, these diodes were all found to degrade gradually with incident RF power. In contrast, silicon diodes failed catastrophically above a critical incident RF power level. The mixer diodes which failed as a result of 5 nsec pulses exhibited substrate filament formation at the edge of the active areas. Metal filament formation resulted in substrate degradation and catastrophic failure.

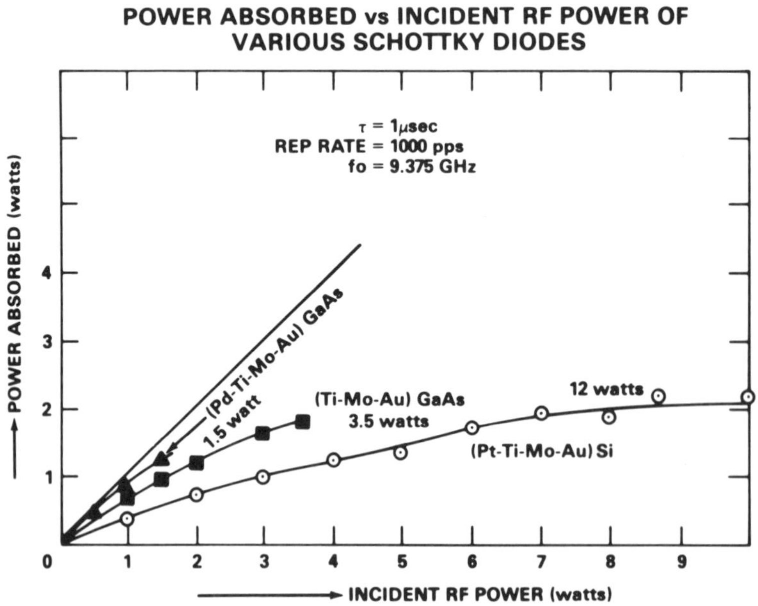

Figure 5.33. Absorbed power *vs.* incident RF power of various GaAs Schottky diodes

With the use of a refractory metallization and a native oxide passivated junction, both high performance from 35–105 GHz and high reliability were achieved.[74] A tantalum–gold metallization system was used with grown GaAs for surface protection against ambient contaminants. Reliability test results to date indicate an MTBF of better than 10^8 hours at 100 °C with no bias and greater than 10^5 hours at 200 °C. In a commercial application of paramps these diodes have successfully completed 2.5 million operational varactor hours with no performance degradation. Pulsed burnout analysis has not been undertaken on these diodes to date, neither has accelerated life tests with DC or RF bias.

The (Ta–Au) structure has been adopted in a beam-lead configuration.[75] Again, at 94 GHz, the mixer conversion loss was below 6.0 dB with no degradation in high temperature reliability observed. The beam-lead configuration for GaAs mixer diodes was also tested for vibration, shock, temperature (400 °K), CW RF power and pulsed RF power.[76] At the conclusion of the above tests, the mixers exhibited no measurable degradation in noise figure or conversion loss. The CW RF power, however, was maintained below 150 mW and the pulsed power was maintained below 1.5 W.

The present status of GaAs mixer diode reliability is that: (1) diode burnout from RF pulses greater than 1 watt remains as a significant problem at frequencies greater than 36 GHz; (2) the mixer diodes have not been evaluated under RF stress–temperature conditions at frequencies above 36 GHz, and in fact there has been no reported elevated temperature–DC bias reliability evaluation of GaAs mixer diodes designed for 36–94 GHz operation; (3) the beam-lead configuration apparently has eliminated shock and vibration reliability problems which were present with whisker contacts.

5.8 MATERIALS RELATED PROBLEMS

5.8.1 Ohmic contacts

The reliability of GaAs microwave devices has been shown to be directly related to the integrity of the ohmic contacts.[77,78,79,80] Industrial suppliers of GaAs devices have standardized on the use of Au–Ge or Au–Ge-Ni ohmic contacts mainly due to their initial low resistance. However, the reliability and long-term stability of these contacts has been questionable. The major problems in GaAs ohmic contacts are: lack of uniform wetting of the metals to the GaAs, metal segregation at the metal–semiconductor interface, an unevenly doped n^+-layer accompanied by a multitude of semiconductor–metal phases, surface roughness, the outdiffusion of Ga to the gold overlay. In this section, the GaAs–ohmic contact interdiffusion phenomenon is emphasized since it has been observed to be a primary degradation mechanism for small signal FETs.

5.8.1.1 Interdiffusion phenomenon

Interdiffusion effects in Au:Ge–Ni contacts has been shown to result in solid phase formation[81,82] with facet growth of Ni–Ga and Ni–Ge having the same orientation as the (100) substrate and aligned in the ⟨100⟩ direction as shown in Figure 5.34. The Au:Ge–Ni contact formed by annealing at 450 °C for 60 seconds has been further analysed by μAES in order to determine the lateral and depth distribution of Au, Ge, Ni, Ga, and As. A schematic of the results is summarized in Figure 5.35 and indicates the presence of Ge in the GaAs substrate. In addition, Figure 5.36 shows Ga outdiffusion to the Au overlay suggesting that Ge has migrated to Ga vacancies in the substrate. It is clear

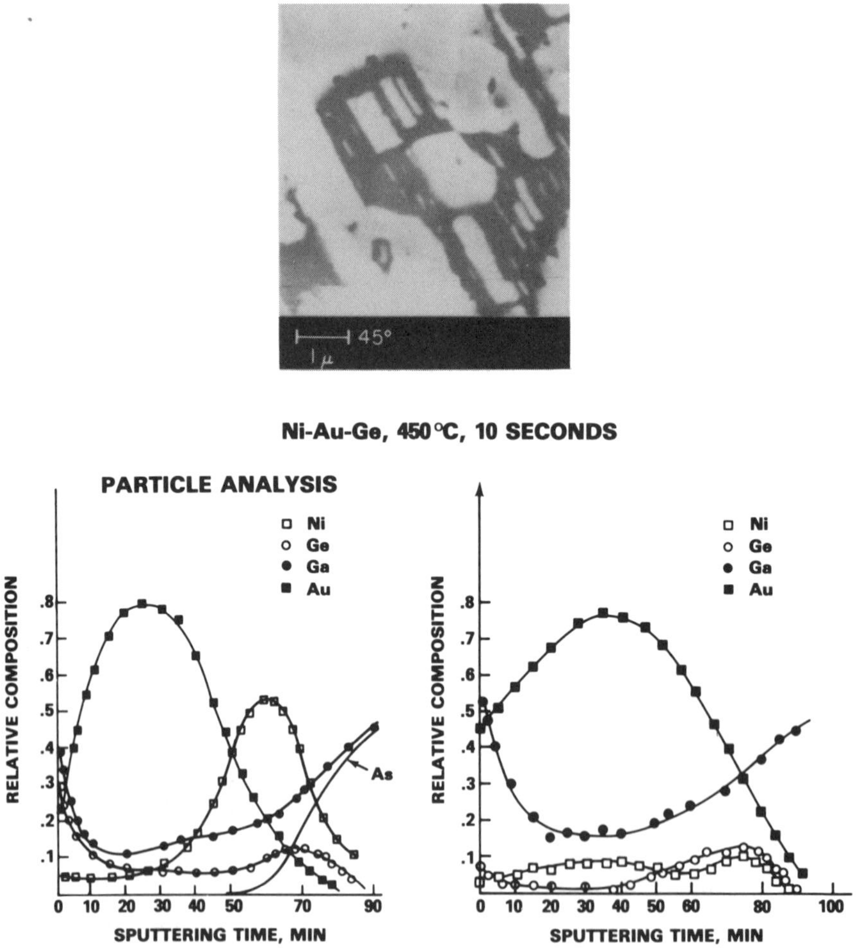

Figure 5.34. Interdiffusion effects in Au:Ge–Ni contacts showing solid phase formation and a variation in composition between particles and matrix

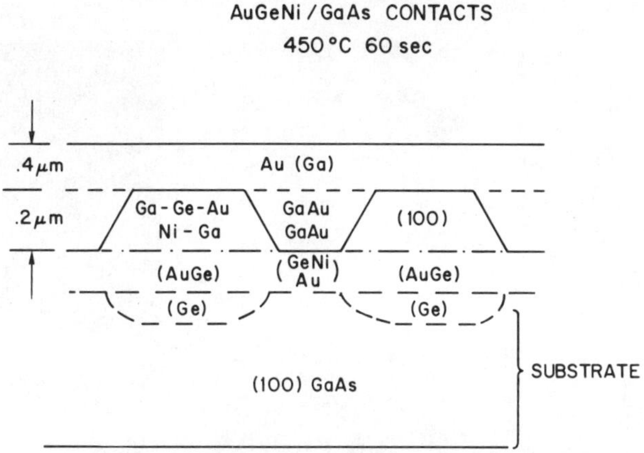

Figure 5.35. A schematic of Au:Ge–Ni interdiffusion with GaAs showing distribution of each constituent of the contact system

that significant outdiffusion through Au:Ge–Ni contacts occurs and that a refractory metallization must be included between the Au:Ge–Ni and the Au overlay.

5.8.1.2 Electromigration and lateral migration

A second serious reliability problem is lateral migration of metallic constituents due to concentration, temperature and field gradients. This phenomenon has been studied by using a 'gateless' GaAsFET test structure which can evaluate both interdiffusion and lateral diffusion under temperature–bias stress.[25] At 200 °C and a channel current density of 10^6 A cm^{-2}, the contacts were observed to degrade rapidly (4–8 hours). The failures were attributed to bulk diffusion of Ga to the surface of the contact and then to lateral migration between the electrodes. In all investigations to date, the outdiffusion of Ga to the gold overlay was observed to accompany contact metal migration under bias. It is well known that gold undergoes electromigration with an activation energy of 2 eV.[83] However, solute effects on gold can drastically reduce the electromigration activation energy to below 1 eV. The effect of solute atoms on gold electromigration has been studied[23] extensively and it is speculated that the same solute effect occurs when Ga atoms dope the gold film.

A solution to the ohmic contact reliability problems as related to Ga outdiffusion was proposed in 1977[24] which would incorporate a combination of refractory films between the Au and the eutectic ohmic contact. The initial refractory metal layers consisted of Ti–Pt, TiW–Cr, Ti–Cr–Pt with the overall effect of preventing the outdiffusion of Ga to the top surface of the gold.

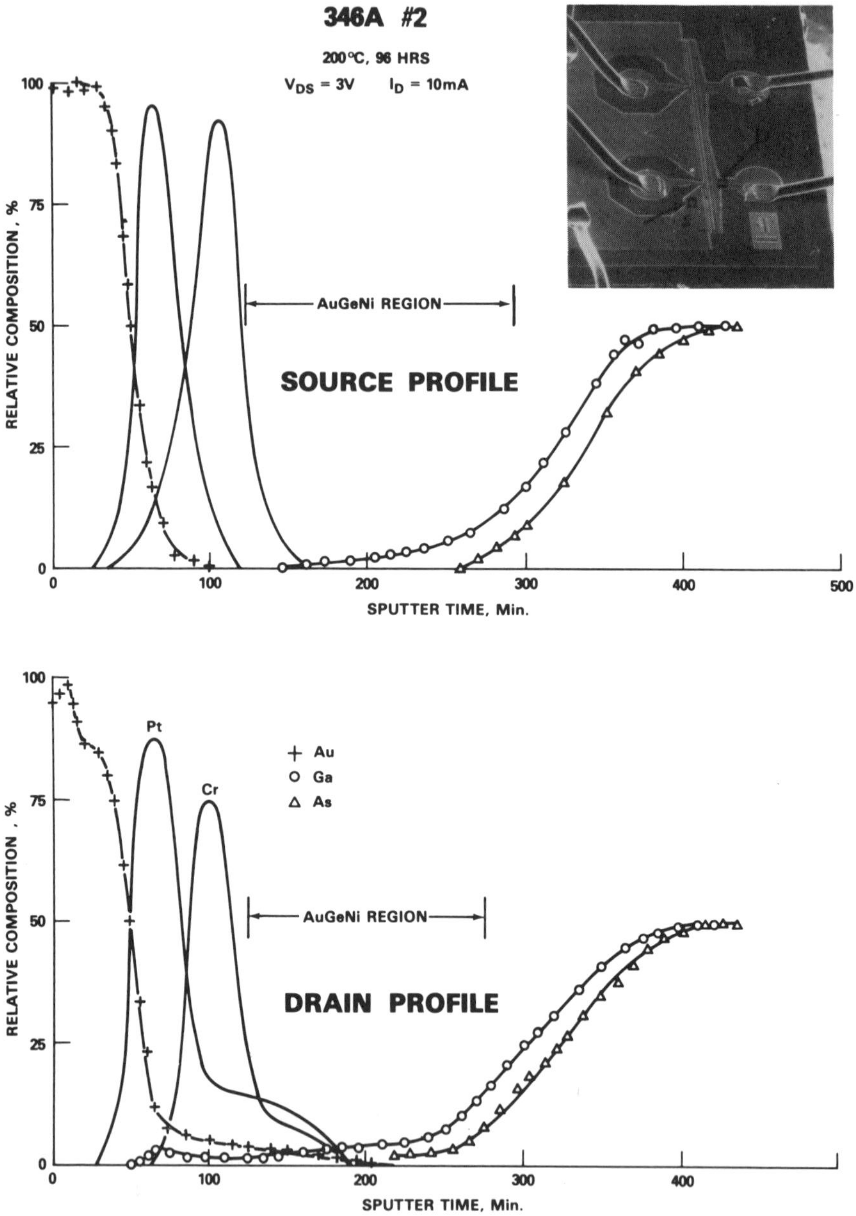

Figure 5.36. A Pt–Cr diffusion barrier for the Au:Ge–Ni system showing that Ga has not migrated to the top surface even after a temperature-bias test at 200 °C

Figure 5.36 illustrates an effective Ga diffusion barrier. In an elevated temperature bias study[24] of the Au-Ge-Ni–Ti–Pt-Cr;Au system, it was shown that the Ti–Pt eliminated metal migration up to 200 °C. Above 250 °C, current crowding or hot spotting was found to occur at high drain currents.

5.8.1.3 New ohmic contact structures and ion-implanted contacts

It has been pointed out in the previous sections that certain advantages accrue from avoiding deleterious metallurgical effects resulting from short critical alloying times. The initial work in this area[48,49] resulted in ohmic contacts to n-type GaAs using epitaxial Ge films on GaAs alloyed with Ni overlays by solid state diffusion at temperatures of 450–550 °C. The contacts were reported to be smooth and continuous showing no evidence of phase separation or other surface structure.

The application of molecular beam epitaxy (MBE) to the development of the GaAs ohmic contact technology has resulted in contact resistance as low as 2×10^{-6} ohm cm^2 for MBE deposited n^+-layers to GaAs. It has been reported[84] that highly reliable contacts can be obtained without alloying by using highly doped n-type GaAs layers grown by MBE. Recently, the Cornell group[85] have employed heavily doped MBE Ge layers to form unalloyed contacts to n-GaAs with contact resistance values below 1.0×10^{-7} ohm cm^2. Non-alloyed ohmic contacts have also been formed[86] by depositing TiPt on highly Sn-doped GaAs layers grown by MBE. Layers with carrier concentrations as high as 6×10^{19} cm^{-3} were obtained resulting in a contact resistance of 1.86×10^{-6} ohm cm^2.

The application of ion-implanted non-alloyed contacts has produced marginal results until recently. However, Mozzi[87] has reported non-alloyed contacts of TiPtAu deposited on pulsed electron beam annealed Se implants which yielded $r_0 \approx 6 \times 10^{-7}$ ohm cm^2. An n^{++}-layer was obtained by a 120 keV Se$^+$ implant of 5×10^{15} cm^2 at room temperature. Electron beam annealing of the layers was accomplished by 10–15 keV electrons with a pulse duration of 150 sec and energy density of 0.71 J cm^{-2}. The application of non-alloyed contacts processed either by ion-implantation or by an epitaxial process will provide a significant impact in improving device reliability both for power FETs and small signal FETs.

5.8.2 Schottky barrier contacts

From the accelerated life tests on small signal MESFETs and power FETs reported to date, it is generally agreed that one important aspect of the reliability of GaAs devices is related to the integrity of the gate–GaAs interface. In the early development of MESFETs, aluminium Schottky barriers were the baseline metallization. Investigations on the Al–GaAs system[8,88] employing SIMS, AES and electrical measurements indicate that interdiffusion occurs between Al and GaAs at temperatures as low as 125 °C, with the presence of Al_2O_3 at the metal–semiconductor interface This outdiffusion will result in a stoichiometric imbalance in the substrate just below the surface of the film. In addition to the Al–GaAs interdiffusion, AuAl phase formation has also been reported as a reliability problem and has been discussed in previous sections. A number of attempts have been made to minimize the Al-related problems[5,6] such as including a Cr barrier or redesigning the FET geometry. However, the survey of FET manufacturers clearly shows a trend toward gold refractory gates. The most successful gold-based systems have been the TiW–Au, the TiPtAu and Ta–Au. To date, the TiW–Au system offers perhaps the most stable system for high temperature operation.

The initial work on TiW–GaAs reported[89] that even after annealing at 600 °C for 15 hours outdiffusion of Ga was not observed. The barrier height was found to increase gradually with annealing temperatures up to 600 °C, but even at 600 °C its value was measured to be 0.85 eV. Similar stability of the TiW–GaAs interface has been reported.[90] Avantek has incorporated this system in low noise FETs. However, as with all refractory metallizations, the thermal expansion difference between TiW and GaAs resulted in the generation of defects just below the gate and a reduction in carrier mobility and associated degradation of g_m. On the positive side, these defects appear to be process related and proper choice of sputter deposition parameters will reduce their magnitude.

The TiPtAu metallization system has been used with great success in the silicon beam-lead technology. Thus, this metallization is also of interest for GaAs Schottky barriers. Auger analysis of TiPtAu Schottky gates,[82] however, has exposed a number of processing and potential reliability problems. These problems are: (1) a temperature–time-dependent interdiffusion phenomenon between GaAs and Ti; (2) formation of a TiO_2 layer above 150 °C. The most dramatic effect is the presence of the non-stoichiometric interface region which may degrade device performance.

Finally, the Ta refractory metallization system has been used successfully for high reliability varactor diodes[74] and for FETs. Initial results with Ta gate FETs[25] has shown that I_{DSS}, g_m, and V_P decreases as a function of anneal temperature. This reduction in FET parameters has been attributed to Ta–GaAs interdiffusion at temperatures between 125–250 °C. In terms of the FET structure, this interdiffusion, in general, means that the effective gate–GaAs interface may be graded resulting in a depletion region which is (i) much closer to the epi–buffer interface or (ii) mostly in the interdiffusion region where it is ineffective in modulating channel conductivity (i.e. low g_m) or (iii) both. The decreased values of I_{DSS} and V_P strongly suggest that the depletion region has moved closer to the channel–buffer interface after annealing. The above observation is not limited to the Ta–GaAs system but may occur with other interfaces such as Ti–GaAs and W–GaAs.

5.9 SUMMARY

In an attempt to establish a baseline for the materials technologies which are currently being employed in industry, in particular for small signal and power GaAs FETs, a questionnaire was sent to 32 industrial laboratories in order to allow the reader the opportunity to relate materials and process factors to device reliability. Of these 32 laboratories, 18 responded and the details are shown in Table 5.5. A number of features are of interest: (a) the preponderant technique for fabricating active channels is VPE for both small signal and power devices; (b) buffer layers are always used except for ion-implanted channels; (c) drift and light sensitivity has not been reported as a problem affecting reliability or performance; (d) most organizations use some form of passivation, SiO_2 SiO_x or Si_3N_4; (e) either channel etch or recessed gates are generally employed; (f) AuGeNi is the standard system for ohmic contacts and usually without a refractory barrier to the Au current carrying contact; (g) about half of the laboratories employ Al gates with most of the remainder using TiPtAu or TiWAu. The small signal FET reliability data indicate MTTFs of 10^7 to 10^9 hours at 100 °C channel temperature with activation energies ranging from 1.0 to 1.8 eV. Power FET reliability data indicates MTTFs of from 10^4 to greater than 10^8 hours at 100 °C channel temperature with activation energies ranging from 0.67 to 2.3 eV. About one-half of the reported failure modes were associated with ohmic contact degradation and gate problems were seen for both Al and refractory systems.

The results of the survey in Table 5.5 are consistent with the observations reported in Sections 5.2 and 5.3. That is, the technology for small signal FETs is the most 'mature' for GaAs devices available and has made significant gains in achieving MTBFs of approximately 10^8 hours at channel temperatures of 100 °C. On the other hand, these devices still exhibit failure modes of source and drain contact degradation, gate–GaAs interdiffusion and electromigration. Although both the AuGe and AuGeNi systems exhibit a number of degradation modes, the survey of industrial suppliers does not indicate a universal enthusiasm for incorporating alternate ohmic contact systems. There is, however, a trend away from aluminium gates and general acceptance of the more reliable Au–refractory Schottky barriers. The presence of high relative humidity and ionic contamination (adverse ambients) has been found to result in graceful degradation or catastrophic failure depending on the type of gate metallization present. Although these effects can be minimized with hermetic packages, it is questionable whether moisture can be kept out of package interiors due to residual package leaks. At present, all investigations

Table 5.5 Summary of industrial suppliers of GaAs MESFETs

Organization	Device	Active Layer	Buffer	Drift Short, Long	Light Sensation sensitivity	Passi-vation	Ohmic	Schottky	Recessed gate or surface etch
NEC (Japan)	SB SM SIG	VPE	YES	No, No	No to very weak	SiO_2	AuGe–Pt AuGe–Ni	Al	Yes (latest)
	SB power	VPE	Yes	No, No	No to very weak	SiO_2 Si_3N_4	AuGe–Pt AuGe–Ni	Al	Yes (latest)
Siemens (W. Germany)	SB SM, SIG	VPE I^2	Yes No	No, No	Weak	SiO_2	GeAuCrAu (Layers, not eutectic)	TiPtAu	Yes
	SB Power	VPE I^2	Yes No	No, No	Weak	SiO_2	GeAuCrAu (Layers, not eutectic)	TiPtAu	Yes
Tektronix	SB SM, SIG	I^2	No	No.—	Yes	SiO_2 Si_3N_4	AuGeNi	TiPdAu	No
	SB power	VPE	Yes	No.—	No	SiO_2 Si_3N_4	AuGeNi	TiPdAu	Yes
	SB IC	I^2	No	No.—	Yes	SiO_2 Si_3N_4	AuGeNi	TiPdAu	No
RSRE (Great Britain)	SB SM, SIG	MOCVD	Yes	—, —	Weak	None	AuGe	TiPdAu	Yes
	SB power	VPE	Yes	—, —	Weak	None	AuGe	TiPdAu	Yes
Hitachi (Japan)	SB SM, SIG	VPE	Yes	No, No	Weak	Sintered organic	AuGeNi	MoAu	Yes
Fujitsu (Japan)	SB SM, SIG	VPE	Yes	No, No	Weak	SiO_2	AuGe	Al	Yes
	SB power	VPE	Yes	No, No	Weak	SiO_2	AuGe	Al	Yes
Avantek	SB SM, SIG	VPE I^2	Yes	Weak, No	Weak	None PGA	AuGeNi	TiWAu	Yes
	SB power	VPE I^2	YES	Weak, No	Weak	None	AuGeNi	TiWAu	Yes
Texas Instruments	SB power	VPE	Yes	No, No	Weak	Si_3N_4	AuGeNi	TiPtAu	Yes/Yes
Plessey (Great Britain)	SB SM, SIG	VPE (etched)	Yes	No, No	Weak	None	InGeAu	Af	Yes
	SB power	VPE	Yes	No, No	Weak	None	InGeAu	Af	Yes/Yes
Teledyne MEL	SB SM, SIG	VPE	Yes	Weak, No	Weak	SiO_2 or none	AuGeNi	TiPtAu	Not uniformly
	SB MED power	VPE	Yes	Weak, No	Weak	None	AuGeNi	TiPtAu	No

Life tests	High temp stress	MTBE (hr) channel temp	ϕ (eV)	Failure mode	Comparison with other devices	Ambient effects studies	Radiation effects studies
Hi temp storage	RF	$10^8 10^9$ 100 °C	1.8	Contact degradation	No	Yes, No problems in PKG	Yes better than bi polar
Hi temp storage	RF	$5–10^7$ 130 °C	1.8	Contact degradation	No	Same as above	No
6V 40 mA 190 C.C.T.	290–400 °C	10^8 100 °C	1.8	Gate leakage	No	No	No
No	No	—	—	—	—	No	No
No	No	—	—	—	—	No	No
No	No	—	—	—	—	No	No
No	No	—	—	—	—	No	No
No	No	—	—	—	—	No	No
No	No	—	—	—	—	No	No
Yes unbiased to 265 °C	Yes unbiased 175, 225, 265 °C	10^7 10^8 80 °C	1.0	Ohmic degrad. Ni indiffusion	No	Yes like US MIL-S-883 No failures	No
—	RF 100 °C 11 GHz 4,000 hr	—	—	—	—	No	Yes
Yes, 199, 227 259 °C	RF, 200 °C 6 GHz 14,000 hr	10^5 175 °C	1.3	Ohmic degrad.	—	No	Yes
Yes	Yes, 225, 275 °C biased SM, SIG	150 hr at 270 °C	1.5	Gain degrad.	No	No	No
Yes, just started	Yes, just started	—	—	—	—	V_{DS} stress to failure also MIL-S-883 tests	No
Yes, 1000 hr V_{DS} 8V 100 mW RF No degrad.	Too early	Too early	—	—	Dexcel NEC MSC	No	No
No	Yes biased NF_{MIN} to 27 °C	Too early	—	—	No	No	Yes, No results yet
Yes, acc.	Yes step stress const. stress 157, 245 °C	10^4 100 °C	0.67	Af gate opens	No	No	Yes. No results yet
Yes 100–125 °C OK after 2000 hr	Yes, in process	Too early	—	—	No	Yes, packaged amps. MIL-S-202 MIL-S-810	Yes No results yet
No	No	No	—	—	No	VIB shock humid temp. alt.	Yes, No results yet

Table 5.5 *(continued)*

Organization	Device	Active Layer	Buffer	Drift Short, Long	Light Sensation sensitivity	Passi-vation	Ohmic	Schottky	Recessed gate or surface etch
Microwave Semiconductor Corp	SB power	VPE	Yes	No, No	No	SiO_2 or none	AuGeNi	TiWAu	Yes
McDonnel Douglass AC	JFET DEPL-ENH	I^2	No	No, No	Weak	Si_3N_4 sputtered neut beam	AuGeNi	NA	No/Yes
Motorola	SB SM, SIG	VPE I^2	Yes/No	Yes, No	Weak	No	AuGeNi TiPtAu	TiPtAu	No
Mitsubishi	SB SM, SIG	VPE	Yes	No, No	Weak	Si_3N_4 sometimes	AuGeNi	Al	Yes
	SB power	VPE	Yes	No, No	Weak	No	AuGeNi	Al	Yes
Texas Instruments	SB power	VPE	Yes	No, No	Weak	Si_3N_4	AuGeNi	TiPtAu	Yes/Yes
Dexcel	SB SM, SIG	VPE (LPE)	Yes	No, No	No prob.	SiO_x	AuGeNi	TiPtAu	Yes
	SB power	VPE (LPE)	Yes	No, No	No prob.	SiO_x	AuGeNi	TiPtAu	Yes
TRW	SB SM, SIG	I^2	No	No, Some	Weak	SiO_2 Si_3N_4	AuGeNi	CrPtAu	Yes
Hughes Aircraft Co.	SB LNFET power	I^2 VPE LPE	Yes	Yes, No No prob.	Weak No prob.	SiO_2 sometimes	AuGeNi	Al	Yes

have shown that gold–refractory gates with silicon nitride or silicon dioxide passivation can withstand exposure to high relative humidity and ionic contamination for prolonged periods of time without degradation.

The status of power GaAsFET reliability, as determined from Section 5.3, is unclear at the present time with reported Arrhenius plot activation energies varying from 0.67 to 2.3 eV and MTTFs from 10^4 to 10^8 hours at 100 °C, in agreement with Table 5.5. However, it can be concluded that: (a) if an aluminium gate is present, voiding or Au–Al reactions is highly probable; (b) the susceptibility to electromigration is very high with Al gates and AuGe source–drain ohmic contacts. There remains the need for more definitive studies of power GaAsFET reliability including gate grain size effects and the comparison of DC and RF failures using the same temperature–stress conditions.

Life tests	High temp stress	MTBE (hr) channel temp	ϕ (eV)	Failure mode	Comparison with other devices	Ambient effects studies	Radiation effects studies
Yes, Hi. temp. stor. at 120 °C		10^8	10	Metal diffusion	No	Yes, detailed of text	Yes, detailed of text
Yes, DC bias 140, 250 °C		10^7 125 °C	185	Gate changes			
No	No	—	—	—	—	No	Yes, detailed of text
No	No	—	—	—	—	—	—
Yes less than 200 FIT	Yes, up to T_{CH} 250 °C	10^7 150 °C	1.8	Au–Ga interdiffusion	No	Yes	Scheduled
Yes	Yes up to T_{CH} 250 °C	10^5, 10^6 150 °C	1.8	Au–Ga interdiffusion	No	Scheduled	Scheduled
Yes, 10^3 hr, 8V V_{DS} 100 mW rF input No degradation	Yes, 200 °C, 8 V_{DS} 100 mW rF input devices degrade	Too early	Too early	I_{DSS}, gain decrease	Dexcel NEC MSC	No	No
Yes, no bias T_1 275 °C data not avail. yet	Yes, T_1 275 °C data not available yet	—	—	I_G leakage Au-Ga interdiffusion	Yes	No	No
Yes, no bias T_1 275 °C data not avail. yet	Yes, T_1 275 °C data not available yet	—	—	I_G leakage Au-Ga interdiffusion	Yes	No	No
No	No	—	—	—	NEC Dexcel	No	Yes, at research triangle inst.
Yes, biased and unbiased	Yes, biased, unbiased and under rF 90 °C to 260 °C LM and PWR	10^9 100 °C LN FET	1.8 LN FET	Ohmic contact degrad, metal changes implied	NEC Dexcel Airtech TI Raytheon MSC Fujitsu Plessey H-P Avantek	No	No

A potentially significant problem for GaAs FETs in discrete or IC format, as utilized in military systems, may be the pulsed radiation susceptibility. GaAs-FETs either ion-implanted or epitaxial have been shown to be susceptible to pulsed radiation with significant long-term transient recovery. A significant materials improvement in GaAs will be required to harden devices and circuits to both transient radiation or positive gate pulsed effects or both.

GaAs IMPATTs although available commercially since 1971 have limited reliability due to metal–GaAs junction motion. Recently, the reliability problems have been complicated by the possibility of interstitial diffusion induced by hot carriers. Transferred electron devices still suffer from anode metal migration induced by the domain's electric field. A new refractory based anode–cathode metallization system must be incorporated in order to minimize the metal migration in two terminal devices. Planar TEDs have also been observed to undergo field enhanced ion migration probably caused by recombination enhanced diffusion. The mechanism of recombination enhanced diffusion requires further study since it is a reliability limiting mechanism for both IMPATTs and TEDs. The assessment of GaAs mixer diodes is at its initial stages. The present status is that diode burnout from RF pulses greater than 1 watt is a significant problem at frequencies above 36 GHz.

Finally, it is emphasized that new approches for GaAs ohmic contacts, Schottky barriers, passivations and substrate materials should be investigated if present reliability levels are to be improved. The problems reviewed in this chapter for standard AuGe, AuGeNi ohmic contacts, Schottky barriers have imposed limits to industry's ability for attaining high device reliability. Defects which are also present in substrates and at interfaces have introduced radiation related problems for GaAsFETs if utilized in a pulsed radiation environment.

REFERENCES

1. D. A. Abbott and J. A. Turner, 'Some aspects of GaAs FET reliability', *1975 IEDM Technical Digest*, pp. 243–246, 1975.
2. H. Kohzu, I. Nagasako, M. Ogawa, and N. Kawamura, 'Reliability studies of one-micron Schottky gate GaAsFET', *1975 IEDM Technical Digest*, pp. 247–250, 1975.
3. T. Irie, I. Nagasako, H. Kohzu, and K. Sekido, 'Reliability study of GaAs MESFETs', *IEEE Trans.,* **MTT-24**, 321 (1976).
4. H. F. Cooke, 'Small signal GaAs FET problems', *Army ECOM Contract DAAB07-76-C-1340-I*, at Avantek, Inc., 1976.
5. R. E. Lundgren and G. O. Ladd, 'Reliability study of microwave GaAs field effect transistors', *16th Annual Proc. Rel. Physics*, pp. 255–269, 1978.
6. R. E. Lundgren, 'Reliability study of GaAs FET', *RADC-TR-78-213*, Oct. (1978).
7. A. Christou, Unpublished research at the Naval Research Laboratory, 1976.
8. A. Christou, 'Low temperature interdiffusion in Al–GaAs', *J. App. Phys.,* **47 (12)**, 5464–5468 (1976).
9. J. Poate, 'Diffusion studies', in *Thin Films Symposium*, Chicago, Ill. Oct. 25, 1977.
10. J. C. Irvin and A. Loya, 'Failure mechanisms and reliability of low-noise GaAsFETs', *Bell System Tech. Journal,* **57**, No. 8, 2823–2846 (1978).
11. K. Mizuishi, H. Kunono, H. Sato, and H. Kodera, 'Degradation mechanisms of GaAsFETs', *IEEE Trans. Electron Devices,* **ED-26**, 1008 (1979).
12. B. L. Walsh, Jr., *Private Communication*, Communications Laboratories, Hughes Aircraft Corp. (1980).
13. E. Calucci, *RADC Contract F30602-78-C-0295* (HRL) (1978).
14. E. Calucci, *RADC Contract F30602-79-C-0037* (TI) (1979).
15. P. M. White, B. L. Hewitt, and J. A. Turner, 'Reliability investigations of GaAs power FETs with aluminum gate', *European Microwave Conference*, Paris, France, 1978.
16. L. Baun, 'Electromigration testing—a current problem', *Microelectronic and Reliability,* **12**, 215 (1974).

17. P. M. White, 'Metallizations for III-V semiconductor devices', *IEEE Professional Group Committee S2, Digest No. 1979/59*, 1979.
18. A. C. Macpherson, K. R. Gleason, and A. Christou, 'Voids in aluminum gate power GaAs FETs under microwave testing', *AFTA Symposium*, Los Angeles, Ca., No. 1978, 1978.
19. R. A. Minasian, 'Simplified GaAs MESFET Model to 10 GHz', *Electronics Letters,* **13**, 18, 549 (1977).
20. E. D. Cohen and A. C. Macpherson, 'Reliability of gold metallized commercially available power GaAs FETs', *17th Annual Proc. Reliability Physics*, 156–160 (1979).
21. I. Drukier, Supplied deliverables on *Navy Contract N00173-78-C-0019*, 1979.
22. B. S. Siegal, 'Use electrical tests for thermal measurements', *Microwaves,* **15**, No. 6, p. 48 (1976).
23. I. Drukier and J. F. Silcox, Jr., 'On the reliability of power GaAsFETs', *Proceedings of the Reliability Symposium*, San Francisco, Ca., April 1979, p. 1 (1979).
24. A. Christou, 'Precipitation and solid phase formation in Au (Ag)/Ge based ohmic contact systems for GaAs FETs', *1977 Cornell Conf.Proc.*, 169 (1977).
25. K. J. Sleger and A. Christou, 'Studies of Ta and Al Gate annealing on GaAs FET structures', *WOCSEMMAD 1977*, New Orleans, La., Feb. 24–25, 1977.
26. H. Fukui, S. H. Wemple, J. C. Hovin, W. C. Niehaus, J. C. M. Huang, H. M. Cox, W. O. Schlosser, and J. V. Dilorenzo, 'Reliability of power GaAs field effect transistors', *IEDM*, Washington, D.C., Dec. 1979.
27. B. Dornan, W. Slusark, Y. U. Wu, D. Pelka, R. Barton, H. Wolkstein, and H. Huang, 'A 4 GHz FET Power Amplifier', *RCA Review*, **41**, 472–501 (1980).
28. R. S. Wagner, A. K. Sinha, and T. T. Sheng, 'Tungsten metallization for LSI applications', *J. Vac. Sci. Tech.,* **11**, 582–589 (1974).
29. R. A. Pucel, D. J. Masse, and C. F. Krumm, 'Noise performance of GaAs field effect transistors', *IEEE J. Solid State Circuits,* **SC-11**, 243 (1976).
30. M. Fukuta, K. Suyama, H. Suzuki, Y. Nakayama, and H. Ishikawa, 'Power GaAs MESFET with a high drain–source breakdown voltage', *IEEE Trans. on Microwave Theory and Tech.,* **MTT-24, 6**, 312 (1976).
31. S. H. Wemple and W. C. Niehaus, 'Source–drain burnout in GaAs FETs', *Inst. Phys. Conf.*, Ser. No. 33B, Chapter 5 (1977).
32. F. Hasegawa, Y. Takayama, A. Higashisaka, T. Furutsuka, and H. Honjo, 'GaAs power MESFETs with a simple recess structure', *ISSCC Dig. Tech. Papers*, p. 118 (1978).
33. T. Furutsuka, T. Tsuji, and F. Hasegawa, 'Improvement of the drain breakdown voltage of GaAs power MESFETs by a simple recess structure', *IEEE Trans. on Electron Devices*, **ED-25**, 6, 563–569 (1978).
34. T. Ianada and S. Kato, 'Ohmic contacts on ion-implanted *n*-type GaAs layers', *J. Appl. Phys.,* **50(6)**, 4466.
35. R. Yamamoto, A. Higashisaka, and F. Hasegawa, 'Light emission and burnout characteristics of GaAs power MESFETs', *IEEE Trans. on Electron Devices,* **ED-25**, 567 (1978).
36. W. C. Ballamy and L. C. Kimerling, 'Premature failure in Pt–GaAs IMPATTs—recombination assisted diffusion as a failure mechanism', *IEEE Trans. on Electron Devices,* **ED-25**, 746–752 (1978).
37. S. P. Bellier, R. F. Haythornwaite, J. L. May, and P. J. Woods, 'Reliability of microwave GaAs field effect transistors', *13th Annual Proc. Rel. Physics*, pp. 193–199 (1975).

38. J. J. Whalen, M. Thom, and M. C. Calcatera, 'Microwave nanosecond pulse burnout properties of GaAs MESFETs', *1979 IEEE MTT-5 International Symposium*, pp. 443–445 (1979).
39. H. L. Hughes and R. R. Giroux, 'Space radiation affects MOSFETs', *Electronics*, Dec. 28 (1964).
40. R. Zuleeg, J. K. Notthoff, and K. Lehover, 'Femtojoale high-speed planar GaAs E-JFET logic', *IEEE Trans. Electron Devices,* **ED-25**, 628 (1978).
41. R. Zuleeg and K. Nothoff, 'Transient response of GaAs ICs to ionizing radiation', *1979 IEEE Radiation Effects Conference*, Santa Clare, Ca., July (1979).
42. D. H. Phillips, 'GaAs digital IC radiation hardness at high dose rates', *GaAs IC Circuit Symposium*, Lake Tahoe, Sept. (1979).
43. M. Simmons and E. E. King, 'Long-term pulsed radiation response characteristics of GaAs FETs', *IEEE Radiation Effects Conference*, Santa Clara, Ca., July (1979).
44. W. T. Anderson, Jr., K. E. Davey, A. Christou, M. L. Bark, and K. J. Sleger, 'Germanium heterojunction gate GaAs FET', *IEEE IEDM*, Dec. (1978).
45. S. V. Bearse, 'GaAsFETs: Device designers solving reliability problems', *Microwaves,* **15**, 32 (1976).
46. D. V. Lang, 'Deep level transient spectroscopy: a new method to characterize traps in semiconductors', *J. Appl. Phys.,* **45**, 3020–3033 (1974).
47. G. P. Nelson, Naval Research Laboratory, Unpublished research, 1978.
48. W. T. Anderson, Jr., A. Christou, and K. J. Sleger, 'Environmental effects on GaAs MESFETs', *IEDM Proceedings*, Dec. 3–5, 1978.
49. W. T. Anderson, Jr., A. Christou, and J. E. Davey, 'Smooth and continuous ohmic contacts to GaAs using epitaxial Ge films', *J. Appl. Phys.,* **49(5)**, 2998–3002 (1978).
50. J. C. Irvin, D. J. Coleman, W. A. Johnson, and I. Tatsuguchi, 'GaAs IMPATT diodes', *Proc. IEEE,* **59**, 1212–1219 (1971).
51. J. C. Irvin, 'GaAs IMPATT diode in perspective', *4th Biennial Cornell Conf.*, Cornell Univ. Press, Ithaca, New York, p. 287, 1973.
52. W. R. Wisseman, D. W. Shaw, R. L. Adams, and T. E. Hasty, 'GaAs Schottky–Read diodes for X-band Operation', *IEEE Trans. Electron Devices*, **ED-21**, pp. 317–326, 1974.
53. H. M. Olsen, 'Thermal runaway of IMPATT diodes', *IEEE Trans. on Electron. Devices,* **ED-22**, April, pp. 165–168 (1975).
54. H. M. Olsen, 'A mechanism for catastrophic failure of avalanche diodes', *IEEE Trans. on Electron Devices,* **ED-22, 10,** 842–849 (1975).
55. S. Su and S. M. Sze, 'Design considerations of high-efficiency GaAs IMPATT diodes', *IEEE Trans. Electron Devices*, **ED-20**, 541–550 (1973).
56. G. E. Mahoney, 'Retardation of IMPATT diode aging by use of tungsten in the electrodes', *Appl. Phys. Lett.,* **27**, 613–615 (1975).
57. T. L. Hierl, J. J. Berenze, and S. I. Long, 'High efficiency GaAs IMPATTs', *Cornell Conf.*, August 1977.
58. J. J. Berenz, G. J. Scilla, V. L. Wrick, L. F. Eastman and G. H. Morrison, 'Interdiffusion in GaAs metallization', *J. Vac. Sci. Technol.,* **13**, 1152–1161 (1976).
59. M. R. Namordi and V. Sokolov, 'The effect of current bias on the reliability of Read-type Schottky barrier GaAs IMPATTs', *1979 IEDM*, Paper 16.6, Washington, D.C. Dec. 3–5, 1979.
60. J. L. Heaton, R. E. Walline, and J. F. Carroll, 'Low–high–low profile gallium arsenide IMPATT reliability', *IEEE Trans. on Electron Devices,* **ED-26**, No. 1, 96–108 (1979).

61. P. Guetin, 'Contribution to the experimental study of the Gunn effect in long GaAs samples', *IEEE Trans. on Electron Devices,* **ED-14**, No. 9, 5522–5531 (1967).
62. C. J. Palmstrom, D. V. Morgan, and M. J. Howes, 'Contact degradation studies on GaAs transferred-electron devices using a focused backscattering technique', *Electronics Lett.,* **13**, 17, 397–398 (1977).
63. C. J. Palmstrom, D. V. Morgan, and M. J. Howes, 'Contact degradation of GaAs transferred-electron devices', *Nucl. Instr. and Methods,* **150**, 305–311 (1978).
64. J. Gyulai, J. W. Mayer, and V. Rodrigues, 'Gunn diode degradation', *J. Appl. Phys.,* **42**, 3578–3589 (1971).
M. Shoji and F. J. D'Alessio 'Reliability of Gunn diodes', *Proc. of IEEE,* **57**, No. 2, 250–263 (1969).
66. T. B. Ramachandran, 'Gunn diode reliability', *Status Report*, Microwave Associates, 12/72, (1972).
67. T. B. Ramachandran, J. L. Heaton, and E. B. Hakim, 'Reliability studies of Gunn diodes', *12th Annual Proceedings, Reliability Physics, IEEE*, pp. 284–292, 1974.
68. B. Jeppsson and I. Marklund, 'Failure mechanisms in Gunn diodes', *Electron. Letters,* **3**, No. 5, 213–214 (1967).
69. W. T. Anderson, Jr., H. B. Dietrich, A. Christou, and M. L. Bark, 'Transferred-electron device development using silicon ion implantation in GaAs', *37th Annual Device Research Conference*, paper WP-A6, June 25–27, 1979, Boulder, Colorado.
70. W. T. Anderson, Jr. and A. Christou, 'GaAs transferred-electron device failure mechanisms', *WOCSEMMAD '80*, San Francisco, Ca., Feb. 11–12, 1980.
71. W. J. Moroney and Y. Anand, 'Low barrier height GaAs Schottky diodes using AuGe alloy', *Proceedings Third Inst. Symp. GaAs*, paper 31, 1970.
72. A. Christou, A. and Y. Anand, 'Burnout mechanisms of GaAs mixer diodes at 36–94 GHz', *1980 Reliability Physics Symposium*, April 1980.
73. Y. Anand and A. Christou, 'High burnout millimeter GaAs mixer diodes', *IEDM*, Washington, D.C., Dec. 3–5, 1979.
74. J. A. Calviello and J. L. Wallace, 'Performance and reliability of a high temperature GaAs Schottky junction and native oxide passivation', *IEEE Trans.,* **ED-24**, 698–704 (1977).
75. J. A. Calviello, J. L. Wallace, and P. R. Bie, 'High performance GaAs beam lead mixer diodes', *Elect. Letters,* **15**, 17, 509 (1979).
76. A. G. Cardiasmenos and P. T. Parrish, 'A 94 GHz balanced mixer using suspended substrate technology', *1979 International Microwave Symposium, Orlando, Fla.,* May 3–6, 1979.
77. A. M. Andrews and N. Holonyak, 'Ohmic contacts for GaAs devices', *Solid St. Electron.,* **15**, 601–612 (1972).
78. G. Y. Robinson, 'Metallurgical and electrical properties of alloyed Ni–Au–Ge films on *n*-type GaAs', *Solid State Electron.,* **18**, 331–334 (1975).
79. P. C. Moutou, J. J. Godart, J. M. Montel, and B. Dixneaf, 'Cathode structures with limited injection current for GaAs TED', *1975 Cornell Conf. Proc.,* pp. 225–230, 1975.
80. M. Omori, 'Metallic contact on GaAs microwave devices', Reliability Physics, *15th Annual Proceedings*, pp. 232–237, 1977.
81. A. Christou, 'Solid phase formation in GaAs ohmic contact systems', *Solid St. Electronic,* **22**, 141–150 (1979).
82. A. Christou, 'Interdiffusion in Au-refractory thin film systems studied by a combination of scattering techniques', *Scanning Electron Microscopy/1979/Vol. 1*, SEM, Inc., AMF O'Hare, Ill., 1979.

83. J. C. Blair and C. R. Fuller, 'Electromigration induced failures in Gold films', *J. Appl. Phys.*, **43**, No. 2, 307–315 (1972).
84. J. V. Dilorenzo, W. C. Niehaus, and A. Y. Cho, 'Nonalloyed and *in situ* ohmic contacts to highly doped *n*-type GaAs layers grown by MBE', *J. Appl. Phys.*, **50**, 951–953 (1979).
85. R. Stall, S. E. C. Wood, K. Board, and L. F. Eastman, 'Ultra low resistance ohmic contacts to *n*-GaAs', *Elect. Lett.*, **15**, No. 24, 800 (1979).
86. P. A. Barnes and A. Y. Cho, 'Ohmic contacts to GaAs by MBE', *Appl. Phys. Lett.*, **33**, 651–653 (1978).
87. R. L. Mozzi, W. Fabian, and F. J. Piekarski, 'Nonalloyed ohmic contacts to *n*-GaAs by pulse-electron-beam-annealed selenium implants', *Applied Phys. Lett.*, **35(4)**, 387–389 (1979).
88. H. B. Kim, G. G. Sweeney, and T. Heng, 'Schottky barriers to GaAs', *Westinghouse Res. Lab. Report*, No. 74-IF6-MPATT-P1, 1974.
89. H. M. Day, A. Christou, and A. C. Macpherson, 'Interdiffusion and Schottky barrier variations in Au–W(Ti–/*n*-GaAs contacts', *J. Vac. Sci. Technol.*, **14**, No. 4, 939–946 (1977).
90. E. Kohn, 'High temperature stable metal GaAs contacts', *1979 IEDM*, Washington, D.C., Dec. 3–5, 1979.

Reliability and Degradation
Edited by M. J. Howes and D. V. Morgan

CHAPTER 6

Reliability and degradation of lasers and LEDs

D. H. NEWMAN and S. RITCHIE

6.1 INTRODUCTION

Semiconductor light-emitting diode (LED) and laser diode sources have become accepted as devices capable of extending the solid-state philosophy beyond the confines of pure electronic circuitry. Visible LEDs, now available freely as sources of red, yellow, and green light have established themselves in an indispensible position as visual indicators. Combinations of LEDs, in a hybrid or monolithic form, are one of the competitors for the lucrative visible alpha-numeric display market. Reliability of such LEDs is now almost taken for granted, mainly as a result of substantial effort during the early part of the 1970s. This effort is reviewed later in the chapter. However the main emphasis will be on the understanding of degradation processes relevant to the more sophisticated high-radiance infrared LEDs and lasers operating in the 0.8–1.6 μm region of the spectrum, developed for application in optical fibre communications. The reliability of these devices can still not be taken for granted even after a decade of study.

Infrared semiconductor lasers and high radiance LEDs are still not readily available and their current prices ($50–$1000+) reflect the fact that, to date, there is neither a large market nor real production capability. The potential market for such devices is showing signs of diversifying with applications for semiconductor lasers in optical video and audio disc reading (and writing) and in rangefinders and infrared illuminators for surveillance purposes. In addition, the original optical fibre communication concept is enlarging to include not only the land-based civil transmission network but undersea transmission and interconnections within telecommunication switching centres. The potential of optical fibre links for control and monitoring purposes in industrial and other environments, which can be electrically noisy or where electrical isolation is necessary, has also been recognized. Potential military markets (e.g. battlefield communications, use in ships and aircraft) are being identified as well. There is thus great incentive for research and development laboratories to perfect designs of high-radiance LED (HRLED) and laser that have the right combinations of characteristics. These include low electrical input power, capability for efficient optical coupling into a particular fibre, correct transient response for the particular application and operation over the required systems temperature range. All these aspects require detailed consideration in their own right,[1] but there is the overriding requirement that these characteristics do not deteriorate during operation either catastrophically or, if gradually, that the device should remain within specification for a time sufficient to make systems operation economically viable.

Semiconductor sources are also being developed in the 3–30 μm region and are assuming some importance in spectroscopy and atmospheric studies. A brief mention of the reliability hazards associated with these devices is also included.

The III–V compound range of materials form the basis of LED and laser devices that have reached the stage of commercial exploitation in the 0.5–1.6 μm region. Additionally, the lead chalcogenides are being developed for use in the wavelength range 3–30 μm. The range of binary III–V materials on which light emitting devices are based can be seen in Table 6.1. In principle, it is possible to cover the whole wavelength region from ~0.5 μm to ~6 μm with ternary and quaternary alloys derived from the binaries GaP, GaAs, AlAs, InP, InAs, InSb, and GaSb.

For visible LEDs, $GaAs_xP_{1-x}$ and GaP are the forerunners,[2] but as GaP is an indirect gap material, efficient radiative recombination is achieved via Zn–O pair emission (red) and via nitrogen iso-electronic centres (green). Nitrogen doping of GaAsP compositions in the indirect gap regime is employed also to give red and yellow emission; red emission can, in addition, be achieved by retaining the direct gap characteristics of GaAs using the compo-

Table 6.1 Binary III–V semiconductors

Material	Band gap energy at 300 K (eV)	Equivalent emission wavelength (μm)
GaP[a]	2.26	0.55
AlAs[a]	2.16	0.57
AlSb[a]	1.58	0.78
GaAs	1.42	0.87
InP	1.35	0.92
GaSb	0.73	1.70
InAs	0.36	3.44
InSb	0.17	7.29

[a]denotes indirect band gap, so low photon emission efficiency in pure material

sition $GaAs_{0.6}P_{0.4}$. Commercially, GaAsP is the favoured material for red-emitting LEDs.

In the infrared, GaAs and $Ga_{1-x}Al_xAs$ have dominated to date. The potential of optical fibres in the 1.2–1.6 μm region, due to their lower loss and smaller materials dispersion in comparison with their characteristics in the 0.8–0.9 μm region, has put emphasis on other combinations of the III–Vs, particularly the $In_xGa_{1-x}As_yP_{1-y}$system.[1]

In order to grow alloys such as InGaAsP in the form of high quality epitaxial layers on a conveniently available binary substrate material, it is usually necessary to ensure lattice matching. This prevents mismatch dislocations and other defects being formed which could affect radiative efficiency and reliability. Figure 6.1 shows a graph of lattice constant against band gap for the material system incorporating Ga, In, As, and P, together with that for Ga, Al, and As. To a first approximation, any composition of the ternary alloy GaAlAs can be grown lattice-matched onto a GaAs substrate, but only one composition of the ternary alloy InGaAs (in fact $In_{0.53}Ga_{0.47}As$) can be grown lattice-matched onto InP. This forces the grower into the quaternary $In_xGa_{1-x}As_{1-y}P_y$ alloy regime for band gaps intermediate between those of $In_{0.53}Ga_{0.47}As$ and InP.

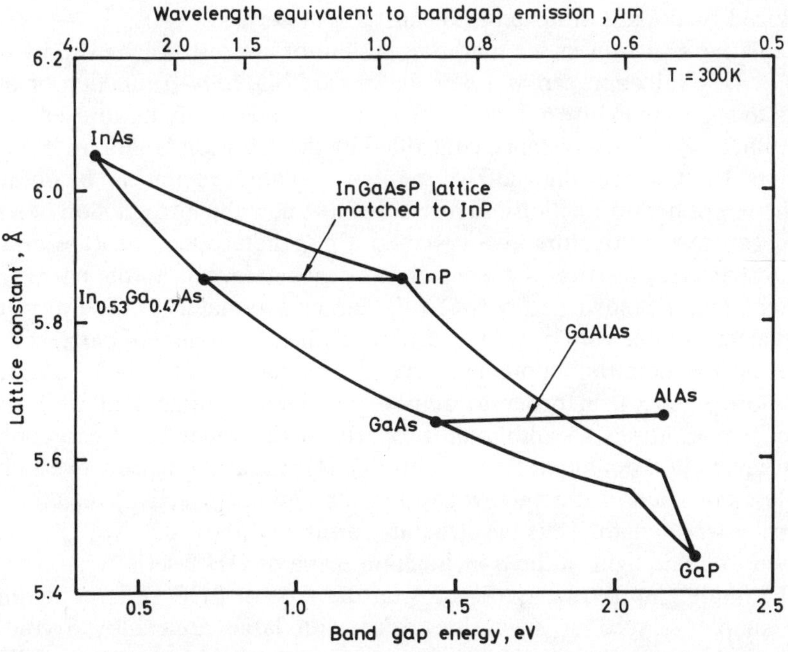

Figure 6.1. Lattice constants and band gaps for the quaternary system InGaAsP, and the ternary system GaAlAs

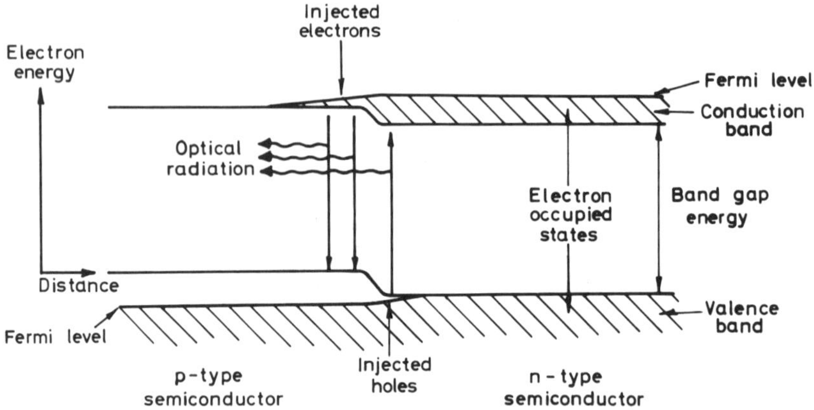

Figure 6.2. Band diagram of a high-doped, forward-biased p–n junction

The most common method of growth for epitaxial layers is liquid phase epitaxy (LPE) although hydride (or chloride) vapour phase epitaxy (VPE), metal organic chemical vapour deposition (MO–CVD) and molecular beam epitaxy (MBE) have been tried with varying success for some of the range of materials.[1] Simple p–n homojunctions in the binary compounds can be produced by dopant diffusion into substrate materials.

Radiative emission in the materials system of interest is achieved by electron or hole injection across a heavily forward-biased p–n junction or both. Thus, as shown in Figure 6.2, radiation is emitted over a volume adjacent to the p–n junction within a distance controlled by the diffusion length for minority carriers. Control over the width of the recombination region can be obtained by the use of heterojunctions. Figure 6.3 shows how the introduction of larger band gap semiconductors at a p–n and a p–p heterojunction (the double-heterostructure) restricts the electron-hole recombination to the narrow gap semiconductor sandwiched between the larger gap materials. The most efficient electron-hole confinement is achieved when the potential barriers introduced at the heterojunctions are several kT (barriers of 0.2–0.3 eV being common practice), in order to achieve adequate confinement well above room temperature. An additional property of the wider band gap confinement layers in a double heterostructure (DH) is that their refractive indices are less than that of the narrow gap recombination (or active) region at the emission wavelength. Thus the structure forms an optical slab waveguide that is used to guide light in the p–n junction plane of DH lasers.[3]

The diode chip geometry depends on the type of LED or laser required. The simple cleaved or sawn-sided dice with large area alloyed contacts (Figure 6.4) may be modified for low current density visible and infrared LEDs. The basic structure is modified in different ways to maximize external radia-

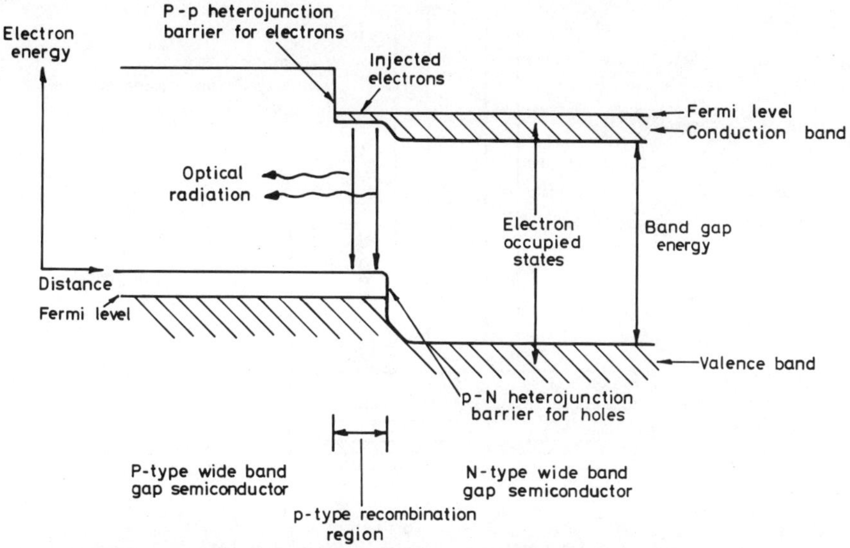

Figure 6.3. Band diagram of a high-doped, forward-biased double heterostructure p–n junction

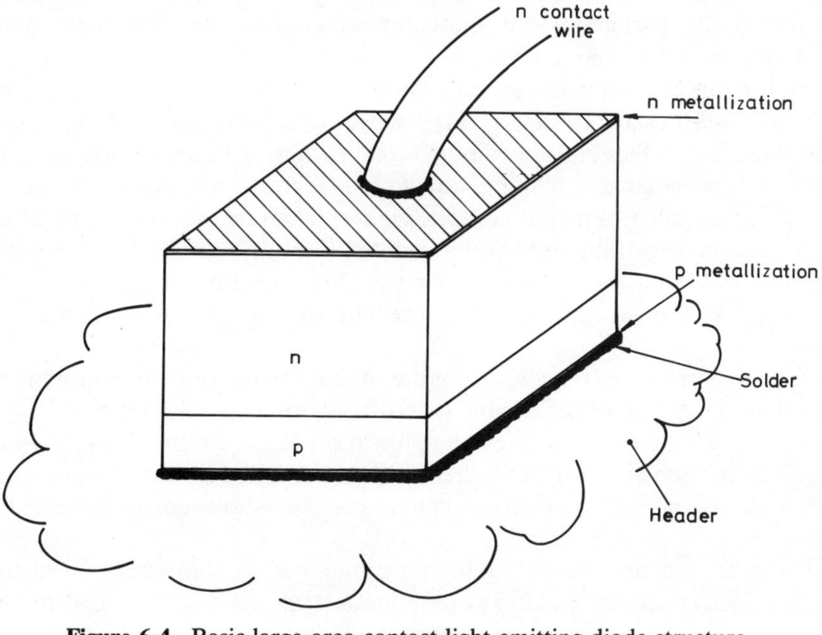

Figure 6.4. Basic large-area-contact light-emitting-diode structure

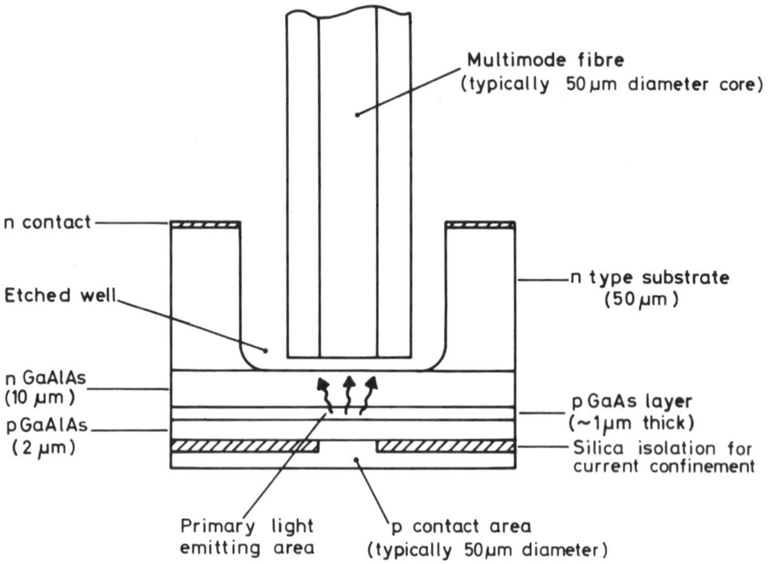

Figure 6.5. Cross-section of a double-heterostructure surface-emitting high-radiance LED, with typical dimensions

tive efficiency[2] depending upon whether the light suffers from self-absorption or not in the particular semiconductor material in use. The light output increases monotonically, in a fairly linear fashion, with injected current.

High-radiance LEDs are specifically designed to give efficient launching into optical fibres and can be of two generic types[1,3]–surface emitting or edge emitting. The surface emitting design achieves high radiance by restricting the emission to a small area within a larger chip (Figure 6.5). This is achieved by the use of a small area p-type contact. Light is taken out through a well etched in the n-type substrate and can be coupled into a fibre-tail. Structures with integral lenses have also been reported. Homojunction and single heterostructure as well as the DH structure depicted in Figure 6.5 have been developed.

The edge-emitting (ELED) configuration makes use of light emitted in the junction plane. Use of the double heterostructure helps with efficient extraction and, at high injection levels, amplification of the emitted photons in the plane of the junction further increases device radiance. A restricted length stripe geometry p-contact arrangement can be employed as indicated in Figure 6.6.

Broad contact area lasers can be fabricated, but for continuous (as distinct from pulsed) room temperature operation a stripe geometry configuration is adopted. The p-contact stripe (typically 3–20 μm wide) extends the full

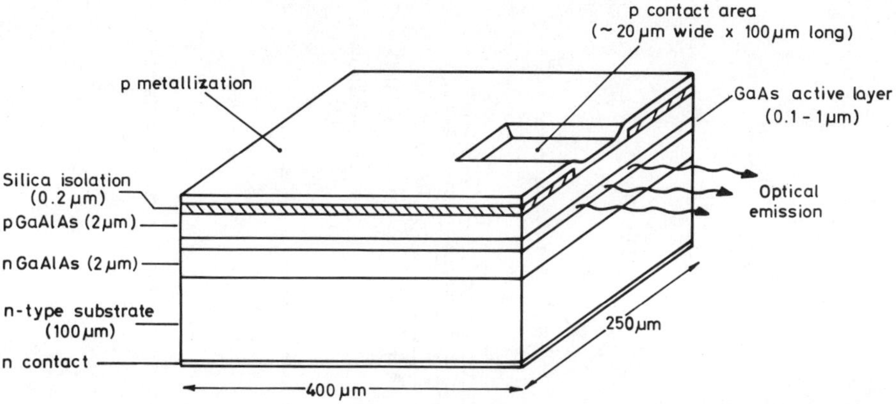

Figure 6.6. Restricted edge-emitting LED structure based on GaAs. Typical layer dimensions are given

length of an optical cavity formed, usually, by cleaving along planes perpendicular to the stripe direction (Figure 6.8). The high refractive index of the semiconductor material ensures (for GaAs and InP) a 30 per cent reflectivity and this is sufficient to achieve the required cavity round-trip gains to give laser action above a critical value of injected current—the threshold current. Thus photons emitted in the plane of the active region and perpendicular to the end facets are amplified as they travel in the cavity and the resulting wave is guided by the slab-waveguide geometry. Minimum threshold currents are achieved for an active region thickness in the range 0.05–0.3 μm.

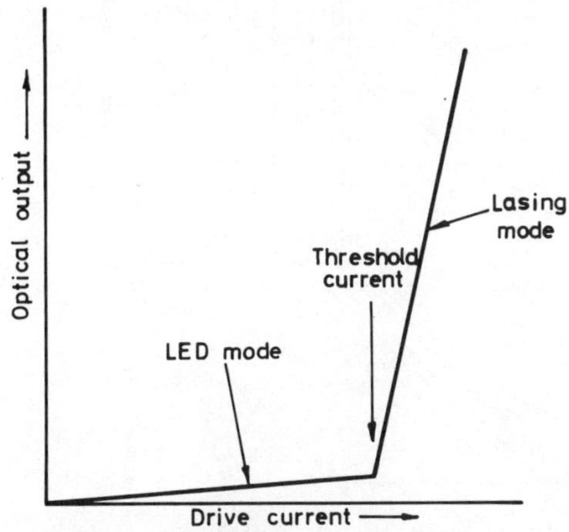

Figure 6.7. Idealized light output–drive current characteristics for a laser

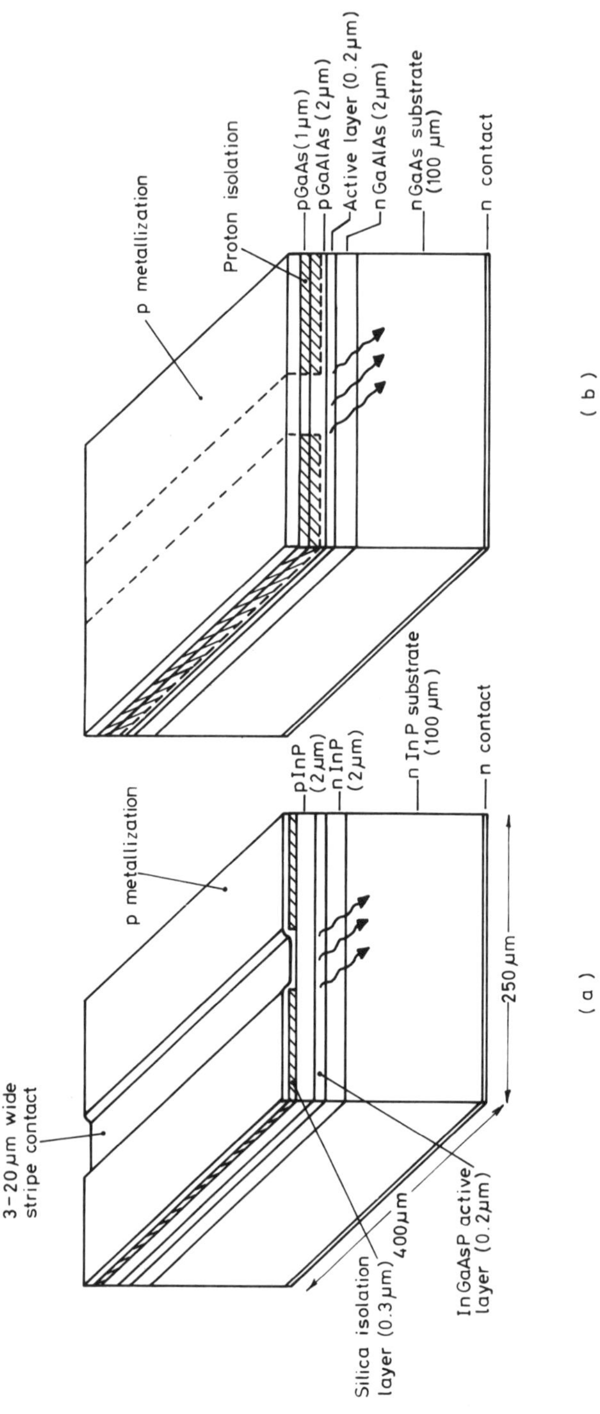

Figure 6.8. (a) oxide isolated InP based and (b) proton isolated GaAs based double-heterostructure lasers. Typical dimensions are given

An idealized optical power output versus injection current characteristic is shown in Figure 6.7. On increasing the injected current a laser initially acts as an LED, but close to the threshold current, stimulated emission begins to dominate and the incremental quantum efficiency becomes much higher above the lasing threshold. Operating currents range typically from a few to several hundred milliamperes as in the case of LEDs.

The most simple stripe-geometry DH lasers are the proton-isolated and oxide-isolated structures (Figure 6.8) and the majority of reliability studies have been associated with these two. These simple laser stripe geometry configurations usually have stripes widths in the range 3–20 μm. The extent of the lasing action in the transverse direction (parallel to the junction plane and perpendicular to the longitudinal cavity) is controlled mainly by the injected carrier profile which peaks at the centre of the stripe. This is termed 'gain-guiding'.[4] In order to achieve stable and reasonably linear optical power output characteristics above the threshold, a stable transverse lasing mode is necessary. The zero-order transverse mode would normally be preferred. This mode has a single peak in intensity at the centre of the stripe. Injected electrons decrease refractive index by the plasma interaction so that, in a

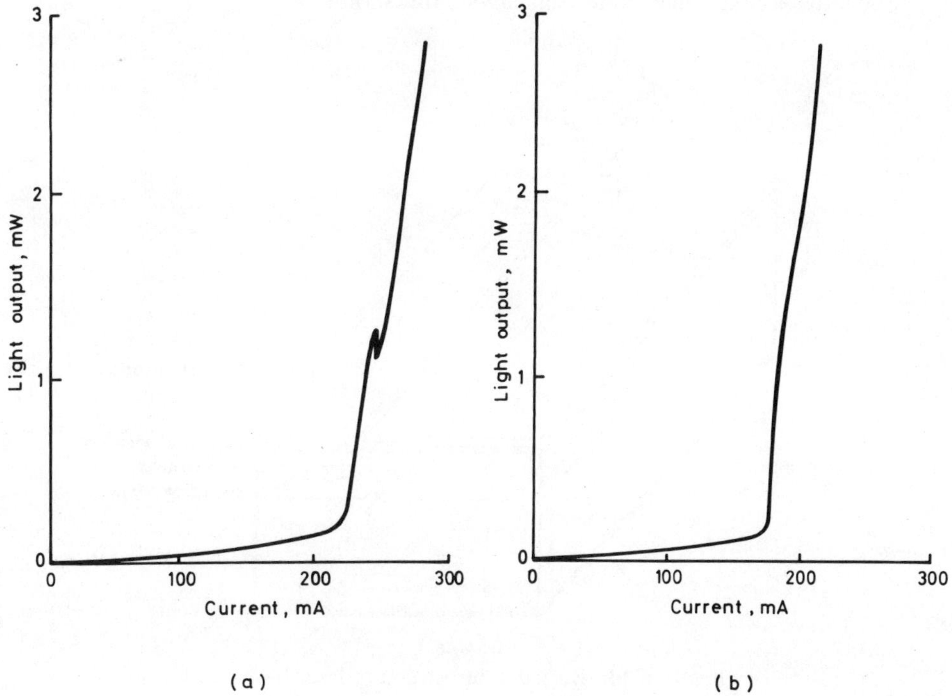

Figure 6.9. Two types of non-linearity in laser light output–drive current characteristics

gain-guided stripe structure, there is no positive refractive-index guiding and the optical wave is able to diffract out in the transverse direction, so increasing the losses. Interaction of the optical field with the injected population can result in changes in transverse mode structure, the appearance of non-linearities ('kinks')[1,4] in the optical power output versus current characteristics above threshold (Figure 6.9) as well as excess noise.[5] There has been much discussion of these problems in the literature, and improved stability has been obtained for some oxide[6] and proton-isolated[7,8] geometries.

There are a number of other configurations that aim to give further improvements in characteristics by stabilisation of the transverse optical field using a positive refractive index guiding structure. The buried heterostructure (BH) laser[1,4] is an example where, by a more complex (two stage) epitaxy process, the laser active region is surrounded by material of larger band gap and lower refractive index (Figure 6.10). Thus a real optical waveguide is formed with the active region as the core material. In order to achieve a zero-order transverse mode configuration, the active region width is restricted to about 1 μm because of the large refractive index steps introduced. For such a geometry very low threshold currents (~10 mA) can be achieved. Other improved stripe geometry structures[1,4] such as the transverse junction stripe (TJS) and channelled substrate planar (CSP) have greater active region widths and so have somewhat higher thresholds.

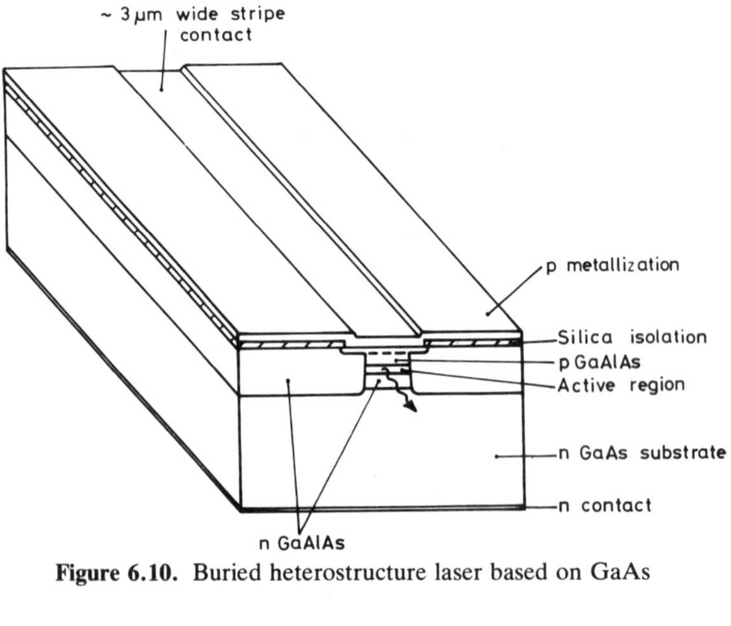

Figure 6.10. Buried heterostructure laser based on GaAs

6.2 DEVICE DETERIORATION AND LIFETESTING METHODS

Soon after the production of the first LEDs and semiconductor lasers in the early 1960s it became clear that there was a reliability problem. This was not, perhaps, too unexpected, as the only other device at that time incorporating a *p–n* junction in a III–V compound semiconductor, the GaAs tunnel diode, also showed poor reliability.[9,10] The reliability of LEDs improved reasonably quickly—however this was because of a general improvement in materials and fabrication techniques, with little understanding of the basic factors determining LED reliability. Lasers have proved much more of a problem, and this is reflected in that most of this chapter is concerned with lasers rather than LEDs.

This section deals with the manifestations of the degradation processes and the lifetesting methods employed.

6.2.1 Deterioration in device parameters

One of the most important parameters of a semiconductor laser is its threshold current. Device structures and materials properties have been optimized over the years, so that threshold currents are now low enough for most applications (less than 100 mA compared with several amperes 15 years ago). During this period also, the gradual increase in threshold current during device operation has been the overriding concern in trying to make more reliable devices. Figure 6.11 gives some idea of the considerable improvement in the reliability of lasers that has occurred during the past decade.

The variation during operation of parameters other than threshold is now seen as very important for specific applications. For example, in optical communications systems, mode stability and noise are two cases in point. However it is the increase in threshold current that has been the major concern in the bulk of the work on the reliability of lasers, and it is the monitoring of this parameter that has given maximum insight into degradation mechanisms.

A fundamental, albeit simplified, equation relating the threshold current (I_t) to more basic properties of the laser is

$$I_t \propto \left(\frac{g_t}{\beta}\right)^{1/m} \frac{qd}{\tau} \exp\left\{\frac{T_j}{T_0}\right\} \tag{6.1}$$

where β is a constant determined by the Einstein stimulated emission coefficient, d is the thickness of the active layer, τ the carrier lifetime and T_0 is a constant which determines the temperature sensitivity of threshold. m is a constant between 1 and 3 and g_t is the optical gain required to achieve

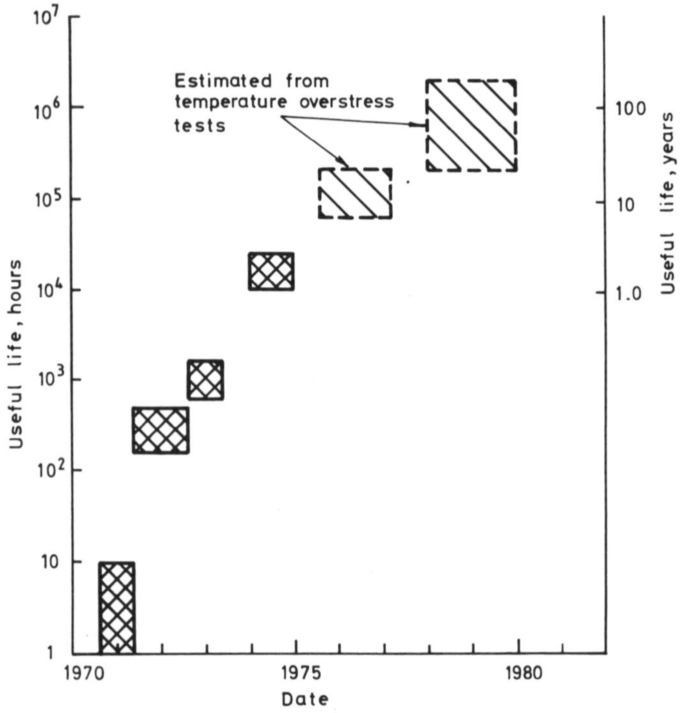

Figure 6.11. An indication of the gradual improvement of laser lives over the past decade

threshold given by:

$$g_t = \alpha + \frac{1}{L} \ln \frac{1}{R} \tag{6.2}$$

where α is the optical loss in the cavity, L its length and R the reflectivity of the cleaved facets. The junction temperature T_j is given by

$$T_j = R_t I_t V + T_h \tag{6.3}$$

where R_t is the thermal resistance and V the voltage over the device. T_h is the temperature of the heat sink. From equations (6.1) to (6.3) it is clear that an increase in threshold may occur as a result of changes in up to 10 parameters. It is indicative of the complexity of the laser reliability problem that for the various degradation mechanisms isolated to date, increases in threshold have been attributed to changes in α, R, τ, T_0 and R_t.

The external differential quantum efficiency, $\Delta\eta_e$ is another parameter which is liable to change during device operation. $\Delta\eta_e$ can be expressed as

$$\Delta\eta_e = \Delta\eta_i\left(\frac{g_t - \alpha}{g_t}\right) \tag{6.4}$$

and is proportional to the slope of the power output–current characteristic above threshold (Figure 6.7). $\Delta\eta_i$ is the internal differential quantum efficiency, which should not be confused with the spontaneous efficiency (η_i).[11] As the external differential quantum efficiency is determined by parameters that also affect threshold current, any change in efficiency is always associated with a threshold change but the converse is not necessarily so.[11,12]

Studies of other optical properties of lasers during lifetests have revealed changes particularly detrimental to the operation of devices in optical communication systems. For example, the onset of kinks in the power output–current curve above threshold, excessive noise and asymmetry between the emission from the front and rear facets have been observed to cause problems.[13] These effects are attributed variously to inadequate transverse mode confinement, inhomogeneities in the gain profile or saturable absorbers. It seems likely that some of these effects may be eliminated by employing suitable device geometry—for example, devices with tight transverse confinement, such as the buried heterostructure laser, do not show kinks (see Section 6.1). The limited data on these phenomena will be discussed in 6.3.2.

Moving to LEDs, the most obvious manifestation of their degradation is the gradual decrease in power output when the device is operated at a constant current, i.e. the spontaneous efficiency decreases with time. This is seen in lasers as well, and is usually attributed to the same cause. In addition, the carrier lifetime is observed to decrease, but this is not a problem for most applications.

6.2.2 The measurement of device reliability

Early lifetest data were difficult to interpret because of the erratic manner in which device parameters deteriorated, and considerable variability in the results from device to device.[14,15] This was due, in part, to the non-uniformity of the degradation process in the broad contact area devices available at that time.

In the case of GaAs lasers, much more consistent behaviour is observed on the more recent stripe-geometry devices, with threshold currents increasing gradually with time, at a rate of, say, 0.2 per cent per thousand hours at room temperature. A 'time to end of life' has become the most favoured parameter with which to measure device reliability. However this time clearly depends on what criteria are used to determine device failure. Although criteria involving noise limits or the onset of spontaneous oscillations may be increasingly relevant, the condition most commonly used is when the laser is unable to give a specified power output, regardless of how high the current is set.[16—21] This is known as thermal runaway; however it may not be a realistic condition for some applications. At room temperature, high currents would be required at thermal runaway, and in many optical communication systems the drive circuitry may not be designed to operate over a wide range of currents. Another shortcoming of the thermal runaway criterion is that because of the very sensitive dependence on temperature of many laser parameters (Equation 6.1), a laser operated and failed at an elevated temperature may operate perfectly satisfactorily at lower temperatures. This can introduce confusion when performing temperature overstress lifetesting. For this reason, some workers have used a specified percentage increase in threshold measured at room temperature as the condition for end of life.[22,23] Yet another method of quantifying the reliability of lasers is the use of a degradation rate[24,25]—for example, the percentage increase in threshold per thousand hours.

LED failure is a gradual process as well, but the problems in quantifying the degradation are not as great as for lasers. The power output decreases with time, although not necessarily in any well behaved manner. Several distinct stages in the power output decrease with time may be observed.[26] The most popular definition of the time to end of life of an LED is the time at which the power output has fallen to either 50 per cent, or 1/e, of its original value.

Although the failure of light emitting devices has become considerably less erratic over the past few years, as with all other semiconductor devices, there is a variation in reliability between devices within a specific batch. A common approach is to consider the lifetest data as a statistical distribution, and to use its characteristic parameters to describe the device population as a whole.

Three distributions have been used to characterise semiconductor components in general; the lognormal, the Weibul and the exponential distributions.[27,28] However it is the lognormal distribution which has proved the most popular for application to reliability data.[29,30] In the few statistical analyses of light emitting devices, both Weibul[20] and exponential[23] have been used, but again the lognormal distribution has had the most success. In the case of lasers, the fact that this distribution has been observed for selected samples,[21] unselected samples,[18] devices made with different structures, and using various methods of measuring life implies a certain degree of similarity between the ageing behaviour of lasers in general.

Figure 6.12 shows a cumulative failure distribution of lasers lifetested at 70 °C.[18] The ordinate is the time to failure on a logarithmic scale, and the abscissa is normalized so that a straight line on the graph corresponds to a lognormal distribution. All but two devices from a population of 90 had failed, and it can be seen that there is a very good fit to a lognormal distribution over much of the range. The important features are the median

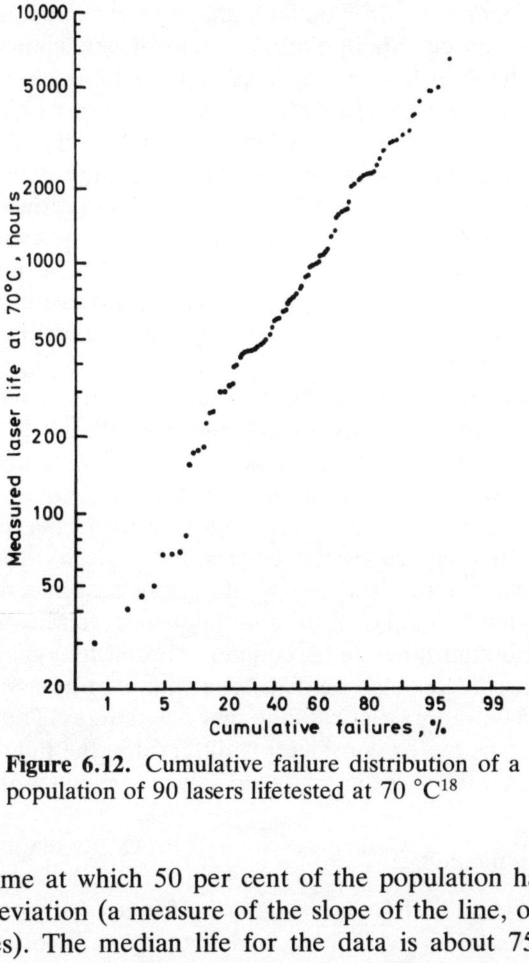

Figure 6.12. Cumulative failure distribution of a population of 90 lasers lifetested at 70 °C[18]

lifetime (the time at which 50 per cent of the population have failed) and the standard deviation (a measure of the slope of the line, or the spread of device lifetimes). The median life for the data is about 750 hours. More recent results have given median lives up to 4,500 hours at 70 °C.[21] With longer lived lasers, it becomes more difficult to fail a high proportion of devices in a reasonable test period and thus confirm that the failure distribution is lognormal. The importance of Figure 6.12 is that it shows a good fit to a lognormal distribution up to high cumulative failures. An additional parameter, incorporating both median and standard deviation, is often used for components in a telecommunications system and is known as the 2 per cent reliability life. This is the time at which 2 per cent of the population will have failed.

In the case of the failure of semiconductor components, it would appear that there are no sound physical reasons for the validity of the lognormal distribution, although there have been suggestions that it could occur as a fundamental consequence of diffusion processes[31] or an Arrhenius type temperature dependence.[29] It is unlikely that a further understanding of the implications of a lognormal failure distribution will yield clues about the basic degradation mechanisms limiting the lives of present day lasers. However the isolation of a particular failure distribution may help to eliminate certain possible mechanisms. For example, Joyce *et al.*[18] have pointed out that device failure caused by the random generation of a particular defect in nominally identical devices would result in a Poisson distribution rather than a lognormal distribution. One method of understanding the implications of the lognormal distribution is to make comparisons with the normal distribution. Whereas the normal distribution results from the additive effects of random variables, the lognormal distribution should result when the random variables interact multiplicatively.

The reason why three different distributions (lognormal, Weibul and exponential) have been used is probably because it is difficult to distinguish between them when only limited lifetest data is available (which is usually the case for lasers). Indeed, the data presented as a Weibul distribution by Furukawa *et al.*[20] can be fitted equally well to a lognormal distribution. As experience on other semiconductor devices has indicated that a lognormal distribution is more appropriate, it would seem better to use it for light-emitting devices until there is a more sound statistical validity for the use of an alternative distribution function. A common approach in characterizing failure distributions is to find the median time to failure by assuming a failure distribution, and extrapolating from the first few failures. The lognormal and Weibul distributions can be considerably different in the tails of the distribution, so that large differences in predicted values of median life could result.

6.2.3 Lifetesting methods

Lifetests on lasers and LEDs are usually performed with two objectives in mind; first to assess their useful lives, and secondly to discover what mechanisms are determining their lives by post mortem examination of failed devices. The latter objective dominated in the early lifetests on rather primitive devices at room temperature. Tests carried out over the past few years have usually employed overstress conditions, particularly elevated temperatures, to accelerate the relatively slow degradation now observed under normal operating conditions. The interpretation of the results from overstress tests will be covered in 6.3.3, but in this section the various problems in establishing the operating conditions will be considered, following a short discussion about lifetesting methods in general.

The requirement for high reliability LEDs and lasers comes mainly from the telecommunications industry, where devices will be carrying digital information between 2 Mbit s^{-1} and several Gbit s^{-1}. Large scale lifetesting under pulsed conditions is complex, and devices are usually tested by operating them continuously, considered to be the worst case condition. Mean power feedback or manual control is often employed to increase the drive current to maintain the power output constant as the device degrades.

Pulsed operation, either from a bias just below threshold (for a laser), or from zero (for a LED or laser) is a more realistic method of testing, as mechanisms associated only with pulsed operation can be envisaged. For example, many lasers show transient oscillations at the beginning of a pulse[1,3,4] which could cause undesirable high optical flux densities at the facets. This effect would not be present under continuous operation thus comparison between continuous and pulsed operation may become important for future reliability assessment.

Operating lasers continuously might seem to require a relatively simple experimental procedure. A problem, probably appreciated fully only by those performing lifetests, is the necessity to suppress all transients and spikes which may be transmitted through the drive circuitry. Even very short (a few 100 ns), very low amplitude (a few mA for high efficiency devices) current spikes can cause threshold increases due to facet damage (see Sections 6.3.1 and 6.3.2). Without great care the reliability of lasers can be more a reflection of the quality of the drive circuitry, a point which is not mentioned in the literature, but which should be considered seriously by those lifetesting lasers.

Lesser problems exist in setting up to test LEDs, because high optical flux densities are not present.

The basic philosophy of accelerated lifetesting is to operate devices at different degrees of overstress, in order to establish a relationship between device life and the amount of overstress. Estimates of device life under normal conditions are then obtained by extrapolation. The best example of this is temperature overstress lifetesting. The degradation process is taken to be thermally activated such that the time to failure (t_f) follows the Arrhenius relationship.

$$t_f \propto \exp\left(\frac{E_a}{kT}\right) \tag{6.5}$$

where E_a is an activation energy for the degradation process and T is the device temperature. An idealized Arrhenius plot is shown in Figure 6.13. Lifetest measurements might be taken, for example, at 150 °C, 125 °C, and 100 °C, and the 25 °C lifetime obtained by extrapolation.

In the case of lasers, three parameters are normally considered as relevant for overstress—junction temperature, power output and operating current. Unfortunately, these cannot be varied independently. For example, the threshold of a GaAs–GaAlAs laser will increase at about one per cent $°C^{-1}$, so to maintain the same power output at 25 °C and 75 °C, the operating current will have to be increased by about 50 per cent. In general, any accelerated lifetest using junction temperature, power output or operating current

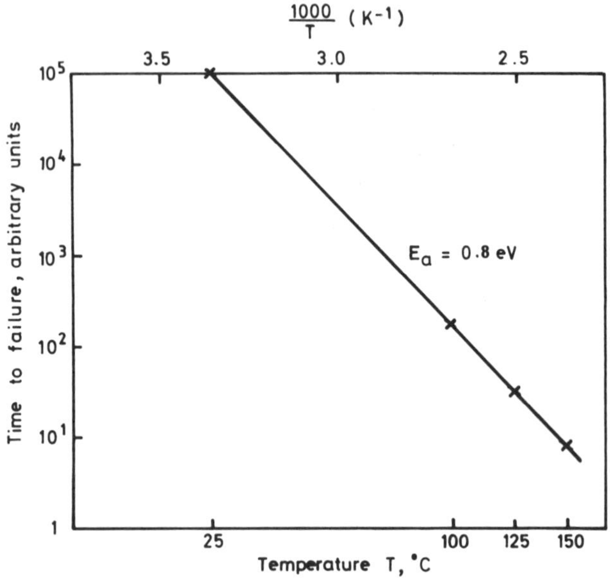

Figure 6.13. An idealized Arrhenius plot

as the prime overstress parameter must involve overstress or understress of one of the other parameters as well. The ability to predict accurately device life using accelerated tests will be influenced by these problems, and is discussed further in 6.3.3.

Most accelerated lifetesting on lasers have used junction temperature as the overstress parameter, and the current has been increased gradually during life to maintain the power output constant.[16–21] Until recently, maximum junction temperatures have been maintained near 100 °C because the laser chips are soldered down to heatsinks with indium (melting point 156 °C). Improved bonding procedures have enabled devices to be held at temperatures up to 140 °C,[23] but at this temperature continuous operation in the lasing regime is usually not possible for conventional devices designed for optical communication purposes. In these circumstances the lasers are held at a constant current below the lasing threshold.

Lifetesting of LEDs can be carried out more easily, with current and temperature being varied independently.

6.3 DEGRADATION OF HIGH CURRENT-DENSITY GaAs AND GaAlAs BASED SOURCES

Reliability studies of GaAs and GaAlAs based emitters can be considered in two categories. Devices can be of a low current density, low output power configuration or of the more specialized high radiance diode or laser types which operate at current densities in the range 10^3 to greater than 10^4 A cm^{-2}. Consideration of low current density GaAs LEDs is delayed until Section 6.4.1 where low current density LEDs in general are discussed. In this section, degradation phenomena in the specialised high current density devices are considered. This has received great impetus because of the need to develop reliable sources for optical fibre systems.

To some extent, reliability problems encountered in high-radiance LEDs and lasers are similar. The high current densities of device operation produce significant amounts of heat which have to be removed efficiently. Thus, as well as the possibility that degradation processes in the bulk of the chip are accelerated by increased temperature, stable bonding is of prime importance. In the case of lasers, there is the added possibility of damage to the emitting facet by the combined action of current, optical flux and the environment.

Some degradation processes are understood to the extent that improvements in technology have been made thought to be capable of limiting the effect of these processes on device life. These degradation processes are reviewed first in this section. Following is a discussion of degradation mechanisms still thought to affect state-of-the-art devices and a review of methods of predicting device life.

6.3.1 Degradation processes thought no longer to limit device life

6.3.1.1 Dark line and spot defects

Early reports of the rather rapid degradation of lasers and HRLEDs in periods of hours to a few hundred hours were concerned, in the main, with the growth of so-called dark-line defects (DLDs)[32,33] and, sometimes, dark spot defects (DSDs).[34] Using the scanning electron microscope in the induced current (EBIC) mode, it has been possible to observe the occurrence of DLDs, such as those shown in Figure 6.14. The dark lines propagate during operation along $\langle 100 \rangle$ crystallographic directions for devices fabricated on (100) orientated material. The dark line is a region of high non-radiative recombination so that carriers injected into that region do not contribute to the luminescent output. In addition, in the case of the laser, light travelling in the junction plane can be absorbed in these regions as they are regions of loss, and this further inhibits laser action.

Transmission electron microscopy (TEM) has identified the defects as arrays of dislocation dipoles. These are formed during device operation by the climb, in the proximity of the device active region, of dislocations threading up through the diffused or grown epitaxial layers from the substrate.[35,36] Figure 6.15 shows an example of such a climb network as seen in the electron microscope. Most workers now agree that the dipoles are of interstitial type,[37]

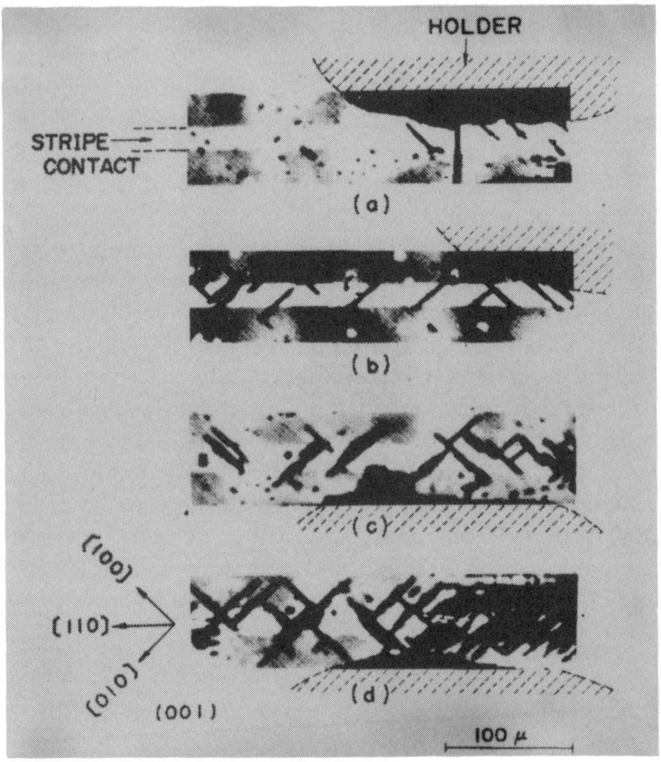

Figure 6.14. Typical example of dark line defect structures induced by ageing a laser. (a) to (d) show the effects with increasing ageing time[33]

that is, they contain extra planes of atoms. However, various aspects of the climb process are still uncertain. It has been suggested that supersaturations of one or both interstitial species (i.e. gallium and arsenic) are present. As a consequence of electron-hole recombination-enhanced diffusion,[38] these defects interact with the dislocation core to produce the observed climb (the phenomenon of recombination-enhanced diffusion is discussed in Section 6.3.2). Others have pointed out that to explain some of the dislocation dipole structures seen,[39] very large ($\sim 10^{19}\,cm^{-3}$) grown-in interstitial concentrations would be required if the source of interstitials was only from the region where electron-hole recombination was occurring. Thus it has also been suggested[39] that energy released by non-radiative electron-hole recombination at the dislocation core causes the dislocation to emit vacancies into the lattice, leaving an interstitial-type climb dipole and no problem concerning the origin of a large number of grown-in interstitials.

In spite of substantial effort, there is as yet no full understanding of the

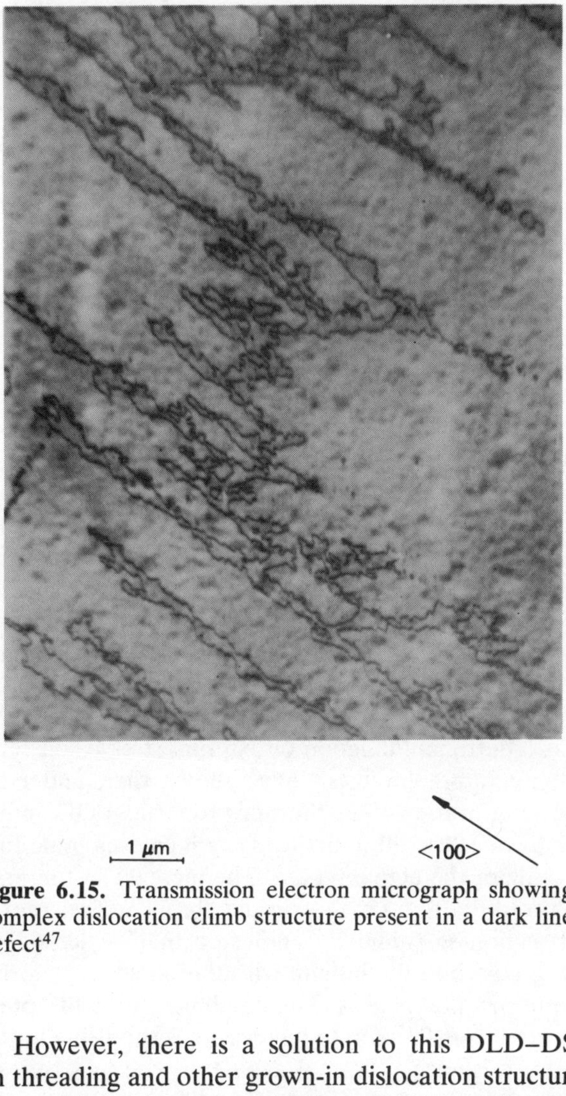

Figure 6.15. Transmission electron micrograph showing complex dislocation climb structure present in a dark line defect[47]

climb process. However, there is a solution to this DLD–DSD problem associated with threading and other grown-in dislocation structures. Elimination of these features is achieved by selection of better quality substrates and by more careful attention to details of epitaxial layer growth or dopant diffusion.

The mechanisms suggested for dislocation dipole growth still have implications if native defects can migrate more easily in the presence of electron-hole recombination. It is possible that the presence of strain or temperature gradients could give a preferred direction to the enhanced diffusion process. The native defects involved might then form clusters and perhaps even spontaneously generated dislocation loops. These could then act as regions of non-radiative recombination. Additionally, if loops were formed, dipole formation by climb might then occur leading to a rapid failure stage in devices made in high quality material. This is an incentive to obtain a much fuller understanding of recombination-enhanced point defect diffusion processes in III–V semiconductors. There is a need to identify the defect species most susceptible to the diffusion process so that methods to minimise its influence in the grown and processed semiconductor material can be formulated.

6.3.1.2 Thermal resistance deterioration

Two dice-to-heatsink bonding schemes are in common use, as shown in Figure 6.16. One involves the plating up of individual integral gold pads directly to the primary *p*-side chip metallization of the HRLED or laser. The plating thickness is usually between 10 and 50 μm. The gold pad spreads the heat so that the pad-to-heatsink soldering process is not critical. Integral heatsinks are more difficult to arrange in the case of lasers. This is because it is necessary for wafer cleaving (to produce the mirror facets) to be carried out in the presence of the individual heatsinks. Nevertheless, such a technology has been employed successfully in the case of lasers[40] as well as HRLEDs.[41] With lasers however, the preference has been to bond with indium solder the *p*-side of the chip, typically metallized with Ti/Au, directly to a gold-plated copper heatsink. Indium is chosen because of its ductility at ordinary temperatures (its melting point is only 156 °C); early experimental evidence[42] had indicated that there was a detrimental action of external stresses on reliability.

Elevated temperature lifetests[24] have shown that, under certain circumstances, rapid deterioration of the thermal properties of the indium bond could occur in periods of 100–200 hours, causing increases in dc threshold due to increased heating of the active region. The increase in thermal resistance of the bond occurred ten times slower in the absence of current flow. Detailed metallurgical sectioning studies[43,44] indicated that device operation induced the diffusion of gold into the indium with the consequent formation of voids and In–Au intermetallic phases. The combination of the presence of intermetallics and voids was thought to be directly responsible for the increases in device thermal resistance. Thermal activation energies of 0.6–0.9 eV were measured for operating devices.[45,46] This implied that significant thermal resistance increases would be expected in devices operating at room temperature over periods of 10,000 to 20,000 hours, as has been observed.[22,48]

Two approaches have been adopted to overcome this problem. One is to minimise or eliminate the gold present on the chip and the header, which removes the possibility of the excessive formation of intermetallics. With less than 100 nm of gold on both surfaces, it has been shown[43] that stable indium based bonds can be produced. Lasers directly bonded with indium not showing this thermal resistance problem probably fall into two categories. There

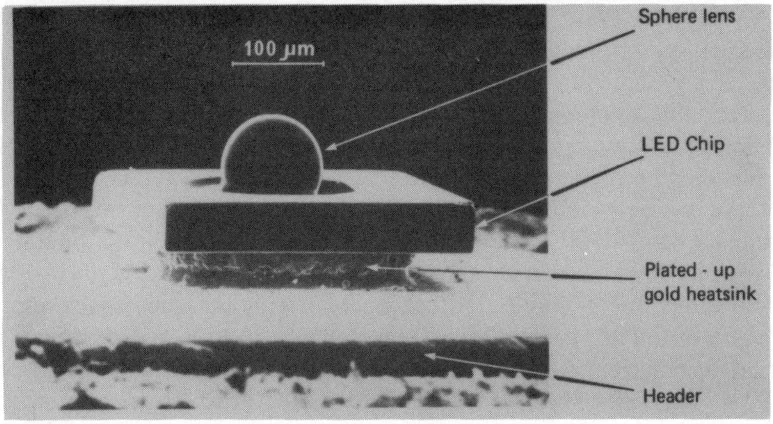

(a)

(b)

Figure 6.16. Examples of device bonding technology: (a) surface-emitting LED with integral plated heatsink (Reproduced by permission of Plessey Research (Caswell) Ltd). (b) DH laser, indium soldered to copper header

are those that have very thin gold layers in their metallization schemes (by design or accident) and those where the heating cycle during dice bonding is prolonged. All the gold will be consumed in the formation of the In–Au intermetallics, and there is the possibility that voids may be infilled. Under these circumstances the bonding cannot deteriorate further during operation, but the initial device thermal resistance will probably be greater than necessary.

The second approach is to question the need for retention of indium as the solder now that improvements in materials and device technology are giving much improved laser reliability. There are encouraging results using AuGe eutectic alloy[45] (melting point 360 °C) to solder laser dice with a thin Ti–Pt–Au *p*-side metallization to metallized diamond heatsinks. Lifetests extending over 10,000 hours at 70 °C on several stripe lasers operated at 4 mW optical power output have demonstrated stable device thermal resistance and only a 6 per cent increase of threshold current. This value compares well with state-of-the-art values being reported for devices using indium as a solder. It is not clear whether similar results could be achieved by soldering with AuGe to copper heatsinks where the thermal expansion mismatch with GaAs is worse.

It is worth bearing in mind that the amount of inbuilt strain due to the bonding will decrease at higher temperature. If there is any suggestion that inbuilt stresses caused by differential contraction can produce degradation in a particular device structure then it is obviously very necessary to be cautious in extrapolating room temperature life from temperature overstress tests conducted on that structure.

6.3.1.3 p-side metallization breakdown

Lasers and HRLEDs have, in the past, usually been required to operate only close to room temperature. Thus, there has been little consideration of the reliability problems associated with metal penetration into the semiconductor material. For example, in GaAs IMPATT diodes, which operate at junction temperatures around 200 °C, gold diffusion into the semiconductor can occur when a simple Ti–Au or Cr–Au two level *p*-metallization is employed.[49] However, with the need to accommodate temperature overstress testing and for devices to operate in environments well above room temperature, it is becoming standard for one of the more reliable three level metallization schemes (e.g. Ti–Pd–Au or Ti–Pt–Au) to be used. In at least one instance[50] it has been shown that penetration of gold through a thin titanium layer within the active area of surface emitting HRLEDs was the primary cause of device failure. In addition, it was shown that the perimeter of the active area was a particularly sensitive region in an oxide-isolated structure. In this case it is particularly difficult to arrange good coverage by the titanium at the oxide step edge. Clearly it is necessary to check for sufficient thicknesses of the Ti–Pt layers for example, especially over oxide steps. Thinner oxides and chamfered steps help avoid the gold penetration problem, as does the use of a planar isolation technology such as proton bombardment. Another point is that indium soldering of thin gold metallized chips may also avoid the problem because the gold will have been consumed during the bonding heat treatment to form In–Au intermetallics. In this respect, it may be significant that the gold penetration problem was reported for a structure with a thick (~50 μm) plated-up gold layer.[50]

6.3.1.4 Catastrophic facet damage in lasers

In the laser structure, the cleaved semiconductor facets form the laser cavity mirrors and can be subjected to high optical power densities. Deterioration of the semiconductor at this interface is capable of producing degradation in device performance. In fact, catastrophic facet failure limits the optical power output in most cases.

Early work on the subject was on homojunction and single heterojunction broad contact lasers. At that time lasing emission across the facet was often non-uniform and scattered results were obtained. However, it became clear that the catastrophic facet damage limit was dependent on both the magnitude and duration of an individual pulse.[3]

Eliseev[51] was first to report measurements with DH lasers and concluded that the critical optical flux density, q_c, had a pulse width dependence proportional to $t^{-\frac{1}{2}}$ for pulse widths, t, in the range 300 ns to 1 μs and with q_c about 4 MW cm^{-2} at 300 ns. The work on homojunction devices had used a measured reduction of q_c with decreasing temperature to argue that stimulated Brillouin scattering was a possible mechanism; more recent work suggests that thermal effects dominate.[52]

Imai *et al.*[53] have reported on catastrophic damage limits at room temperature in 15 μm wide stripe geometry DH lasers as a function of pulse length (Figure 6.17) and obtained an approximate empirical relation:

$$\begin{cases} q_c \propto t^{-1/4} (100 \text{ ns} < t < 10\text{–}50\ \mu\text{s}) \\ q_c \simeq \text{const}\ (t > 10\text{–}50\ \mu\text{s}) \end{cases} \tag{6.6}$$

For 300 ns pulses, q_c was about 3 MW cm^{-2}, which is comparable with Eliseev's value. The reason for the difference in the power law compared with the results of Eliseev is not clear, but Imai *et al.* suggest that different thermal characteristics of the early broad-contact samples and the state-of-the-art stripe lasers with good heatsinking may be important. However, the general tendency for q_c to saturate with increasing pulse length is an important feature of this damage mechanism.

Imai *et al.* also showed that the effects of catastrophic facet damage can be mitigated by the use of half-wavelength dielectric facet coating—such thickness coatings have minimal effect on laser threshold (as opposed to quarter-wavelength coats that are anti-reflecting). They suggest that Al_2O_3 as a coating material is superior to SiO_2 (see Figure 6.17). Also, there is evidence that optical flux rather than current density is the critical parameter.

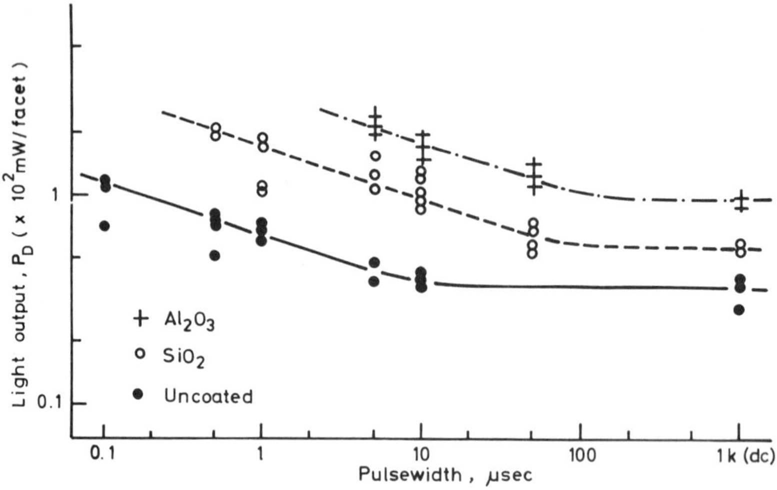

Figure 6.17. Light output required for catastrophic damage as a function of pulse width. Results are shown for devices with different facet coatings[53]

As half-wavelength facet coatings do not change the reflectivity (and hence the peak optical field) of the facets, it might seem surprising that they influence catastrophic damage limits so significantly. However, the studies of Henry *et al.*[52] may help to explain this beneficial effect. Their model is based on the idea that extra non-radiative electron-hole recombination occurs at the facets via surface states. This reduces the carrier concentrations close to the facets (as shown in Figure 6.18) and will produce lossy regions there. Absorption of the stimulated radiation will result in increased heating. This heating reduces the band gap and the local absorption coefficient then increases

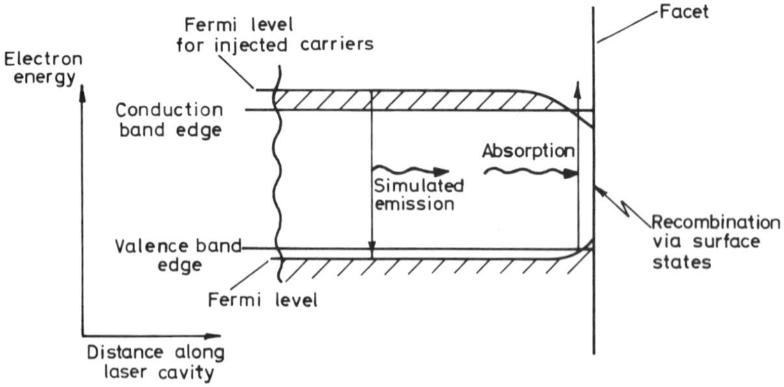

Figure 6.18. An indication of how extra non-radiative recombination near the facet, and thus a lower carrier concentration, results in extra absorption

rapidly to greater than 10^4 cm^{-1} as the local temperature rises to about 300 °C. Calculations show that, in a typical DH laser, such heating can occur in 10 ns, for an incident flux of 15 MW cm^{-2} at the facet (equivalent to an output power of 5 MW cm^{-2}—the approximate catastrophic damage flux for several hundred ns pulses). At this time thermal runaway begins due to the large optical absorption coefficient, and 1500 K, the melting point of GaAs, can be reached in about 100 ns.

Several workers[52–55] report ⟨110⟩ dark line defects (in electroluminescence or photoluminescence) running perpendicular to the cleaved facets many micrometres into the bulk of the device after catastrophic damage. TEM shows these dark line defects (DLDs) to be different to the ⟨100⟩ DLDs related to climb of threading dislocations and responsible for the early GaAlAs laser failure already discussed. Henry *et al.*[52] conclude that the ⟨110⟩ dark line is associated with a recrystallized molten region that propagates into the bulk during the damaging optical pulse. They detected a variation in the aluminium content of the active layer in the dark line region as proof that the material had gone through a molten stage.

Ueda *et al.*[55] have shown, also by TEM studies, that several types of dislocation structure can be produced. They suggest that one type (a simple dislocation dipole) can be generated at temperatures well below the melting point of GaAs as a result of stresses present in the heterostructure layers due to lattice mismatch effects. The implication is that hetero-interface stresses should be minimized in order to maximize the catastrophic damage limit.

If this surface recombination–absorption–heating mechanism is correct, then the initial amount of surface recombination is important in determining the rate at which the initial heating due to absorption can occur. Possibly Al_2O_3, as deposited by Imai *et al.*[53] is a better inhibitor of surface states than SiO_2, or it is a better encapsulant for minimizing the loss of arsenic from the surface, so controlling surface deterioration on heating.

The insensitivity of flux levels for catastrophic damage to pulse length for long (greater than 10 μs) pulses is explained qualitatively by the setting up of a quasi-steady-state heat flow away from the device active region. Measured catastrophic damage limits for long pulse or continuous operation are around $1\text{–}5 \times 10^5$ W cm^{-2} for uncoated devices. For example, for a 15 μm wide stripe DH laser, this corresponds to a catastrophic damage power output limit of 30 mW for an uncoated facet.[53] For an Al_2O_3 coated facet a limit of 100 mW has been suggested.[53]

For some of the more advanced stripe geometry devices, such as buried heterostructure or narrow stripe lasers, which have a near-field optical emission width restricted to only a few micrometres, the catastrophic optical damage power limit is less.[56] The possibility of long-term facet erosion at power densities below the catastrophic damage limit will be considered later (in Section 6.3.2).

For some applications lasers have to be optimized for high optical power pulse operation. For these applications, the use of quarter-wavelength anti-reflection coatings gives the added advantage that the power density at the facets can be reduced for equivalent output powers. The associated increase in threshold is relatively unimportant as high power output devices will, in any case, be operated at current densities well in excess of threshold. Another approach to increase the catastrophic optical damage level is to modify the optical guiding in order to decrease the power density at the facets. Variants of the large optical cavity (LOC) structure can be used.[57]

A novel development reported by Yonezu *et al.*,[58] decreases the influence of the semiconductor output facet. Catastrophic facet damage can be eliminated by use of a zinc-diffused 'window-stripe' laser. The maximum available optical power is now determined by catastrophic bulk damage and is at least one order of magnitude higher than obtained in conventional structures under pulsed operation—optical power densities up to 10^7W cm^{-2} for 100 ns pulses being reported. The 'window-stripe' laser is made by diffusing zinc into the

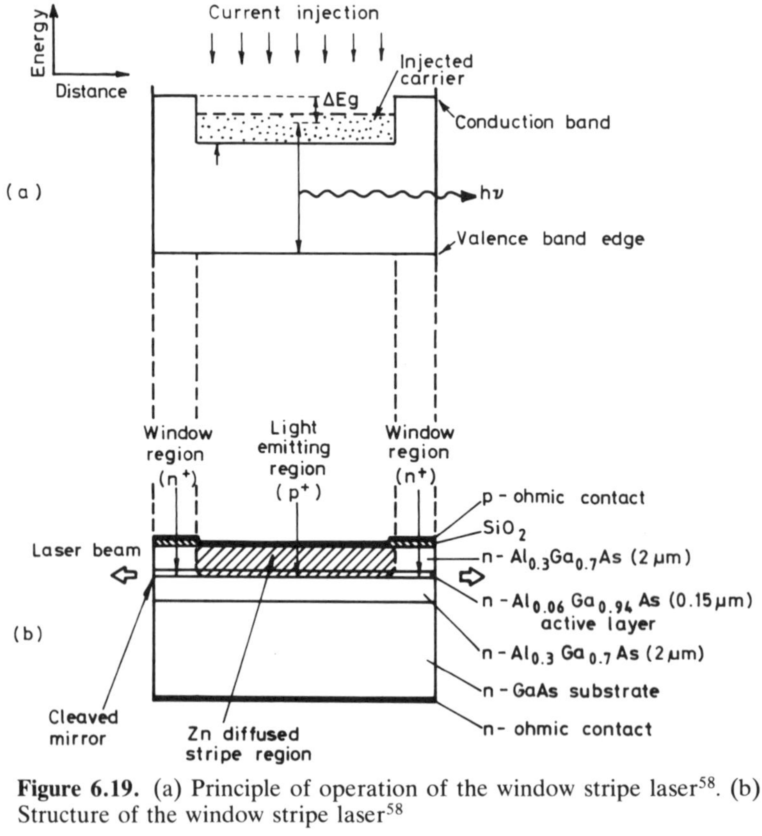

Figure 6.19. (a) Principle of operation of the window stripe laser[58]. (b) Structure of the window stripe laser[58]

heavily n-doped active region of a GaAlAs DH wafer, except close to the proposed position of the mirror facets as shown in Figure 6.19. The absorption edge of GaAs (or GaAlAs) is shifted to higher energy by heavy n-type doping and shifted to lower energy in overcompensated p-type material. This is sufficient to produce low absorption (at the lasing wavelength) window layers adjacent to the facets. On increasing the power density above $\sim 10^7 W\ cm^{-2}$ instantaneous catastrophic optical bulk damage occurred, evidenced by long dark line defects in the whole of the light-emitting part of the active region, but not in the window regions. A peak output of 1.8 W (100 ns pulse) has been obtained from a 10 μm wide window-stripe laser before catastrophic optical bulk damage.

6.3.2 Possible mechanisms determining the reliability of state-of-the-art devices

In the previous section, discussion centred on the mechanisms that are thought to be understood well enough so that action can be taken to prevent them limiting device lives. For example, although the process of growth of a $\langle 100 \rangle$ dark line defect (DLD) from a threading dislocation is not understood fully at present, the elimination of such dislocations from devices prevents the occurrence of this particular failure mechanism. Thermal resistance deterioration is understood to the extent that different metallizations and bonding procedures are resulting in improvements. Catastrophic degradation in lasers can be avoided by operating devices at sufficiently low power densities.

This section discusses mechanisms thought to limit the lives of the best present day devices, and the steps being taken to overcome them.

6.3.2.1 The role of native defects, and homogeneous degradation

Some of the longest lived lasers show no dark line defects, no thermal resistance deterioration, and are not thought to have facet damage, yet still degrade. This has prompted speculation about failure mechanisms associated with homogeneous effects in the light emitting region of the device. The most popular idea is that the radiative efficiency decreases gradually during operation as the density of non-radiative recombination centres increases with respect to the radiative centres. Alternatively, the centres introduced by the degradation process may be radiative, emitting long wavelength photons associated with radiative recombination at a centre within the band gap.

The rôle of native defects in DLD formation has already been discussed. Point defect creation is also raised in a discussion of non-catastrophic facet damage later in this section. Thus the generation and movement of defects seem particularly important in the degradation of light emitting devices. However the detection and identification of recombination centres in semiconductors is difficult and the results on light emitting devices are variable—in some cases contradictory. Instead, there has been great interest in the experiments of Lang and Kimmerling[38,59] showing that the mobility of defects in GaAs and other materials can be enhanced in the presence of electron-hole recombination. In their earliest work,[38] Lang and Kimmerling investigated defects introduced by electron bombardment which annealed out with an activation energy of 1.4.eV at a temperature in the range 175–225 °C. In the presence of electron-hole recombination, annealing took place below 100 °C with an activation energy of 0.34 eV. In its simplest interpretation, the enhanced mobility of defects in the presence of electron-hole recombination is caused by the energy released when recombination occurs non-radiatively. This energy is localized at the defect for a short time before being transferred to the lattice in the form of phonons. This extra energy may allow the defect to surmount the potential barrier limiting its movement. This is also a very attractive way of explaining the movement of defects to or from a dislocation during the growth of a dark line (Section 6.3.1).

Clearly such ideas may be extended to homogeneous modes of degradation. A model involving the recombination-enhanced motion of defects has been used by Kobayashi and Furukawa.[61] Experimenting with remote junction heterostructure (RJH) lasers, in which the active region is separated from the pn junction by a fraction of a micrometre, they found that threshold current **decreased** during operation (on all devices). Their model considers defects in the active region migrating to the *p-n* junction under the influence of electron-hole recombination. In a conventional DH laser, where the active region is adjacent to the *p-n* junction, the accumulation of defects in the recombination region would give degradation. In the RJH laser the radiative efficiency of the active region improves as defects are removed. Electroluminescence and EBIC studies confirmed that the defect density was increasing at the 'remote' *p-n* junction. These results would seem to be particularly significant, not only because of the obvious implications on device lives, but also because of the insight into the behaviour of defects in semiconductors that further studies on such structures should reveal. It is to be hoped that structures with similar ageing properties can be produced by other workers, with the advent of a new generation of particularly long-lived devices.

Temperature gradients, electric or stress fields are expected to influence the direction of motion of defects. For example there is evidence that the mechanical stress fields in devices affect defect mobility. It has been known for some time that stresses caused by bonding can be deterimental to life.[42] There have also been suggestions that the strain caused by the small lattice mismatch between the GaAs active layer and the GaAlAs passive layer would enhance degradation. The addition to the active layer of a small amount (<5 per cent) of aluminium improved device lives,[62] although it was not clear whether this was due to a decrease in the stress, or other effects—for example the chemical action of aluminium in gettering oxygen during LPE growth. The lives of short wavelength (0.73–0.82 μm) LEDs are improved by the use of a thick GaAlAs buffer layer between the substrate and the high aluminium content layer.[63] This buffer layer is thought to 'take up' the lattice mismatch. In the case of oxide-isolated stripe geometry lasers, lives have been found to depend on the oxide thickness.[48] This was attributed to the stresses set up by opening a stripe window in oxide deposited at high temperatures.

All these observations have been in the absence of any clear idea as to how stress fields are related to the degradation process. Recent results[64] on oxide-isolated stripe geometry lasers have shown that stress can influence the motion of defects. Investigation of the radiative efficiency of the substrate of degraded lasers using a photoluminescence technique, showed two parallel dark lines beneath the stripe contact (Figure 6.20). The position of these lines

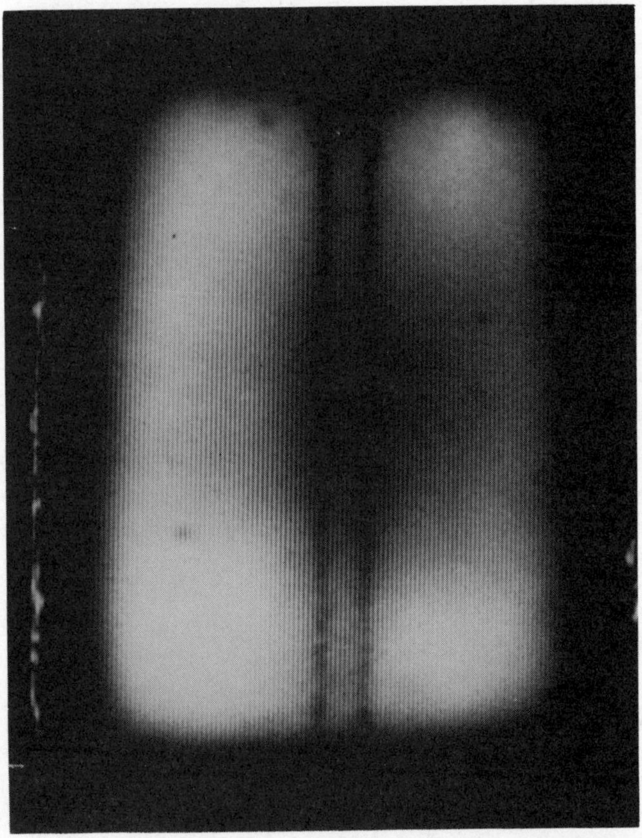

Figure 6.20. Photoluminescence of the substrate of a degraded oxide stripe laser[64]

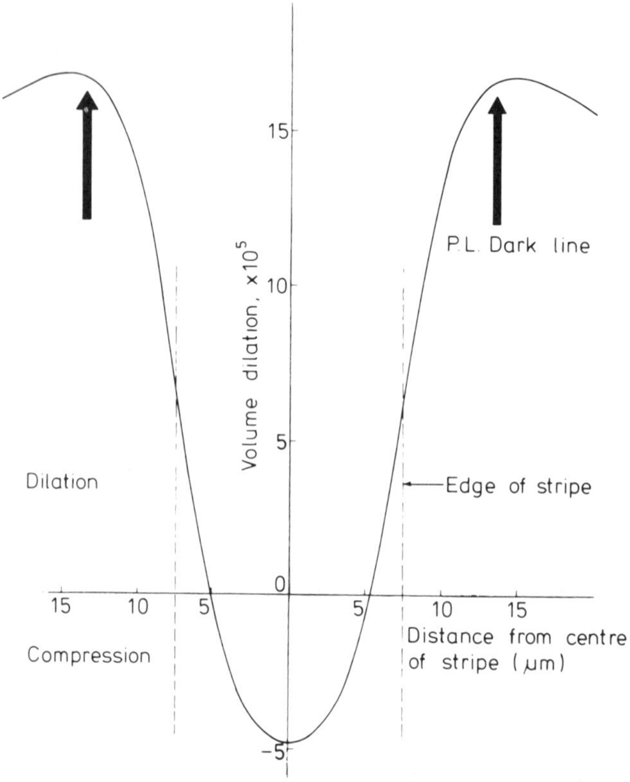

Figure 6.21. Calculated volume dilation induced in the substrate of an oxide stripe laser[64]

corresponded to the calculated position of the maximum volume dilation of the lattice (Figure 6.21). Degradation of the luminescence from the substrate is due to the absorption of spontaneous emission from the active region. This absorption creates electron-hole pairs which may enhance the mobility of any defects present. It is proposed that these defects will migrate to the region of maximum volume dilation. Further work[65] using the SEM in the cathode luminescence and EBIC modes have shown similar effects in the active layers of devices and both in the bulk and near the facets. In oxide-isolated lasers, the combination of stresses resulting from the oxide and heteroepilayers gives particularly complex stress fields at the facets. Defects, depending on their type, will tend to migrate to or from the facets. This provides a link between degradation in the bulk of the device and near the facet. It may also explain the observation[66] that degradation often occurs preferentially near the facet, even in the absence of facet damage.

An important factor in any assessment of recombination-enhanced diffusion as the basis of a degradation mechanism is the identification of a specific defect responsible for the extra recombination.

Sophisticated techniques to detect very small concentrations of centres within the band gap have been developed over the past decade. These include thermally stimulated current (TSC), thermally stimulated capacitance (TSCap) and photocapacitance methods. All these techniques are based on the measurement of the change in occupancy of traps in the space charge region of a *p–n* junction or a Schottky barrier. The occupancy is altered by thermal or optical processes, and the change in occupancy is detected by monitoring current or capacitance variations.

Perhaps the most popular technique is deep level transient spectroscopy (DLTS).[67] This gives spectroscopic information about traps within the band gap. Initially the occupancy of the traps is set by appropriate biassing of the junction, and capacitance changes are observed as the non-equilibrium trap occupancy is adjusted by thermal processes. By monitoring how the time constants of these capacitance transients change with temperature, information about the position of the traps within the band gap can be deduced. Capture cross-sections and trap concentrations can also be found using these methods. There are two main limitations to the use of DLTS in the study of the degradation of light emitting devices. First, traps are detected only in the space-charge region. Unless the active region of the device has a much lower doping than the other side of the *p–n* junction, the traps detected will not be in the region where the majority of carrier recombinations occurs in forward bias. Secondly, and more important, distinction has to be made between a trap and a recombination centre. This is determined by the ratio of capture cross-sections for electrons and holes. Centres thought to be responsible for spontaneous efficiency decreases are non-radiative recombination centres. These have relatively large electron and hole capture cross-sections, and give rise to short lifetime transitions. However this type of centre may be difficult to observe with DLTS, because the short lifetime makes changing the occupancy of the centre difficult. It is possible that modifications to the basic DLTS technique, such as optical DLTS, may alleviate this problem. In optical DLTS, the trap occupancies are determined by pulses of light rather than by carrier injection.

Deep levels have been investigated extensively in GaAs[68,69] and other III–V compounds, and there have been studies of defects in the GaAs/GaAlAs double heterostructure.[40,70,71] In addition, some workers have monitored changes in concentration of deep levels during device ageing.[40,71] One such centre, the E3 centre, is an electron trap 0.89 eV below the conduction band and thought to be a gallium vacancy.[72] Investigation of proton-isolated devices lifetested at 70 °C, showed the concentration of this trap increasing by over an order of magnitude in the early stages of the test.[40] However in lasers with very long lives, the concentration of this trap decreased. Although the traps were monitored in the *n*GaAlAs layer, rather than the active region, the properties of this trap are such that it would act as a very fast recombination centre—a 'killer' centre—in *p* material. It is also known to be mobile under conditions of electron-hole recombination. This centre has only been observed in material damaged by 1 MeV electron irradiation, and proton-isolated material. It is not clear whether the E3 centre would be associated with degradation in device structures not using proton isolation. However, it has been suggested that a deep radiative centre (emitting at 1.0 eV), observed in all LPE GaAs may be associated with the E3 centre.[73] This radiative centre has been observed to increase[74] and decrease[73] during ageing. Another electron trap (at 0.35 eV below the conduction band) has also showed changes during device life[71]—again erratically, with both increases and decreases in concentration being observed.

Another centre, which seems common to all *n*-type GaAs doped with Te or Sn, the DX centre, has been associated with the dislocation climb process. Using DLTS with spatial resolution (scanning DLTS), Petroff *et al.*[75] showed that the concentration of this centre decreased in the vicinity of a climbing dislocation, implying that the DX centre is the point defect needed for dislocation climb. The DX centre is thought to be a complex of the *n*-dopant and some other unknown defect, and is present with a concentration comparable with that of the donor ($>10^{17}$ cm^{-3}). The number of point defects required to enable the climb process to proceed is large, and the DX centre is the only centre to have been detected at such a high concentration.

To summarize, it appears that there are no generally accepted views on the defects responsible for degradation of GaAs devices. It seems most probably that the centres involved are native defects—possibly simple complexes even incorporating dopant atoms—rather than contaminants. Even if a centre responsible for degradation is isolated, it may well be difficult to eliminate it from devices. In this respect the development of growth techniques, other than LPE, which in principle may offer greater control over stoichiometry and native defect concentration, could result in better devices. Lasers have been made from material grown by molecular beam epitaxy (MBE) and metal-organic chemical vapour deposition (MO–CVD), and recent reports indicate that their lives are approaching the standard of those made by LPE.

6.3.2.2 Non-catastrophic facet deterioration of lasers

As has been discussed previously, catastrophic damage of lasers can result if too high a power is incident on the laser facets. The power density for this catastrophic damage tends to be constant at about 1–5 $\times$ 10^5 W cm^{-2} for pulse widths greater than 10 μsec (for uncoated devices). For lasers in optical communications systems it can be assumed that catastrophic damage will not be a problem as the advantages to be gained by operating the source at high powers are slight. Although there will be benefit to the extinction ratio, a doubling of launched power will only provide an extra 3 dB of signal, which is small compared to the overall system losses of about 50 dB for 8 Mbit s^{-1} and about 40 dB for 140 Mbit s^{-1}.

On a much longer time scale another problem associated with the facets becomes apparent and is known as non-catastrophic facet damage. Part of the gradual threshold increase over several thousand hours at moderate optical power levels (about 10^4 W cm^{-2}) is thought to be associated with such a mechanism.

Early work showed that the lives of DH lasers were improved significantly if facet coatings were applied.[76,77] Chinone *et al.*[78] found that the degradation rates of these devices were proportional to the power output up to 1.5 mW μm^{-1} of facet width. Optical microscopy and SEM studies detected apparent erosion at the facet of a different type from that seen in devices with catastrophic damage. This erosion was less significant in devices with facet coatings. In addition Chinone *et al.* found that the GaAs active region at the facet could not be etched by a GaAs preferential etch after only a short period of operation. All the evidence pointed to some photochemical reaction at the facet which was being enhanced by the ambient conditions and optical flux densities.

Another observation[78,82] is that the differential external quantum efficiency decreases with time. This is not compatible with a threshold current increase caused by a reflectivity decrease alone (see Equations 6.1 and 6.2) when differential external quantum efficiency would increase). A detailed investigation of the facets of degraded DH lasers by Auger electron spectroscopy[80] has shown that the facets can be oxidized during operation even when in a nominally oxygen free environment (dry nitrogen). Typical oxide thicknesses were given as 60 nm on the GaAs layer and 4.5 nm on the $Ga_{0.7}Al_{0.3}As$ passive layers. Oxidation did not occur to nearly the same extent on devices with facet coatings. Oxidation of the cleaved surfaces of semiconductors has also been produced by excitation with the 488 nm emission from an argon ion laser.[81] This work showed that the oxidation resulted in a decrease in photoluminescence efficiency attributed to an increase in non-radiative recombination in the semiconductor–oxide interface region.

Based on these observations and more detailed investigations of the characteristics of degraded devices, Yonezu *et al.*[82] have provided a quantitative basis for a model where increases in threshold current occur because of current being diverted through extra non-radiative recombination centres. These centres are formed in association with the oxide layer. Gallium and arsenic may diffuse from near the surface to combine with the oxygen in the environment. The resulting vacancies remain as recombination centres. The diffusion process will be enhanced by electron-hole recombination at the facet.

The regions of lower carrier concentration near the facets introduce extra losses (α in Equation 6.2) thus decreasing the differential efficiency. It is interesting to note that changes in the reflectivity of the facets are not needed to explain the observations. The conclusion is that even very thin oxide layers may cause threshold increases if the associated recombination is significant. However it is to be expected that oxide layers of optical thickness will affect reflectivity as well as recombination rate.

Yonezu *et al.* also observed (in EBIC) $\langle 100 \rangle$ dark lines extending from the facets after extended periods of operation. The growth of these defects were attributed to the increasing defect mobility as both absorption and temperature continue to increase in the lossy regions. Thus, from this model it could be deduced that the role of facet coatings on the facet is to prevent the build up of oxide which causes non-radiative recombination centres.

There seems to be almost general agreement that facet coatings improve the life of devices, and the model of Yonezu *et al.* is not inconsistent with the experimental results. However there have been reports suggesting that the lives of good lasers are not improved by the application of facet coatings,[66] and that degradation is not dependent on optical flux up to 5 mW per facet.[17,23] Dixon and Hartman[17] have operated devices at 70 °C under either spontaneous conditions or between 3 and 5 mW per facet and observed little difference in life. Similar conclusions have been reached on a limited number of samples operated at room temperature.[24] In a particularly comprehensive study by Nash *et al.*[66], devices both with and without protective coatings were subjected to lifetests under various ambient conditions. The conclusion was that facet coatings did not improve the lives of their devices under the chosen operating conditions (70 °C). The results suggested however, that the residual degradation may still be associated with the facets in some way. In their best devices a structureless darkening occurred near one or both of the mirrors, and in their shorter lived devices DLDs were observed near the facet.

Another interesting point to emerge from this study was that the facet erosion of degraded devices observed with an optical microscope could be suppressed considerably by using facet coatings, but without any effect on device life. This implies that the facet erosion may be merely a 'cosmetic' effect.

Thus there seems a difference in opinion as to whether facet coatings promote long lives in good GaAs–GaAlAs lasers. This rather important point will have to be resolved by further experimentation, but it is worth mentioning some aspects which may help to explain the differences in results.

A particularly important aspect is a comparison of the power density incident on the facet in the tests carried out by the different workers. However this comparison is made difficult for several reasons. The near-field profile is usually narrower than the stripe width because of various guiding mechanisms, so even if the total power from the facet and the stripe width are known, it is difficult to estimate the maximum power per micrometre. This is complicated further by higher order transverse modes that may occur at high powers in certain laser structures. For example it has been observed[83] that the peak power density at the facet of a 20 μm oxide isolated laser operated at a total output of 3 mW (zero order transverse mode) can be considerably higher than that of the same device operated at 5 mW (higher order transverse mode). The optical distribution perpendicular to the layers can depend critically on the active layer thickness and could vary significantly from device to device. In addition, the effect on threshold of damage near, or at the facet may depend on the strength of the optical guiding. This is because in a weakly guiding structure the mode may move, retaining a similar threshold, whereas a strongly guided mode may not be able to move and threshold will increase considerably. In the absence of any better method, values of the average power output per micrometre of stripe width may be compared. In general the maximum values range from 0.7 mW (μm^{-1})[80] to 1.2 mW (μm^{-1})[77] compared with the maximum of 0.6 mW (μm^{-1})[66] in the work where there was no improvement with the application of facet coatings. However, bearing in mind all the other uncertainties mentioned previously it is not clear that this constitutes a significant difference.

A specific feature of the work of Nash *et al.*[66] is that they carried out the lifetests at 70 °C, whilst other workers have employed room temperature testing. Saturable initial-stage degradation (often over a few 1000 hours) is sometimes observed in long-lived devices and there is the possibility that room temperature lifetests may be assessing the effect of the facet coatings on this initial-stage process. They may not reflect the usefulness of coating in preventing any long-term degradation effects that are the ultimate limit of device life. On the other hand it could be argued that the degradation mechanism(s) limiting life at high temperatures may be different from those at room temperature. If for example, the facet oxidation mechanism was limiting laser life at room temperatures, and it was an athermal process, then a different, more thermally activated mechanism could be responsible for the degradation at 70 °C. This point will be discussed in more detail in 6.3.3, but it is interesting to note that the oxidation of GaAs facets in the absence of excitation is in fact athermal.[81] However tests on lasers immersed in water (not operated) showed that the degradation rate was particularly dependent on temperature.[84]

Another point that could be important is the composition of the active regions. In experiments investigating the rise in threshold after immersion in water,[84] the amount of degradation was found to depend on the aluminium concentration and the doping levels (or dopant) in the active region. Also, the thickness of the oxide induced by an argon ion laser was much less in $Ga_{0.7}Al_{0.3}As$ than in GaAs.[81]

Thus there seems no generally accepted views about the usefulness of facet coatings, but to date there is no evidence that their use has a deleterious effect on device reliability. In general, exposed semiconductor surfaces are undesirable in any device, so it is worth considering the coatings that have been used.

As in the case of protection against catastrophic damage, the coatings most commonly used are SiO_2[80,85,86] and Al_2O_3[76,77,85,87]. Comparative studies[85] have indicated that Al_2O_3 is superior to SiO_2 which is similar to the findings (by different workers) for catastrophic degradation (Section 6.3.1). SiO has been used,[76] and more recently, carbon films[88] and Si_3N_4.[89] Coatings are applied as half wavelength films (thickness equals $\lambda/2n$, where n is the refractive index of the film). This should not alter the reflectivity when the refractive index of the semiconductor equals n^2; the properties of the film should be primarily as a barrier between the semiconductor and the environment. In addition, if the oxidation model of Yonezu *et al.*[82] is relevant, the film should not introduce extra defects at the interface, and preferably act in such a way as to immobilize defects that exist before deposition. The method of deposition of the film will be important in this respect. Al_2O_3 has been applied by electron-beam evaporation,[77] chemical vapour deposition,[85] and Si_3N_4 by plasma deposition.[89] However the trend seems to be towards lower temperature processes and recently carbon ion-beam deposition[85] and SiO_2[86] sputter deposition have been carried out near room temperature.

Another approach that looks very promising for decreasing facet effects related to reliability problems, is the use of the 'window stripe' laser as proposed by Yonezu *et al.*[58] It has been mentioned that this structure has shown greatly increased catastrophic damage levels (Section 6.3.1). Preliminary results also show good long-term reliability.[58,82] If gradual facet degradation is induced by electron-hole recombination at the facets (rather than as a direct consequence of high optical flux levels) then further improvements can be envisaged by maximizing the effective band gap difference between the window region and the light-emitting region.

To summarize this discussion of non-catastrophic facet damage, it would seem that the majority of workers find that facet coatings improve device life, at least over the short term. Degradation results on devices without facet coatings are not inconsistent with a model where gradual oxidation introduces extra non-radiative recombination centres near the facets, causing threshold to rise and introducing losses resulting in a decrease in external quantum efficiency. The application of a half wavelength thickness of facet coating helps to prevent this oxide forming. Alternatively, the window stripe laser structure, where there is little electrical activity at the facets, may be the best approach to prevent long term degradation associated with the mirrors.

More empirically based work is needed to establish the most effective coating material and method of deposition, but this will take a long time unless a valid method of overstress testing for this failure mechanism is found. This will probably require a detailed knowledge of the failure process, such as its dependence on optical flux and temperature.

6.3.2.3 *Intensity pulsations in lasers*

A phenomenon which can be very serious for laser operation in high bit-rate optical communication systems is the onset of pulsations or oscillations in the light output during device operation.[3,4] These intensity pulsations occur at frequencies between 200 MHz and 2 GHz, with a modulation depth approaching 100 per cent. Lifetests on some types of devices giving long lives, as measured by threshold increases, showed that pulsations could occur after only a very small threshold increase.[13] In lifetests on some conventional oxide and proton isolated devices, 30 per cent of the samples showed these effects.[13,98] The operation of a high bit-rate optical communication system would be affected severely by these pulsations—in fact the end of life could be determined by their onset rather than the more usual criteria associated with the increase in threshold.

The reason for these intensity pulsations is still under discussion. There have been suggestions that second order mode locking,[90] or switching between two lateral filaments[91] may be responsible. However, it seems most probably that effects associated with electron–photon interactions in the presence of non-uniform gain or loss are the cause.[92]

Two non-linear rate equations give a simplified description of the variation in photon density (P) and electron density (N) with time:

$$\frac{dP}{dt} = \left(g(N) - \frac{1}{\tau_p}\right) P + k\,\frac{N}{\tau_e} \tag{6.7}$$

$$\frac{dN}{dt} = \frac{J}{qd} - \frac{N}{\tau_e} - g(N)P \tag{6.8}$$

τ_e and τ_p are the lifetimes of electrons and photons, J the applied current density and d the active region width. $g(N)$ is the optical gain and k is a constant defining the fraction of carriers that provide spontaneous photons into the lasing mode. By making further approximations, for example that $g(N)$ is a linear function of N, and kN/q is small, a solution giving damped oscillations is obtained. These relaxation oscillations, as distinct from the intensity pulsations, are well understood.[3] Although undesirable for high bit-rate optical communication systems, their effect may be minimized by careful design of the drive circuitry.[93] Also, a more rigorous solution of Equations 6.7 and 6.8 indicate that device parameters can be adjusted to decrease the damping time constant. In particular, the damping time constant decreases with increasing k. Devices with a high degree of spontaneous coupling into the lasing mode show heavily dampled oscillations, and lasers showing no relaxation oscillations[6,94] have been made.

Under certain circumstances the solution of Equations 6.7 and 6.8 can be repetitive intensity pulsations rather than damped oscillations. Calculations have indicated that the solutions to Equations 6.7 and 6.8 are very sensitive to the linearity of the gain and loss functions[95] and to whether the gain saturates at threshold.[96] A very small non-linearity may give rise to these pulsations. For example, a change in mode structure because of inadequate transverse guiding may provide the non-linearity required. This may be the reason why such pulsations are usually observed in devices showing kinks in their light output–current characteristics (Figure 6.9).

More specifically, intensity pulsations have been attributed to the effect of saturable absorbers in the active region.[91,92,97,98] Such an absorber may result from a localized region of lower population inversion, caused perhaps by a region of non-radiative recombination, surface recombination at the facets or the unpumped regions at the edges of the stripe. The absorption in these areas will decrease with increasing pumping, giving rise to a Q switching effect. It has been suggested that loss decrease due to a saturable absorber of only one per cent may result in continuous oscillations in conventional devices.[99] This may explain why intensity pulsations can occur during device life tests with only a minimal increase in threshold current.

In summary, it appears that there is no general agreement about the exact mechanism for these pulsations. It may be that the origins vary between different laser structures. If the effects are associated with small variations in uniformity, then this may present a considerable problem. However, it is probable that devices with better guiding properties will be less likely to show intensity pulsations. Fuller understanding may allow the development of structures in which the optical properties are less sensitive to non-uniformities.

6.3.3 Prediction of device life

Accelerated lifetesting to predict device life has been used because of the absence of any other technique. There are several problems in the use of data from such lifetests—not least the difficulty in proving that a life prediction for a device under normal operating conditions, sometimes several decades, will indeed come true. Overstress testing relies on a relationship between the device life and the degree of overstress being extrapolated down to normal operating conditions. If this extrapolation can be justified by experimental results under normal operating conditions, then the accelerated lifetesting techniques may be used with confidence, but in general this is not possible. Justification is usually based on an understanding of the failure mechanisms involved, but can be proved unfounded by an unexpected mechanism not sensitive to the particular overstress parameter used.[100] There are still failure mechanisms in semiconductor light emitting devices which are not understood, as discussed in Section 6.3.2 (especially so for lasers) and so accelerated lifetests must be treated with even more caution than those for other semiconductor devices.

6.3.3.1 Overstress testing of GaAlAs DH lasers

As well as the general reservations mentioned above, other difficulties particular to semiconductor lasers must be considered. Two of these have been discussed in Section 6.2—the definition of device failure and the difficulties in overstressing one particular operating parameter without affecting others.

A general outline of the results presented by various groups of workers is given in Table 6.2. Some of the work has been carried out in two stages. First an activation energy is evaluated—in most of the work this has been from limited data and many of the plots have shown considerable scatter. Secondly, a full statistical analysis of a relatively large sample operated at elevated temperatures can be carried out. The result of this, for example the median time to failure, is used in conjunction with the activation energy and Equation 6.5 to give the predicted median life under normal operating conditions.

It must be stressed that the life figures in Table 6.2 depend on many factors, such as operating conditions, definitions of end of life and statistical presentation of the data. The table should not be considered as a league table of reliable lasers.

Perhaps the most significant feature of these studies is the variety of operating conditions and methods of defining end of life. This is reflected in the variety and uncertainty of activation energies that have been obtained. The most common definition of end of life is when a specified power output, say 1 mW, can no longer be obtained at the operating temperature[16–18] (however high the current is set). This method has the disadvantage that devices failed at one temperature will, most likely, still operate satisfactorily at a lower temperature. The activation energy obtained from this approach could more accurately be called an 'extrapolation' energy. A more systematic approach is to monitor device degradation by measuring the device parameters at a fixed temperature (room temperature)[20,22–25] even though the devices are being aged at elevated temperatures. A specified increase in room temperature threshold would then define the end of life. Alternatively, a degradation rate may be used (i.e. mA kh^{-1} or per cent khr^{-1}). These methods will give activation energies reflecting more accurately the thermal activation of the degradation process. However, activation energies obtained in this way have been used to extrapolate from data based on the power output definition of end of life,[18,23] and it is not clear that this procedure is valid.

An Arrhenius plot obtained by one set of workers[23] is shown in Figure 6.22. In these tests, devices operated at the higher temperatures were in the spontaneous mode, and the devices at the lower temperatures were operated at constant optical power in the lasing mode. End of life was defined as when pulsed threshold, measured at room temperature, had increased by 50 per cent.

The criterion that a device has only failed when the required stimulated

Table 6.2 Activation energy

Worker	Hartman *et al.*	Ettenberg *et al.*	Furukawa *et al.*	Ishikawa *et al.*	Mizuishi *et al.*	Ritchie *et al.*
References	16, 21	19, 22	20	23	25	24
Operating Conditions	Constant optical power output	Constant current (spontaneous)	Constant optical power output	Most devices at constant current (spontaneous)	Constant current and constant optical output	Constant current (spontaneous)
Temperature range	50–90 °C	60–100 °C	70–110°C	70–140 °C	50–100 °C	25–90 °C
End of life	Inability to emit 1 mW	20 per cent increase in pulsed threshold at room temperature	Degradation rate in pulsed threshold at room temperature	50 per cent increase in pulsed threshold at room temperature	Degradation rate in pulsed threshold at room temperature	Degradation rate in pulsed threshold at room temperature
Activation energy	0.7–0.9 eV	0.95 eV	0.7–0.9 eV	0.74 eV	0.8 eV	0.5 eV

Lifetime prediction

Operating Conditions	Constant optical power output	Constant optical power output	Constant optical power output	Constant optical power output	Constant optical power output	—
Temperature range	70 °C	70 °C	70–110 °C	70 °C	50–100 °C	—
End of life	Inability to emit 1 mW	Inability to emit 1 mW	100 per cent increase in pulsed threshold at room temperature	When lasing is not possible	Drive current increased by 100 mA	—
Life-time at room temperature	3.10^5 h (0.7 eV)	10^6 h	1.310^6 h (0.7 eV)	3.10^5 h (at 30 °C)	$3–4 \times 10^5$ h	—
Statistics	Median of 100 unselected devices from 10 selected slices	Median of 40 devices	77 devices over 4 temperatures. Rates averaged	50 samples from 8 slices		—

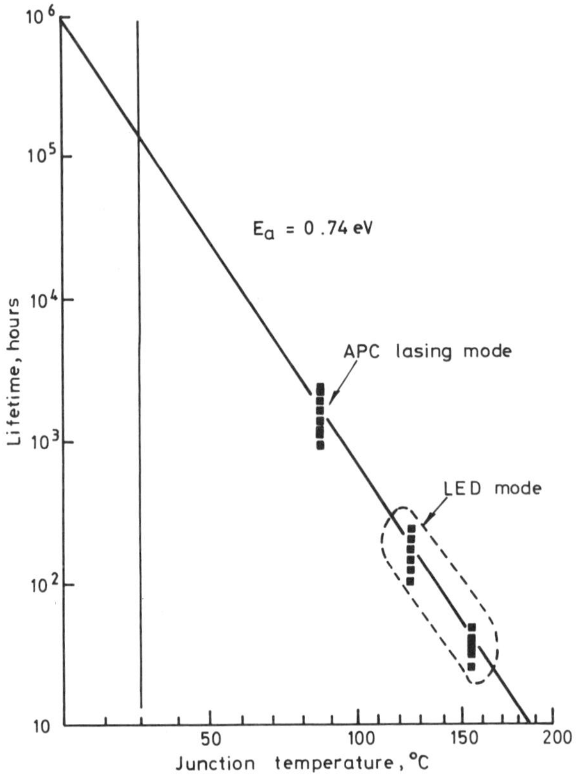

Figure 6.22. Arrhenius plot for lasers aged at elevated temperatures. End of life is when threshold current, measured at 25 °C, has risen by 50 per cent[23]

power output can no longer be obtained is not necessarily appropriate in an optical communications system, where drive current may be at a premium. In this respect a specified current increase is particularly useful as a definition of failure.[25] As long as devices are operating well away from thermal runaway, those failed at any temperature will have suffered the same amount of degradation.

A final point about these definitions of laser failure is whether degradation of other parameters can be more important than threshold increases. As has been mentioned in 6.3.2, the onset of intensity pulsations may be a much more serious limitation, and this can occur with very little threshold increase. There is insufficient statistical data on this aspect yet, although Channin *et al.*[98] have reported that about 30 per cent of a sample of long lived devices tested at 70 °C showed oscillations.

In an optical communication system it is most probable that feedback will be used to keep the peak optical power output constant. This is the main reason why the definition of end of life based on the power output criterion has evolved. It is also the reason why feedback, or manual control, to maintain the light output power constant, is used in lifetests. However there are the disadvantages that devices may be at different currents and, additionally, that the operating currents will rise during device life. The role of current overstress in lasers is not clear, but could have important implications for the interpretation of thermally accelerated lifetest data. Work on early lasers suggested a J^2 dependence[12,15] whilst DH LEDs have been found to degrade at $J^{3/2}$.[101] More recently, for CSP lasers, an exponential relationship was found.[25] In general, devices do not degrade 'on the shelf' so unless any current induced degradation mechanisms saturate at low currents, it seems likely that increasing the current will accelerate degradation to some extent. In view of these problems associated with current overstress, some thermal overstress tests aimed at obtaining an activation energy have been carried out at constant current. This has meant that some, or all of the devices have been operated in the spontaneous mode. The validity for this approach has been based on the claim that degradation rates are not affected by relatively low levels of stimulated emission,[17,22,23] especially if facet coats are used.[22] However, the tests on which these claims are based were all carried out at elevated temperatures—usually 70 °C. Care must be exercised before extending these conclusions to room temperature operation. There is a possibility that a mechanism associated with optical power levels (probably at the facets) may have a lower thermal activation than the mechanism causing failure at 70 °C. If so, the mechanism induced by optical power might dominate at room temperature. This is an example of a major problem in overstress testing—the necessity for a single mechanism to be dominant over the temperature range of interest. A study of oxide-isolated devices by Ritchie *et al.*[24] showed that the standard deviations of the lognormal failure distributions were different for devices tested at different temperatures. This implied that the mechanisms dominating at each temperature were not the same. This view was confirmed by using a photoluminescence technique which showed a homogeneous darkening in the active region of devices operated at room temperature, but more non-uniform features in devices aged at high temperatures. A further example of different failure mechanisms dominating at different temperatures is the work of Furukawa *et al.*[20] Devices tested at the maximum temperature (110 °C) failed by contact deterioration, whilst devices at lower temperatures showed dark line defects. With the mechanism(s) controlling device life in present day lasers still not understood fully, it is particularly important to be sure that a unique extrapolation energy exists over the whole temperature range of interest.

It may be difficult to assess whether the same degradation mechanism is dominant over the whole temperature range. One method would be to check statistically lifetest data at different temperatures. If the failure distributions have the same standard deviation, it would give confidence that the mechanisms were the same. A superior method, although probably impractical with the present state of knowledge of laser reliability, would be to prove by post-mortem examination of failed devices that the mechanisms were the same.

Another problem is the validity of using an activation energy obtained from one batch of lasers on a different, newer, batch. In many cases the activation energy is more an extrapolation energy, incorporating such features as the thermal properties of the device and possible current or optical power overstress effects as well as the basic thermal activation of the degradation process. Improved devices with, say, better thermal properties may well give a different energy.

In summary then, there are considerable difficulties in carrying out accelerated lifetests on lasers. The factors that have to be considered are:

—operating conditions
—definition of end of life
—current overstress
—optical flux effects
—validity of Arrhenius extrapolation over whole temperature range
—validity of using the same activation energies for different batches of lasers

It is very important to be fully aware of all these problems before placing too much confidence in lives predicted by accelerated lifetesting.

Even with these uncertainties, it can be seen from Table 6.2 that predicted lives fall between 10^5 and 10^6 hours. From the statistical distributions of Hartman *et al.*[21] and Ettenberg,[19] the 2 per cent failure time can be estimated at 1–5 $\times$ 10^4 hours at room temperature. This compares favourably with the lives of silicon MOS integrated circuits, but is still several orders of magnitude less than the lives of conventional silicon bipolar transistors.[102]

6.3.3.2 Overstress testing of GaAlAs HRLEDs

Choice of operating conditions and the definition of end of life are much easier with LEDs. In the work of Yamakoshi *et al.*[103] devices showing dark line defects were screened out at an early stage, and the results were claimed to represent a homogeneous bulk mechanism. The activation energy obtained was 0.57 eV, and lives of 5.10^6 hours at 40 °C were predicted.

6.3.4 Additional comments

For certain applications, resistance to degradation in the presence of ionizing radiation is of interest. Ionizing radiation can create a range of defect types in semiconductor lattices which may act as non-radiative recombination centres. There is a large body of literature on radiation-damage-induced defects.[104] The creation of such defects close to the *p–n* junction of an LED or laser is detrimental to efficient device operation. It has been shown that the degree of resistance to the formation of radiation-induced defects is dependent on both the III–V alloy composition and also on the dopants and doping levels employed. For example, $Ga_{1-x}Al_xAs$ diodes with x as little as one per cent, have been shown to be 5 times more resistive to 1.25 MeV electron or X-ray induced degradation than GaAs diodes similarly fabricated.[105] The effect of dopant species is highlighted by the ten times greater resistance to X-rays of Zn diffused GaAs diodes in comparison with amphoterically silicon-doped devices.[106] No doubt results on HRLEDs and DH lasers will be reported in due course.

GaAs and GaAlAs HRLEDs and lasers are now reaching a stage of development where there is a requirement for much more attention to be paid to the packaging of the device chip, as the device environment becomes more varied. Full hermetic packaging is now being specified so that testing and screening of devices suitable for optical communications and other applications is becoming more complex. In addition, there is a growing requirement for the LED or laser to be presented to the systems' application engineer as a 'black box', with electrical input pins and an optical fibre output.[107] For some applications, electronic drive and control circuitry has also to be incorporated within the package. Full reliability assessment of the whole package will present a major problem.

It remains to be seen whether current research into integrated optical technology will lead eventually to the monolithic integration of semiconductor electronic and optical devices on a single chip giving the sort of improvement in reliability achieved in silicon-based, monolithic electronic integrated circuitry.

6.4 OTHER SEMICONDUCTOR SOURCES

This section is concerned with various types of semiconductor source not covered in Section 6.3. Visible LEDs and low current density GaAs sources are discussed briefly, followed by a review of 1.0–1.6 μm sources. The section is completed by a mention of lead chalcogenide devices.

6.4.1 GaP, GaAsP and low current density GaAs LEDs

The recent literature on the reliability of these devices is sparse in comparison with that on lasers. This is due in part to the fact that, even in the early 1970s, various workers[108–110] were predicting constant current room temperature lifetimes, ranging from 10^5 up to 10^8 hours. This range is adequate for most, if not all, applications. The large scale production of red and green GaP–GaAsP LEDs and their widespread use for indicators and in displays has been sufficient to give confidence to the user that the early predictions, based on accelerated lifetesting, were justified.

Because of the much lower current densities of operation of these devices in comparison with high radiance LEDs and lasers, the early work in this area was very much preoccupied with the detrimental effects of surface leakage. This can result in a significant proportion of the current being diverted from radiative processes. As an example regarding the importance of surface effects, treatment of red emitting GaP diodes in hot hydrogen peroxide followed by an anneal to remove excess water from the oxidized surfaces was shown to improve dramatically device degradation resistance.[108]

In addition, contaminants were also thought to affect adversely device performance. For example, the intentional contamination of GaAs and GaP LEDs by copper was shown to increase degradation rates.[109,111] This implied that it was necessary to eliminate potential mobile impurity ions such as copper from the processing environment.

Stresses induced by contact alloying were also thought to be important. This led to the use of alloyed contact areas of small dimensions compared to the *p–n* junction area itself.[112]

The realization of the sensitivity of LEDs to surface preparation, contamination and fabrication technology led to a consolidation of their performance. Lifetesting at elevated temperature became necessary in order to begin to estimate failure times at room temperature.

6.4.1.1 GaP LEDs

There have been a number of studies of red GaP (Zn:O) LEDs, but there has not been complete clarification of the dominant degradation mechanism. There have been proposals that a reduction in the bulk *p*-side radiative recombination efficiency[113] or a reduction in concentration of the Zn–O luminescent centres[114] could be responsible. However, Mizuta *et al.*[115] have pointed out that the initial values of minority carrier diffusion length and the Zn–O pair concentration of the device material is a critical factor that could determine the relative importance of the different mechanisms.

Whatever the mechanism, lives in excess of 10^6 hours for operation around room temperature are now claimed for some devices, as the results of Ralston and Mann[113] demonstrate (Figure 6.23). Note from the figure that an activation energy of $\sim$0.7 eV is obtained independent of stress current. The figure also demonstrates that there is also a current dependence of degradation.

Jordan[31] has presented statistical information on GaP red LEDS—basically one log-normal plot of devices degraded at 155 °C and 10 mA drive current. His results indicate a median time to failure of about 6000 hours, in agreement with the Arrhenius plot in Figure 6.23.

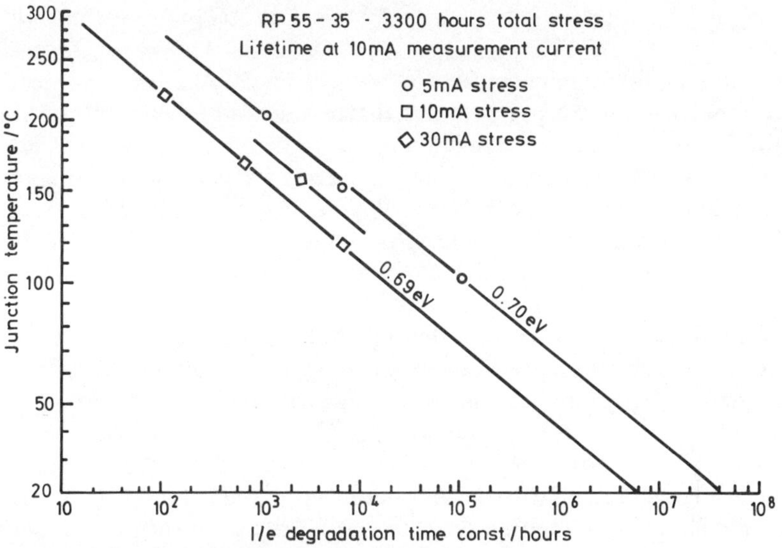

Figure 6.23. Arrhenius plot for red GaP LEDs aged at elevated temperatures and several values of drive current[113]

GaP red LEDs are no longer commercially preferred because, although they are efficient LEDs, most of their emission is in the infrared. In addition, their emission saturates at relatively low currents due to the restricted Zn–O pair population that it is possible to achieve in practice in the diode material.

The technology developed for GaP red diodes has, however, been applied to the green (GaP:N) variety. The lack of any apparent reliability problem (lives of 10^3 hours at 205 °C having been measured[116] that compare with the reported lives of GaP red LEDs—Figure 6.23) has resulted in a minimum of published literature in this area. However, Petroff *et al.*[116] have suggested that green LED degradation is associated with the growth of dislocation dipole structures formed by the climb of threading dislocations in a process similar to that responsible for the early rapid failure of GaAlAs DH lasers and HRLEDs as discussed in Section 6.3.1. Again, however, the mechanism of dipole growth is uncertain. The climb rate of the dislocation network in the green LEDs is considerably slower than in the case of GaAlAs DH lasers. This is explained qualitatively on the basis that the injected current densities are ~100 times smaller in the LED case and also that there is minimal optical flux.

Another investigation of green GaP LED degradation (by Kaneko[117]) concludes that the migration of interstitial zinc towards and across the junction in devices prepared by liquid phase growth is responsible for the slow degradation of the green emission.

6.4.1.2 GaAsP and GaAs LEDs

There have been comments that, in red $GaAs_{0.6}P_{0.4}$ LEDs, zinc diffusion source type, arsenic overpressure and *n*-type dopant and concentration govern the initial quantum efficiency and degradation rate of these devices. There have also been reports that native defects are generated during operation and that these are responsible for degradation.[118,119] This has a similarity with mechanisms suggested for GaAlAs DH laser and GaAs HRLED degradation. However, as in the case of GaAlAs lasers, there are inconsistencies in the reported behaviour of defects and their effect.

Reliability data on low current density GaAs and GaAlAs emitters is very limited, but the very great advances made in understanding degradation of high radiance devices in these materials must have had a substantial influence on the quality of low radiance devices that can now be produced.[143]

Most of the low current density LED types discussed in this section, need to be encapsulated, often with the plastic encapsulation moulded to form an integral lens to help direct the emitted light. The use of such encapsulations obviously limit the temperature range of operation and commercial specifications often give limits for both storage and operation of −55 °C to +100 °C. Early problems associated with encapsulation included bond wire failure due to differential thermal contraction, but modern plastic encapsulants do not suffer in this way.

6.4.2 1.0–1.6 μm sources

The reliability of sources operating in the 1.0–1.6 μm wavelength range is assuming an increasing importance as a result of the decreased optical fibre losses in this region of the spectrum. Particular interest has developed in wavelengths near 1.3 μm (where silica based fibre materials have zero dispersion) and around 1.55 μm (where fibre loss is lowest). Now that the basic materials and device technology is more firmly under control, assessment of reliability will become increasingly important. With the probability that submarine optical fibre systems will operate in this wavelength region (in order to maximize repeater separation), there will be a need for submerged repeaters capable of operating for 20 years or more without attention. This will put more pressure on reliability qualification than has been regarded as necessary for most 0.8–0.9 μm sources.

Although reliability assessment of 1.0–1.6 μm sources is in its infancy, it is generally possible to take an optimistic view for the future, based on the few results available on InGaAsP HRLEDs and lasers. The majority of results are on devices emitting in the 1.1–1.3 μm region; devices emitting close to 1.55 μm are less well developed. The first results were reported in 1977 on InGaAsP 1.2–1.3 μm surface emitting double heterostructure HRLEDs by

Dentai *et al.*[120] and on InGaAsP 1.2 μm lasers by Shen *et al.*;[121] lives in excess of 1000 hours of room temperature operation were obtained. This was viewed as very promising since the threading dislocation content of the active layers was thought to be high due to the 10^4–10^5 cm^{-2} dislocation density in InP substrates on which the LPE layers were grown. Such a dislocation content in a GaAlAs device would have resulted in very quick failure due to DLD formation (Section 6.3.1). These results led to speculation that the probability of recombination-enhanced diffusion of defects was less in InGaAsP because of the smaller bandgap and hence energy available for non-radiative recombination. An alternative possibility is that the threading dislocations are pinned more effectively in InGaAsP by the presence of the indium ion.[122] Yet another reason could be associated with the types of defect present in these quaternary alloys which could be fundamentally different from those in GaAs. Also, short range variations in periodicity (alloy disorder) of the quaternary active region lattice may decrease defect mobility in comparison with the near-binary GaAlAs active region case.

More recent results, although still sparse, give further encouragement. In the case of HRLEDs operating in the 1.1–1.3 μm range, the most significant results are those of Wada *et al.*[123] They reported operation of 1.2–1.3 μm DH InGaAsP surface emitters in the temperature range 20 °C up to 170 °C for periods of between 7000 and 9000 hours without any measurable degradation of light output at 100 mA (8 kA cm^{-2}) drive current. It is interesting to compare these results with those obtained by Yamaoka *et al.*[103] on DH GaAlAs surface emitters operating in a similar regime and discussed in Section 6.3.3. In that case devices degraded to 20 per cent of the initial light output in 9000 hours at 180 °C. This comparison underlines the apparent superiority of the InGaAsP devices.

Olsen *et al.*[124] have reported useful results with hydride vapour phase epitaxially grown DH InGaAsP edge emitting LEDs (essentially broad contact laser structures run below threshold). No measurable reduction in light output was seen after 10^4 hours of DC operation at 60 °C for 1.12 μm emitters run at 1 kA cm^{-2}.

For lasers, Olsen *et al.*[124] also reported 4500 hours of DC operation with minimal degradation at 20 °C for oxide stripe devices emitting at 1.25 μm and run at 4 mW output per facet. These lasers were prepared by VPE. Liquid phase grown oxide stripe lasers, emitting at 1.31 μm have been operated, also at 20 °C, for periods up to 10^4 hours at 2–4 mW output.[125]

Horikoshi *et al.*[126] have reported temperature accelerated lifetests of 1.1 μm oxide-isolated InGaAsP lasers. Devices were aged in the spontaneous mode at 100 mA (2.5 kA cm^{-2}) drive at 75 °C, 110 °C, and 180 °C. An activation energy of 0.78 eV was obtained, giving an extrapolated 1.1×10^7 hour potential lifetime at 25 °C. Lasers were fabricated on both low and high dislocation density substrate material. It was suggested that localized degradation could sometimes occur when the latter type of substrate is used, so that low dislocation substrates are preferred, as in the case of GaAlAs lasers.

Buried heterostructure lasers have been operated at 70 °C for periods in excess of 4500 hours with no increase in threshold, when run at 5 mW output per facet.[131]

Results on 1.5 μm InGaAsP are at a much earlier stage, but surface emitting DH LEDs have been reported[128] with no degradation in 1500 hours when operated at almost 10 kA cm^{-2} at a temperature of 120 °C. For 1.5 μm 17 μm wide oxide stripe lasers, the absence of degradation after 2000 hours of room temperature operation has also been reported.[129]

To date, there has been little direct attention paid to laser facet erosion as such, except for reports[130,131] that catastrophic burn-off limits are a factor of 3 or more higher for the InGaAsP system than for the GaAlAs system. This may be because the thermal conductivity of InP is about 50 per cent higher than GaAs and about 5 times higher than GaAlAs so that heating of the facets by surface non-radiative recombination and absorption is less pronounced.

If long-term non-catastrophic facet damage is caused by the oxide growth mechanism as proposed by Yonezu,[82] then there is hope that it will be not nearly as severe in InGaAsP lasers. This may be so because photoexcitation induced oxidation of InP was reported to be considerably less than for GaAs. Also, both catastrophic and non-catastrophic damage should be helped by the much lower surface recombination velocity of InP.[132]

With the small number of devices tested and the low degradation rates observed there have been inadequate results to formulate possible degradation mechanisms. However, there are some comments in the literature regarding possible indium solder–gold metallization interdiffusion effects[125] that result in increased thermal resistance or local dissolution[133] of InP. As with GaAs devices, these types of effects should not limit device lives, as improved metallization and bonding procedures are developed. For example, low resistance, non-alloyed *p*-contacts have been obtained by evaporating a Ti–Pd–Au metallization onto an InGaAs *p*-capping layer; the conductivity at the surface of the InGaAs having been raised by a zinc diffusion treatment.[134] Reliability studies of alloy systems other than InGaAsP giving emission in the 1.0–1.6 μm wavelength range have also been reported. Olsen *et al.*[124] have investigated InGaAs–InGaP DH ELEDs and lasers emitting close to 1.1 μm and grown by VPE. These incorporated an InGaP graded layer onto GaAs substrates. Such devices showed appreciable degradation in 10^4 hours at 22 °C and substantial degradation in 100 hours at 60 °C. This result suggests that the InGaAs–InGaP DH devices compositionally graded onto GaAs substrates may have inferior reliability to either VPE or LPE InGaAsP–InP devices lattice matched onto InP substrates. Comments on the reliability of LPE GaAlAsSb DH devices emitting around 1.06 μm compositionally graded onto GaAs substrates also suggest inferior performance. Thus it seems that lattice matched structures are necessary for reliable device operation in the 1.0–1.6 μm spectral region. To date there have been no reports on the reliability of InGaAsSb[135] or GaAlAsSb[136] (lattice matched to GaSb substrates) devices. It still remains to be seen whether the resistance to degradation of InGaAsP–InP heterostructures is related to the lower energy of non-radiative recombination available (because of smaller band gap energy) or is related to its specific ion and lattice configuration. The method of growth appears not to be of overriding importance.

The major problem with InGaAsP DH devices, particularly lasers, and one not encountered with GaAlAs devices, is the strong dependence of internal quantum efficiency on temperature.[137–139] This is thought to be due to a temperature dependent non-radiative recombination process[138,139] occurring in the bulk of the InGaAsP active layer or at either of the adjacent heterojunction interfaces. Whether this effect is fundamental to the particular materials system is not yet clear. Thus the temperature sensitivity of increase in laser threshold or reduction in LED optical power output at constant current is rather more at the present time than for equivalent GaAs or GaAlAs devices. For example T_0 in Equation 6.1 is around 50–70 K as opposed to 120–150 K for GaAlAs lasers.

In passing, it is of interest to note that a T_0 of 112 K has been reported[135] close to room temperature for one of the other quaternary alloy system double heterostructure—InGaAsSb–AlGaAsSb lattice matched to GaSb substrates—and, in this case, emitting at 1.8 μm.

For the present, the low value of T_0 means that CW degradation rates in InGaAsP lasers are more sensitive to a given increase in pulsed threshold. Hence there is incentive to minimize thermal resistance by ensuring non-deterioration of the solder bond and low values of *p* contact resistivity[134] in stripe lasers. With good reliability being reported for InGaAsP devices, it is now important to look critically at these areas to ensure that the effect of any internal chip degradation has the minimum possible effect on CW characteristics.

6.4.3 Lead chalcogenide sources

Lead chalcogenide diode lasers and LEDs have been developed because of their usefulness as sources of relatively monochromatic radiation over the whole spectral region from 3 μm to 30 μm. They are being used increasingly in spectroscopy, where the tuning of the emission wavelength of an individual diode is possible over a range by temperature variation as well as, less usually, by pressure and magnetic fields. Because of their narrow band gaps, laser operation is usually confined to below 100 K for CW operation even with double-heterostructure configurations.[140] Pulsed laser operation up to around

200 K is possible however and continuous 4.6 μm LED operation at room temperature has been reported.[141]

The reliability problems encountered are somewhat different to those in the materials systems discussed earlier in this chapter. Cycling of the diodes down to low temperature puts constraints on the packaging technology. Pressure contacts using thick indium layers as buffers are often employed to take up the differential thermal contraction of the device and packaging materials on repeated temperature cycling. In this context silver is better than copper as a heatsink material because of its closer match to the thermal expansion coefficients of the lead chalcogenides. There is a need to protect the essential parts of the lead-salt chip from direct contact with the indium contact layers because indiffusion is significant at room temperature and so affects shelf-life. Indium, however, is used routinely as the *n* contact metal. Gold and platinum are useful *p* contact metals. One scheme that eliminates the formation of brittle Au–In alloys during the bonding operation (which can cause problems on repeated temperature cycling), is to use a Au–Pt–In contact sandwich.[142] Here the platinum acts as a barrier layer. PbSSe layers with this metallization have not shown contact degradation after being thermally cycled 50 times, coupled with several hundred hours continuous operation.

Gradual internal chip degradation of the type seen in devices emitting in the red and near infrared region, has not been reported from lead chalcogenide devices. It is not at all clear whether their low operating temperature is important in this respect. The advent of room temperature lead salt LEDs should be watched with interest from this point of view.

6.5 CONCLUSIONS

Degradation processes in semiconductor light emitters are, as yet, not fully understood. However, with low radiance visible and near infrared LEDs, attention to processing environment and fabrication technology has created fully acceptable products.

High radiance GaAs based LEDs and lasers are becoming generally acceptable, but devices still show degradation effects over long periods. Enhanced defect migration phenomena are now thought to play a dominating role in residual bulk and, possibly, also surface degradation, and are probably intrinsic in nature.

Longer wavelength (1.0–1.6 μm) emitters are giving encouraging results. It remains to be seen whether such devices are intrinsically more reliable than GaAs based devices and what extent such a finding would have on the relative emphasis given in, for example, optical fibre systems' applications to the two wavelength regions.

Lead-salt lasers and LEDs, up until now operated at cryogenic temperatures, have apparently not been plagued by internal degradation processes.

REFERENCES

1. For a review, see R. C. Goodfellow and R. Davis, 'Optical source devices' in *Optical Fibre Communications* M. J. Howes and D. V. Morgan (Eds.), 27–106, Wiley, 1980.
2. For a review, see A. A. Bergh and P. J. Dean, *Light-emitting Diodes*, Clarendon Press, 1976.
3. For a review, see H. Kressel and J. K. Butler, *Semiconductor Lasers and Heterojunction LEDs*, Academic Press, New York, 1977.
4. For a review, see H. C. Casey and M. B. Panish, *Heterostructure Lasers, Part B*, Academic Press, New York, 1978.
5. T. L. Paoli, 'Noise characteristics of stripe-geometry double-heterostructure junction lasers operating continuously–I. Intensity noise at room temperature', *IEEE J. Quantum Electron.*, **QE-11**, 276, 1975.
6. A. R. Goodwin, A. W. Davis, P. A. Kirkby, R. E. Epworth, and R. G. Plumb, 'Narrow stripe semiconductor laser for improved performance of optical communication systems', *Optical Communication Conference Proceedings*, Amsterdam, 1979.
7. A. G. Steventon, P. J. Fiddyment, and D. H. Newman, 'Low threshold current proton-isolated (GaAl) As DH lasers', *Opt. and Quant. Electronics*, **9**, 519–525 (1977).
8. R. W. Dixon, F. R. Nash, R. L. Hartman, and R. T. Hepplewhite, 'Improved light-output linearity in stripe-geometry double-heterostructure junction lasers', *Appl. Phys. Lett.*, **29**, 372–374. (1976).
9. R. D. Gold and L. R. Weisberg, 'Permanent degradation of GaAs tunnel diodes', *Solid State Electronics*, **7**, 811–821 (1964).
10. S. A. Steiner and R. L. Anderson, 'Degradation of GaAs injection devices', *Solid State Electronics*, **11**, 65–86 (1968).
11. D. H. Newman, S. Ritchie, and S. O'Hara, 'Experimental tests of proposed mechanisms for gradual degradation of GaAs double-heterostructure injection lasers', *IEEE J. of Quantum Electron.*, **QE-8**, 379–382 (1972).
12. D. H. Newman and S. Ritchie, 'Gradual degradation of GaAs double-heterostructure lasers', *IEEE J. of Quantum Electron.*, **QE-9**, 300–305 (1973).
13. T. L. Paoli, 'Changes in the optical properties of cw (AlGa)As junction lasers during accelerated lifetesting', *IEEE J. Quantum Electron.*, **QE-13**, 351–359 (1977).
14. H. Kressel and N. E. Byer, 'Physical basis of non-catastrophic degradation in GaAs injection lasers', *Proc. IEEE*, **57**, 25–33 (1969).
15. N. E. Byer, 'Role of optical flux and of current density in gradual degradion of GaAs injection lasers', *IEEE J. Quantum Electron.*, **QE-5**, 242 (1969).
16. R. L. Hartman and R. W. Dixon, 'Reliability of DH GaAs lasers at elevated temperatures', *Appl. Phys. Lett.*, **26**, 239–242 (1975).
17. R. W. Dixon and R. L. Hartman 'Accelerated aging and a uniform mode of degradation in (AlGa)As double-heterostructure lasers', *J. Appl. Phys.*, **48**, 3225–3229 (1977).
18. W. B. Joyce, R. W. Dixon, and R. L. Hartman, 'Statistical characterization of lifetimes of continuously operated (AlGa)As double-heterostructure lasers', *Appl. Phys. Lett.*, **28**, 684–686 (1976).
19. M. Ettenberg, 'A statistical study of the reliability of oxide-defined stripe CW lasers of (AlGa)As', *J. Appl. Phys.*, **50**, 1195–1202 (1979).
20. Y. Furukawa, T. Kobayashi, K. Wakita, T. Kawakami, G. Iwane, Y. Horikoshi,

and Y. Seki, 'Accelerated lifetest of AlGaAAs–GaAs DH lasers', *Japan. J. Appl. Phys.*, **16**, 1495–1496 (1977).
21. R. L. Hartman, N. E. Schumaker and R. W. Dixon, 'Continuous operated AlGaAs double-heterostructure lasers with 70 °C lifetimes as long as two years', *Appl. Phys. Lett.*, **31**, 756–759 (1977).
22. H. Kressel, M. Ettenberg and I. Ladany, 'Accelerated step-temperature aging of $Al_xGa_{1-x}As$ heterojunction laser diodes', *Appl. Phys. Lett.*, **32**, 305–308 (1978).
23. H. Ishikawa, T. Fujiwara, K. Fujiwara, M. Marmoto, and M. Takusagawa, 'Accelerated aging test of $Ga_{1-x}Al_xAs$ DH lasers', *J. Appl. Phys.*, **50**, 2518–2522 (1979).
24. S. Ritchie, R. F. Godfrey, B. Wakefield, and D. H. Newman, 'The temperature dependence of degradation mechanisms in long-lived (GaAl)As DH lasers', *J. Appl. Phys.*, **49**, 3127–3132 (1978).
25. K. Mizuishi, N. Chinone, H. Sato, and R. Ito, 'Acceleration of the gradual degradation in (GaAl)As double-heterostructure lasers as an exponent of the value of driving current', *J. Appl. Phys.*, **50**, 6668–6674 (1979).
26. S. D. Hersee, 'The reliability of high radiance GaAs LEDs', Proceedings AGARD Symposium, *CPP219*, paper 41, 1977.
27. N. A. J. Hastings and J. B. Peacock, *Statistical Distributions*, Butterworth and Co.
28. N. R. Mann, R. E. Schafer, and N. D. Singpurwalla, *Methods for Statistical Analysis of Reliability and Life Data*, Wiley, New York, 1974.
29. D. S. Peck and C. H. Zierdt, 'The reliability of semiconductor devices in the Bell system', *Proc. IEEE*, **62**, 185–211 (1974).
30. F. H. Reynolds, 'Thermally accelerated aging of semiconductor components', *Proc. IEEE*, **62**, 212–222 (1974).
31. A. S. Jordan, 'A comprehensive review of the lognormal failure distribution with application to LED reliability', *Microelectron. Reliab.*, **18**, 267–278 (1978).
32. B. C. De Loach, B. W. Hakki, R. L. Hartman, and L. A. D'Asaro, 'Degradation of CW GaAs double-heterostructure lasers at 300 K', *Proc. IEEE*, **61**, 1042–1044 (1973).
33. H. Yonezu, M. Ueno, T. Kamejima, and J. Sakuma, 'Lasing characteristics in a degraded GaAs–GaAlAs double heterostructure laser', *Japan J. Appl. Phys.*, **13**, 835 (1974).
34. D. H. Newman, R. F. Godfrey, A. R. Goodwin, and D. Lovelace, 'Enhanced localized degradation and anomalous emission spectra of $Ga_{1-x}Al_xAs$ double heterostructure lasers induced by fabrication processes', *Appl. Phys. Lett.*, **29**, 353–355 (1976).
35. P. W. Hutchinson, P. S. Dobson, S. O'Hara and D. H. Newman, 'Defect structure of degraded heterojunction GaAlAs–GaAs lasers', *Appl. Phys. Lett.*, **26**, 250–252 (1975).
36. P. Petroff and R. L. Hartman, 'Defect structure induced during operation of heterojunction GaAs lasers', *Appl. Phys. Lett.*, **23**, 469–471 (1973).
37. P. W. Hutchinson and P. S. Dobson, 'Defect structure of degraded GaAlAs–GaAs double heterojunction lasers', *Phil. Mag.*, **32**, 745–754 (1978).
38. D. V. Lang and L. C. Kimmerling, 'Observation of recombination-enhanced defect reactions in semiconductors', *Phys. Rev. Lett.*, **33**, 489–492 (1974).
39. S. O'Hara, P. W. Hutchinson and P. S. Dobson, 'The origin of dislocation climb during laser operation', *Appl. Phys. Lett.*, **30**, 368–371 (1977).
40. D. V. Lang, R. L. Hartman, and N. E. Schumaker, 'Capacitance spectroscopy studies of degraded $Al_xGa_{1-x}As$ DH stripe-geometry lasers', *J. Appl. Phys.*, **47**, 4986–4992 (1976).

41. R. C. Goodfellow, 'High-radiance, small area GaAs lamps', *Proc. Electro-Optics International Conference*, Brighton, 1974.
42. R. L. Hartman and A. R. Hartman, 'Strain-induced degradation of GaAs injection lasers', *Appl. Phys. Lett.*, **23**, 147–149 (1973).
43. R. G. Plumb, A. R. Goodwin, and R. S. Baulcomb, 'Thermal impedance aging characteristics of CW stripe lasers', *Solid State and Electron Devices*, **3**, 206–209 (1979).
44. K. Fujiwara, H. Imai, T. Fujiwara, K. Hori, and M. Takusagawa, 'Analysis of deterioration in In solder for GaAlAs DH lasers', *Appl. Phys. Lett.*, **35**, 861–863 (1979).
45. K. Fujiwara, T. Fujiwara, K. Hori, and M. Takusagawa, 'Aging characteristics of $Ga_{1-x}Al_xAs$ double-heterostructure lasers bonded with gold eutectic solder', *Appl. Phys. Lett.*, **34**, 668–670 (1979).
46. S. Ritchie (unpublished work).
47. P. W. Hutchinson and P. S. Dobson, 'Climb asymmetry in degraded GaAs lasers', *Phil. Mag. A*, 1980 to be published.
48. A. R. Goodwin, P. A. Kirkby, I. G. A. Davies, and R. S. Baulcomb, 'The effects of processing stresses on residual degradation of long lived $Ga_{1-x}Al_xAs$ lasers', *Appl. Phys. Lett.*, **34**, 647–649 (1979).
49. J. D. Speight and K. Cooper, 'Interlayer diffusion phenomena in Ti–Au metallizations on *n* type GaAs at 250–450 °C', *Thin Solid Films*, **25**, S31–S37 (1975).
50. P. D. Augustus and S. D. Hersee, 'Precipitate-induced generation of $\langle 110 \rangle$ dislocations in GaAs LEDs', *Proceedings 9th Int. Congress Electron Microscopy*, Toronto, 1978.
51. P. G. Eliseev, 'Degradation of injection lasers', *J. Luminescence*, **7**, 338–356 (1973).
52. C. H. Henry, P. M. Petroff, R. A. Logan, and F. R. Merritt, 'Catastrophic damage of $Al_xGa_{1-x}As$ double-heterostructure laser material', *J. Appl. Phys.*, **50**, 3721–3732 (1979).
53. H. Imai, M. Morimoto, H. Sudo, T. Fujiwara, and M. Takusagawa, 'Catastrophic degradation of GaAlAs DH laser diodes', *Appl. Phys. Lett.*, **33**, 1011–3 (1978).
54. B. W. Hakki and F. R. Nash, 'Catastrophic failure in GaAs double-heterostructure injection lasers', *J. Appl. Phys.*, **45**, 3907–3912 (1974).
55. O. Ueda, H. Imai, and T. Kotani, 'TEM observation of catastrophically degraded $Ga_{1-x}Al_xAs$ double-heterostructure lasers', *J. Appl. Phys.*, **50**, 6643–6647 (1979).
56. N. Chinone, K. Saito, N. Shige, and K. Aiki, 'High-efficient (GaAl)As buried-heterostructure lasers for analog communications', *Optical Communications Conference*, Amsterdam, 1979.
57. G. D. Henshall, 'Gallium aluminium arsenide heterostructure lasers: factors affecting catastrophic degradation', *Solid-State Electron.*, **20**, 595–602 (1977).
58. H. Yonezu, I. Sakuma, T. Kamejima, M. Ueno, K. Iwamoto, I. Hino, and I. Hayashi, 'High optical power density emission from a window stripe AlGaAs double heterostructure laser', *Appl. Phys. Lett.*, **34**, 637–639 (1979).
59. L. C. Kimmerling, 'Recombination enhanced defect reactions', *Solid State Electronics*, **21**, 1391–1401 (1978).
60. P. M. Petroff and L. C. Kimmerling, 'Dislocation climb model in compound semiconductors with zinc blende structure', *Appl. Phys. Lett.*, **29**, 461–463 (1976).

61. T. Kobayashi and Y. Furukawa, 'Recombination enhanced annealing effect in AlGaAs–GaAs remote junction heterostructure lasers', *IEEE J. Quantum Electron.*, **QE-15**, 674–684 (1979).
62. M. Ettenberg, H. Kressel, and H. F. Lockwood, 'Degradation of $Al_xGa_{1-x}As$ heterojunction electroluminescent devices', *Appl. Phys. Lett.*, **25**, 82–85 (1974).
63. I. Ladany, T. R. Furman, and D. P. Marinelli, 'Internal stress and degradation in short wavelength AlGaAs double-heterojunction devices', *Electronics Lett.*, **15**, 342–343 (1979).
64. B. Wakefield, 'Strain-enhanced luminescence degradation in GaAs–GaAlAs double-heterostructure lasers revealed by photoluminescence', *J. Appl. Phys.*, **50**, 7914–7916 (1979).
65. M. J. Robertson and B. Wakefield (unpublished work).
66. F. R. Nash, R. L. Hartman, N. M. Denkin, and R. W. Dixon, 'GaAs laser reliability and protective facet coatings', *J. Appl. Phys.*, **50**, 3122–3132 (1979).
67. D. V. Lang, 'Deep level transient spectroscopy: A new method to characterise traps in semiconductors', *J. Appl. Phys.*, **45**, 3023–3032 (1974).
68. G. M. Martin, A. Mitonneau and A. Mircea, 'Electron traps in bulk and epitaxial GaAs crystals', *Electronics Lett.*, **13**, 191–193 (1977).
69. A. Mitonneau, G. M. Martin, and A. Mircea, 'Hole traps in bulk and epitaxial GaAs crystals', *Electronics Lett.*, **13**, 666–668 (1977).
70. B. Balland, G. Vincent, D. Bois, and P. Hirtz, 'Donor levels analysis in GaAlAs double heterostructure', *Appl. Phys. Lett.*, **34**, 109–110 (1979).
71. T. Uji, 'Deep levels in the n-$Al_{0.3}Ga_{0.7}As$ layer of AlGaAs double heterostructure lasers', *Japan. J. Appl. Phys.*, **17**, 727–728 (1978).
72. D. V. Lang, R. A. Logan, and L. C. Kimmerling, 'Identification of the defect state associated with a gallium vacancy in GaAs and $Al_xGa_{1-x}As$', *Phys. Rev.*, **15**, 4814–4822 (1977).
73. S. Metz, 'Deep centre luminescence (1.02 eV) in GaAs–(GaAl)As epitaxial layers and double heterostructure lasers', *Appl. Phys. Lett.*, **33**, 198–200 (1978).
74. H. Imai, K. Isozumi, and M. Takasagawa, 'Deep level associated with the slow degradation of GaAlAs DH laser diodes', *Appl. Phys. Lett.*, **33**, 330–332 (1978).
75. P. M. Petroff, D. V. Lang, R. A. Logan, and W. D. Johnson (Jr), 'Deep level-dislocation interactions in $Ga_{1-x}Al_xAs$ (DH)', *Inst. Phys. Conf. Ser.*, **46**, 427–432 (1979).
76. H. Kressel and I. Ladany, 'Reliability aspects and facet damage in high power emmission from (AlGa)As CW laser diodes at room temperature', *RCA Review*, **36**, 230–239 (1975).
77. I. Ladany, M. Ettenberg, H. F. Lockwood and H. Kressel, 'Al_2O_3 half-wave films for long life CW lasers', *Appl. Phys. Lett.*, **30**, 87–88 (1977).
78. N. Chinone, H. Nakashima, and R. Ito, 'Long term degradation of GaAs–$Ga_{1-x}Al_xAs$ DH lasers due to facet erosion', *J. Appl. Phys.*, **48**, 1160–1162 (1977).
79. T. Kajimura, K. Saito, N. Shige, and R. Ito, 'Dependence of degradation rate of $Ga_{1-x}Al_xAs$ buried hetero-structure injection lasers on optical flux density', *Japan. J. Appl. Phys.*, **18**, 693–694 (1979).
80. T. Yuasa, M. Ogawa, K. Endo, and H. Yonezu, 'Degradation of (AlGa)As DH lasers due to facet oxidation', *Appl. Phys. Lett.*, **32**, 119–121 (1978).
81. T. Suzuki and M. Ogawa, 'Degradation of photoluminescence intensity caused by excitation-enhanced oxidation of GaAs surfaces', *Appl. Phys. Lett.*, **31**, 473–475 (1977).

82. H. Yonezu, K. Endo, T. Kamejima, T. Tonkai, Y. Yuasa, and T. Furuse, 'Mirror degradation in AlGaAs double-heterostructure lasers', *J. Appl. Phys.*, **50**, 5150–5156 (1979).
83. S. Ritchie (unpublished work).
84. K. Wakita and M. Fukuda, 'Degradation of (GaAl)As–GaAs DH laser in water due to facet deterioration', *Japan. J. Appl. Phys.*, **17**, 2175–2176 (1978).
85. Y. Shima, N. Chinone, and R. Ito, 'Effects of facet coatings on the degradation characteristics of GaAs–$Ga_{1-x}Al_xAs$ DH lasers', *Appl. Phys. Lett.*, **31**, 625–627 (1977).
86. T. Yuasa, K. Endo, T. Torikai, and H. Yonezu, 'Facet protection of (AlGa)As lasers using SiO_2 sputter deposition', *Appl. Phys. Lett.*, **34**, 685–687 (1979).
87. T. Kajimura, K. Saito, N. Shige, and R. Ito, 'Dependence of the degradation rate of $Ga_{1-x}Al_xAs$ buried heterostructure injection lasers on optical flux density', *Japan. J. Appl. Phys.*, **18**, 693–694 (1979).
88. T. Furuse, T. Suzuki, S. Matsumoto, K. Nishida, and Y. Nannichi, 'Insulating carbon coating on (AlGa)As DH laser facets', *Appl. Phys. Lett.*, **33**, 317–318 (1978).
89. H. Namizaki, S. Takamiya, M. Ishii, and W. Sukaki, 'High-power-density single-mode operation of GaAs–GaAlAs TJS lasers utilizing Si_3N_4 plasma deposition for facet coating', *J. Appl. Phys.*, **50**, 3743–3745 (1979).
90. T. L. Paoli and J. E. Ripper, 'Coupled longitudinal mode pulsing in semiconductor lasers', *Phys. Rev. Lett.*, **22**, 1085–1088. (1969).
91. G. Arnold and K. Petermann, 'Self-pulsing phenomena in (GaAl)As double heterojunction injection lasers', *Optical and Quantum Electronics*, **10**, 311–322 (1978).
92. R. W. Dixon and W. B. Joyce, 'A possible model for sustained oscillations (pulsations) in (Al, Ga)As double-heterostructure lasers', *IEEE J. Quantum Electron.*, **QE-15**, 470–474 (1979).
93. Y. Suematsu and T.-H. Hong, 'Suppression of relaxation oscillations in light output of injection lasers by external resonance circuit', *IEEE J. Quantum Electron.*, **QE-13**, 756–762 (1977).
94. M. R. Matthews and A. G. Steventon, 'Spectral and transient response of low-threshold proton-isolated (GaAl)As lasers', *Electronics Lett.*, **14**, 649–650 (1978).
95. J. P. Van der Ziel, J. L. Merz, and T. L. Paoli, 'Study of intensity pulsations in proton-bombarded stripe-geometry double-heterostructure $Al_xGa_{1-x}As$ lasers', *J. Appl. Phys.*, **50**, 4620–4637 (1979).
96. B. W. Hakki, 'Instabilities in output of injection lasers', *J. Appl. Phys.*, **50**, 5630–5637 (1979).
97. R. L. Hartman, R. A. Logan, L. A. Koszi, and W. T. Tsang, 'Pulsations and absorbing defects in (Al, Ga)As injection lasers', *J. Appl. Phys.*, **50**, 4616–4619 (1979).
98. D. J. Channin, M. Ettenberg, and H. Kressel, 'Self-sustained oscillations in (AlGa)As oxide-defined stripe lasers', *J. Appl. Phys.*, **50**, 6700–6706 (1979).
99. H. Kuwahara, 'Simulation of intensity self-pulsation in CW semiconductor lasers', *Appl. Phys.*, **20**, 67–73 (1979).
100. D. Baker, 'High reliability transistors for submarine systems', *Inst. Phys. Conf. Ser.*, **40**, 87–105 (1978).
101. H. Kressel, M. Ettenberg, and H. F. Lockwood, 'Effect of edges on the reliability of GaAs heterojunction LEDs', *J. Electronic Mats.*, **6**, 467–481 (1977).

102. F. H. Reynolds and J. W. Stevens, 'Semiconductor component reliability in an equipment operating in electromechanical telephone exchanges', *IEEE 16th Annual Reliability Physics Proceedings*, 1978.
103. S. Yamakoshi, O. Hasegawa, H. Hamaguchi, M. Abe, and T. Yamaoki, 'Degradation of high-radiance $Ga_{1-x}Al_xAs$ LEDs', Appl. Phys. Lett., **31**, 627–629 (1977).
104. See, for example, 'Defects and radiation effects in semiconductors, 1978', *Inst. Phys. Conf. Ser.*, **46**, 243 (1979).
105. A. S. Epstein, S. Share, R. A. Polimadei, R. J. Lynch, and D. Sullivan, 'The role of aluminium in irradiated $Ga_{1-x}Al_xAs$ light-emitting-diodes', *IEEE Specialist Conference on the Technology of Electroluminescent Diodes*, Atlanta, 1974 (unpublished).
106. K. J. Soda, C. E. Barnes, and R. A. Kiehl, 'The effect of gamma irradiation on optical isolators', *IEEE Trans. Nucl. Sci.*, **NS-22**, 2475–2481 (1975).
107. K. H. Cameron, M. R. Matthews, K. R. Preston, and D. W. Smith, 'A hybrid laser transitter module for use in fibre systems', *Electronic Components Conference, San Francisco*, 1980.
108. R. L. Hartman, B. Schartz, and M. Kuhn, 'Degradation and passivation of GaP light-emitting diodes', *Appl. Phys. Lett.*, **18**, 304–307 (1971).
109. A. A. Bergh, 'Bulk degradation of GaP red LEDs', *IEEE Trans. Electron Devices*, **ED-18**, 166–170 (1971).
110. S. I. Koyama, T. Koshisa, and S. Suzuki, 'Reliability of $GaAs_{1-x}P_x$ light emitting diodes', *IEEE 10th Annual Reliability Physics Proceedings, Las Vegas*, 203–207 (1972).
111. A. Bahraman and W. G. Oldham, 'Role of copper in the degradation of GaAs electroluminescent diodes', *J. Appl. Phys.*, **43**, 2383–2387 (1972).
112. R. L. Hartman and M. Kuhn, 'The effects of device configuration in the degradation of GaP red light-emitting diodes', *IEEE 10th Annual Reliability Physics Proceedings, Las Vegas*, 137 (1972).
113. J. M. Ralston and J. W. Mann, 'Temperature and current dependence of red-emitting GaP LEDs', *J. Appl. Phys.*, **50**, 3630–3636 (1979).
114. C. H. Henry and P. D. Dapkus, 'Deep-level changes associated with the degradation of gallium phosphide red-light-emitting diodes', *J. Appl. Phys.*, **47**, 4067–4072 (1976).
115. M. Mizata, J. Yoshino, and H. Kukimoto, 'Degradation of high efficiency GaP red LEDs', *IEEE Trans. Electron Devices*, **ED-26**, 194–197 (1979).
116. P. M. Petroff, O. G. Lorimor, and J. M. Ralston, 'Defect structure induced during forward-bias degradation of GaP green-light-emitting-diodes', *J. Appl. Phys.*, **47**, 1583–1588 (1976).
117. K. Kaneko, 'Degradation of GaP green LEDs', *Japan. J. Appl. Phys.*, **15**, 1287–1296 (1976).
118. C. Lopez, A. Garcia, F. Garcia, and E. Minoz, 'Native levels and degradation in $GaAs_{0.6}P_{0.4}$ LEDs', *Solid-State Electronics*, **22**, 81–85 (1979).
119. H. Schade, C. J. Nuese, and J. J. Gannon, 'Direct evidence for generation of defect centres during forward-bias degradation of $GaAs_{1-x}P_x$ electroluminescent diodes', *J. Appl. Phys.*, **42**, 5072–5075 (1971).
120. A. G. Dentai, T. P. Lee, C. A. Burrus, and F. Buehler, 'Small-area, high-radiance CW InGaAsP LEDs emitting at 1.2–1.3 μm', *Electron. Lett.*, **13**, 484–485 (1977).
121. C. C. Shen, J. J. Hsieh, and T. A. Lind, '1500-h continuous CW operation of double-heterostructure GaInAsP–InP lasers', *Appl. Phys. Lett.*, **30**, 353–354 (1977).

122. P. A. Kirkby, 'Dislocation pinning in GaAs by the deliberate introduction of impurities', *IEEE J. Quantum Electron.*, **QE-11**, 562–568 (1975).
123. O. Wada, S. Yamakoshi, M. Abe, K. Akita, and Y. Toyama, 'A new type of InGaAsP–InP DH LED for fibre optical communication systems at 1.2–1.3 μm', *Proceedings of the Optical Communication Conference*, Amsterdam, 1979.
124. G. H. Olsen, C. J. Nuese, and M. Ettenberg, 'Reliability of vapour grown InGaAs and InGaAsP heterojunction laser structure', *IEEE J Quantum Electron.*, **QE-15**, 688–693 (1979).
125. T. Yamamoto, K. Sakai, and S. Akiba, '10,000 h continuous CW operation of $In_{1-x}Ga_xAs_yP_{1-y}$–InP DH lasers at room temperature', *IEEE J. Quantum Electron.*, **QE-15**, 684–687 (1979).
126. Y. Horokoshi, T. Kobayashi, and Y. Furukawa, 'Lifetime of InGaAsP–InP and AlGaAs–GaAs DH lasers estimated by the point defect generation model', *Japan. J. Appl. Phys.*, **18**, 2237–2244 (1979).
127. J. N. Walpole, T. A. Lind, J. J. Hsieh, and A. G. Foyt, 'Self-sustained pulsations in GaInAsP diode lasers', *Appl. Phys. Lett.*, **36**, 240–242 (1980).
128. H. Nagai and Y. Noguchi, 'InP–$Ga_xIn_{1-x}As_yP_{1-y}$ double heterostructure for 1.5 μm wavelength', *Appl. Phys. Lett.*, **32**, 234–236 (1978).
129. S. Akiba, K. Sakai, Y. Matsushima, and T. Yamamoto, 'Effects of double-cladding structure on LPE-grown InGaAsP–InP lasers in the 1.5 μm range', *Japan. J. Appl. Phys.*, **19**, L79–L82 (1980).
130. Y. Suematsu, 'Sources for long wavelength optical communications systems', *Proceedings of the Optical Communication Conference*, Amsterdam, 1979.
131. M. Nakamura, 'Semiconductor injection lasers for long wavelength optical communications', *Integrated and Guided-wave Optics Conference Technical Digest, Lake Tahoe*, 1980.
132. H. C. Casey and E. Buehler, 'Evidence for low surface recombination velocity on *n*-type InP', *Appl. Phys. Lett.*, **30**, 247–249 (1977).
133. Y. Horikoshi, H. Saito, and M. Kawashima, 'Degradation of InGaAsP–InP DH lasers by In solder', *Japan. J. Appl. Phys.*, **18**, 1623–1624 (1979).
134. A. G. Steventon, R. E. Hobbs, and P. J. Fiddyment, (unpublished work).
135. N. Kobayashi, Y. Horikoshi, and C. Uemura, 'Room temperature operation of the InGaAsSb–AlGaAsSb DH laser at 1.8 μm wavelength', *Japan. J. Appl. Phys.*, **19**, L30–32 (1980).
136. L. R. Tomasetta, H. B. Law, K. Nakano, and J. S. Harris, 'GaAlAsSb–GaSb lattice matched semiconductor lasers operating at 1.25–1.40 μm', *6th International Semiconductor Laser Conference*, San Francisco, 1978.
137. Y. Horikoshi and Y. Furukawa, 'Temperature sensitive threshold current of InGaAsP–InP double heterostructure lasers', *Japan. J. Appl. Phys.*, **18**, 809–815 (1979).
138. R. E. Nahory, M. A. Pollack, and J. C. DeWinter, 'Temperature dependence of InGaAsP double heterostructure laser characteristics', *Electron. Lett.*, **15**, 695–696 (1979).
139. G. H. B. Thompson and G. D. Henshall, 'Non-radiative carrier loss and temperature sensitivity of threshold in 1.27 μm (GaIn)(AsP)–InP DH lasers', *Electron. Lett.*, **16**, 42–44 (1980).
140. H. Preier, 'Recent advances in lead-chalcogenide diode lasers', *Appl. Phys.*, **20**, 189–206 (1979).
141. W. Lo and D. E. Swets, 'Room-temperature, 4.6 μm light-emitting diodes', *CLEOS/ICF 1980*, San Diego, 1980.

142. W. Lo and D. E. Swets, 'Diffused homojunction lead-sulphide-selenide diodes with 140 K laser operation', *Appl. Phys. Lett.*, **33**, 938–940 (1978).
143. L. R. Dawson, V. G. Keramidas, and C. L. Zipfel, 'Reliable, high-speed LEDs for short-haul optical data links', *BSTJ*, **59**, 161–168 (1980).
144. J. A. F. Peek, 'Water vapour facet erosion and the degradation of (Al, Ga)As DH lasers operated at CW output powers of up to 3 mW/μm stripewidth', *IEEE J. of Quantum Electron.*, **QE-17**, 781–787, 1981.

Reliability and Degradation
Edited by M. J. Howes and D. V. Morgan

CHAPTER 7

Reliability and degradation of microwave integrated circuits

A. G. VAN NIE, W. GOEDBLOED, and G. KERSUZAN

7.1 INTRODUCTION

Electronic circuits for frequencies below, say, 1 GHz contain elements such as resistors, inductors, and capacitors, whose dimensions are much smaller than the wavelength of the frequency in use. The phase shifts due to wave propagation can be neglected for such 'lumped' elements. The elements in these circuits are interconnected by wires whose capacitances and inductances can often be neglected. In microwave circuits, however, 'distributed' elements are used and these elements are connected by transmission lines. Distributed elements, e.g. directional couplers, hybrid rings, and resonators, have dimensions of the same order of magnitude as the wavelength and use is often made of phase shifts due to wave propagation. The transmission lines have a characteristic impedance and relevant parameters of the wave propagation are wavelength attenuation and dispersion.

In microwave integrated circuits (MICs), planar transmission lines and distributed elements are integrated on the same substrate. To realize such a MIC, a metallic pattern is formed on the substrate and 'add-on' components, e.g. Schottky barrier diodes, FETs, Gunn or Avalanche diodes, are assembled on the planar part of the microwave circuit. After component assembly, the MIC is mounted in a package provided with suitable (coaxial) connectors. Most of the discussions in this chapter will relate to MICs for use in the frequency range from 1 to 20 GHz. Such MICs are applied in telecommunication systems, in TV receivers for satellite broadcasting, in radar and navigation systems, etc.

The properties of the planar circuit are determined by the geometry of the pattern(s) and the substrate and by the properties of the materials used. Several suitable planar transmission lines are discussed in Section 7.2.

The term 'reliability' stands for the complex of functional accuracy, stability and life. The present chapter does not deal with the reliabilities of complete

MICs. It rather focuses on physical and chemical mechanisms that cause the inaccuracies and degradations of MICs properties. As a convenient vehicle, microstrip is used for the study of these phenomena. The properties of microstrips are, therefore, treated in detail in Section 7.2. The technologies and materials, that are suitable for MIC manufacturing, are described and evaluated in Sections 7.3.1 and 7.3.2, respectively.

The accuracy of the performance of a MIC depends on pattern manufacturing and on the spread in thickness and properties of the substrate material. The geometrical accuracies, obtainable in different technologies, are discussed and compared in Section 7.3.3.

The stability and life of several substrate/layer combinations are dealt with in Section 7.4, and models of relevant physical or chemical phenomena are discussed. If one of these phenomena predominates, then the corresponding model can be used for the extrapolation of the aging behaviour from results of accelerated life tests.

Assembly techniques for add-on components are described in Section 7.5. At microwave frequencies ($f > 1$ GHz), the parasitic impedances of the connections between add-on components and planar circuits can cause deterioration in the desirable properties of the components and therefore special precautions are necessary. Finally, MIC packaging and connector provisions are treated in Section 7.5.

The requirements imposed on MICs strongly depend on the application. For a given application, it is not always a simple matter to justify the choice from the available technologies and materials.

7.2 PLANAR TRANSMISSION LINES

As indicated in the introduction, planar transmission lines are used in MICs. In what follows, several suitable transmission lines are presented. Microstrip transmission lines are discussed at length, because they are generally applied in MICs and emphasis will be put on the characteristic impedance, wavelength and propagation losses, because these parameters are well-suited for the study of the consequences of technology for the accuracy and stability of MICs. Distributed elements used in MICs are extensively discussed in the literature,[1] but are outside the scope of the present chapter.

7.2.1 Various planar transmission lines

7.2.1.1 Microstrip

The microstrip shown in Figure 7.1(a) consists of a conductor strip with width w, made on a dielectric substrate with thickness h. The underside of the substrate is metallized. In the microstrip both the electric and magnetic lines

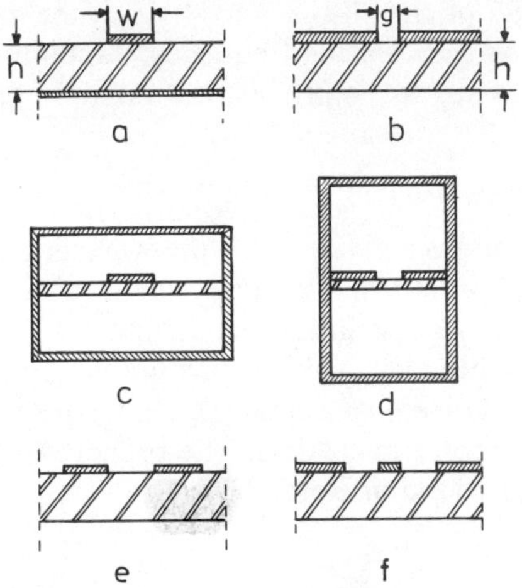

Figure 7.1. Several kinds of transmission lines: (a) microstrip; (b) microslot; (c) suspended microstrip; (d) finline; (e) coplanar stripline; (f) coupled slotline

of force are practically perpendicular to the longitudinal axis (see Figure 7.2) and the wave is propagated in a quasi-TEM (transverse electromagnetic) mode. The characteristic impedance of the microstrip and the phase velocity depend both on the geometry of the microstrip and the properties of the dielectric material, however these propagation parameters are nearly independent of frequency. In Section 7.2.2, properties of microstrip are discussed at length.

7.2.1.2 Slotline

The cross-section of a slotline is shown in Figure 7.1(b) and its field configuration in Figure 7.3. Due to the fact that the electric field appears across the slot, the slotline is well suited for parallel connections of semiconductor elements. The presence of longitudinal components in the magnetic field implies that a rotating magnetic field vector is present at some points. This effect can be used for integrating non-reciprocal devices such as circulators in the circuit. The field configuration of a slotline shows some resemblance to that of a hollow waveguide.

The propagation properties of the waves in slotlines have been investigated by Cohn.[2] The slotline exhibits a marked dispersion and its characteristic impedance is strongly dependent on frequency (Figure 7.4). Cohn has also treated the field extension beside the slot[3] and it turns out that at 10 GHz the extension in the plane of the slot is 3 to 4 times the slot width. This field extension decreases with increasing frequency. A combination of microstrip and slotline has been used for improving the isolation of 3 dB couplers.[4]

7.2.1.3 Suspended microstrip

In a suspended microstrip (Figure 7.1(c)) the wave is propagated in a quasi-TEM mode. This kind of transmission line has the advantage over the microstrip that the substrate has less influence, so that for the same characteristic impedance the strip is broader, which is favourable in view of the attainable impedance range. A disadvantage is that the field extension beside and below the strip is larger than for a microstrip. The properties of suspended microstrips have been investigated in References 5, 6, 7.

7.2.1.4 Finline

The electrical performance of finlines (Figure 7.1(d)) has been investigated by Saad and Bates.[8,9,10] The field configuration of a finline is similar to that of a slotline: dispersion occurs and its characteristic impedance is strongly dependent on frequency. A disadvantage is that the cross-section must be limited to avoid spurious mode propagation. An advantage is that (semiconductor) elements can easily be mounted in such a way that the parasitic impedances are small. Finline circuits are successfully applied in mixers at frequencies from 10 GHz to 75 GHz.

7.2.1.5 Coplanar strip and coupled slotline

The coplanar strip and coupled slotline configurations (Figure 7.1(e) and Figure 7.1(f)) have been investigated by Knorr, Hofman, Becker.[11,12,13] If the substrate thickness is much smaller than the widths of the slots and strips, the waves will propagate in a quasi-TEM mode and the dispersion will be low. Like with finlines, the cross-section of these kinds of transmission lines must be limited to avoid spurious modes.

7.2.2 Propagation parameters of microstrips

In microstrip with a homogeneous dielectric, the electric and magnetic lines of force are perpendicular to the direction of the wave propagation, so that a pure TEM mode occurs. The characteristic impedance Z of this microstrip depends on the relative dielectric constant ε_r and the relative permeability μ_r. In general, one can write.

$$Z = Z_0 \sqrt{\frac{\mu_r}{\varepsilon_r}} \tag{7.1}$$

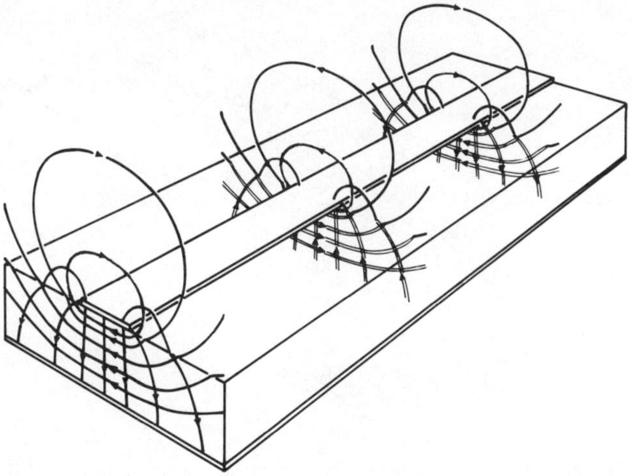

Figure 7.2. Pattern of the electromagnetic field of a microstrip

in which Z_0, being the impedance for $\mu_r = \varepsilon_r = 1$, is a function of the form factor $p = w/h$. The wavelength of this transmission line is given by

$$\lambda = \frac{\lambda_0}{\sqrt{\varepsilon_r \mu_r}} \tag{7.2}$$

in which λ_0 is the free-space wavelength. The ratio λ/λ_0 of this kind of transmission line turns out to be independent of frequency; in other words there is no dispersion.

If the medium is inhomogeneous, as shown in Figure 7.2, the wave is no longer propagated in the pure TEM mode. At lower frequencies, however, the deviation from the TEM mode is relatively small and in first approximation, this effect is therefore neglected. Such a quasi-TEM approach, also called a quasi-static analysis, has already been widely used for the case that $\mu_r = 1$.[14,15]

Schneider's relations, in which the influence of the thickness of the strip has been neglected give

$$Z = \frac{Z_0}{\sqrt{\varepsilon_{\text{eff}}}} \tag{7.3}$$

$$\lambda = \frac{\lambda_0}{\sqrt{\varepsilon_{\text{eff}}}} \tag{7.4}$$

For $p \leqslant 1$,

$$Z_0 = 60 \cdot \ln(8/p + p/4) \tag{7.5}$$

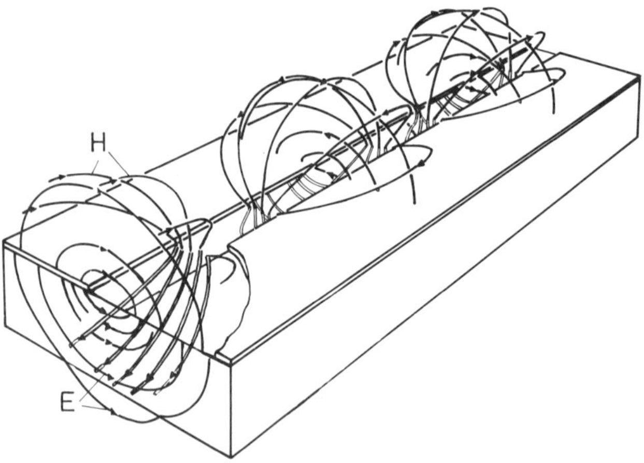

Figure 7.3. The pattern of the electromagnetic field of a slot-line. A part of the metallic layer has been removed for demonstrating the field lines in the dielectric

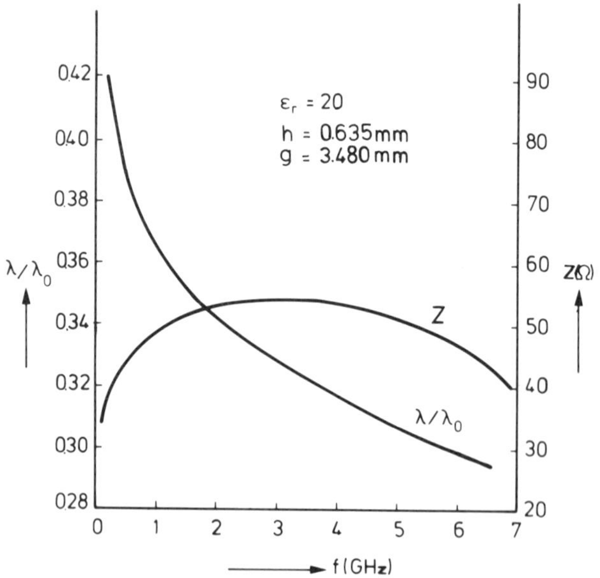

Figure 7.4. The dispersion expressed by λ/λ_0 and the characteristic impedance Z of a microslot versus frequency according to Cohn; h is the thickness of the dielectric and w is the slot width.

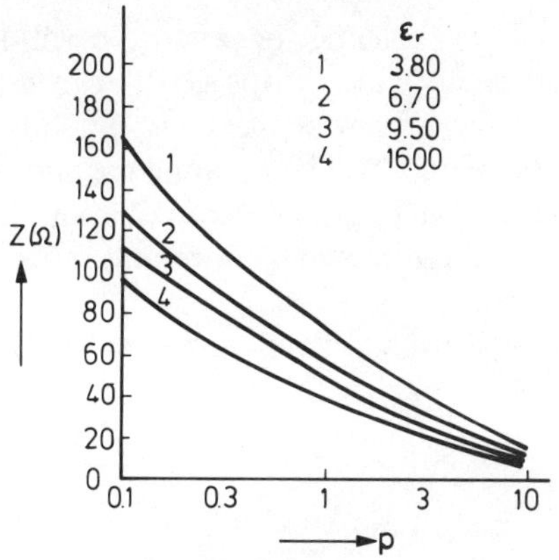

Figure 7.5. The characteristic impedance Z as a function of $p = (w/h)$ for several values of ε_r

For $p > 1$,

$$Z_0 = \frac{120\pi}{p + 2.42 - 0.44/p + (1 - 1/p)^6} \tag{7.6}$$

$$\varepsilon_{eff} = \frac{\varepsilon_r + 1}{2} + \frac{\varepsilon_r - 1}{2}(1 + 10/p)^{-1/2} \tag{7.7}$$

in which $p = w/h$. The accuracy is claimed to be ±0.25 per cent for $0 < p \leqslant 10$ and ±1 per cent for $p > 10$. Figure 7.5 shows the characteristic impedance Z as a function of p for several values of ε_r. This graph gives an

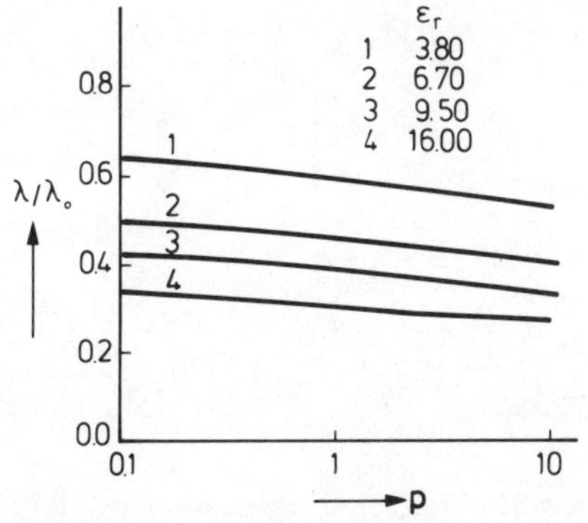

Figure 7.6. The ratio of the wavelength λ in microstrip and the free space wavelength λ_o as a function of $p = (w/h)$ for several values of ε_r

idea of the values that Z can assume; in practice p will lie within the region given in the graph, since the wider strips ($>\frac{1}{2}\lambda$) can give rise to undesired propagation modes and the narrower ones are difficult to make accurately enough, see Section 7.3.3. The wavelength along the microstrip, expressed as a fraction of the free-space wavelength λ_0, decreases with increasing p, because the electromagnetic field then becomes more and more concentrated in the dielectric (see Figure 7.6).

The properties of microstrips with magnetic substrates have been found from the TEM approach[16] to be:

$$Z = Z_0\sqrt{\frac{\mu_{\text{eff}}}{\varepsilon_{\text{eff}}}} \tag{7.8}$$

$$\lambda = \lambda_0\sqrt{\frac{1}{\mu_{\text{eff}} \cdot \varepsilon_{\text{eff}}}} \tag{7.9}$$

The μ_{eff}, the effective value of the relative permeability, can be obtained from equation 7.7 by replacing ε_r by $1/\mu_r$ and taking the inverse value of the function, so that

$$\mu_{\text{eff}} = \frac{2\mu_r}{(1+\mu_r) + (1-\mu_r)\cdot(1+10/p)^{-1/2}} \tag{7.10}$$

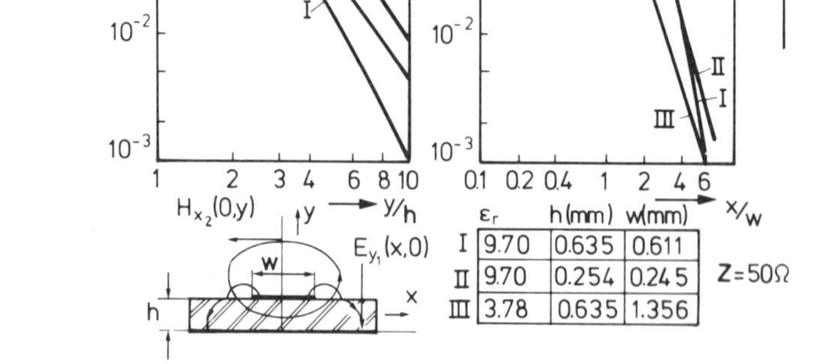

Figure 7.7. The normalized field extensions at 10 GHz for several microstrips. (a) magnetic field above the strip; $H_{x2}(x, y)$ is the component in the x-direction of the magnetic field outside the dielectric; (b) electric field beside the strip; $E_{y1}(x, y)$ is the component in the y-direction of the electric field inside the dielectric.

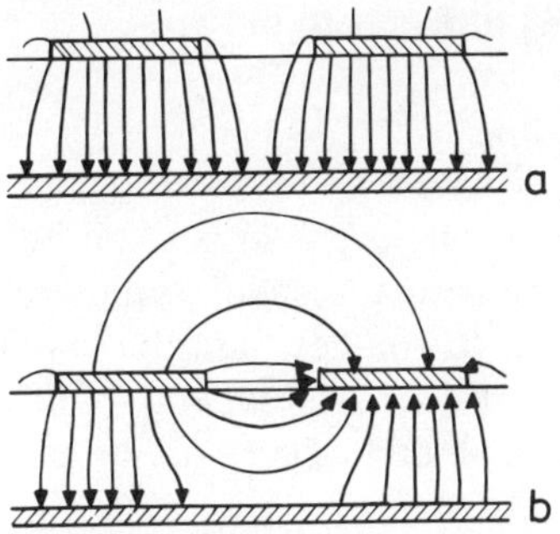

Figure 7.8. Pattern of the electric field of coupled microstrips: (a) even mode; (b) odd mode

7.2.2.1 Electromagnetic field of a microstrip

The field extensions above and beside the strip have been calculated by van Nie.[17] Figure 7.7 shows the field extensions which are important for packaging and integration of several microstrip circuits without coupling on the same substrate.

7.2.2.2 Coupled microstrips

Microstrip filters and couplers are made of adjacent strips with part of the electromagnetic field in common. Figure 7.8 shows the odd and the even mode of the field of two coupled microstrips. The coefficient of coupling strongly depends on the gap width, the substrate thickness and the strip width. More information about microstrip circuits can be found in the literature.[1]

7.2.3 Microwave losses in microstrip

The attenuation of microwaves in microstrip circuits is caused by losses in conductors and dielectric materials and further by parasitic radiation. A summary of the state-of-the-art information on losses in single and coupled microstrips with dielectric or magnetic substrate materials has been given by Denlinger.[18] In the following the relations derived by Schneider will be used.

Radiation losses can be substantial[19] but parasitic radiation due to objects in the environment of the strip and currents in the strip at the strip–air interface can be sufficiently reduced by taking the strip thickness to be at least five times the skin depth. The radiation due to the geometry can be reduced to an acceptable level by taking design precautions to avoid discontinuities where possible.

The dielectric losses α_d turn out to be:

$$\alpha_d = \frac{27.3}{\lambda} \cdot \tan\delta \ \mathrm{d}B/m \tag{7.11}$$

where λ is the wavelength. Dielectric losses can be reduced significantly by using materials of a high purity and small grain size. If $\tan\delta < 0.0001$, then the dielectric losses are much smaller than the conductive losses.

The conductive losses in microstrip depend mainly on the resistivity of the conductors and on their geometry. These losses can be obtained from equations derived by Schneider:

$$\alpha_c = \frac{R_c \cdot A\sqrt{\varepsilon_{\mathrm{eff}}}}{h} \ \mathrm{d}B/m \tag{7.12}$$

in which for $p \leq 1$:

$$A = \frac{10}{\pi \ln(10)} \cdot \frac{(8/p - 0.25p) \cdot (1 + [1 + q] \cdot 1/p)}{Z_0 \cdot \exp(Z_0/60)}$$

and for $p > 1$:

$$A = \frac{Z_0 \cdot (1 + 0.44/p^2 + [1 - 1/p]^5 \cdot 6/p^2) \cdot (1 + p + q)}{720 \cdot \pi^2 \cdot \ln(10)}$$

with

$$q = \frac{1}{\pi} \cdot \ln\left(\frac{4\pi w}{t}\right) \quad \text{for} \quad p \leqslant \frac{1}{2\pi}$$

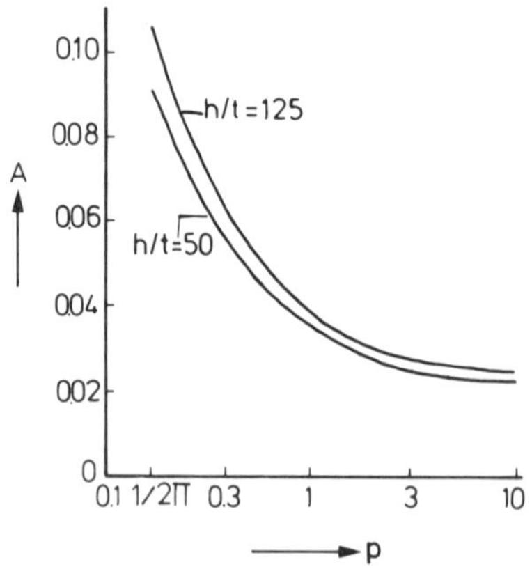

Figure 7.9. The normalized conductor attenuation A in dB Ω^{-1} as a function of $p = (w/h)$ for two values of h/t and $\varepsilon_r = 1$

and

$$q = \frac{1}{\pi} \cdot \ln\left(\frac{2h}{t}\right) \quad \text{for} \quad p > \frac{1}{2\pi}$$

where t is the thickness of the strip and R_c is the surface resistivity defined below. Figure 7.9 shows A as a function of p for two values of the ratio h/t. It turns out that the conductive losses decrease with increasing p and that these losses are slightly dependent on the ratio h/t.

7.2.3.1 Surface resistivity

For thick-film conductors, the relationship between the surface resistivity and the resistivity of the conductor material is difficult to determine in practice, because the conductor material is inhomogeneous and this inhomogeneity depends on the process used. To enhance the adhesion of the conductor to the substrate, the percentage of the glass at the conductor–substrate interface is higher than elsewhere.

For thin-film conductors the surface resistivity is:

$$R_c = \xi_1 \times \xi_2 \times R_{sc} \tag{7.13}$$

in which R_{sc} is the surface resistivity due to the skin effect alone, ξ_1 takes account of the increase of R_{sc} due to the surface roughness of the substrate and ξ_2 the increase due to the adhesion layer. The surface resistivity is:

$$R_{sc} = \rho_c/\delta_c \tag{7.14}$$

in which ρ_c is the resistivity of the conductor and δ_c is the skin depth in the conductor material:

$$\delta_c = \sqrt{\frac{\rho_c}{\pi \mu f}} \tag{7.15}$$

Table 7.1 shows the relation between the skin depth δ_c and the surface resistivity R_{sc} for electrolythic Au and Cu.

Table 7.1 Relations for skin depth δ_c and the surface resistivity R_{sc}

	$\rho_c (\mu\Omega\ \text{cm})$	$\delta_c\ (\mu\text{m})$ f in GHz	$R_{sc}\ (\Omega)$ f in GHz
Au	2.55	$\frac{2.54}{\sqrt{f}}$	$1.00\ 10^{-2}\sqrt{f}$
Cu	1.83	$\frac{2.15}{\sqrt{f}}$	$0.85\ 10^{-2}\sqrt{f}$

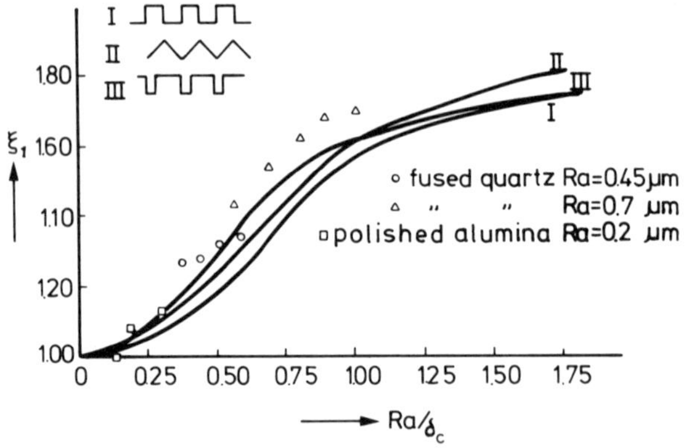

Figure 7.10. The relative increase of the surface resistivity ξ_1 with the surface roughness expressed by R_a/δ_c; the curves I, II, III correspond to different types of surface roughness, the indicated points have been measured using an open-ended straight microstrip resonator

Figure 7.10 shows the measured increase in attenuation due to the surface roughness R_a of the substrate compared with that of polished ones. Curves derived by Morgan[20] are also shown. The agreement between the measured values and the theoretical curves is surprisingly good considering the rather poor accuracy of roughness measurements and the simplifications of the theory. The conductive losses increase rapidly when the ratio R_a/δ_c exceeds 0.25.

The surface resistivities of bilayered structures have been discussed.[21] In Figure 7.11, the pertinent configuration is shown: the adhesion layer has a thickness d and a resistivity ρ_a. The surface resistivity of a bilayer conductor turns out to be:

$$(R_s)_{\text{bilayer}} = R_{sa} \cdot \text{Re}\left[(1 + \text{j}) \cdot \left(\frac{\sinh \tau d + \{R_{sc}/R_{sa}\}\cosh \tau d}{\cosh \tau d + \{R_{sc}/R_{sa}\}\sinh \tau d}\right)\right] \tag{7.16}$$

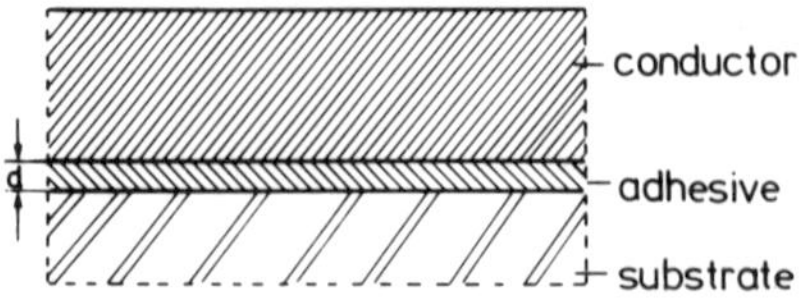

Figure 7.11. The multilayer configuration; d is the thickness of the adhesion layer

where

$$\tau = \frac{1 + j}{\delta_a}, R_{sa} = \rho_a/\delta_a$$

in which δ_a and δ_c are the skin depths for adhesion and conductor materials, respectively. The relative increase of the surface resistivity due to the adhesion layer is given by

$$\xi_2 = \frac{(R_s)_{bilayer}}{R_{sc}} \tag{7.17}$$

It turns out that the maximum values of ξ_2 occur when: $2.6 < R_{sa}/R_{sc} < 4$. At higher values of R_{sa}/R_{sc}, the current leaves the lossy layer, so that the total conductive losses decrease.

7.2.3.2 Example

As an example, the microwave attenuation at 10 GHz will be calculated for a 50 Ω microstrip consisting of 10 μm gold conductors on an alumina substrate with a thickness $h = 635\ \mu$m, a surface roughness $R_a = 0.2\ \mu$m, a relative dielectric constant $\varepsilon_r = 9.5$ and a loss tangent $\tan\delta = 0.0001$. Further, a NiCr adhesion layer has been applied with a thickness $d = 0.040\ \mu$m and a resistivity $\rho_a = 400\ \mu\Omega$cm, so that the DC surface resistivity is 100 Ω per square. From Figure 7.5, it follows that $p \sim 1$, and from Equation 7.7 it can be deduced that $\varepsilon_{eff} = 6.5$, so that the wavelength $\lambda = \lambda_0/\sqrt{6.5} = 1.2$ cm. The dielectric losses are according to Equation 7.11:

$$\alpha_d = \frac{27.3}{\lambda} \cdot \tan\delta = 0.23 \text{ dB m}^{-1} \tag{7.18}$$

Figure 7.9 reveals that $A \sim 0.036$ for $h/t = 63.5$ and $p = 1$. From Equation 7.14 and Equation 7.15 can be calculated that the conductor has a surface resistivity $R_{sc} = 0.032\ \Omega$ and a skin depth $\delta_c = 0.8\ \mu$m, so that $R_a/\delta_c = 0.25$. From Figure 7.10 it can be seen that $\xi_1 \sim 1.08$. From Equations 7.16 and 7.17, it follows that in this case $\xi_2 \sim 1$, and consequently the adhesion layer hardly contributes to the conductive losses.

From Equation 7.12 it follows that the conductive losses are given by

$$\alpha_c = \frac{\xi_1 \xi_2 A R_{sc} \sqrt{\varepsilon_{eff}}}{h} \sim 5 \text{ dB m}^{-1} \tag{7.19}$$

In this case, the conductive losses indeed exceed the dielectric losses.

7.2.3.3 *Experimental*

The microwave attenuation in microstrip can be experimentally determined by measuring the quality factor Q_0 of an unloaded microstrip resonator. The equation between the attenuation α and Q_0 is given by

$$\alpha = \frac{27.3}{Q_0 \cdot \lambda} \text{ dB m}^{-1} \tag{7.20}$$

Two methods for determining the quality factors will be discussed. One method is suitable for fast in-line determination of the quality factor Q_1 of a loaded resonator, whereas the other method, which is more elaborate, can be used for establishing the unloaded quality factor Q_0.[22]

In the former method, use is made of two pairs of $\lambda/2$ resonators: one pair at 8.20 and 8.38 GHz and the other pair at 12.0 and 12.18 GHz, as shown in Figure 7.12. The open-ring resonators of one pair have slightly different diameters, which results in a difference in their resonance frequencies. It can be shown that in the case of quadratic detection of the microwave signal, the loaded quality factor is

$$Q_1 = \frac{f_r}{2|f - f_r|} \cdot \sqrt{\frac{1 - F(f)}{F(f)}} \tag{7.21}$$

where f is the frequency, $F(f)$ the output signal of the detector and f_r the resonance frequency of the relevant resonator.

Both pairs of resonators have the same difference in resonance frequency, namely $\Delta f = 0.180$ GHz. The output signal at the intercept f_c of the resonance curves, belonging to a resonator pair, is governed by the Q_1 of the relevant resonator (see Figure 7.13). Superposition of the curves, belonging to the same pair, then results in a dip of -10 dB.

Reference cases, for which $F(f_c) = -13$ dB, have been chosen with

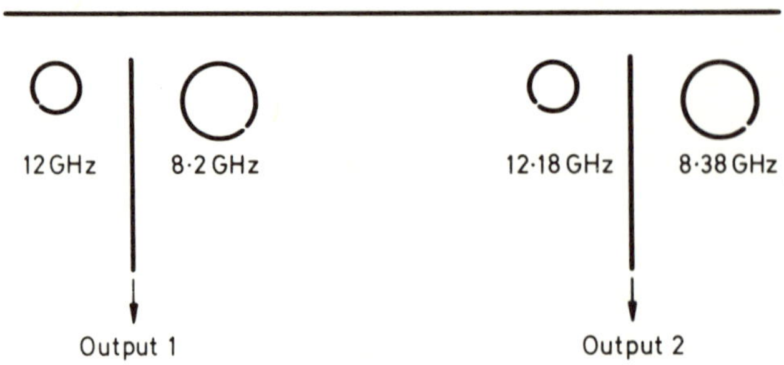

Figure 7.12. Two pairs of open-ring resonators used for a fast quality control during manufacturing

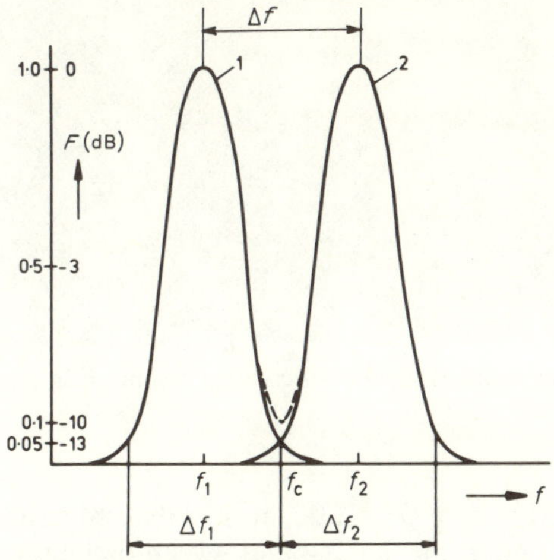

Figure 7.13. Frequency response F (dB) of one pair of open-ring resonators

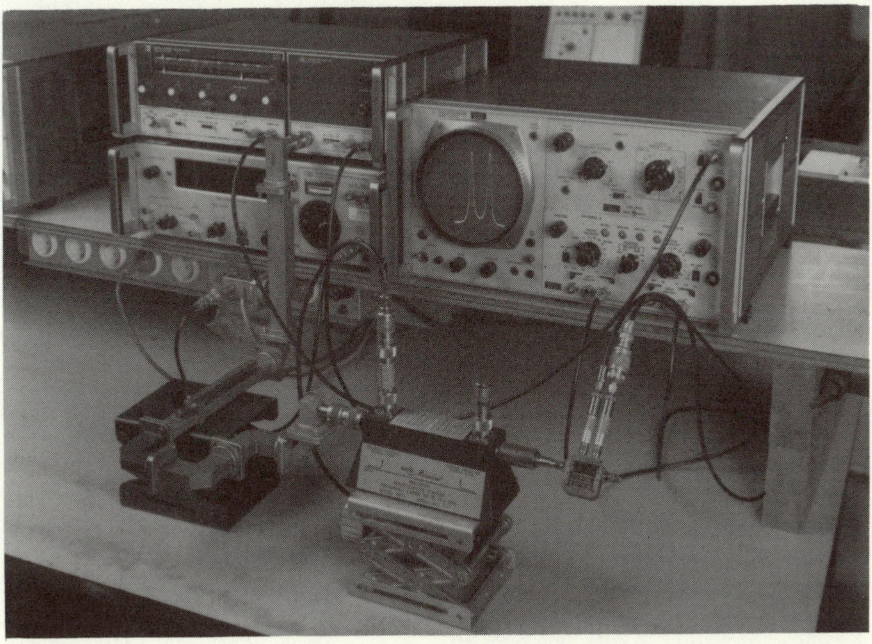

Figure 7.14. Equipment for the in-line determination of the loaded quality factors Q_1 of open-ring resonators

Figure 7.15. An open-ended straight microstrip resonator in a waveguide below cut-off

$Q_1 = 200$ at 8.2 GHz and $Q_1 = 300$ at 12 GHz. Measurements of actual cases are carried out, using an X-band sweep-oscillator. The outputs of the detectors are applied to a dual-trace oscilloscope (Figure 7.14). After equalizing the peak values of the output signals, that correspond to the resonance frequencies f_1 and f_2 of the pair, and adding the two signals, one can quickly check if the dip corresponds to Q_1s that lie close to the respective reference values.

In the second measuring method, the resonator to be measured consists of an open-ended straight microstrip, coupled to the measuring circuit at each end, as shown in Figure 7.15. The microstrip resonator is located in a waveguide which operates below its cut-off frequency in order to prevent radiation losses.[23] The loaded quality factor Q_1 is measured by determining the resonance frequency and the bandwidth of the resonance curve. The unloaded quality factor Q_0 can be derived from Q_1 and the insertion loss at the resonance frequency. The Q_0-factors of resonators can in this way be determined with an accuracy of 5 per cent in the frequency range from 2 to 12 GHz.

7.2.4 Deviations from simplified microstrip theory

7.2.4.1 Dispersion

The deviations from the TEM field, see Figure 7.16, increase with frequency. One effect of this increase is that the phase velocity decreases with rising frequency, causing the line to be dispersive. Dynamic analyses of infinitively thin strip configurations have already been made by Denlinger and Mittra.[24,25] From these analyses, the dispersion of a microstrip has been calculated.[17]

In Figure 7.17, the relative deviation of the wavelength with respect to the wavelength λ_{TEM}, associated with the pure TEM mode, is plotted as a function of frequency.

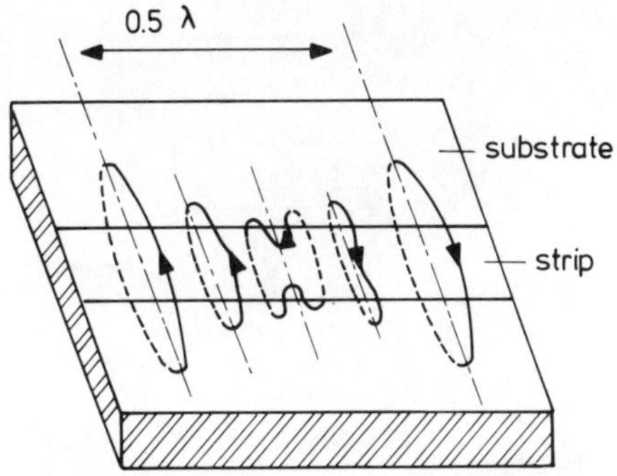

Figure 7.16. The magnetic field pattern of a quasi-TEM mode

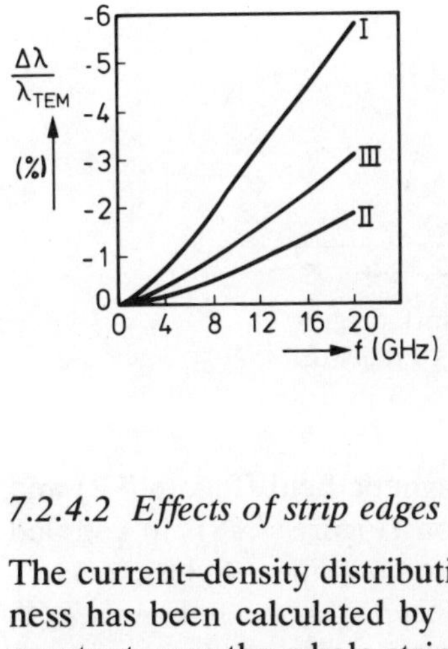

Figure 7.17. Dispersion of several microstrips versus frequency. λ_{TEM} is the wavelength calculated using the quasi-static analysis and $\Delta\lambda = \lambda - \lambda_{TEM}$; see Figure 7.7 for the indications I, II, III

7.2.4.2 Effects of strip edges

The current–density distribution at the surface of a strip with a certain thickness has been calculated by Wiesbeck.[26] This current distribution is nearly constant over the whole strip width except near the edges. The current distribution near the edges shows peak values, which depend on the strip thickness and the form of the strip edges. For a practical strip thickness, the peak values can be about 8 times larger than the value at the middle of the strip, see Figure 7.18. The narrower the strips, the larger the influences are on the values of Z, λ and the propagation losses. In the cases where these effects are important, Schneider's relations cannot be applied.

Figure 7.6 shows an example of a strip with mushrooming edges. It illustrates that in the case of narrow strips, both the thickness of the strip and the shape of its edges must be taken into account when designing for specified propagation parameters. The influence of thickness and shape can hardly be calculated, hence, when this influence is strong, a design that is based on Schneider's Equations 7.3 to 7.7 must be corrected experimentally.

On the other hand, the thickness and the mushroom shape of the edges will

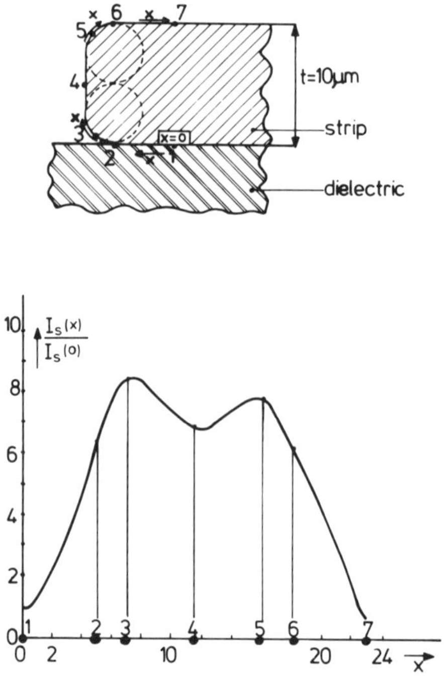

Figure 7.18. The normalized surface current density $I_s(x)$ near the edge of the strip

favourably affect the shielding of the electromagnetic field (Figure 7.2) and the spreading of the current-density distribution (Figure 7.18). In coupled microstrips, an even stronger effect of the mushroom shape is to be expected. An idealized geometry, with a finite strip thickness t, is shown in Figure 7.19, together with a more realistic geometry with mushroom shape.

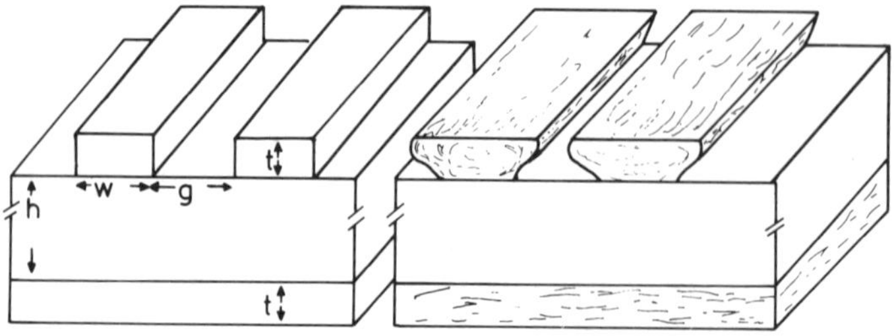

Figure 7.19. Coupled microstrips. To the left an idealized example, to the right a more realistic one

7.3 ACCURACY IN RELATION TO MATERIALS AND TECHNOLOGIES

Different techniques are used to make conductive patterns on dielectric materials. In industrial applications, the great majority of the MICs are at present made in thin-film or thick-film processes, while combinations of both technologies are now under investigation.

The terms 'thin-film' and 'thick-film' refer to the method of metal deposition and not to the thickness of the layers. In thin-film deposition, thin layers are deposited by evaporation or sputtering, followed by electroplating to the required thickness of the circuit pattern, while in thick-film deposition, the conductive and resistive materials are screen-printed through aperture masks and fired at a temperature around 850 °C.

Electroless processes are successfully applied in several laboratories but hardly used in industrial applications. In such processes, the substrate is first roughened by a combination of chemical and mechanical treatments for achieving a good adhesion of the conductor pattern to the substrate.[27] The increase of the conductive losses due to this process must be kept within acceptable limits. A seed layer, usually consisting of copper or nickel, is deposited in an electroless way and afterwards electroplated to the required thickness with either copper or gold.

In general, the manufacture of microstrip does not allow for any intermediate adjustment of circuit parameters. It, therefore, demands an accurately controlled process.

The pattern inaccuracies can be analysed in terms of position and detail inaccuracies and edge acuities. For instance, the detail inaccuracy determines the spread Δw in the strip width w of a microstrip. The edge acuity, denoted by r, is related to local deviations from the mean edge line, e.g. an edge acuity of 2 μm means that 90 per cent of the edge lies in a 2 μm wide band with the mean edge line in the middle. The mean edge line has in its turn a certain positional inaccuracy.

The spreads in substrate thickness Δh and dielectric constant ε_r depend on the material and available quality grade. These spreads Δh and ε_r are listed in Table 7.2 for available substrates.

7.3.1 Technologies

7.3.1.1 Thin-film technology

Thin-film patterns on substrates can be manufactured by using either an 'additive' or a 'subtractive' method. The additive method consists of the following stages:

(a) overall metallization of both surfaces of the substrate, implying the deposition of an adhesion layer and a thin initial gold layer;

Table 7.2 Properties of MIC substrate materials

Material			10 GHz 25 °C		Thickness tolerance	Surface roughness (μm)	Thermal conductivity ($W/cm^{-1}\,°C^{-1}$)	Thermal expansion ($°C^{-1}\,10^{-6}$)	Young's modulus ($N\,m^2$)	density ($g\,cm^3$)
			ε_r	Tan δ 10^{-4}						
Sintered Al_2O_3		96%	9.2	4	10%	0.3–1.3	0.25	6.0	3.1×10^7	3.7
		99.5%	9.7	1	10%	0.1–0.3	0.35	6.3	3.6×10^7	3.9
Ferrite (YIG)		I > 7 GHz	15.5	2	5 μm	polished	0.06	10	2×10^{11}	5.1
		II > 3.4 GHz	14.8	2	5 μm	polished	0.06	10	2×10^{11}	5.0
Sintered BeO		99.5%	6.7	3	10%	0.5–1.3	2.4	6.4	3.6×10^7	2.9
Fused quartz			3.8	1	5 μm	polished	0.014	0.55	7×10^{10}	2.2
Sapphire		// C-axis	11.5	0.3	5 μm	polished	0.42	9.3	4.7×10^{11}	4.0
		// A-axis	9.3	0.9	5 μm	polished	0.42	8.8	4.7×10^{11}	4.0
Filled PTFE	3M	K 6089	2.5	18	50 μm		0.001	9–10[a]	0.5×10^6	2.2
		Epsilam 10	10.3	18	50 μm		0.004	20–30[a]		3.0
	Duroid	5880	2.3	12	50 μm		0.003	20–70[a]		2.2
		6010	10.5		50 μm					2.9
Polyguide			2.3	5	60 μm			108	0.7×10^6	0.9

[a]The linear coefficients of thermal expansion in *X*-, *Y*-, and *Z*- directions greatly depend on temperature

(b) formation of a photoresist layer on one of the metallized surfaces;
(c) exposure of the photoresist to ultraviolet radiation through a mask and subsequent development to obtain a negative circuit pattern;
(d) growth of the conductor pattern and the layer on the rear side of the substrate by electroplating to a thickness of 10 μm;
(e) removal of the photoresist and subsequent etching away of the initial thin layers outside the thick layers.

In the subtractive method, stage (a) is followed by overall electroplating of both sides of the substrate to the final thickness. Subsequent photoresist steps lead to an imprint of a positive circuit pattern on one side and an overall covering of the other side of the substrate. Etching away the bare parts of the thick gold layer results in the desired conductor pattern.

The subtractive method is not so attractive as the additive method, because of a waste of gold and a less accurate definition of the pattern. On the other hand, gold layers with low internal stresses can be obtained by proper deposition of homogeneous layers so that the danger of peel-off will be small. Both methods for realizing conductor patterns are depicted in Figure 7.20. As can be seen, the forms of the cross-sections of the conductors are different. A growth over the edges of the photoresist pattern results from the additive method (right) and undercutting of this pattern occurs using the subtractive method (left). The deviations of a 10 μm thick conductor pattern from the mask dimensions are illustrated in Figure 7.34 for both methods.

For the first step (a), ultra high-vacuum evaporation or RF sputtering is the metallization technique currently in use. The choice of the technique depends in principle on the adhesion layer employed. An important factor governing conductor losses at high frequencies is the substrate finish. Highly polished surfaces are preferred, for reasons of low losses (see Section 7.2), but sufficient adhesion to such surfaces is a difficult problem. Much research

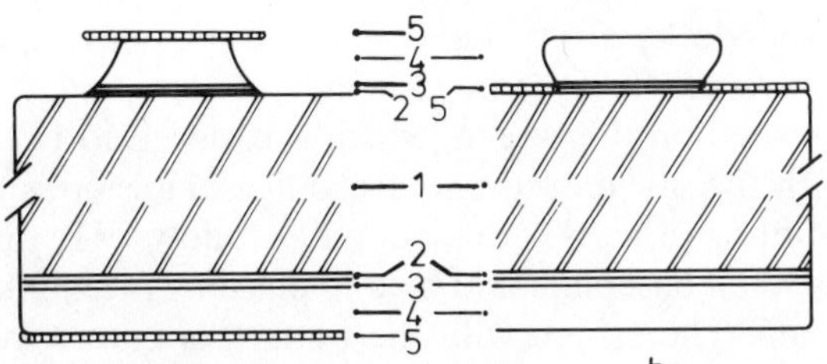

Figure 7.20. Layers configuration of the conductor patterns in subtractive process (a) and additive process (b): (1) substrate; (2) adhesion layer; (3) conductor layer; (4) electrodeposited layer; (5) photoresist layer

has therefore been carried out to find satisfactory adhesion layers that do not themselves introduce higher losses than would otherwise occur in layers of better conductivity, due to the rougher substrate surfaces that they need.

For a good adhesion to smooth surfaces, chemical bonds are required. Materials that are rather poor conductors but do have good adhesive properties like chromium, nickel chromium, tantalum and titanium are used. Aluminium is the only adhesive material with a low resistivity. In addition to the adhesive properties of the first metal layers, their purity is of importance. A very thin (20–50 nm) adhesion layer with poor conductive properties is subsequently coated by a well-conductive layer e.g. 200–300 nm gold. For purity, the sequence of both depositions is carried out in a non-interrupted vacuum.

Because the resistivity of the electroplated gold is strongly influenced by impurities, very pure gold bathes are used with very small additives to influence crystalline structure, adhesion and the current efficiency.

An adhesion layer of proper resistivity can also be used to integrate resistive elements into the conductive pattern. The resistors can be realized by masking corresponding parts of the adhesion layer during electroplating and etching. The resistors, obtained in this way, can accurately be laser-trimmed to the desired values.

7.3.1.2 Thick-film technology

In thick-film technology, the conductive patterns are obtained by a screen-printing method in which various types of ink are used. In general, the inks contain metallic and glass particles, oxides or metallic components, organic binder and solvents. These inks have a viscosity of about 25,000 PAs and are thixotropic. Using this technique, conductors, resistors and capacitors can be printed onto ceramic substrates. The adhesion between the printed patterns and the substrates can be achieved by means of a glass frit or a binder based on copper oxide.

The screen is made by stretching a fine stainless steel mesh over a cast aluminium-alloy frame, fixing it by glueing and clamping. A photosensitive emulsion is deposited on the screen, which is exposed to UV light via a film mask, after which the unexposed photosensitive emulsion is removed.

The screen-printing process is carried out as follows. The substrate is accurately positioned on a baseplate and held in place by mechanical means or by the use of vacuum. The screen with the pattern is placed in parallel to the substrate at a small distance (snap-off). This done, a small quantity of ink is spread on the screen. A squeegee is thereafter moved over the screen, pressing the screen against the substrate and forcing the ink through the openings of the pattern. The process can be carried out manually or automatically.

After each deposition process, the solvents in the ink are removed by heating the substrate in a furnace or infrared tunnel. The printed substrates are then fired in a furnace of which the temperature profile is adjustable within the limits 600 and 1000 °C. The profile should, of course, always be selected according to the recommendations for the resistor ink used. The total firing time is usually 45 min. It is necessary to stabilize the furnace temperature to within 2 °C over the whole temperature profile. The gas used in the furnace must be very clean and must, in particular, neither contain oil nor contaminants like freon and other detergents.

7.3.1.3 Plated-through holes

In MIC circuits, it is often required to have plated-through holes to minimize the length of the connections to the earth plane. Holes in ceramic substrates can be made before or after sintering. After sintering, holes can be made by ultrasonic drilling or by laser beam. A disadvantage of using a laser beam is that structural changes will result in the ceramic material around the hole. A too high laser energy will vitrify the wall of the hole, leading to microcracks and to poor adhesion for the layer that has to be deposited to this wall. The surface condition of the wall of the hole can be improved by a heat treatment at a temperature just below the melting point of the ceramic material.

In thin-film technology, the holes can be metallized by sputtering the seed layer onto both sides of the substrate, so that the layer thickness in the holes becomes rather uniform. When, however, the seed layer is to be vapour-deposited, the holes must be conical in shape, or the substrates must be rotated under an angle with the evaporation beam.

In thick-film technology, the holes are metallized in the following way (Figure 7.21). The substrate is placed on a plexiglass counterplate, having the holes corresponding to those of the substrate, but with slightly larger diameters. An under-pressure is applied to the counterplate so that the ink is sucked into the holes. However, the parameter setting is rather critical and difficult. One has to optimize the under-pressure, accounting for the viscosity of the ink and the number of holes in the substrate.

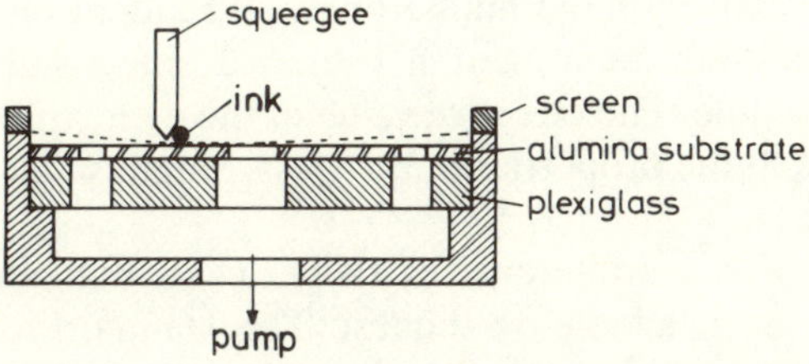

Figure 7.21. Method for metallizing holes in thick-film technology

7.3.2 Material properties

7.3.2.1 Substrates

Various materials can be used for the dielectric substrate. The choice depends on the required performance and frequency band, as well as on the kind of transmission to be applied. In Table 7.2 several relevant properties of dielectric materials are listed. The choice of the dielectric constant depends on the required frequency band. Because the dimensions of circuit parts are governed by the wavelength, reductions of wavelength and size by means of a large dielectric constant are preferable at lower frequencies. At higher frequencies, dielectrics with smaller dielectric constants are applied. In general, the dielectric losses must be low. It should be noted that the loss tangent increases with frequency.

The permitted tolerance in substrate thickness is determined by the required accuracy of the MIC performance (as will be discussed in Section 7.3.3). The surface roughness R_a, mentioned in Sections 7.2.3 and 7.3.1, is important in view of additional conductive losses and adhesion. The surface roughness can also influence the edge acuity of the resulting photoresist patterns, as will be dealt with in Section 7.3.4.

Apart from the spreads of the geometric dimensions of the substrate, Young's modulus and the coefficient of thermal expansion of the substrate material are important parameters for the assembly and for the design of the interconnections and transitions (see Section 7.3.5). The coefficients of thermal expansion are given in Table 7.2 for the temperature range from 25 °C to 200 °C.

Sintered alumina substrates are often used in microstrip circuits up to 18 GHz. They meet many specifications and are also relatively inexpensive. Such substrates are currently available in three grades, based on the surface quality. These grades concern as-fired, ground, and polished substrates, respectively. Single-sided grounding or polishing gives rise to camber, but is nevertheless usual. Therefore, the substrates of the higher grades of quality will normally exhibit more camber than the lowest grade with as-fired substrates. Alumina with 96 per cent purity is also available. The loss tangent of that material is relatively large and strongly dependent on the additives. In addition to the materials mentioned in Table 7.2, single-sided glazed alumina substrates are available. The effective value of the dielectric constant depends on the thickness and the properties of the glass. Single-sided glazed substrates also exhibit camber.

The standard thickness tolerance is 10 per cent, but substrates with narrower tolerances are available on request. This standard tolerance is 15 per cent for glazed alumina substrates. The standard length and width tolerances are ± 100 μm. By laser scribing, tolerances of ± 25 μm can be met with. The closer the tolerance the higher the prices are.

Various types of ferrite are used at microwave frequencies. Each type has a minimal operating frequency. Two types of ferrite are mentioned in Table 7.2. Ferrite substrates are made from bulk material by machining, two-sided lapping and polishing, so that the dimensional accuracy is higher than with alumina. Further, these substrates have no camber. They are especially suitable for integration of non-reciprocal circuit functions such as isolators and circulators.

The disadvantages of ferrite substrates are their somewhat brittle behaviour and their low thermal conductivity in combination with a rather large coefficient of thermal expansion, which results in a high sensitivity to thermal shocks.

Sintered beryllium oxide is used for circuits with a high heat dissipation. It is worthwhile noting that BeO dust is toxic.

Fused quartz substrates are used for higher frequencies, say beyond 18 GHz. These substrates can be made within small tolerances. A drawback is their brittleness and low thermal conductivity.

Sapphire substrates are available with polished surfaces. They have the advantage of a rather high thermal conductivity, but are anisotropic, which is complicating the EM field and the design process.

Copper clad PTFE reinforced with glass and Polyguide are used, for example, in finline configurations. The coefficient of thermal expansion of PTFE depends on the geometric directions and is strongly dependent on temperature. The relevant data, mentioned in Table 7.2, are given by the manufacturers. Conductive patterns can be obtained by selective etching of the copper layer.

7.3.2.2 Conductors

The conductors of a MIC must have a low resistivity in order to obtain low transmission losses and high performances of the distributed elements. A good bondability or solderability is important as well. Finally, the conductors must have sufficient adhesion to the substrates and they must be insensitive to environmental influences.

The conductor layer, that is electroplated on to the adhesion layer(s) of a thin-film circuit, is usually of gold with a purity of 99.99 per cent. Pure gold has an excellent bondability and a low resistivity: ρ_{Au} is between 2.50 and 2.55 $\mu\Omega$ cm for annealed layers. Electrolytic copper has even a lower resistivity, namely $\rho_{Cu} = 1.83$ $\mu\Omega$ cm, but copper layers must be protected against environmental influences.

As already mentioned in Section 7.3.1, the seed or adhesion layer can be used to integrate resistors. For these layers, NiCr or Ta (either Ta_2N or TaO_xN_y) are usually applied, resulting in a surface resistivity that is ranging from 50 to 200 Ω per square. For the resistors, the temperature coefficient (TCR) and the resistor stability during life are the crucial parameters.

Table 7.3 Surface resistivity of conductors used in thick-film MICs

Conductors	Weight ratio	Surface resistivity (mΩ/square)
Au	—	5–10
Au/Pd	1.25/1	40–100
Au/Pt	—	80–100
Ag	—	2–3
Ag/Pt	(20–40)	3–10
Ag/Pd	(2–3)	20–80
Cu	—	1.3–2
Ni	—	50

In thick-film MICs, the conductors are usually made of gold and sometimes of copper or silver. After processing, the thickness of the conductors normally lies in the range from 10 to 15 μm. Copper has the lowest resistivity, see Table 7.3, and seems very promising. However, precautions must be taken during processing, because copper is rather reactive. Pure silver has also a low resistivity, however additions to the silver are necessary for reduction of silver diffusion into the resistors and to obtain good solderability and bondability.

The resistor inks are defined, for a thickness of about 15 μm after processing, by their surface resistivity ρ per square. In general, ρ per square lies in the range from 10 Ω per square to 1 MΩ per square, so that resistors can be made in the range from 3 Ω to 10 MΩ. A resistor accuracy of $\frac{1}{2}$ per cent can be obtained by laser trimming. The temperature coefficient TCR of these resistors mostly lies in the range from − 100 p.p.m./°C to + 100 p.p.m./°C. The differences between the TCRs of resistors, made with the same ink, are within ± 20 p.p.m./°C.

At present, inks of the following kinds are available:

(a) RuO_2-based, after processing one obtains some $Pb_2Ru_2O_6$, because the ink contains Pb-based glasses;
(b) RuO_2-based + dopant, e.g. a dopant with niobium oxide for controlling the TCR;
(c) RuO_2 + thallium oxide;
(d) $Pb_2Ru_2O_6$ for large values of ρ per square.

For the insulators and low-ε_r dielectrics, inks with vitreous glasses are used. However, vitreous glasses melt during each firing cycle, so that ceramic powder is added in order to limit their flow during subsequent firing steps. Recrystallizable glasses, that remain solid during the subsequent firing steps, are also used.

7.3.3 Inaccuracies from processing

A variety of permutations of manufacturing steps are used to produce MICs by thin-film or thick-film techniques. For a discussion of all possible sequences with inherent accuracies, the reader is referred to relevant handbooks.[28,29] However, to promote attention for the limitations of these techniques, when applied to microstrip circuitry, and to make the reader familiar with accuracy problems of MICs, relevant aspects of some commonly used manufacturing sequences will be dealt with.

In thin-film technology, both for additive and subtractive methods (Section 7.3.1), the accuracy is mainly determined by:

(a) causes of mask inaccuracies;
(b) inaccuracies from photolithography;
(c) inaccuracies from electroplating and etching.

The position and detail inaccuracies and edge acuities of the mask(s) are added to the inaccuracy contributions from the photolithography. The final contributions are from electroplating and etching. How the contributions will accumulate to the total detail and position inaccuracies, will depend on whether the subsequent contributions are systematic or random.

In conventional thick-film technology, accuracy is largely determined by the screen manufacturing techniques and by the printing and firing procedures of the deposits.

7.3.3.1 Mask inaccuracies

Forming a circuit pattern involves imprint of mask images in photoresist layers. In practice, this is mostly done by contact printing using either film or glass masks.

The film mask is flexible and is preferred where the flatness of the substrates can be insufficient for glass masks, e.g. in cases of large substrates, rough surfaces or camber. On the other hand, film masks are sensitive to deformation due to changes of environmental conditions like temperature and humidity.

Film masks are usually made from a 10 times enlarged studnite model of the desired pattern. This model is imaged onto the film with the aid of an optical system. The position accuracy is ± 25 μm and the minimum obtainable line and gap width are 1 mm and 0.5 mm, respectively. These accuracy values and minimum widths translate into ten times smaller values for the film masks. The resulting mask accuracy also depends on the precision of the optical system.

Glass masks are more rigid and form-retaining. These masks are usually made with the aid of a pattern generator. In such a system, an optical or UV

Table 7.4 Accuracy figures for several glass masks

	High resolution emulsion (μm)	Metal film mask Cr (μm)	Metal film mask FeO (μm)
Position accuracy	0.5	0.5	0.5
Detail accuracy	2–2.5	0.5	0.5
Minimal obtainable strip/gap width	5	2.5	2.5

beam directly writes the pattern from input data specifying the geometric dimensions on the glass. Different types of glass masks are found, e.g. glass masks with a high-resolution emulsion (HR plates) or glass masks with a metal-film mask (Cr, FeO). The latter are made by using a 0.6 μm photoresist layer deposited on the metal-clad glass. After photolithography, the pattern is obtained by etching. Table 7.4 presents some accuracy figures for these masks. The glass masks with a metallic pattern have the best durability for reproduction techniques, based on contact exposure.

Film masks for manufacturing can also be made by contact printing from glass masks that have been made in a mask-making centre. A high accuracy can be maintained, if the copying of this mask is carried out in the same environment as the film mask is used.

Multiple masks for making an array of MICs on one substrate may have an additional inaccuracy to deviations from exact step and repeat actions during mask making. In that case, the accuracy of separating the multiple MICs by sawing or laser scribing is also important, because the performance of a MIC will also depend on the position of the pattern with respect to the edges of the substrate.

7.3.3.2 Inaccuracies from photolithography

The obtainable accuracies in the transformation of mask images into photoresist images depend on the photoresist properties, the surface quality (camber and surface roughness) of the metallized substrates and the performance of the UV exposure sources. If instead of a single mask, a set of masks are necessary to make a MIC, then the effects of alignment inaccuracies are added to the position and detail inaccuracies.

The deposition of photoresist layers is mostly carried out by spinning, dipping, spraying, or roller coating. Sometimes rims occur in the photoresist layer at the edges of the substrate. In MICs, the metallic pattern extends to the substrate edges, where the rims are situated. The UV exposure must be prolonged because of these rims. Possible overexposure of the other parts may cause dimensional changes in the pattern.

For photoresist patterns with a thickness of 2 to 16 μm, the obtainable accuracies and edge acuities have been reported in literature. Maes[30] preferred spraying or roller coating to obtain reproducible photoresist layers of uniform thickness. Rims at the edges could practically be avoided by using Shipley AZ 119 photoresist of appropriate viscosity, deposited by means of a roller coater.

Optical phenomena during UV exposure, like diffraction in the photoresist layer and reflection at the (metallized) surface of the substrate, can make the photoresist pattern deviate from the mask pattern.

The calculated diffraction of light around a straight edge in a homogeneous medium is represented in Figure 7.22 by the normalized curve of the relative intensity distribution I/I_0 as a function of the normalized parameter V, where $V = X\sqrt{2/D \cdot \lambda}$, X is a distance, λ the wavelength of the light in the photoresist and D the thickness of the photoresist layer. In Figure 7.23 the diffraction effects are demonstrated: (a) for an exposure time just sufficient to remove the photoresist from the centre of the gap and (b) for a proper exposure time.

The influence of the reflection is shown in Figure 7.24, where an optically polished, gold-covered ferrite substrate reflects the incident light. The resulting interference effects will cause strong intensity fluctuations, referred to as 'standing waves'. The stepped shape of the photoresist edges is caused by these standing waves. This effect can be reduced by postbaking of the photoresist or by using substrates with a sufficiently rougher surface.

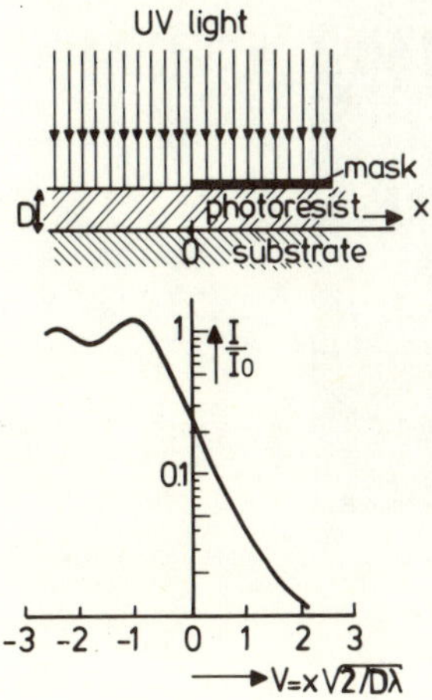

Figure 7.22. Normalized intensity distribution due to diffraction around a straight edge

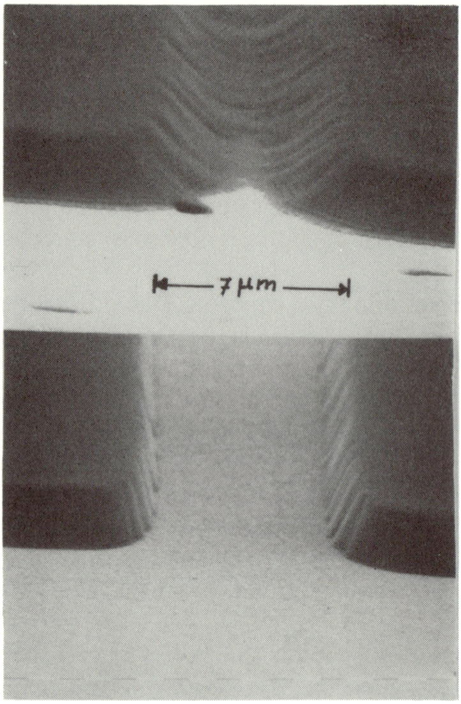

Figure 7.23. Diffraction effects demonstrated in developed patterns of Shipley AZ 1350 photoresist with a thickness of 2.5 μm: (a) under exposure time $t_{UV} = 6$ sec; (b) under exposure time $t_{UV} = 24$ sec

Figure 7.24. An edge of an AZ 1350 H pattern deposited on an optically polished gold-covered surface; the influence of standing waves in the UV light beam is clearly observable

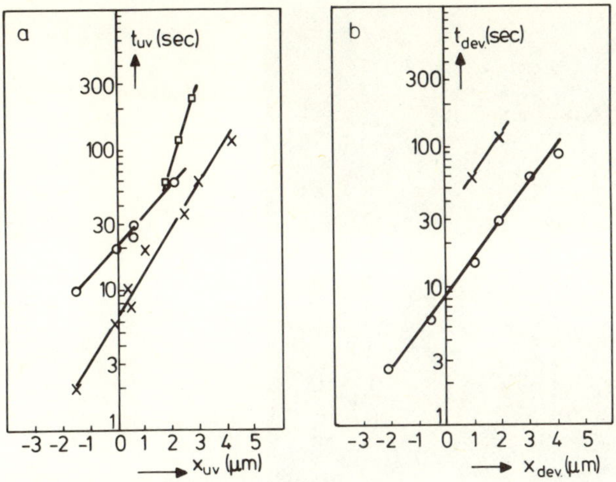

Figure 7.25. (a) Edge displacement X_{UV} versus exposure time;
□ 2 µm AZ 119, optical polished surface;
× 4 µm AZ 340 } surface roughness
○ 8 µm AZ 340 } $R_a \sim 0.2 - 0.4$ µm.
(b) Edge displacement X_{dev} as a function of the development time

Edge displacements are possible, depending on the times taken for exposure and development of the photoresist. Apart from these times, the surface roughness, denoted by R_a, can strongly influence the spreads of the manufactured patterns. Figure 7.25(a) and (b) show the edge displacement as a function of the times for exposure and development, respectively. These graphs correspond to different thicknesses of the photoresist layers and to a given surface finish of the substrates. They are used for the correction of the mask dimensions in order to compensate for effects of edge displacements.

When MIC patterns are made by means of the additive method, sufficiently thick photoresist layers with steep edges are important to reduce mushrooming in electroplating. Prebake and postbake procedures influence the slope of the edge. For Shipley AZ 1350H photoresist, Figure 7.26 shows an example of deformations due to postbaking at 110 °C. The slope of the edges after postbaking is 40°. Figure 7.27 shows the edge of a Shipley AZ 119 photoresist pattern, which does not need to be postbaked. This photoresist was applied by a roller coater and prebaked at 80 °C. The edge slope is 75°.

7.3.3.3 Inaccuracies from plating and/or etching

The selective electroplating of the additive thin-film process has to result in a pattern with a final thickness in the range of 5 to 10 µm. The final process

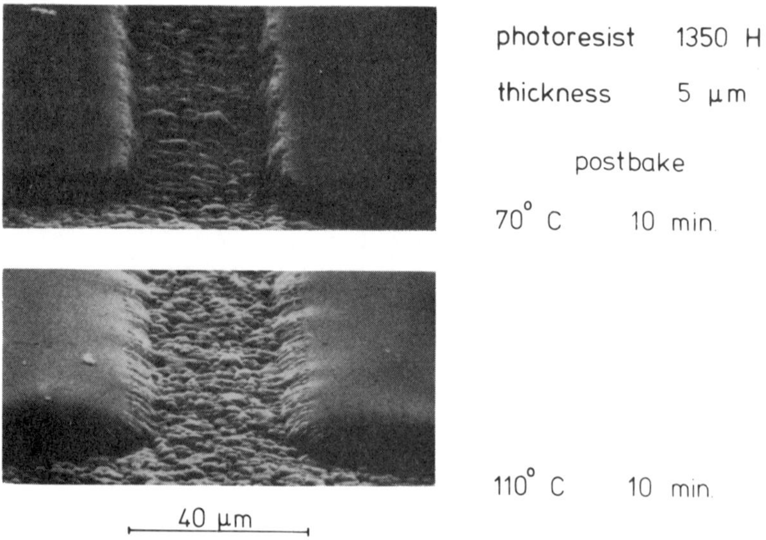

Figure 7.26. Example of a postbake deformation of a 5 μm photoresist Shipley AZ 1350 H layer:
(a) before postbaking;
(b) after postbaking treatment at 110 °C for 10 min

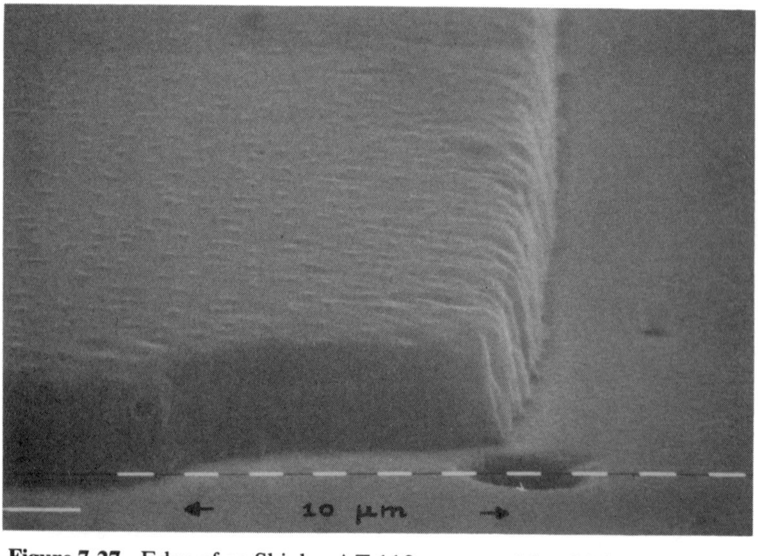

Figure 7.27. Edge of an Shipley AZ 119 pattern with a thickness of 4 μm

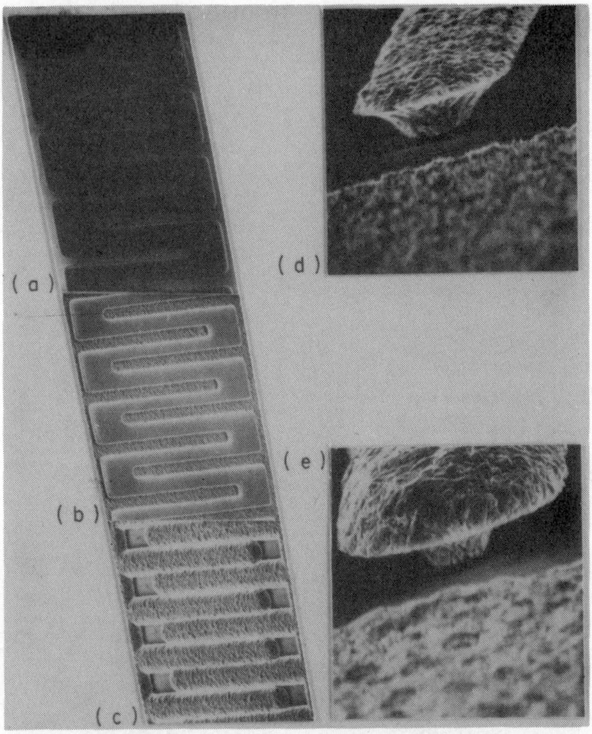

Figure 7.28. Patterns made using a part of the test pattern (see Figure 7.36): (a) resist pattern (620×); the photoresist has been applied by a spinner; (b) 5 μm electrodeposited patterns (620×); (c) 12 μm electrodeposited patterns (620×); (d) as (b) side view (6200×); (e) as (c) side view (6200×)

steps are removal of the photoresist and etching away of adhesion and seed layers outside the electroplated pattern.

Figure 7.28 and Figure 7.29 show electroplated test patterns obtained by using different photoresist layers.[31] The finger structures to the left are used for determining the thickness of the gold patterns. The fingers are 5 μm wide and the gaps between these fingers are alternately 13 μm and 17 μm. The photoresist patterns are applied by spinning using Shipley AZ 1350H (Figure 7.28) and by roller coating using AZ 119 (Figure 7.29). Magnification of the finger tips reveals strong mushrooming in the case of spinning. The mushroom shape is influenced by the thickness of the photoresist pattern. If the gold growth is not impeded by the edges, the growth rates are nearly equal in horizontal and vertical directions. Consequently, the widening of the strips,

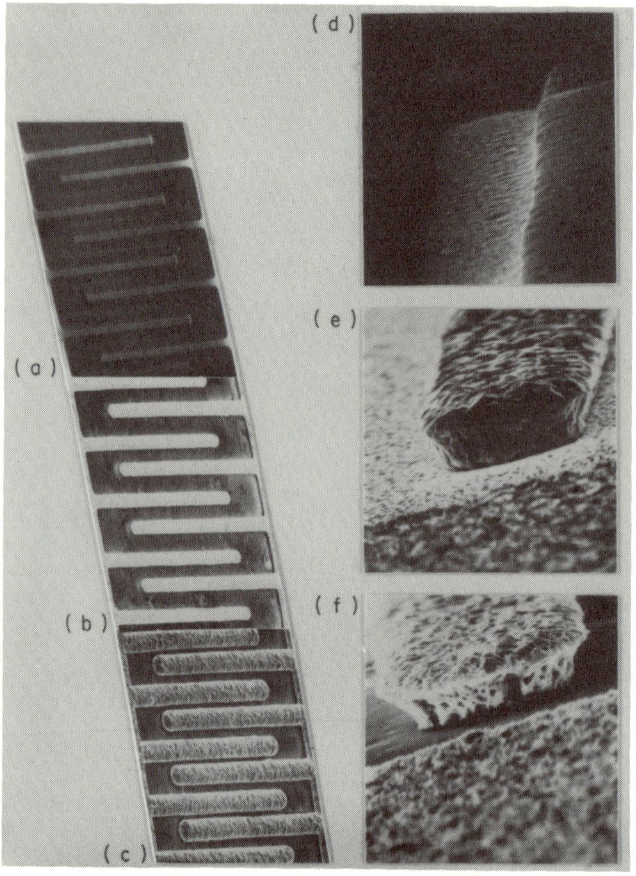

Figure 7.29. Patterns made using a part of the testpattern (see Figure 7.36): (a) resist pattern (620×); the photoresist has been applied by a roller coater; (b) 5 μm electrodeposited patterns (620×); (c) 10 μm electrodeposited patterns (620×); (d) as (a); detail of the photoresist pattern (6200×); (e) 5 μm Au pattern in which the evaporated Au layer is present; (f) as (e), but evaporated layer has been etched away

i.e. the narrowing of the gaps, is twice the difference between the thicknesses of the electroplated and the photoresist layers.

As has been mentioned in Section 7.2.4, a certain mushroom shape has favourable effects on the microwave performance (better shielding and current-density distribution), but results in pattern dimensions, that are different from the corresponding mask dimensions. The narrower the strips or gaps, the larger the resulting shifts in electrical parameters will be. Strong mushrooming is poorly controllable and will result in inaccurate shifts. The shape of the edges may also give rise to mechanical peel-off stresses in narrow strips and give rise to failures (as dealt with in Section 7.4). Use of photoresist layers, several μm thick, results in slightly mushrooming strips which are preferable for transmission of microwaves (Figure 7.29).

For the same substrate material, higher microwave frequencies will require smaller MIC dimensions and higher accuracies will be needed. On the other hand, however, the thickness of patterns may then be reduced thanks to the smaller skin depth. This thickness reduction is in turn favourable for obtaining a higher accuracy and less mushrooming.

Anisotropic plating techniques can also be used for diminishing the mush-

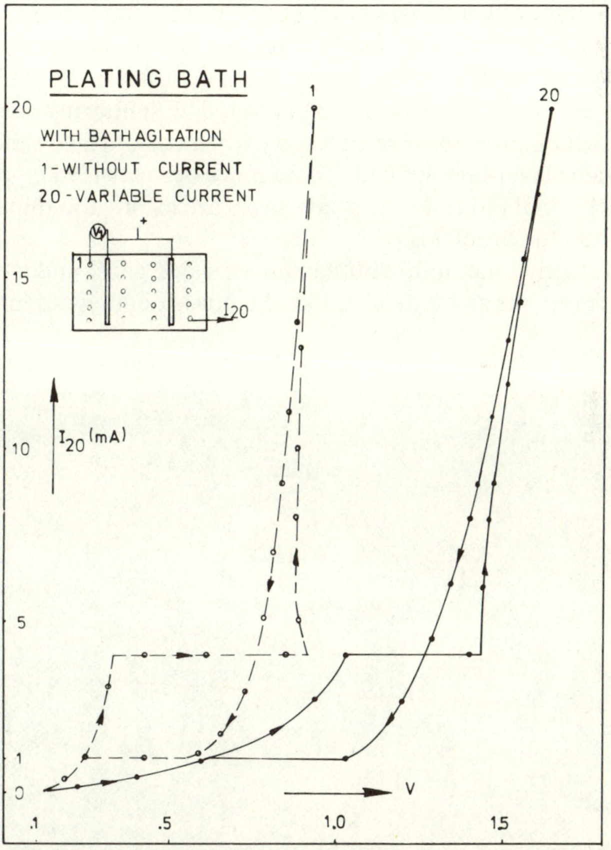

Figure 7.30. (a) Variation of the cathode potential during Au-plating of a substrate in position 20; (b) local variations of the bath potential recorded in position 1 during plating of a substrate in position 20

room shape. The latter will be influenced by adding chemicals to the plating bath or by periodically reversing the plating potentials. Pulse plating by means of an 'on–off' direct current is entering the field,[32] because it offers electro-deposits of superior properties, e.g. increased density and decreased resistivity. In the current-on period the cathode potential stays lower than with other methods, consequently the impurity content will also be lower, whereas during the current-off period hydrogen and other absorbed impurities will be released again. The cathode-diffusion potential that is building up during DC plating is shown in Figure 7.30. The MIC substrate in position 20 is supplied with a varying current. The substrate in position 1, being without current supply, is used for recording of the local bath potential during the plating. The potentials of both substrates are recorded as functions of the plating current. The hysteresis effect is due to bath polarization. In pulse plating, the sudden increase to the steady-state cathode-diffusion potential can be avoided if the 'on' and 'off' periods are properly chosen.

In the additive method, the edge profile may also be affected by etching away the initial layers, that have been deposited by sputtering or evaporation. Successive etching processes are necessary when complicated sandwich structures of metallic layers are applied. These processes may lead to undercutting. However, the initial (adhesion or seed) layers are mostly too thin (200 nm) to give rise to serious problems.

In the subtractive method, with narrow strips or gaps, under-etching may lead to inaccurate parameter shifts, like the mushrooming did in the additive

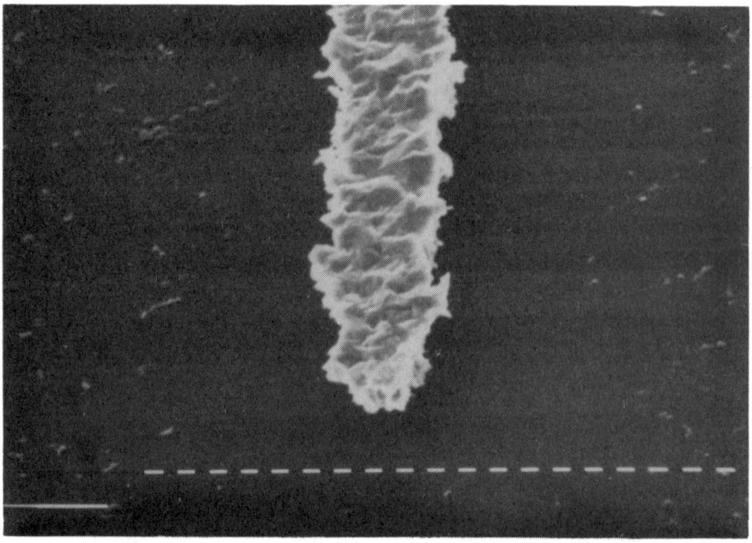

Figure 7.31. Edge acuity demonstrated on narrow strip made by subtractive method

method. Both mushrooming and under-etching have unfavourable effects on the acuity. Figure 7.31 shows a part of a strip, obtained by etching a 10 μm gold layer. The width of the strip was designed to be 10 μm, but the applied subtractive method led to heavy under-etching. Consequently, the detail accuracy (2 μm) and the edge acuity (2 μm) are both moderate.

Alternatives to the chemical way of etching are sputter etching and plasma etching, which have anisotropic effect. Both are carried out by means of accelerated ions. In plasma etching, surface atoms are removed by a chemical reaction in vacuum. This technique[33] is still in its infancy but is quite promising. An example of a pattern realized by sputter etching is shown in Figure 7.32. The mushrooming edges of the gold pattern act as an umbrella, that

Figure 7.32. Initial layers of a 10 μm electrodeposited pattern removed by: (a) chemical etching, (b) sputter etching

prevents etching away the (lossy) layer underneath along the edges. Sputter etching has been compared to chemical etching by making $\frac{1}{2}\lambda$ ring resonators in both ways (see Figure 7.31). The decrease of the effective width of the gap, due to the lossy layer that remained after the sputter etching, resulted in a 130 MHz shift of the resonance frequency.

The anisotropic effect of sputter etching, allows for finer line geometries. Chemically inert material can also be removed. But, sputter etching and plasma etching will also erode masking material, which may cause contamination problems.

7.3.3.4 Thick-film versus thin-film

Thick-film circuits are more and more used for higher frequencies up to 10 GHz. However, these circuits must be made by means of a special technology, because current factory processes will not result in sufficient accuracies of edges and line widths. Figure 7.33 shows a drawing of a part of the test mask composed of strips and gaps of different widths. This pattern has been deposited by means of a current factory process. The first and the tenth sample of the batch have been photographed. The strip width increased and the gap width decreased due to the wetting of the screen on the substrate side during running-in. Figure 7.34 shows the results as compared to thin-film patterns, realized by means of additive and subtractive techniques, respectively.

Better line definitions can be obtained in thick-film technology by first depositing and firing an overall layer and, thereafter, applying photolithogra-

Figure 7.33. Detail of test mask composed of strips, gaps and squares of varying dimensions (measures in μm)

Figure 7.34. Patterns made in different technologies using the detail shown in Figure 7.33: (a) thin-film in additive technique; (b) thin-film in subtractive technique; (c) thick-film first sample of a batch; (d) thick-film tenth sample of a batch; (e) mask

phy to obtain the desired pattern like with the subtractive thin-film method. However, inhomogeneities (interdispersed glass) and spread in thickness of the thick film layer give rise to dimensional variations of the etching results. Nevertheless, the obtainable accuracy is only slightly inferior to the accuracy that has been obtained using the subtractive thin-film method.

The subtractive method becomes more and more attractive the thinner the gold patterns may be. Instead of chemical etching, sputter, or plasma etching can again be used for improved definition of lines and edges.

Q-factors of thick-film and thin-film resonators for the X- and Ku-band have been discussed in literature.[34]

7.3.4 Inaccuracies and their electrical effects

The parameter spreads which can result for MICs, depend on the planar transmission lines used, on the parameter spreads of the applied materials, on the geometry and on the obtainable dimensional accuracies. These accuracies, in turn, depend on applied materials and technologies.

The spreads in substrate thickness Δh and dielectric constant ε_r were said to depend on the material and its quality grade. The spreads Δh are listed in Table 7.2 for available substrates.

As already mentioned, the final geometrical inaccuracies of thin-film pattern are accumulated from inaccuracy contributions from the masks, the photolithography and the electroplating and etching. Frequently, position inaccuracies are mainly determined by the masks and the alignment, while the detail inaccuracies and edge acuities are mainly due to the photolithography.

As has been mentioned in Section 7.2.3, the thickness of the pattern, and the shape and acuity of the edges, are of significant influence on the electrical MIC parameters if the strip width is narrow enough. In particular, for microstrip filters and couplers, an accurate definition of the narrow gap between adjacent strips is often required for accuracy in MIC performance.

But, for a single microstrip that is wide enough, a simplification to a planar pattern geometry will allow for acceptable approximations of propagation parameters by means of Schneider's Equations 7.3 to 7.7. Hence it should also be possible to derive acceptable approximations of the spreads of these parameters.

7.3.4.1 Electrical effects of inaccuracies in single microstrips

The effects of the spreads in w, h, and ε_r on the characteristic impedance Z and the effective dielectric constant ε_{eff} of microstrips have been investigated.[35,36] A set of sensitivity coefficients have been derived, however, the effects of the strip thickness and the edge form were left out of account. The following equation resulted for the deviations in Z and ε_{eff}.

$$\frac{\Delta Z}{Z} = S_{Zw} \cdot \frac{\Delta w}{w} + S_{Zh} \cdot \frac{\Delta h}{h} + S_{Z\varepsilon} \cdot \frac{\Delta \varepsilon_r}{\varepsilon_r} \tag{7.22}$$

$$\frac{\Delta \varepsilon_{eff}}{\varepsilon_{eff}} = S_{\varepsilon w} \cdot \frac{\Delta w}{w} + S_{\varepsilon h} \cdot \frac{\Delta h}{h} + S_{\varepsilon\varepsilon} \cdot \frac{\Delta \varepsilon_r}{\varepsilon_r} \tag{7.23}$$

in which the sensitivity coefficients:

$$S_{Zw} = \frac{p}{Z} \cdot \left(\frac{\partial Z}{\partial p}\right)_{\varepsilon_r}; \quad S_{\varepsilon w} = \frac{p}{\varepsilon_{eff}} \cdot \left(\frac{\partial \varepsilon_{eff}}{\partial p}\right)_{\varepsilon_r}$$

$$S_{Zh} = -\frac{p}{Z} \cdot \left(\frac{\partial Z}{\partial p}\right)_{\varepsilon_r}; \quad S_{\varepsilon h} = -\frac{p}{\varepsilon_{eff}} \cdot \left(\frac{\partial \varepsilon_{eff}}{\partial p}\right)_{\varepsilon_r}$$

$$S_{Z\varepsilon} = \frac{\varepsilon_r}{Z} \cdot \left(\frac{\partial Z}{\partial \varepsilon_r}\right)_p; \quad S_{\varepsilon\varepsilon} = \frac{\varepsilon_r}{\varepsilon_{eff}} \cdot \left(\frac{\partial \varepsilon_{eff}}{\partial \varepsilon_r}\right)_p$$

In Figures 7.35(a), (b), (c), (d), these sensitivity coefficients, derived from Schneider's equations, are plotted as functions of $p(= w/h)$ for several values of ε_r. The coefficients S_{Zw} and S_{Zh} appear to be nearly independent of ε_r. S_{Zw} decreases with decreasing p, but for a constant edge displacement Δw, and a constant thickness h, the ratio $\Delta w/w$ increases with decreasing w, so that the overall effect of the edge displacement increases.

For $p \to \infty$, $-S_{Zw} = S_{Zh} \to 1$, because in that case the impedance

$$Z \to \frac{120\pi}{p \cdot \sqrt{\varepsilon_r}} \tag{7.24}$$

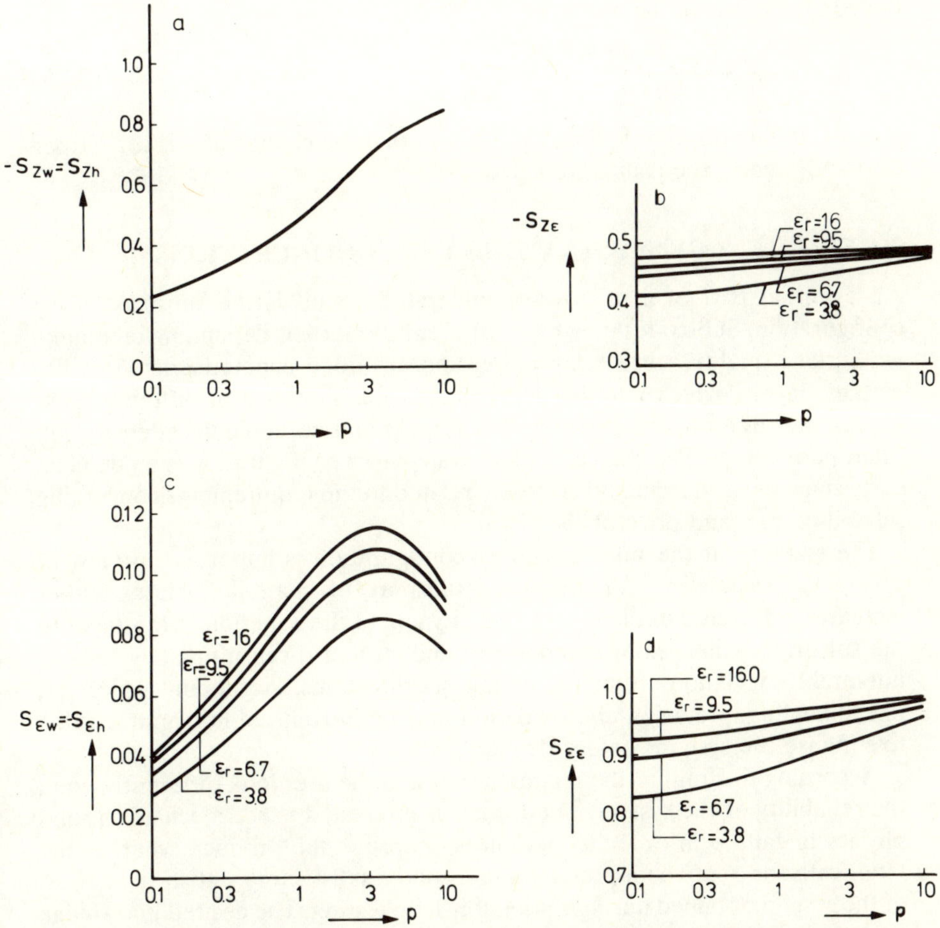

Figure 7.35. Sensitivity coefficients used in tolerance formulae plotted as function of $p = w/h$ for different values of ε_r

Further:

$$-S_{Z\varepsilon} \rightarrow 05 \quad \text{for } p \rightarrow \infty.$$

It appears that $|S_{\varepsilon w}| \ll |S_{Zw}|$, in other words: the effect of the width spread is much smaller on ε_{eff} than on Z. The coefficient $S_{\varepsilon w}$ shows a maximum at $p \sim 3$. Finally, if $p \rightarrow \infty$, then $S_{\varepsilon w} \rightarrow 0$ and $S_{\varepsilon\varepsilon} \rightarrow 1$.

For illustration, one can consider a microstrip on an alumina substrate with a thickness $h = 635$ µm and a dielectric constant $\varepsilon_r = 9.5$. If $p = 1$ and $Z \sim 50\ \Omega$, then $-S_{Zw} = S_{Zh} = 0.48$, so that a spread $\Delta h/h = 5$ per cent results in $\Delta Z/Z = 2.4$ per cent.

The ε_{eff} is changed by glazing the surfaces of the substrate. If the glass has a dielectric constant $\varepsilon_{rg} = 5$ and a thickness h_g, then the dielectric constant ε_{rt} of the combination will be:

$$\varepsilon_{rt} = \frac{\varepsilon_r \cdot \varepsilon_{rg} \cdot (h_g + h)}{\varepsilon_r \cdot h_g + \varepsilon_{rg} \cdot h} \tag{7.25}$$

If $h_g = 10$ µm and $\varepsilon_{rg} = 5$, then $\varepsilon_{rt} = 9.4$, so that the change of ε_r is *ca.* -1 per cent and, hence, the change of ε_{eff} *ca.* -0.95 per cent.

7.4 STABILITY OF SUBSTRATE–LAYER CONFIGURATIONS

An essential part of a microwave integrated circuit is the substrate–layer configuration. Substrates may have different properties, depending on material, surface quality, surface structure, and absorbed contaminants. Suitable metallic layers, which may be deposited on the substrate in different techniques, may differ in composition, structure, grain size, and defect density and other parameters. The choice of a substrate–layer configuration is made in an early stage of the design and is closely related to the requirements concerning reliability, life, and price of the circuits.

The stability of the substrate–layer configuration is important in view of reliability. Instability, for instance transport of material, causes either increase or decrease of the layer resistivity and of the adhesion of this layer to the substrate. Changes in composition and structure can ultimately lead to intolerable deviations from the circuit specifications. Additional processes, like annealing, pre-aging and oxidation may be introduced in manufacturing to stabilize the systems before delivery.

According to Braun,[37] the manufacturer and the user look differently upon the reliability of the system. The former is interested in the mechanisms and physics of failure in order to be able to improve the product, whereas the latter calls for a sufficient product quality and a satisfactory statistical analysis of the system obtained during reasonably long testing. The desired knowledge for both cases can be obtained by life testing of samples with special patterns. Such patterns should be carefully chosen to test the potential failure sources that are inherent of the materials used and the technology applied.

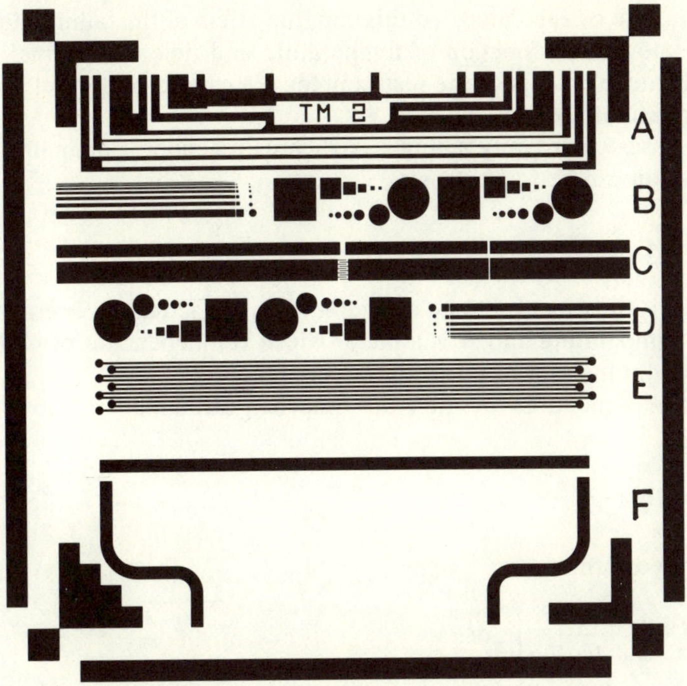

Figure 7.36. Test pattern used for MIC testing.

An example of a test pattern for MIC testing[22] is shown in Figure 7.36. It can be used to test geometrical accuracy and electrical behaviour, and to investigate effects of electromigration, electrocorrosion, diffusion, and adhesion losses that are related to the physics of failure.

The reproducibility of the manufacturing can be checked by means of the mentioned pattern, if realized in test samples which have been processed together with batches of MIC substrates in production. In doing so, the behaviour of the test sample is supposed to remain representative for the properties of the other circuits throughout processing. Destructive (adhesion) tests can be carried out on the relevant parts of the test samples, while the (microwave) properties can be checked using other parts. The (life) testing results of the test-samples can be subjected to statistical analyses and used for investigation of failure mechanisms.

7.4.1 Material transport

The aging behaviour of the substrate–layer combinations is usually investigated using tests in which the failure mechanisms are accelerated by thermal activation. For predicting life under operational conditions, the measured results must be extrapolated. To this end, the effect of the failure mechanism must be known as a function of temperature and time. Isothermal and isochronal heat treatments of the test samples are carried out to determine the wanted relationships.

In the cases where only a single Arrhenius reaction[38] dominates, the life extrapolation can be found from:

$$\frac{\mathrm{d}p}{\mathrm{d}t} = A \cdot \mathrm{e}^{-E/kT} \tag{7.26}$$

in which p is the property to be measured, E is the activation energy, T is the absolute temperature and A is a factor which is independent of p. E can be deduced from the plot of $\ln(\mathrm{d}p/\mathrm{d}t)$ versus $1/T$.

Should A depend on p, then the Meechan–Brinkman relation[39] can be used:

$$\frac{\mathrm{d}p}{\mathrm{d}t} = f(p) \cdot \mathrm{e}^{-E/kT} \tag{7.27}$$

In integrated form:

$$\theta(p) = \int_{p_0}^{p} \frac{\mathrm{d}p}{f(p)} = \int_0^t \mathrm{d}t \cdot \mathrm{e}^{-E/kT} \tag{7.28}$$

or

$$\theta(p) = t \cdot \mathrm{e}^{-E/kT}. \tag{7.29}$$

The activation energy of a Meechan–Brinkman process can be derived from an isothermal and an isochronal characteristic. The isothermal characteristic is obtained from a series heat treatments carried out at a constant temperature T_a for different storage times t, whereas the isochronal characteristic is obtained from a series treatments at different temperatures T for a constant storage time t_a, see Table 7.5. The property p is always measured at the same (room) temperature before and after each heat treatment and the change in p is determined. The isothermal and isochronal characteristics are associated with the functions

$$\theta_t(p) = t \cdot \mathrm{e}^{-E/kT_a} \tag{7.30}$$

Table 7.5 Process parameters in the heat treatments

	Time	Temperature
Isothermal	t (variable)	T_a (constant)
Isochronal	t_a (constant)	T (variable)

$$\theta_T(p) = t_a \cdot e^{-E/kT} \tag{7.31}$$

respectively. The Meechan–Brinkman time–temperature characteristic can be found by taking $\theta_t(p) = \theta_T(p)$, see Figure 7.37, so that:

$$\ln t = \frac{E}{k} \cdot \left(\frac{1}{T_a} - \frac{1}{T}\right) + \ln t_a \tag{7.32}$$

This characteristic shows the trade-off between the storage time t of an isothermal treatment and the temperature T of an isochronal treatment for equal changes in $\theta(p)$. This characteristic will be a straight line if $1/T$ is plotted against $\ln t$. The thermal activation energy E can then be derived from the directional coefficient of the line. In the cases where several processes take place simultaneously, the characteristic will deviate from a straight line. If the activation energies differ sufficiently, they can be found by decomposing the curve into straight lines.

For these measurements, several samples are needed, which should have equal activation energies, identical functions of p and the same starting values

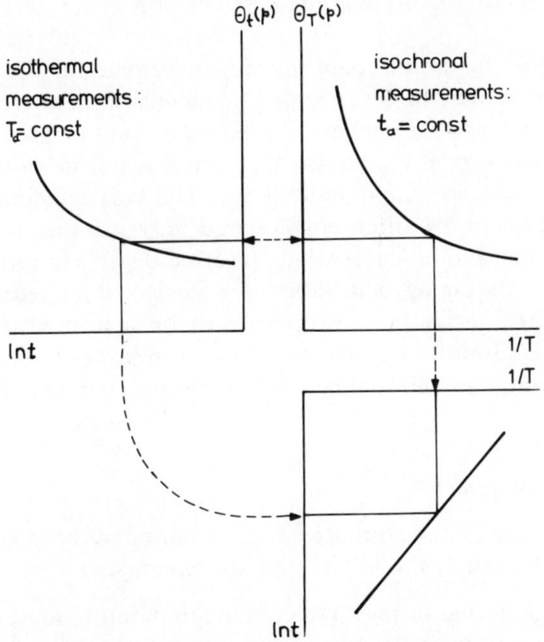

Figure 7.37. Meechan–Brinkman time–temperature characteristic derived from isothermal and isochronal measurements

p_0 of p. One of these samples is needed for the series of the isothermal treatments, whereas each isochronal treatment needs its own sample. However, if treatments are carried out with increasing temperature and the temperature increments are large enough to make the previous changes negligible, then the same sample can be used for all isochronal treatments.

In the Meechan–Brinkman analysis, the functions $\theta_t(p)$ and $\theta_T(p)$ are used. Because of Equation 7.28, heat treatments resulting in the same value of p will also result in the same $\theta(p)$. Hence, $\theta_t(p)$ and $\theta_T(p)$ in Figure 7.37 can be replaced by the values of p obtained during the isothermal and isochronal treatments, respectively.

In practice, the samples are not equal, but their properties exhibit a certain spread, as opposed to what has been assumed above. In order to reduce this the effect of this spread, a set of samples is taken instead of a single sample. Sufficiently large sets of samples must be taken in order to obtain an acceptable confidence interval for the value of the activation energy.

The life of a device can be defined as the time at which a tolerance limit of a relevant property is reached. If the device is liable to only one failure mechanism and if the corresponding process obeys Arrhenius' law or the law of Meechan–Brinkman, then the life of the device can indeed be found by life-test extrapolations as mentioned above. Aging processes, caused by diffusion, recovery, and electromigration are relevant examples of such failure mechanisms.

However, other failure mechanisms may determine the life of a MIC, and these mechanisms must be investigated by means of other treatments. Tests like thermal shock and thermal cycling may represent an accelerated simulation of some operational conditions that could result in more rapid failure than by the degradation mentioned above. The corresponding stresses and temperature gradient are often encountered in transitions like terminations and bonding spots, and in places where inserted devices introduce hot spots.

To guarantee the aging behaviour of a device, it is necessary to use a reproducible MIC technology. Properties to be measured and tests to be applied must be carefully selected to cover the relevant failure mechanisms. Changes in technology imply time consuming investigations of aging consequences.

7.4.1.1 DC measurements

Properties, suitable for a quantitative determination of the aging behaviour of multilayers, must satisfy the following requirements:

(a) their change during storage at a certain temperature must be measurable with a high accuracy;
(b) their Meechan–Brinkman time–temperature (trade-off) characteristic should be continuous and monotonous.

DC resistances often fulfil these requirements. Part A of the test mask (Figure 7.36) contains wide strips for investigation of electromigration phenomena and narrow strips for DC resistance measurements.

In thin-film MICs, gold patterns of 5–10 μm with adhesion layers are normally used. Sometimes barrier layers are sandwiched between adjacent layers to reduce the speed of degradation processes and thus increase the life.

MICs with substrates of alumina, ferrite or fused quartz, and with metal patterns of NiCr–Au, Ti–Pt–Au, Al–Au, Ti–Au, Cr–Au, NiCr–Au, Ni–Au or Mo–Au are commonly used. Test circuits with several of possible combinations have been made and subjected to isochronal and isothermal heat treatments. Results of isochronal treatments, see Figure 7.38, can be grouped into the following categories of aging behaviour:[22]

(1) A steady decrease of resistance;
(2) A slow increase of resistance;
(3) A steady decrease of resistance with a sudden jump in the temperature interval between 460 and 500 K;
(4) A slope discontinuity from a strong initial increase to a decrease in resistance.

Bearing in mind that several reactions take place simultaneously in all samples and that the measured DC resistances of the narrow strips also depend

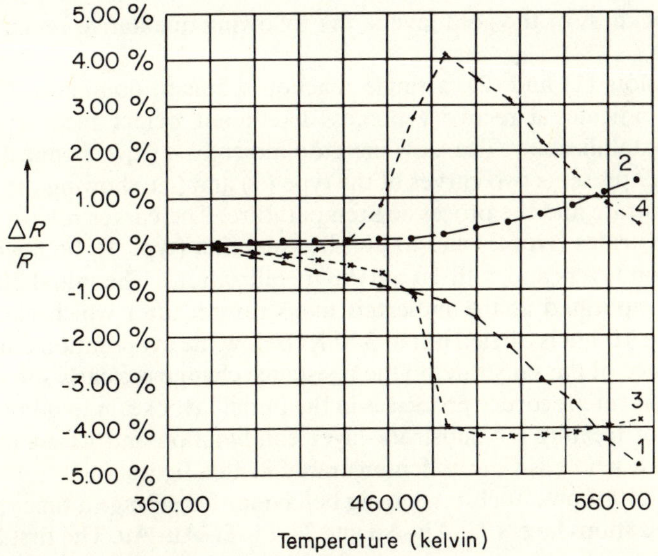

Figure 7.38. Four types of aging behaviour of tested MIC metallizations observed via deviations of DC resistance during isochronal heat treatments (storage time ~ 1 hour)

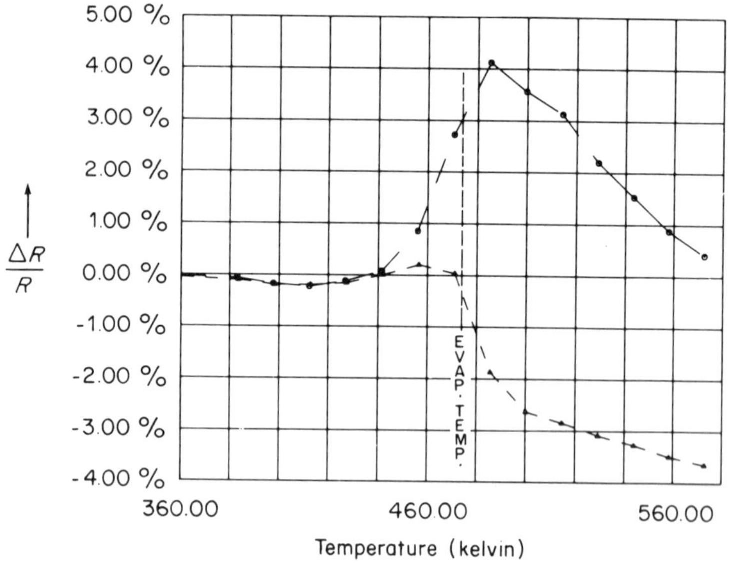

Figure 7.39. Discontinuous isochronal aging characteristics. 'Evaporation temperature' is the temperature at which the seed layers are evaporated

on the thickness of the gold layers, the following qualitative remarks can be made.

In the plots (1) and (2), a single reaction seems to dominate. The steady decrease indicates a recovery process like point defect recovery or grain boundary stabilization. The slow increase indicates a slow chemical reaction. Figure 7.39 presents two curves of the type (3) and (4), showing the influence of the substrate and the processing temperature. The curves refer to polished ferrite substrates (type 3) and to polished alumina (type 4), both with 15 nm Al adhesion layers and with 10 μm Au metallizations. The initial 200 nm Au layer is evaporated at the indicated temperature, after which electrodeposition up to 10 μm is carried out at 340 K. Below the evaporation temperature the influence of the substrate on the resistance change is clearly shown; above this temperature recovery processes in the initial layer seem to dominate. The necessity of treating the substrate–layer combination and its manufacturing process as a whole is clearly demonstrated in this figure.

Figure 7.40 shows isochronal aging behaviour for storage a time $t_a = 4$ h of the metallizations layers Ti–Au–Au and Ti–Pt–Ti–Au–Au. The first layer (Ti) is the adhesion layer sputtered on polished ferrite substrates. The thicknesses of the metal layers of the Ti–Au–Au combination are 20 nm for Ti, 200 nm for the sputtered Au and 10 μm for the electrodeposited Au. For the second

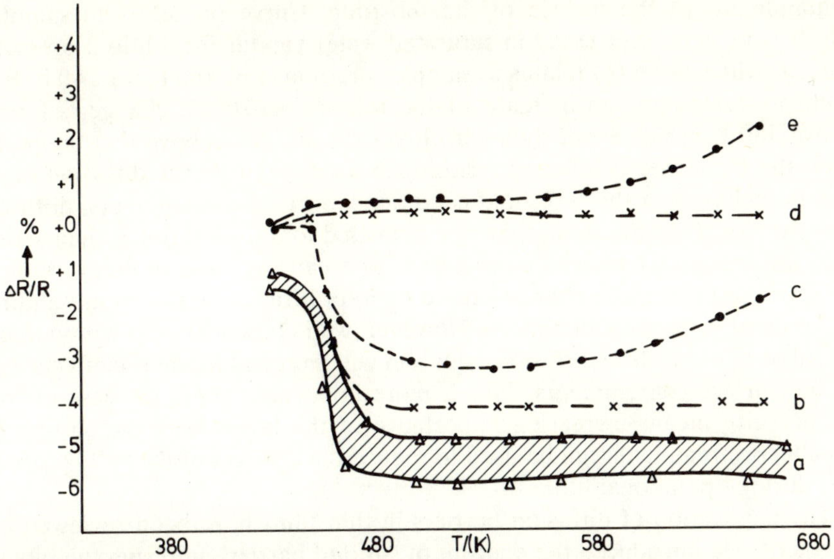

Figure 7.40. Isochronal aging characteristics of metallic multilayers on polished ferrite substrates:
(a) spread of the results of Ti–Au–Au batch directly after manufacturing;
(b) representative curve of Ti–Au–Au sample pretreated in a pressure cooker at 390 K for 4 h;
(c) as (b) for Ti–Pt–Ti–Au–Au sample;
(d) representative curve of Ti–Au–Au sample subjected to heat pretreatment at 480 K for 140 hours.
(e) as (d) for Ti–Pt–Ti–Au–Au sample

combination Ti–Pt–Ti–Au–Au, the thicknesses are 20 nm for Ti, 20 nm for Pt, 10 nm for Ti, 200 nm for the sputtered Au and 10 μm for the electrodeposited Au.

The region (a) indicates the spread of the isochronal characteristics of a batch of samples of the ferrite–Ti–Au–Au combination, investigated immediately after their manufacture. Samples that have been stored for a long time, or have been pretreated, show isochronal characteristics which are markedly different from region (a). What is of interest, however, is only the deviation from the initial value. Pretreatment may result in smaller deviations. Curve (b) presents the behaviour of ferrite–Ti–Au–Au samples, pretreated in saturated water vapour at 390 K for 4 h in a pressure cooker. The deviations are found to be one per cent less than those of the fresh substrates.

Curves (c) and (e) refer to the metal combination Ti–Pt–Ti–Au–Au, as mentioned, and show the influence on the relative change in resistance due to the very thin Pt barrier within the Ti layer. The Pt barrier influences the rate of metal interdiffusion and the rate of diffusion of oxygen and other gas contaminants of the surface of the substrate. Curve (c) refers to samples which have been pretreated in saturated water vapour for 4 h in a pressure cooker, while curve (e) relates to samples pretreated at 480 K for 140 h. For both pretreatments, an increase of the relative resistance change is found above 480 K, which is not shown in curves (b) and (d). Above that temperature, the Pt barrier is affected resulting in an increase of the diffusion rates and in a change of the threshold conditions. Surface contaminants, diffused into the first Ti layer, are in that case activated to diffuse through the barrier into the second Ti layer. In conclusion one can say, that multilayers may exhibit a fast resistance change caused by interdiffusion of the metals and by diffusion of surface contaminants. However, well-chosen barriers within these multilayers will reduce diffusion rates and will limit and locate the effects due to the surface contaminants. This is more important where increase of frequency also means increasing importance of the layers near the substrate! Finally, pretreatments can be used to reduce the changes of the MIC properties during operational life.

The application of diffusion barriers in thin films have been reviewed by Nicolet.[40] He introduced the concept of 'stuffed barrier', implying the blocking of high-rate diffusion pathes by stuffing the barrier with appropriate atoms and molecules. Figure 7.40 shows such an example, in which substrate contaminants may block diffusion at the barrier. The number of high-rate diffusion pathes formed by grain boundaries and dislocations is also reduced by the process interruptions which are necessary for the deposition of thin-film multilayers.

7.4.1.2 Microwave measurements

At microwave frequencies, the current is concentrated in the conductor layers at the substrate–layer interface due to the skin effect. The current density increases with frequency. Changes of the material composition in multilayers at this interface can result in changes of microwave losses, which can be determined from measurements of the change of the Q_0-factor of a resonator, as indicated in Section 7.2. Shifts of resonance frequencies can also give an indication of the changes in composition. Part F of the test pattern (Figure 7.36) containing a straight microstrip resonator circuit was aged together with part A to compare the aging behaviour of the DC resistance with the resistance of the metallic multilayer at microwave frequencies. The quality factor Q_0 of the test resonators are measured prior to and after the isothermal or isochronal treatments. Figure 7.41 shows Q_0-factors of straight resonators of alumina–NiCr(20 nm)–Au(10 μm), measured before and after isochronal aging with temperatures up to 620 K. The shifts of the Q_0s and the increase of the resonance frequencies due to this treatment are small. Despite a substantial change of DC resistance, the Q_0-factors remain surprisingly constant for

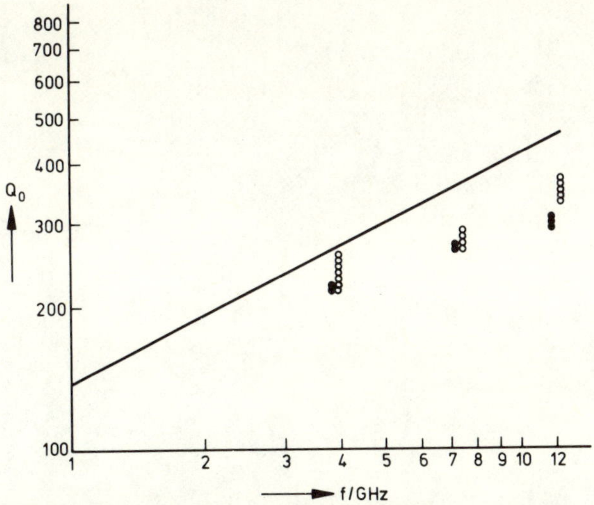

Figure 7.41. The Q_0-factors of samples (alumina–NiCr–Au) measured by using straight-line microwave resonators;
● spread of Q_0 before the heat treatments.
○ spread of Q_0 after the heat treatments.
For clarity the measured spreads of Q_0 after the heat treatments are shifted slightly to the right. The solid line indicates the theoretical Q_0-factors

temperatures up to destructive values. The change of Q_0 due to heat treatments become larger the thinner the gold layers are.

7.4.1.3 Grain size

The grain structure of the metal layers depends on the kind of deposition process and its parameters, such as temperature and surface condition of the substrate or the rate and angle of the metal deposition. During deposition, high dislocation densities are formed near the grain boundaries. Dislocations strongly affect the recrystallization process during the heat treatment. In combination with stresses, dislocations deteriorate the stability of metal layers.

Figure 7.42(a) and Figure 7.42(b) show the surfaces of samples with ferrite–Ti–Au (10 μm) and ferrite–Ti–Pt–Ti–Au (10 μm) after a isothermal heat treatment at 530 K for 400 h. The heat treatment resulted in remarkedly different changes in surface structures of both 10 μm Au layers. The crystalline transitions are attributed to thermodynamic instabilities, which may give rise to anomalous crystal growth e.g. whiskers, large single crystals, and 'cannibals'. The morphology of the transitions in thin-film Au layer has been

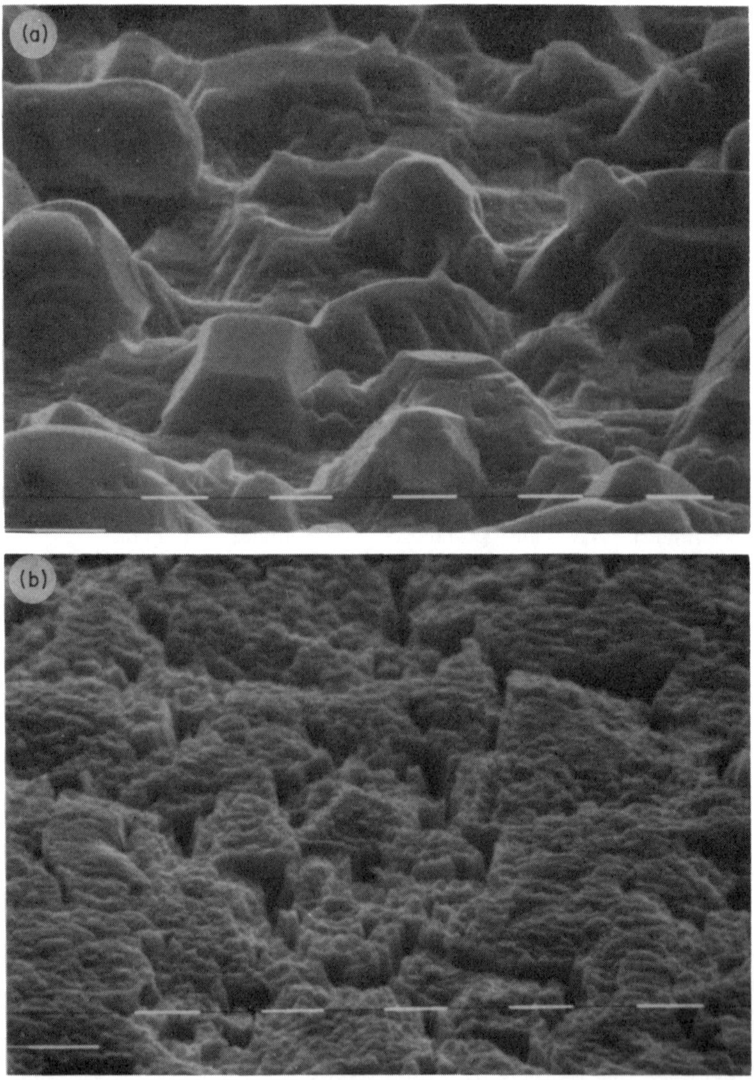

Figure 7.42. Surface of (a) ferrite–Ti–Au–Au (10 μm) sample; (b) ferrite–Ti–Pt–Ti–Au–Au (10 μm) sample after isothermal treatments at 530 K; total storage time was 400 hours

discussed in the literature.[41] An example of whisker-like extrusions, formed on a surface scratch after a heat treatment at 664 K for 200 h, is shown in Figure 7.43. Stresses generated by damages of the surface lead to material transport. Therefore, the circuits must be carefully handled. Furthermore, changes in structure and surface conditions may degrade the bondability and the solderability of the layers.

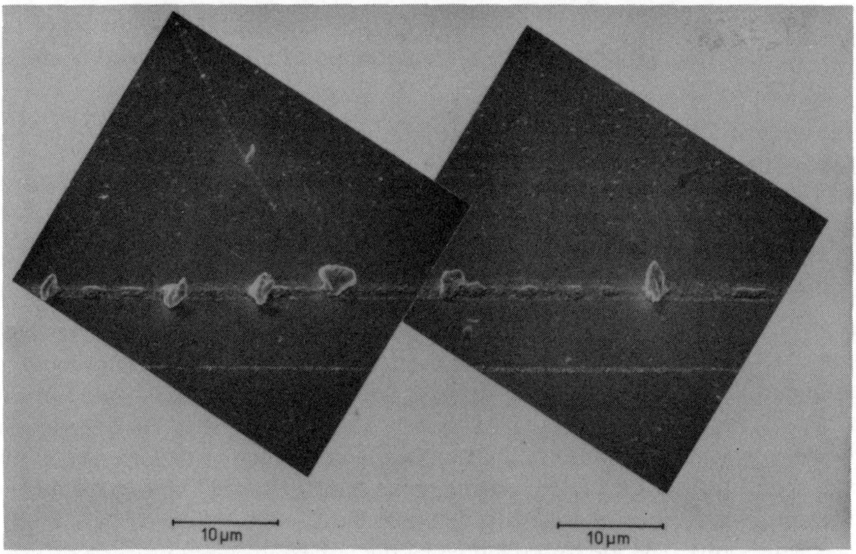

Figure 7.43. Whisker-like extrusions in a scratch on a sputtered 0.5 μm Au layer after a heat treatment at 664 K for about 200 hours

7.4.2 Adhesion

As already mentioned, the adhesion between the metallic thin-film layers and the dielectric substrates is achieved by an adhesion layer, whereas the adhesion of thick-film layers is obtained by means of a binder, containing glass and/or other oxides, which had been included in the ink. The adhesion of metallic layers has already been widely investigated. Comprehensive surveys are given in literature.[42,43,44] In the following, only an outline of problems concerning adhesion will be given and some measuring methods will be discussed.

7.4.2.1 Basic adhesion

The basic adhesion between two materials in close contact follows from the nature and the strength of the bonding forces between them. In principle, one can distinguish between three types of basic adhesion, i.e. basic adhesion achieved by:

(a) physical interactions, which are usually referred to as van der Waals' interactions. The bonding energy is of the order of 0.1 eV;
(b) chemical interactions, e.g. solid-state interdiffusion between the metallic layer and the substrate. In this case the bonding energy is 10 eV at the most;

(c) mechanical interlocking, for which a favourable surface roughness of the substrate can be obtained by a combination of mechanical and chemical roughening.[27]

The adhesion forces due to chemical interaction are much stronger than the forces due to physical interaction or mechanical interlocking.

7.4.2.2 Actual adhesion

The actual adhesion is a combination of the three types of basic adhesion. In general, the basic adhesion can be made sufficiently strong for MIC applications. In view of reliability, it is often desirable that the chemical interaction is dominant and that the bonding strength, which is the measured value of the adhesion, ranges from 10 to 100 N mm^{-2}. The bonding strength depends on the basic adhesion but also on extraneous factors, such as the stresses in the layers, and further on the measuring techniques. It should be noted that the measured values of the adhesion may not be directly related to any of the basic adhesions. The measured values of the adhesion can in practice be used for a comparison between adhesions obtained by different technologies.

The stresses in the metallic layers can generate peel-off forces at the edges of the patterns. Further, the adhesion decreases with life. Kadereith[45] found that the bonding strength can decrease to about half its initial value after storage at 150 °C for 5000 h. The adhesion loss[46] can be caused by:

(a) diffusion of the adhesion layer into the adjacent metallic layer and diffusion that can be enhanced by stresses in the multilayer;
(b) recovery of atomistic defects;
(c) a chemical reaction.

7.4.2.3 Measuring methods

There are many methods for measuring the bonding strength. However, each method has its own merits and limitations. In what follows, four methods will be briefly discussed.

A simple method for testing the adhesion is the 'Scotch tape test'. Adhesive tape is pressed onto the film and then torn off. There are, of course, three possibilities: the film is not removed or is partly or completely removed. It is clear that this test is a highly qualitative one and can only be used for distinguishing cases with a very poor adhesion from those where the adhesion is appreciable.

The adhesion is often investigated by peel tests, see Figure 7.44(a). The metallic strip is peeled off from the substrate at a constant rate and the needed force is continuously monitored. This method can only be used in the cases of a relatively weak adhesion, which may be due to van der Waals'

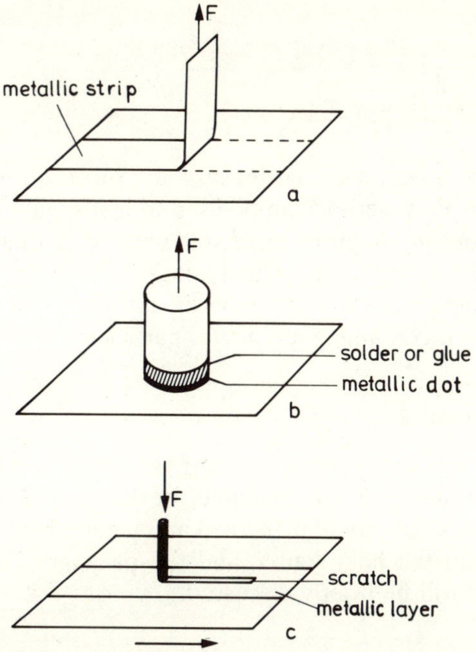

Figure 7.44. Three examples of adhesion tests: (a) peel test; (b) pull-off test; (c) scratch test

interactions or mechanical interlocking. A disadvantage of this method is that the peel-off force depends on the thickness and the ductility of the layer.

In Figure 7.44(b), the principle of the direct pull-off test is shown. A pulling device, e.g. a brass pin, a beam, a rod, or a wire, is attached to the film by soldering or glueing. The device is pulled in the perpendicular direction using a tensile tester and the force needed for detaching the layer is measured. The dots in the parts B and D of the test pattern are intended to be used for this measuring method. A problem is that the pull direction must be exactly perpendicular in order to avoid a non-uniform stress distribution over the interface. Unwanted stresses at the interface of the multilayer and substrate can nevertheless be generated during glueing or soldering, and adhesive or solvent can penetrate into the multilayer and affect the adhesion.

The scratch method has been reported in literature.[47] A smooth round (chrome-steel) tip is drawn over the metal surface (Figure 7.44(c)). A vertical load is applied to this tip and increased until the film is completely stripped off from the substrate surface, leaving a clean channel. This critical force W is determined. The adhesion is expressed by a shearing force F per unit area according to:

$$F = \frac{P \cdot \sqrt{W}}{\sqrt{\pi \cdot P \cdot r^2 - W}} \tag{7.33}$$

in which P is the indentation hardness of the substrate material and r is the radius of the tip. This method can be used to investigate the change of adhesion due to diffusion. A problem of this method is that there is no simple general theory for deducing the actual adhesion from the measured values. It is neither a simple matter to determine the minimum force needed to draw a channel that is not containing any metal particles.

7.4.3 Thermal analysis

To improve the reliability and to prolong the life of a MIC, it is desirable to keep its temperature as low as possible. In the case of heat dissipation, the outside of the package should be cooled and a good heat transport should be provided between the heat source and the package. The principles of the thermal analysis will be briefly discussed.

7.4.3.1 Heat conduction

The heat conduction of an infinitesimally thin slab of material is governed by Fourier's law (Figure 7.45):

$$\Phi = \lambda \cdot \frac{\Delta T}{\Delta x} \cdot S \tag{7.34}$$

in which Φ is the rate of heat flow, λ the thermal conductivity, ΔT the temperature increment across the slab, Δx is the thickness of the slab and S

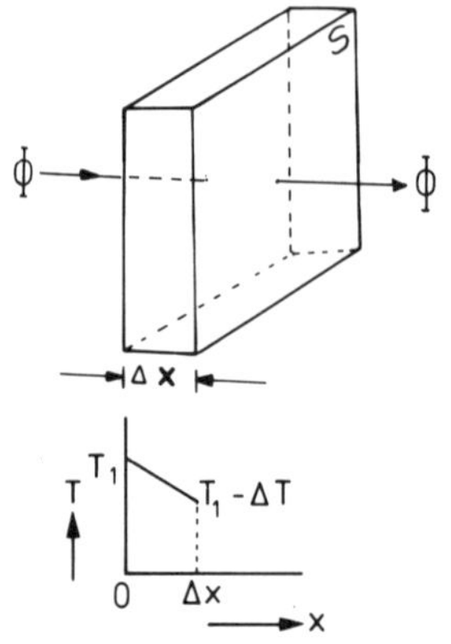

Figure 7.45. Heat flow Φ in a thin slab of material; the correlated temperature gradient is also shown

Table 7.6 Analogy between thermal and electron flow

Heat flow	Electron flow
Heat flow rate [W]	Electric current [I]
Temperature gradient $[\mathrm{T.L^{-1}}]$	Potential gradient $[\mathrm{V.L^{-1}}]$
Thermal conductivity $[\mathrm{W.T.^{-1}L^{-1}}]$	Electric conductivity $[\Omega^{-1}.\mathrm{L}^{-1}]$

the surface of its cross-section. The heat conductance G_{th} of the slab is given by:

$$G_{\mathrm{th}} = \lambda \cdot \frac{S}{\Delta x} \tag{7.35}$$

so that

$$\Phi = G_{\mathrm{th}} \cdot \Delta T \tag{7.36}$$

This equation is equivalent to Ohm's law. A complete analogy exists between the thermal and the electron flow, see Table 7.6.

7.4.3.2 Thermal behaviour of a body

The thermal behaviour of a body can be analysed by establishing the heat equation for every infinitesimal part of that body. For an immobile and undeformable body one can obtain:

$$\Phi - \lambda \cdot \nabla^2 T = \rho_{\mathrm{m}} \cdot c_{\mathrm{p}} \cdot \frac{\partial T}{\partial t} \tag{7.37}$$

in which t is the time, Φ the applied power per unit of volume, ρ_{m} the density and c_{p} the specific heat capacity, so that $\rho_{\mathrm{m}} \cdot c_{\mathrm{p}} \cdot \partial T/\partial t$ is the power stored per unit of volume. Furthermore, $\lambda \cdot \Delta^2 T$ is the power transferred per unit of volume by the conduction, in which the temperature Laplacian is:

$$\nabla^2 T = \frac{\partial^2 T}{\partial x^2} + \frac{\partial^2 T}{\partial y^2} + \frac{\partial^2 T}{\partial z^2} \tag{7.38}$$

In the steady state, the temperature is constant, so that $\partial T/\partial t = 0$ and consequently:

$$\Phi - \lambda \cdot \nabla^2 T = 0 \tag{7.39}$$

This equation can be used for the heat exchange Φ between surface parts and their environment, where $\lambda \cdot \nabla^2 T$ is the heat flow into the body owing to conduction.

At normal working temperatures of the MIC, the heat radiation inside the body can be neglected in comparison with the heat transport due to conduction, so that for parts inside the body $\Phi = 0$. Consequently:

$$\nabla^2 T = 0 \tag{7.40}$$

Consider e.g. the slab shown in Figure 7.45. Assume that the heat only flows in the x-direction, so that:

$$\nabla^2 T - \frac{\partial^2 T}{\partial X^2} = 0 \tag{7.41}$$

After integration:

$$T = a \cdot x + b \tag{7.42}$$

in which a and b must be determined from the boundary conditions. For $x = 0$, one has $T = T_1$ and for $x = \Delta x$ one gets $T = T_2 = T_1 - \Delta T$, thus:

$$T = -\frac{\Delta T}{\Delta x} \cdot x + T_1 \tag{7.43}$$

i.e. the temperature across the slab decreases linearly with x.

7.4.3.3 *Boundary conditions*

To solve the heat equation, one must first determine the boundary conditions of which several kinds exist:[48]

(a) the considered body is in close contact with another body; Figure 7.46(a) represents the situation without, and Figure 7.46(b) with a contact resistance;

(b) the body irradiates to another body (Figure 7.46(c)). The rate of heat flow due to radiation is:

$$\Phi = \varepsilon \cdot \sigma \cdot (T_1^4 - T_2^4) \tag{7.44}$$

in which ε is the transmission coefficient ($0 < \varepsilon < 1$), where the Stefan–Boltzmann constant $\sigma = 5.67\ 10^{-8}$ W m^{-2} k^{-4}, and where T_1 and T_2 are the temperatures of the bodies 1 and 2, respectively.

(c) the heat is drawn by convection according to Newton's law (Figure 7.46(d)):

$$\Phi = H \cdot (T_1 - T_2) \tag{7.45}$$

in which H is Newton's transfer coefficient, T_1 the temperature of the body and T_2 the ambient temperature.

Several values of the thermal conductivity λ and the transfer coefficient H are listed in the Tables 7.7 and 7.8, respectively.

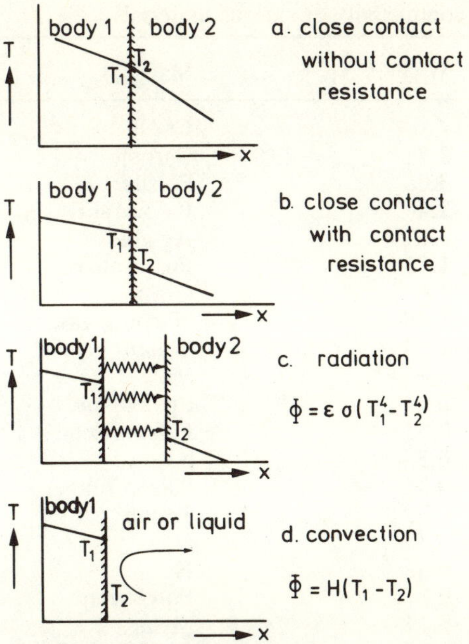

Figure 7.46. Different cases of heat exchange: (a) close contact without contact resistance (b) close contact with contact resistance; (c) heat exchange by radiation; $\Phi = \varepsilon \cdot \sigma \cdot (T_1^4 - T_2^4)$; (d) heat exchange by convection $\Phi = H \cdot (T_1 - T_2)$

As an example, a square heat source with side w, located on a substrate with thickness h and a thermal resistivity ρ, will be treated, see Figure 7.47. The substrate is glued or soldered to a heatsink and the heat flowing through the substrate from source to sink is considered. In the case of steady state, one has to solve Equation 7.40, which is rather complicated to solve exactly. Therefore, an approximate solution will be tried by assuming that the heat is flowing in a pyramidal with angle α. The increase ΔT of the temperature can be obtained in a first approximation by substituting the mean value of the cross-section of the frustum for S in Equation 7.34:

$$\Delta T = \Phi \cdot \rho \cdot \frac{h}{(w + h \cdot \tan\alpha)^2} = \Phi \cdot \frac{\rho}{w} \cdot \frac{1/\tau}{(1/\tau + \tan\alpha)^2} \tag{7.46}$$

in which $\tau = h/w$.

An alternative method makes use of integration:

$$\Delta T = \Phi \cdot \rho \cdot \int_0^h \frac{dx}{(w + 2x \cdot \tan\alpha)^2} = \Phi \cdot \frac{\rho}{w} \cdot \frac{1}{1/\tau + 2\tan\alpha} \tag{7.47}$$

Table 7.7 Thermal conductivity of various materials

Material	W cm^{-1} °C^{-1}	Material	W cm^{-1} °C^{-1}
Ag	4	Cr Ni	0.13
Cu	3.7	Carbon	0.07
Au	3.08	Ferrite	0.064
BeO	2.4	Pyroceram	0.034
AuGe (88/12)	2.27	Ag glue	0.016
Al	1.75	Fused quartz	0.015
Diamond	1.6	Glass	0.011
Mo	1.48	Silicon grease	0.0087
Brass	1.15	Water	0.0067
Si	0.85	Mica	0.005
Pt	0.74	Epoxy glue	0.0035
NiSn	0.65	Polyethylene	0.0033
SnPb (62/38)	0.53	Nylon	0.0021
Al_2O_3	0.35	Silicon rubber	0.0021
Steel	0.35	Teflon	0.0021
Pb	0.35	H_2	0.001
Borium nitride	0.32	He	0.001
Stainless steel	0.2	Polystyrene	0.001
Silicon carbide	0.17	Mylar	0.0003
Kovar	0.17	Air	0.00017

How to find the most appropriate value for α is the next problem. The temperature rise of a square heat source has been calculated from Equation 7.40 using Fourier transforms. David[49] found that for $\tau \leqslant 2$, the angle $\alpha = 26.6°$ can be taken, so that the 'mean value' formula Equation 7.46 becomes:

$$\Delta T = \Phi \cdot \frac{\rho}{w} \cdot \frac{\tau}{(0.25\,\tau^2 + \tau + 1)} \tag{7.48}$$

Table 7.8 Heat transfer coefficients for normal and forced convection

	H (W m^{-2} °C^{-1})	
	Normal convection	Forced convection
Air	6–28	10–300
Water	100–900	300–10.000
Oil		50–2000

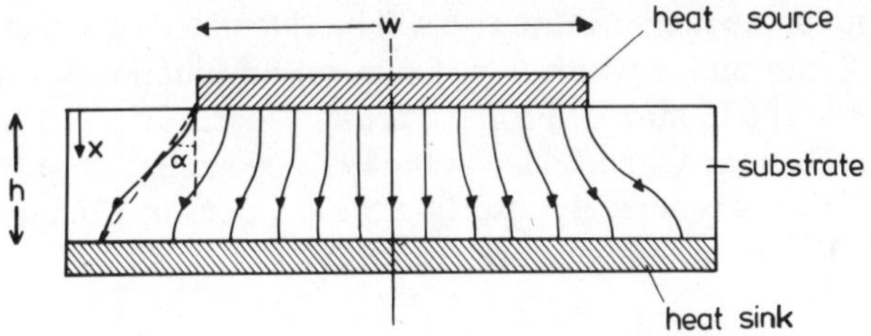

Figure 7.47. Heat flow through substrate

For $2 < \tau < 32$, it turns out that a better approximation can be obtained by using the 'integration' formula Equation 7.47 with $\alpha = 40°$. Consequently:

$$\Delta T = \Phi \cdot \frac{\rho}{w} \cdot \frac{1}{1/\tau + 1.7} \qquad (7.49)$$

In Figure 7.48, the normalized temperature rise θ is given as a function of τ for several values of α. The solid line refers to David's calculations.

In the case of a rectangular heat source with length 1 and width w, the latter is replaced by w_{eff}. If $1/w \leqslant 4$, one can take as a first approximation $w_{\text{eff}} = \sqrt{w \cdot 1}$. The approximated increase in the temperature will then be about 10 to 20 per cent larger than the actual value.

Only single sources have been discussed, however, non-interfering heat sources can also be treated in the same way. For interfering heat sources, software programs are commercially available.

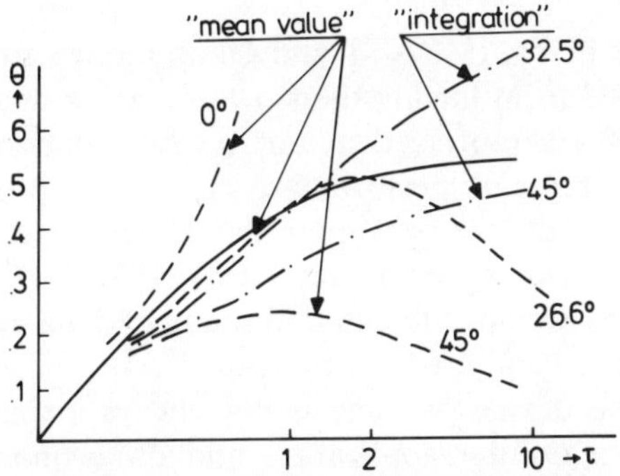

Figure 7.48. Normalized temperature rise Θ as function of τ; solid curve refers to David's calculations; 'mean value' curves are calculated with Equation 7.46 for $\alpha = 0°$, 26.6°, and 45°; 'integration curves are calculated with Equation 7.47 for $\alpha =$ 32.5° and $\alpha = 45°$. ($\theta = \Delta T \cdot w/(\Phi \cdot \rho)$).

As an example, the temperature rise will be calculated of a heat source with a length $l = 2$ mm and a width $w = 1$ mm on an alumina substrate with a thickness $h = 0.635$ mm. From these data, it follows that $w_{eff} = \sqrt{2} = 1.414$ mm and $\tau = 0.449$. Further, $\rho = 1/\lambda = 2.86$ (cm Cdeg W^{-1}) for alumina. By inserting these values in Equation 7.48, one obtains for $\Phi = 1$ W, a temperature rise of $\Delta T = 6.1$ °C.

7.5 MIC ASSEMBLY

After forming the conductor pattern, elements or functions can be added to the circuit. The assembly techniques for manufacturing MICs are in principle the same as those applied for hybrid integrated circuits designed for lower frequencies. However, at microwave frequencies, the performance of add-on components can be seriously affected by parasitic inductances and capacitances. Therefore, at these frequencies, the connection wires must be as short as possible. Hence, naked chips are preferred over encapsulated ones and earth connections are preferably made on the spot via plated-through holes.

Many elements need provisions in the conductor pattern for their optimal functioning. The enormous progress in MIC-system performance is actually due to the progress in the performance of the active devices at higher frequencies and the possibility for the MIC designer to create an optimum operational environment for these devices, e.g. by applying matching elements and filters that can be realized in the conductor pattern.

7.5.1 Add-on components

Active devices like FETs, IMPATT and Gunn diodes are inserted as naked chips or in packaged form having their own heatsink and connection provisions. Naked chips are preferred at microwave frequencies, although test conditions and guaranteed performance are more problematic than for encapsulated devices. Before packaging the active devices are tested on the wafer (static tests) with dynamic tests carried out after packaging. Unencapsulated devices are not subjected to the final dynamic check, so that the relevant performance cannot be guaranteed.

The choice of the device package is not always a simple matter since it depends very much on the application and on economic considerations. Encapsulations, specially designed for use in hybrid ICs, like the leadless inverted device (LID) package, are sometimes used at frequencies just below the microwave frequency range.

An intermediate form between the naked and packaged device is the device to which gold beams are attached. Beam-lead devices offer the advantage of a highly predictable MIC behaviour due to the straightforward gold-on-gold bonding to the microstrip circuit. Mixer diodes in beam-leaded form are used up to very high frequencies. In addition to the possibility of manufacturing beam-lead devices with very small parasitics the reproducibility of these parasitics is much better than for gold-wire connected devices.

An add-on component, recently introduced into the MIC area, is the ceramic-disc resonator. It is made from a ceramic material such as barium titanate ($\varepsilon_r \approx 40$) and functions as a dielectric resonator in oscillator circuits. Its temperature characteristics are favourable to realize stable oscillators. The disc itself is glued to the substrate close to the oscillator circuit to which it is coupled electromagnetically.

Circulators and isolators are often used in MICs. If a proper ferrite substrate can be applied, then the conductor pattern of the non-reciprocal devices can be part of the whole pattern. A pair of permanent magnets can then be mounted on either side of the substrate to provide the magnetic field (Figure 7.49). If non-ferrite substrates are used, then the non-reciprocal devices, made on ferrite substrates, are inserted in the assembly (Figure 7.55).

The transformation of the circulator into an isolator can be effected by terminating one of the terminals with a matching load in the form of an integrated or add-on resistor.

Figure 7.49 shows an example of a MIC with add-on components. The right part of the circuit is a balanced 7 GHz down-converter with a bandwidth of 10 per cent. The circuit consists of an isolator and a balanced mixer. An add-on resistor and a magnet are used in the isolator. The balanced mixer consists of a 3 dB-branch coupler and two add-on Schottky-barrier diodes. In addition, plated-through holes are used for earth contacts. The ferrite substrate is pressed to the pins of the connectors with the aid of a spring loaded system beneath the substrate, while two clamps fix the substrate in the right position. The left part of the circuit is a 70 MHz pre-amplifier.

Figure 7.50 shows a survey of currently used assemblies. The contact between the connector pin and the strip can be brought about by pressing, but sometimes a bonded ribbon is used for this purpose (a). The earth contact near the connector is mostly achieved by pressing. A beam-leaded device is only fixed by bonding its beams, (b), while a plated-through hole is often used for a local earth contact (c). Further, a naked chip (d), e.g. a FET, is usually mounted to a post in order to minimize the length of the bonding wires. This post also serves as earth contact and heatsink. The connection (e) between two substrates is achieved by pressing contacts or bonded ribbons. In addition, encapsulated devices (f), dielectric resonators or magnets (g) must often be bonded to the circuit. Finally, (Gunn) diodes (h) are usually mounted just beside the edges of the substrate because of their heat dissipation. In the case of heat dissipation in the circuit on the substrate, the latter can be soldered or glued to the package. Often the substrates are mounted in the packages with the aid of a mounting plate.

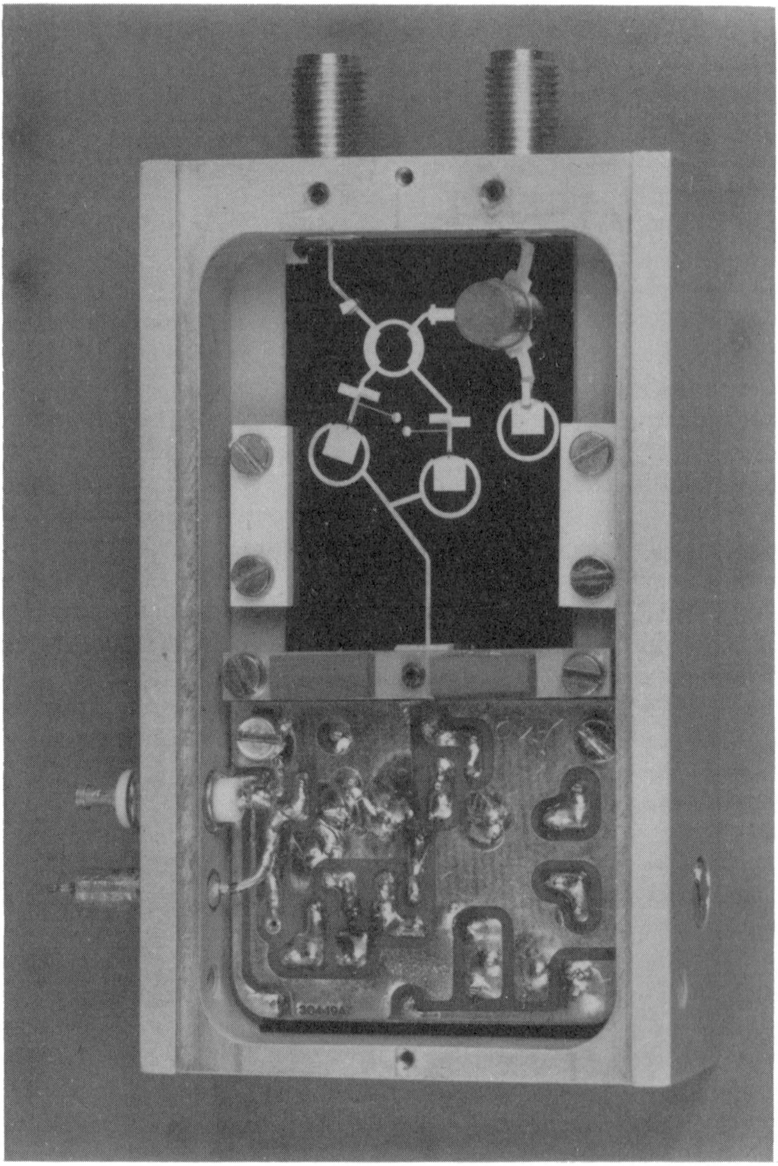

Figure 7.49 A 7 GHz down-convertor in MIC technology with a 70 MHz pre-amplifier; added on are a magnet, a resistor, and two beam-lead Schottky barrier diodes bonded on the squares in the ring

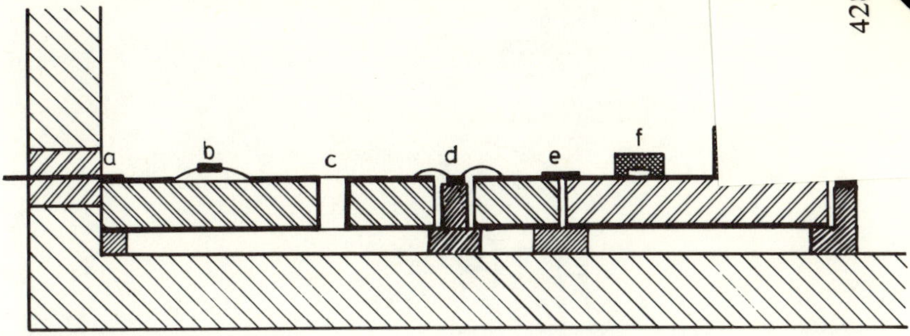

Figure 7.50. Illustrative view of several connections and add-on components: (a) connector contact; (b) beam-lead device; (c) metallized hole; (d) three terminal device; (e) substrate to substrate connection; (f) LID packaged device; (g) dielectric resonator or magnet; (h) packaged diode on stud connected to substrate

7.5.2 Bonding, glueing, and soldering

Current assembly techniques for add-on components make use of thermo-compression (TC) or ultra-sonic (US) wire and beam bonding, while eutectic bonding, glueing, and soldering are used for die bonding.

7.5.2.1 Thermo-compression bonding

TC bonding is a process in which the energy needed for bonding is applied by pressure and heat. The required temperature for making a bond is lower than that for interface melting. The pressure effects a plastic deformation at the interface, while a diffusion occurs during the bonding cycle. Plastic deformation destroys interface films of oxides and dirt, while it achieves an optimum contact between the metal surfaces. The required pressure and temperature strongly depend on the material used. For thin-film circuits, pure gold (99.99 per cent) has the best bonding properties; for thick-film circuits, inks with good bonding properties are commercially available.

There are two methods to supply heat to the bond: first, by a heat source located under the substrate, and second by a heated bonding tip. In the former method, the temperature of the substrate and its holder is rather high and the heat transport through the substrate may present a problem. In the latter case, a ceramic bonding tip must be used in order to prevent corrosion. Mostly, a combination of both methods is applied. Bonding equipment with a pulse-heated bonding tip is also used. The heat is only supplied during bonding, so that the overall temperature of the substrate can be relatively low.

The current techniques for accomplishing TC bonds are 'wedge' bonding (see Figure 7.51(a)) and 'nailhead', or 'ball' bonding (see Figure 7.51(b)).

For TC wedge bonding, Au wires are used, whose ductility is very impor-

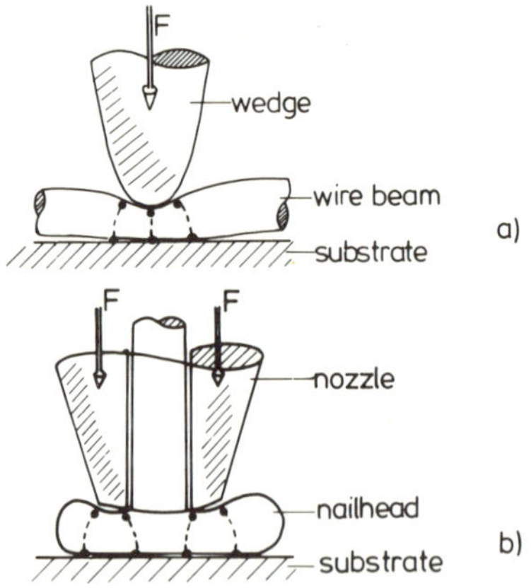

Figure 7.51. Principle of thermo-compression bonds (TC) by: (a) wedge; (b) nailhead; F is the applied force

tant. Any bending or cold working of these wires is sufficient to cause poor bonds. Such wires are very weak because of their ductility and are thus difficult to handle.

Ball bonding is a TC wire-bonding technique in which at the end of the wire a small ball is formed and thereafter deformed, by pressing it against the surface of the metallic layer. Such bonding can only be applied to Au wires. This method is often preferred because of its relatively high-bonding speed and the strength of the bonds. However, in comparison to wedge bonding, ball bonding needs larger bonding pads and longer connection wires, so that for microwave applications, wedge bonding is often preferred.

For a chip-to-circuit connection, the second bond must always be a wedge bond. If first a ball bond has been used, then the direction for the second bond can be arbitrarily chosen. On the other hand, if a wedge bond has been used first, then the direction is fixed.

7.5.2.2 Ultra-sonic wire bonding

US wire bonding makes use of the same mechanism as TC bonding, except for the fact that the energy, needed for the bonding, is applied mechanically. Both wedge and ball bonding are possible in this way. US wire bonding can be used for Al wires, but US bonds can also be made between Au wires and pads of Al, Au, Ni etc. For MICs, bonds between Au wires and Au pads are preferred.

The reliability of US bonds is significantly enhanced by heating the bonding area during the bonding. This combination is called 'thermo-sonic' bonding.

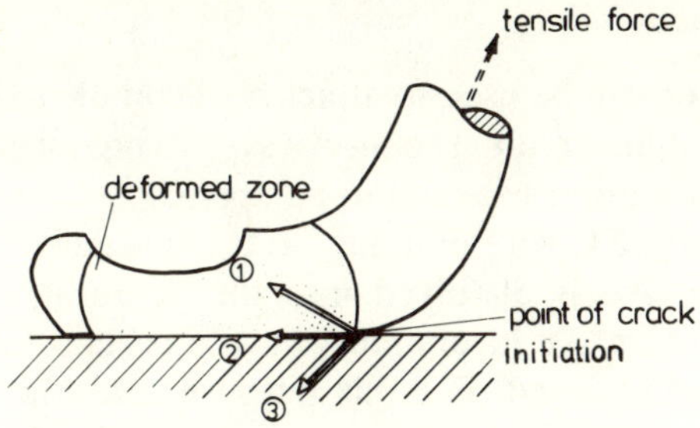

Figure 7.52. Crack initiation and propagation directions into: (1) the deformed wire or beam material; (2) the bond interface; (3) the substrate–metallization system

7.5.2.3 Failure mechanisms of wire bonding

An important failure mechanism of wire bonding is cracking. Cracks, which ultimately lead to fracture of the bond, can be initiated at different locations in the bond area. If a force is applied to the wire, the crack often starts at the edge of the bonded area in the neighbourhood of the bonded wire and penetrates into the ductile material (Figure 7.52). Zero-hour destruction is frequently caused by plastic deformation which leads to cracking. Mechanical loads may shorten the life of the bonds. The expected life of a bond on a thin-film layer as a function of the load is shown schematically in Figure 7.53.

There are three mechanisms that affect the long-term stability of the bonds:[50]

— interface reaction by diffusion;
— diffusion via the grain boundaries;
— stress-enhanced diffusion in the neighbourhood of the extremity of the crack.

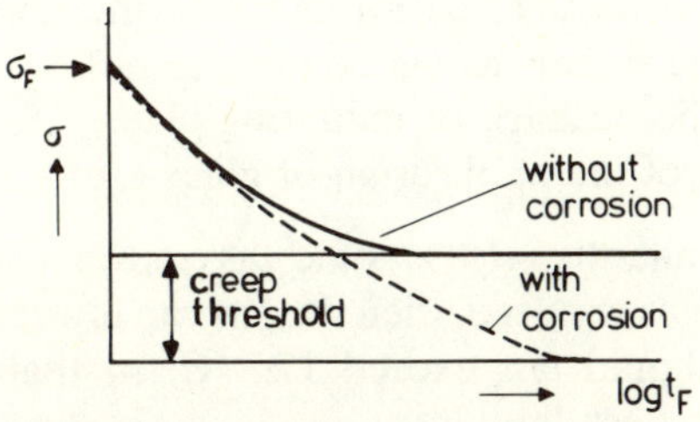

Figure 7.53. Schematic diagram of bond stress versus time to fracture during creep tests with constant loads at a constant temperature

7.5.2.4 *Eutectic die bonding*

Eutectic die bonding can be used to attach a silicon die to a metallization of a thin-film or a thick-film circuit. The Si–Au system has an eutectic temperature at 370 °C, which is neither too high to be applicable, nor to low for allowing other techniques, e.g. TC wire bonding, to be subsequently used in assembly of the MIC. Some Au is dissolved into the Si during the bonding cycle, therefore, the thin-film Au layers must have a certain thickness in order to obtain a good eutectic bond. For this purpose, use can be made of Au–Si preforms, which are melted on a heated substrate. The silicon die is placed on the molten alloy and the substrate is then cooled to form the bond.

Without the preform, the silicon die must be 'scrubbed' into the surface of the gold layer in order to achieve a contact good enough for a reliable bond.

7.5.2.5 *Glueing and soldering*

Adhesives are often used to attach dies and add-on components to the conductive pattern. These adhesives or glues are most commonly epoxies, filled with Ag flakes or Au grains for conducting purposes. The glue can be deposited by screen printing, so that this method fits well into the system of thick-film processing. The properties of epoxy adhesives have been extensively evaluated.[51,52,53]

A good thermal conductivity λ of the glue is important for heat dissipation. The Ag-filled epoxy conducts heat slightly better than an Au-filled glue. However, Au-filled glues, used for Au–Au contacts, show better performance during life tests than Ag-filled glues.

The thermal resistivity of Ag-filled glue is much higher than that of solder (see Table 7.7). On the other hand, elastic properties of glues are much more favourable than those of solder, so that glue can better be used for the realization of a 'graded seal' between materials with different coefficients of thermal expansion. For instance, a die can directly be glued on a Cu heatsink, whereas in soldering a Kovar heatsink is often used. Kovar is a poor heat conductor, so that the thermal performance of the glue–Cu combination is better than that of the solder–Kovar combination. Glues can also be used to fix the substrate to the package or mounting plate.

The disadvantages of the application of glues are:

—the glues will permanently release some gas contaminants, hence, if hermetically sealed packages are used, then the gas can lower the life of the MICs;
—the temperature should not exceed 125 °C, so that the possibilities for accelerated life tests are limited;
—input-quality tests are difficult to specify for commercially available glues;
—the backside of the die must be gold plated.

Table 7.9 Properties of solders used for Au–Au soldering

	Fusion temperature (°C)	Tensile strength (kg cm^{-2})	Shear strength (kg cm^{-2})	Hardness (BHN)
Pb In Ag (15/80/5)	149	179	150	5.2
In Pb (50/50)	183	469	427	0.6
Au Sn (20/80)	280	—	—	—

Table 7.9 shows the properties of three solders which are almost the only ones that can be used for Au–Au soldering. The normal solders, e.g. SnPb (60/40), dissolve too quickly in Au for making a good solder contact on thin-film or thick-film metallizations. During soldering, the scavenging phenomenon appears, which can deteriorate the adhesion between the metallization and the substrate.

Once the solder has been solidified, a solid-state diffusion occurs. This diffusion becomes the more pronounced the higher the temperature and may decrease the adhesion, especially in the case of thin-film metallization. The thickness of the intermetallic layer, formed between Au and In based solders, can be obtained for several combinations from the following relations.

For electrolytic Au–In-Pb (50/50), 1000 h tests:[54]

$$d_{\mu m} = 3.1 + 1.91 \cdot 10^7 \cdot t \cdot e^{-7100/T};$$

for Au 9791–In-Pb (50/50), 1600 h tests:[55]

$$d_{\mu m} = 6.3 + 55.3 \cdot 10^6 \cdot t \cdot e^{-7900/T};$$

for Au 9910–In-Pb (50/50), 1600 h tests:[55]

$$d_{\mu m} = 2.8 + 135.09 \cdot 10^7 \cdot t \cdot e^{-9600/T};$$

for Au 9791–Pb-In-Ag (15/80/5), 5000 h tests:

$$d_{\mu m} = 17260 \cdot t^n \cdot e^{-3400/T}.$$

where $d_{\mu m}$ is the intermetallic thickness in μm;
t is the time in hours and
T is the temperature in kelvins.

The exponent n is dependent on the temperature (Figure 7.54).

The first three combinations obey Arrhenius' law, the activation energies being 0.61, 0.68, and 0.83 eV, respectively. However, the diffusion is slowed

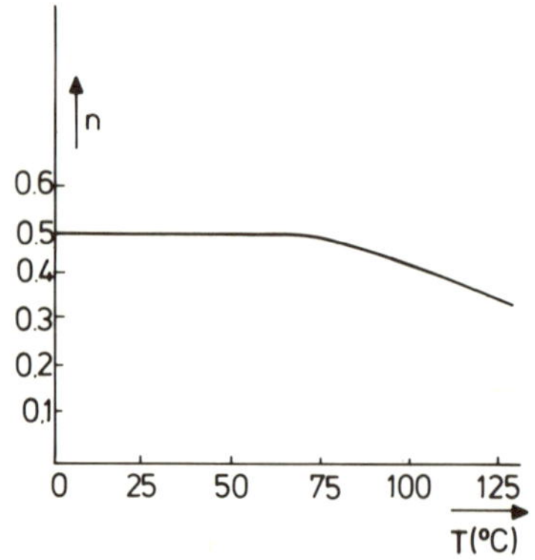

Figure 7.54. Influence of temperature on the exponent n found in the relation for calculation of the thickness of the intermetallic layer formed between Au and In based solders

down for the last combination. For temperatures below 50 °C, this diffusion can be treated with the Meechan–Brinkman relation, see Equation 7.27. It turns out that in that case $n = 1$, so that $\theta(p)$ is proportional to p^2 and the activation energy

$$E = 2.3700 \cdot k = 0.59\,\mathrm{eV}.$$

Au–Sn (80/20) solder has a very low diffusion rate, but a relatively high melting temperature in comparison with other solders.

7.5.2.6 *Compatibility of die and wire bonding*

The question arises which combinations of die bondings and wire bondings can be used together. Appropriate combinations are:

Die bonding	*Wire bonding*
eutectic Au–Si	TC and US
Au–Sn	TC and US
Au–Ge	TC and US
Pb–Sn (185 °C)	US
glue	US and TC (pulsed heat)

7.5.2.7 *Assembly of substrates and packages*

Substrates can be secured on mounting plates and in packages by clamping, glueing or soldering. A good earth contact is required, especially in the neighbourhood of the connectors and the plated-through holes, where slits and slots introduce parasitics. In general, it is recommended, that the entire edge underside of the substrate has a good earth contact. During mounting, the edge of the substrate must be kept free from solder or other conductive materials. Gold plated substrates can be soldered using a sheet of indium. A sheet having a sufficient thickness can be used as a graded seal between materials with different coefficients of thermal expansion.

7.5.3 Packaging and connectors

MICs are usually shielded from the environment by metallic packages in order to prevent parasitic radiation to and from the environment. Further, the MICs must often be protected against the constituents of the ambient atmosphere. Water vapour, in particular, may cause corrosion of the contacts of the active devices and may promote migration of silver and even of gold. Water vapour can influence the performance of resonance circuits, because the part of the microwave field that is in the air is sensitive to moisture. If water condenses in slots of the microwave circuits, the MIC performance will

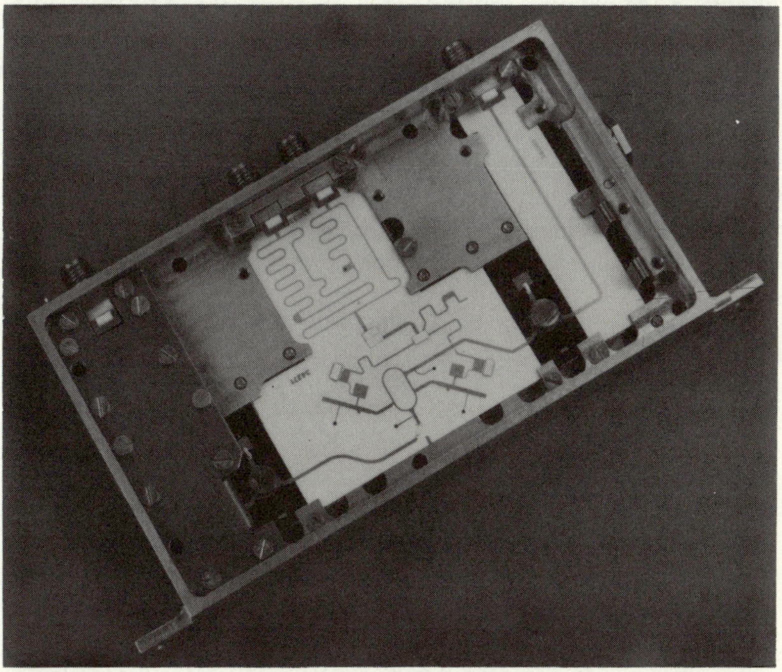

Figure 7.55. A 8.5–9.0 GHz up-convertor with an intermedium frequency (IF) of 1 GHz; several MICs are assembled in one package and interconnected by ribbon bonds

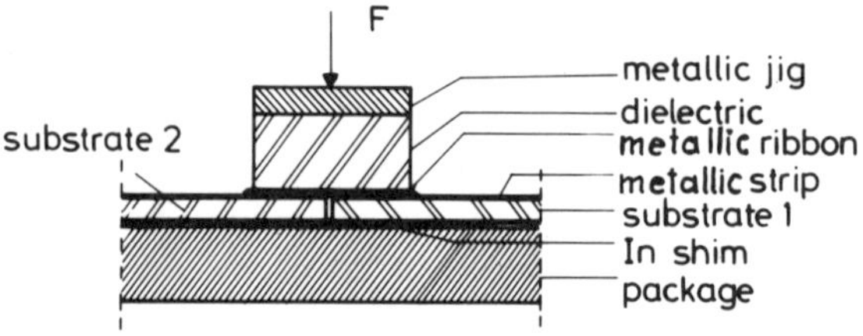

Figure 7.56. Interconnection between MIC substrates, designed for minimum parasitic radiation

deteriorate. To reduce these effects, it is often required to seal the MICs from the ambient atmosphere.

A hermetically sealed package is an expensive part of a MIC. Therefore, if possible, several MICs are assembled in a common package to reduce the cost. Moreover, the reliability is then also enhanced due to the fact that the number of interconnections between the packages is reduced.

Figure 7.55 shows several MICs assembled in one package. The MICs have different substrate materials. Ferrite substrates have been applied for MIC circulators, while alumina substrates are used for other MIC functions. The microstrip transitions between MICs are achieved by bonded ribbons. Figure 7.56 shows a special interconnection between MICs located in the same package. This interconnection consists of a ribbon pressed to the strips on both substrates by means of a dielectric spacer designed in such a way that the radiation losses of the interconnection are minimal.

A hermetically sealed package is much more expensive that a non-hermetically sealed one. In some cases, it is possible to protect only the sensitive parts of the circuit, while a non-hermetically sealed package is used for the entire MIC. Local protection can be achieved by a cap which is hermetically sealed to the substrate. Figure 7.57 shows a local protection of a MIC made by using thick-film technology. For some systems, the costs of

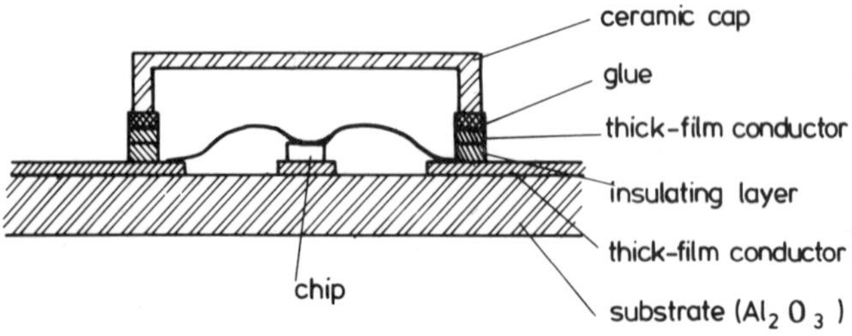

Figure 7.57. Local protection in MIC thick-film circuits by a hermetically sealed ceramic cap

packaging can be lowered by using a hermetically sealed box for the entire system, while non-hermetically sealed packages are applied for the MIC subsystems.

Usually, MICs are connected via 50 Ω coaxial cables or waveguides to other systems. The properties of the microstrip–coax and the microstrip–waveguide transitions are important for the performance of the MIC. These transitions may have a hermetic seal. Different microstrip–coax transitions have been discussed in the literature.[56,57] In the following, some microstrip–coax transitions will be briefly considered.

The performance of a microstrip–coax transition depends on the form of the connection between the inner conductor and the strip, on the earth contact and on the gap between the substrate and the wall of the package. For a non-hermetically sealed transition, a reflection coefficient $\rho \leqslant 0.05$ can be obtained up to 10 GHz and by matching one can obtain $\rho = 0.01$.[58,59] In general, the reflection coefficient of hermetically sealed transitions lies in the range from 0.1 to 0.2 up to 10 GHz.

Figure 7.58 shows a cross-section of a non-hermetically sealed transition. The shape and length of the extended part of the inner conductor can be dimensioned so as to minimize reflection. It is preferable to use a strip width nearly equal to the diameter of the inner conductor. The latter is pressed to the strip to achieve a reliable contact. Connectors whose extended part of the inner conductor can slide, are available. That extended end of the inner conductor can follow the displacements of the substrate caused by temperature fluctuations.

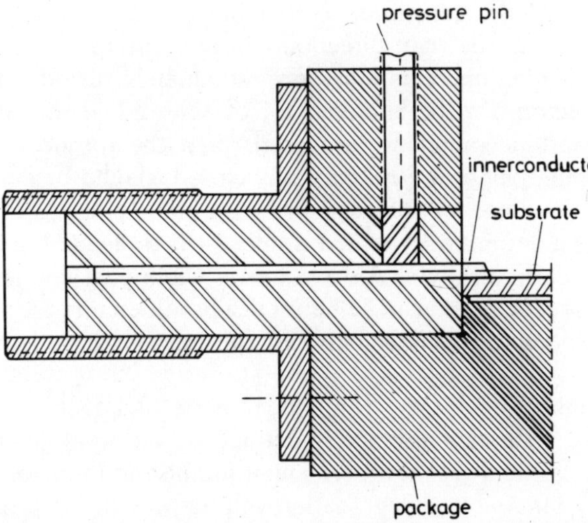

Figure 7.58. Cross-section of a transition of microstrip to coaxial connector

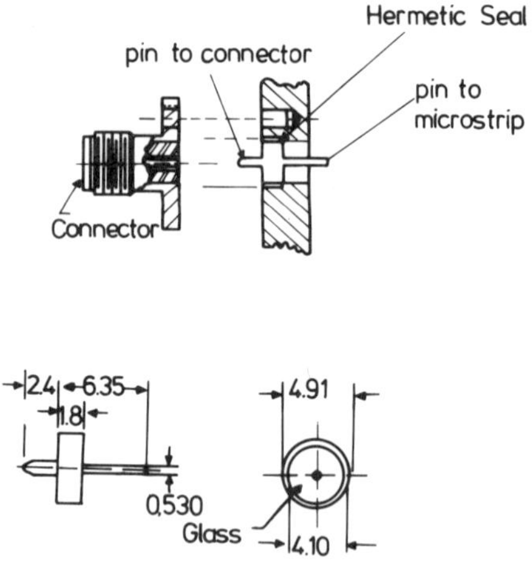

Figure 7.59. A microstrip–coax transition with an accessory pin. Hermeticity is obtained by soldering the pin in the wall of the package (measures in mm)

In order to lower strains and stresses in the transition, expansion bends are sometimes used in the coaxial cables just in front of the connectors. To reduce the gap between the package wall and the substrate, it is desirable to design the microwave circuit in such a way that the transitions are situated at one or two adjacent sides of the substrate, so that the gap can be minimized by pushing the substrate in the right direction before securing.

Hermetically sealed connectors must have adequately sealed glass insulation. These connectors can be soldered to, or screwed in the wall of the package. If screwed-in connectors are used, then the torque required to ensure that matching planes are hermetically closed, should be substantially higher than the torque needed for unscrewing the cable plug. Another possibility for realizing a hermetically sealed connector is to divide the connector into two parts, a hermetically sealed pin and a cable plug connection, see Figure 7.59. The accessory pin can be hermetically soldered in the wall of the package and provides the contact between the microstrip and the coaxial cable.

MIC subassemblies are mostly component parts of larger systems. The reliability of these systems is ultimately defined by the weakest link in the chain of subsystems. As a result, a permanent impulse to improve reliability and to reduce the influence of degradation will be present for future MICs. Therefore, much effort will be devoted to the subjects discussed in this chapter.

REFERENCES

1. J. Frey, 'Microwave integrated circuits', *Modern Frontiers in Applied Science*, Artech. House, Inc., Denham, Massachusetts, USA, 1975.
2. S. B. Cohn, 'Slot line on a dielectric substrate', *IEEE Trans. MTT*, **17**, no. 10, 768–778 (1969).
3. S. B. Cohn, 'Slot-line field components', *IEEE Trans. MTT,* **20**, No. 2, 172–174 (1972).
4. B. Schieck and J. Köhler, 'Improving the isolation of 3-dB couplers in microstrip-slotline technique', *IEEE Trans. MTT,* **26**, no. 1, 5–7 (1978).
5. D. L. Gish and O. Graham, 'Characteristic impedance and phase velocity of a dielectric-supported air strip transmission line with side walls', *IEEE Trans. MTT,* **18**, no. 3, 131–148 (1970).
6. T. Itoh and A. S. Hebert, 'A generalized spectral domain analysis for coupled suspended microstriplines with tuning septums', *IEEE Trans. MTT,* **26**, no. 10, 820–826 (1978).
7. R. Mittra and T. Itoh, 'Charge and potential distributions in shielded striplines', *IEEE Trans. MTT,* **18**, no. 3, 149–156 (1970).
8. A. M. K. Saad and G. Begemann, 'Electric performance of finlines of various configurations', *IEE Microwaves, Opt., Acoust.*, **1**, no. 2, 81–88 (1977).
9. A. M. K. Saad and K. Schünemann, 'A simple method for analysing fin-line structures', *IEEE Trans. MTT*, **26**, no. 12, 1002–1007 (1978).
10. R. N. Bates and M. D. Coleman, 'Milimetre wave finline balanced mixers', *9th Eur. Micr. Conf. Brighton U.K. Sept., Proc.,* pp. 721–725, 1979.
11. J. P. Becker and D. Jäger, 'Electrical properties of coplanar transmission lines on dielectric substrate', *IEEE Trans., MTT*, **23**, no. 7, 541–548 (1975).
12. H. Hofmann. 'Dispersion of planar waveguides for millimetre-wave application', *Archiv für Electronik und Ubertragungstechnik*, **31**, no. 1, 40–44 (1977).
13. J. P. Becker and D. Jäger, 'Electrical properties of coplanar transmission lines on lossless and lossy substrates', *Electr. Letters,* **15**, no. 3, 88–90 (1979).
14. H. A. Wheeler, 'Transmission-line properties of parallel strips separated by a dielectric sheet', *IEEE Trans. MTT,* **13**, no. 2, 172–185 (1965).
15. M. V. Schneider, 'Microstrip lines for microwave integrated circuits', *The Bell Syst. Techn. J.,* 1421–1444 (1969).
16. R. A. Pucel and D. J. Massé, 'Microstrip propagation on magnetic substrates—Part I: design theory', *IEEE Trans. MTT,* **20**, 304–308 (1972).
17. A. G. van Nie, 'On fields and currents in microstrip', *J. Applied Science and Eng. A,* no. 2, 205–212 (1977).
18. E. J. Denlinger, 'Losses of microstrip lines', *IEEE Trans. MTT,* **28**, no. 6, 513–522 (1980).
19. J. H. C. van Heuven, 'Conduction and radiation losses in microstrip', *IEEE Trans. MTT*, **22**, no. 9, 841–844 (1974).
20. S. P. Morgan, 'Effect of surface roughness on eddy current losses at microwave frequencies', *J. Applied Physics,* **20**, 352–362 (1949).
21. S. Ramo, J. R. Whinnery, and T. van Duzer, *Fields and Waves in Communication Electronics*, John Wiley and Sons Inc., New York, London, Sydney, pp. 298–301, 1965.
22. W. Goedbloed, H. Hieber, and A. G. van Nie, 'Ageing tests on microwave integrated circuits', *The Radio and Electronic Engineer,* **48**, no. 1/2, 13–22 (1978).
23. J. H. C. van Heuven and A. G. van Nie, 'Microwave integrated circuits', *Philips Tech. Rev.,* **32**, no. 9/10/11/12, 292–304 (1971).

24. E. J. Denlinger, 'A frequency dependent solution for microstrip transmission lines', *IEEE Trans. MTT,* **19**, no. 1, 30–39 (1971).
25. R. Mittra and T. Itoh, 'Analysis of microstrip transmission lines', *Advances in Microwaves*, Vol. 8, Academic Press, New York, pp. 111–121, 1974.
26. W. Wiesbeck, 'Berechnung und Messung der Stromverteilung auf ungeschirmten Streifenleitungen, *Nachrichtentechnische Zeitscrift NTZ,* **25**, no. 1, 1–6 (1972).
27. A. G. van Nie, 'Electroless NiP processing for hybrid integrated circuits', *Microelectronics and Reliability,* **15**, 221–226 (1976).
28. L. I. Maissel and R. Glang, *Handbook of Thin-Film Technology*, Mc-Graw-Hill Inc., 1970.
29. C. A. Harper, *Handbook of Materials and Processes for Electronics*, McGraw-Hill Inc., 1970.
30. J. J. Maes, A. G. van Nie, and G. B. A. Hut, 'Thick photoresist patterns for selective electroplating', *Microelectronics and Reliability,* **17**, 325–332 (1978).
31. W. Goedbloed, 'Thin film MIC activities in respect to reproducibility and reliability', *Internepcon/Europa '75, Int. Microelectronics Conf.*, Session III, Stuttgart Germany, Nov. 11–13 (1975).
32. G. Perger and P. M. Robinson, 'Pulse plating—retrospects and prospects', *Metal Finishing,* **77**, no. 12, 17–19 (1979).
33. P. S. Burggraaf, 'Plasma etching technology', *Semiconductor International*, **2**, no. 10, December, pp. 49–58, 1979.
34. J. P. Ramy, M. Th. Cotte, J. P. Bolloch, R. Schnitzler, J. J. Guena, and C. Thebault, 'Optimization of the thick- and thin-film technologies for microwave circuits on alumina and fused silica substrates', *IEE Trans. MTT,* **26**, no. 10, 814–820 (1978).
35. R. Garg, 'The effect of tolerances on microstripline and slotline performances', *IEEE Trans. MTT,* **26**, no. 1, 16–19 (1978).
36. R. Garg, A microstrip design guide', *Int. J. Electronics,* **46**, no. 2, 187–192 (1979).
37. L. Braun, 'Electromigration testing—a current problem', *Microelectronics and Reliabiilty,* **13**, 215–228 (1974).
38. S. Glasstone, *The Elements of Physical Chemistry*, D. van Nostrand Co. Inc., New York, 2nd Ed., pp. 606–608, 1946.
39. C. J. Meechan and J. A. Brinkman, 'Electrical resistivity study of lattice defects introduced in copper by 1.25–MeV electron irradiation at 80 K', *Phys. Rev.,* **103**, no. 5, 1193–1202 (1956).
40. M. A. Nicolet, 'Diffusion barriers in thin films', *Thin Solid Films,* **52**, 415–443 (1978).
41. H. Hieber and K. Pape, 'Umkritallisation dünner Goldschichten', *Zeitschrift für Metallkunde,* **70**, no. 7, 459–466 (1979).
42. D. S. Campbell, 'Mechanical properties of thin films', *Handbook of Thin Film Technology*, Eds. L. I. Maissel and R. Glang, McGraw-Hill, New York, pp. 12-6–12-21, 1970.
43. B. N. Chapman, 'Thin-film adhesion', *J. Vac. Technol.,* **11**, no. 1, 106–113 (1974).
44. K. L. Mittal, 'Adhesion measurement of thin films', *Electrocomponent Science and Technology,* **3**, 21–42 (1976).
45. H. G. Kadereit, 'Adhesion measurements of metallizations for hybrid microcircuits', *Proc. Eur. Hybrid Microelectronics Conf. (ISHM)*, Bad Homburg, Germany, Session IX, 1977.
46. H. Hieber, F. Betke, and K. Pape, Ageing tests on gold layers and bonded contacts', *Proc. Eur. Hybrid Microelectronics Conf. (ISHM)*, Bad Homburg Germany, Session IX, 1977.

47. P. Benjamin and C. Weaver, 'Measurements of adhesion of thin films', *Proc. Roy. Soc.*, **254A**, 163–176 (1960).
48. G. Kersuzan, 'Comment résoudre les problèmes thermiques en microélectronique hybride', *Electron, et. Microelectron. Ind.*, **218**, 24–29 (1976).
49. R. F. David, 'Computerized thermal analysis of hybrid circuits', *IEEE Trans. PHP*, **13**, no. 3, 283–290 (1977).
50. H. Hieber, *Behaviour of the Material of Electronic Microcontacts*, to be published.
51. P. J. Planting, 'An approach for evaluating epoxy adhesives for the use in hybrid microelectronic assembly', *Trans. IEEE PHP*, **11**, 4, 305–311, (1975).
52. J. Curabet, 'Adhesive choice for die attach in microcircuits: properties, characterization and reliability', *Eur. Hybrid Microelectronics Conf. (ISHM), Gent Belgium, Proc.* pp. 433–4440, 1979.
53. C. W. L. Kooring and D. Riphagen, 'Application of conductive adhesives in microcircuits for "long-life" equipment', *Eur. Hybrid. Microelectronics Conf. (ISHM), Gent Belgium, Proc.*, pp. 425–432, 1979.
54. F. G. Yost, 'Aspects of lead–indium solder technology', *Proc. Int. Microelectron. Symp.*, pp. 61–66, 1976.
55. S. J. Horowitz, D. J. Gerry, and R. E. Coté, 'Alloy element additions to gold thick film conductors: effects on indium–lead soldering and ultrasonic aluminum wire bonding', *Electron. Components Conf.*, **27**, 358–366 (1977).
56. J. H. C. van Heuven, 'A new integrated waveguide-microstrip transition', *IEEE Trans. MTT*, **24**, no. 3, 144–146 (1976).
57. P. Röschmann, 'Mikrostripleitunsübergang auf andere wellenleiter', *Deutsche Patentanmeldung*, **P26**, 16, 907–9 (1976).
58. E. H. England, 'A coaxial to microstrip transition', *IEEE Trans.*, **24**, no. 1, pp. 47–48 (1976).
59. R. L. Eisenhart, 'A better microstrip connector', *IEEE MTT-S Int. Microwave Symposium, Ottawa Canada, Digest*, pp. 318–320, 1978.

Index